Architectural Detail
Analysis of Famous Buildings

名家建筑细部详解

佳图文化 主编

中国林业出版社
China Forestry Publishing House

图书在版编目（CIP）数据

名家建筑细部图集 / 佳图文化 主编 . -- 北京：中国林业出版社，2016.3
ISBN 978-7-5038-8397-2
Ⅰ.①名… Ⅱ.①佳… Ⅲ.①建筑结构－细部设计－世界－图集 Ⅳ.① TU22-64

中国版本图书馆 CIP 数据核字 (2016) 第 021229 号

中国林业出版社·建筑家居出版分社
责任编辑：李 顺 唐 杨
出版咨询：（010）83143569

出 版：中国林业出版社（100009 北京西城区德内大街刘海胡同 7 号）
网 站：http://lycb.forestry.gov.cn/
印 刷：广州中天彩色印刷有限公司
发 行：中国林业出版社
电 话：（010）83143500
版 次：2016 年 3 月第 1 版
印 次：2016 年 3 月第 1 次
开 本：889mm×1194mm 1 ／ 16
印 张：17
字 数：200 千字
定 价：298.00 元

preface

This book specially chooses the highly representative architectures with detailed designs from countries like the UK, Italy, Japan, Spain, and Switzerland etc. to comprehensively introduce and display the excellent works with professional analysis, photos, diagrams and drawings. It introduces every case with especially enlarged details, which is believed as a quite important element for the architecture, and sometimes even the key to assess the value or success of the building. It is an important reference book for related professionals and those that are practicing and researching on architecture and details. With this reference book, their design level will be greatly improved.

CONTENTS

COMMERCE

CULTURE & ARTS

EDUCATION

OFFICE

TRANSPORTATION

SPORTS, HEALTH & OTHERS

Commerce

Calypso

Location: Rotterdam, Netherland
Architect: Alsop Architects
Total Floor Area: approx. 72,500 m^2
Completion Date: 2013
Photography: Ben Blossom

The project is part of the urban strategy scheme Rotterdam Central undertaken for Rotterdam between 2000 and 2001 by Alsop Architects. The development of this site afforded the opportunity to establish the beginnings of a quality public realm on the Westersingel leading from the station, the desired "cultural route" of the city. It was also desirable that the development reinforces and enhances pedestrian routes across the city which has been identified as positive contributions to the city mesh in this quarter. Pedestrian permeability is a priority, with the main route across the site connecting Mauritsweg itself with the Schounbergtplein and Pathe cinema.

Currently occupied by a Hotel, Retail space and St. Paul's church, the site is a key gathering point of a number of primary city routes. A new hotel, new shops, a replacement church and new residential accommodation above are planned, preserving a mix of public and private realms and accommodating a broad social spectrum. Consultation and workshop exercises with the community of Pauls Kerk church (which incorporates a drug rehabilitation scheme) in particular, were central to the development of new designs. Representatives of the Kerk continue to be involved in detail development.

The quality of the public areas in particular have been considered with a two-tiered circulation space below the 200 bed hotel building; the lower below the transparent pavement of the upper, and connecting directly with car parking for 350 cars. The development takes the form of five forms - the "rocks" - which accommodate hotel, apartment, retail, commercial space and the church. The replacement "arthouse" cinema has three screens and forms part of the subsurface development along with the car park.

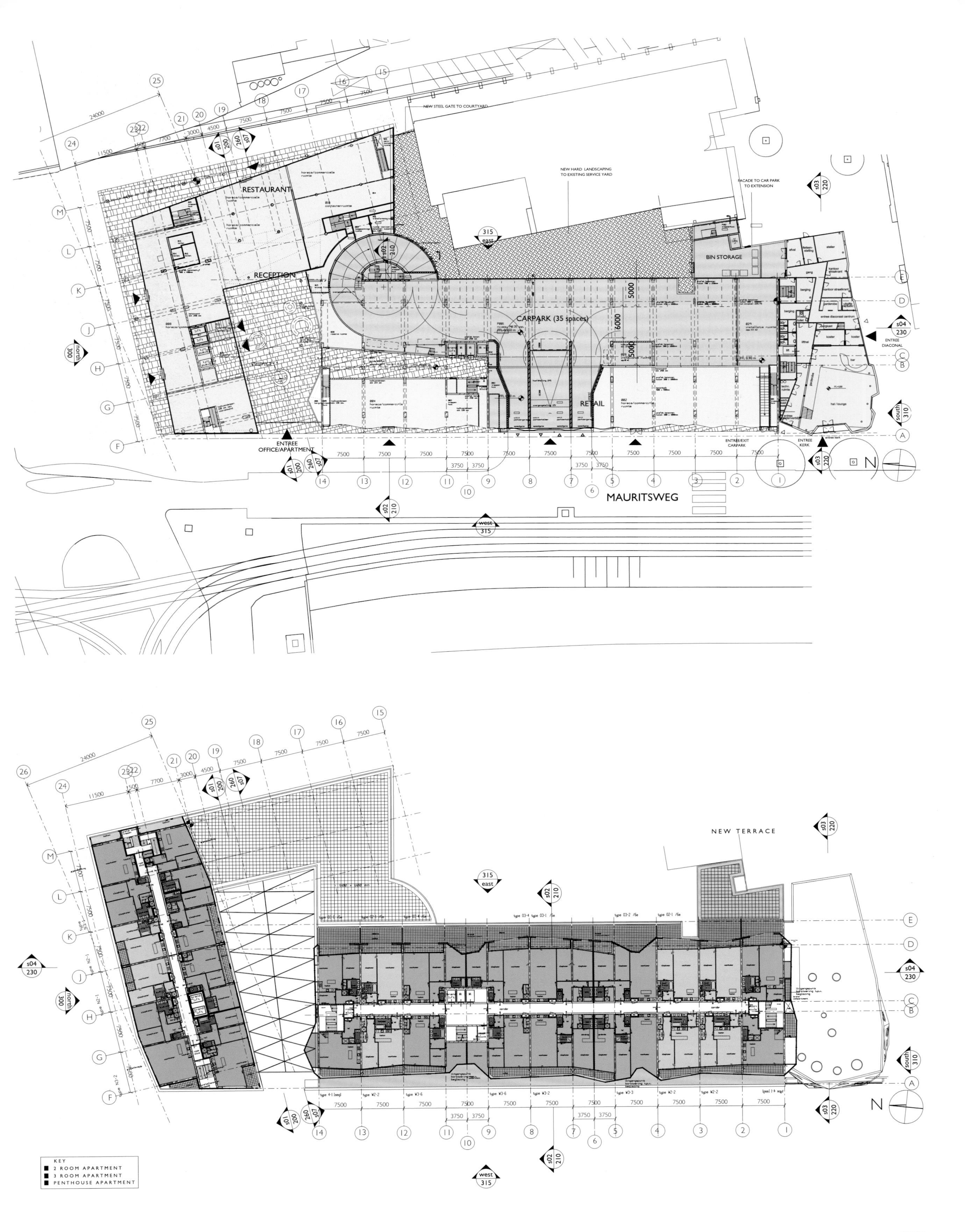
NEW STEEL GATE TO COURTYARD
NEW HARD LANDSCAPING TO EXISTING SERVICE YARD
FACADE TO CAR PARK TO EXTENSION
RESTAURANT
RECEPTION
BIN STORAGE
CARPARK (35 spaces)
RETAIL
ENTREE OFFICE/APARTMENT
ENTREE/EXIT CARPARK
ENTREE KERK
ENTREE DIACONAAL
MAURITSWEG
NEW TERRACE
KEY
2 ROOM APARTMENT
3 ROOM APARTMENT
PENTHOUSE APARTMENT

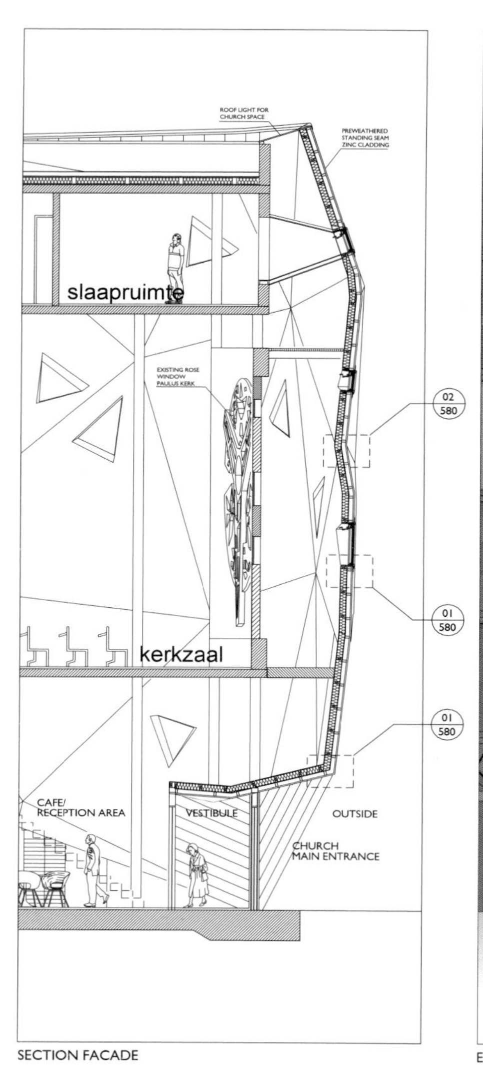

SECTION FACADE

ELEVATION FACADE

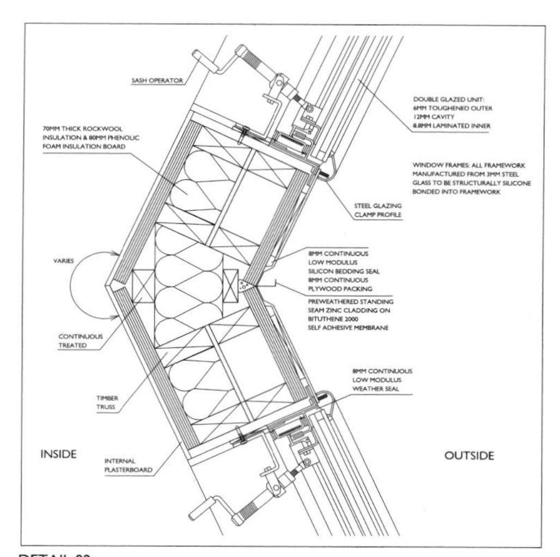

DETAIL 02

0918/VO/580

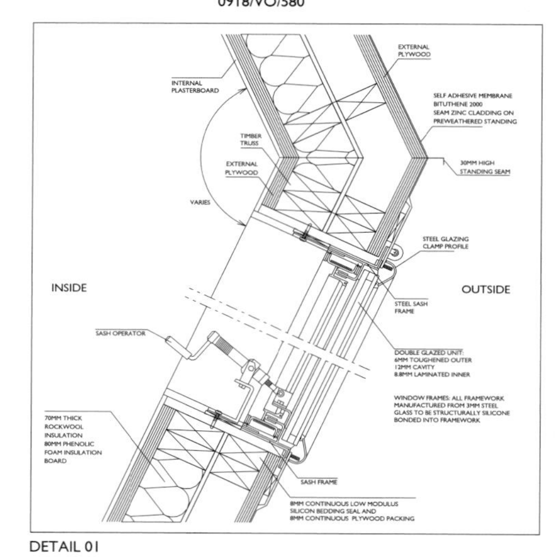

DETAIL 01

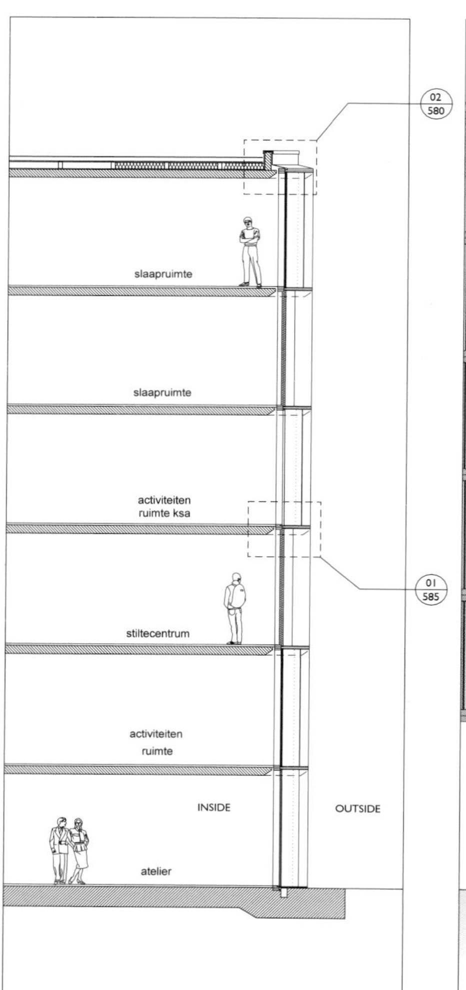

SECTION FACADE

ELEVATION FACADE

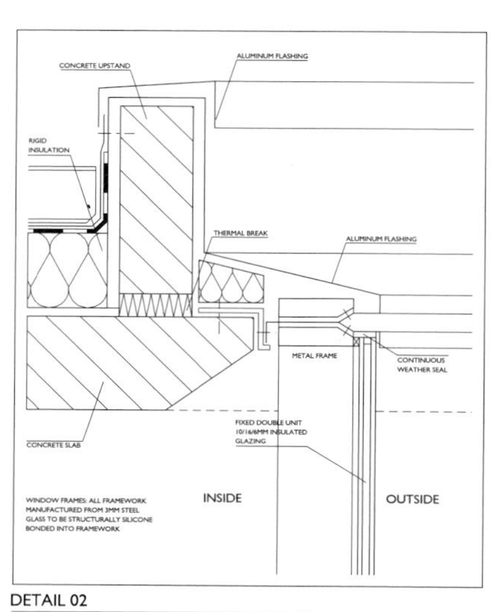

DETAIL 02

0918/VO/585

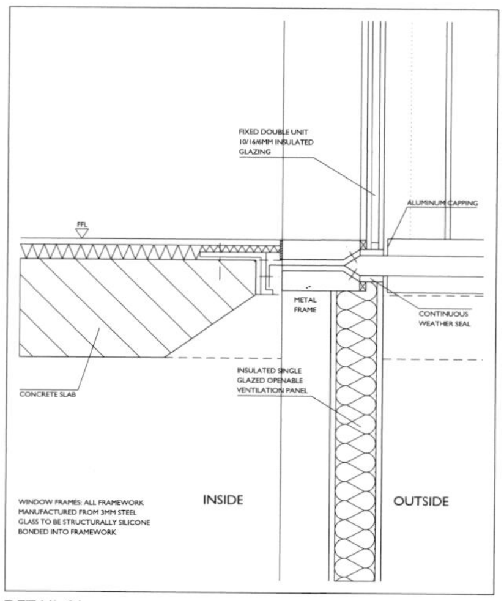

FIXED DOUBLE UNIT 10/16/6MM INSULATED GLAZING
ALUMINUM CAPPING
FFL
METAL FRAME
CONTINUOUS WEATHER SEAL
CONCRETE SLAB
INSULATED SINGLE GLAZED OPENABLE VENTILATION PANEL
WINDOW FRAMES: ALL FRAMEWORK MANUFACTURED FROM 3MM STEEL GLASS TO BE STRUCTURALLY SILICONE BONDED INTO FRAMEWORK
INSIDE
OUTSIDE

DETAIL 01

M
L
K
J
H
7500
7500
7500
7500
PARTITION
STRUCTURAL WALLS AT GRID LINES
AP. 3000
BALCONY
LIVING AREA
BEDROOM
BEDROOM
BEDROOM
LIVING AREA
LIVING AREA
BEDROOM
11 FFL +34.000
10 FFL +31.050
09 FFL +28.100
08 FFL +25.150
300
2650
2950
BALCONY
BALCONY
BALCONY
FACADE CRANK
LIVING AREA
FACADE AREA 19M2
PANEL AREA 5.5M2
GLASS AREA 13.5M2 (70%)
DOUBLE GLAZED UNITS WITH DIFFERING WIDTHS
750, 900, 1200 AND 1500 MM
DOUBLE GLAZED PANEL
BEDROOM AREA
FACADE AREA 19M2
PANEL AREA 10.8M2
GLASS AREA 8.2M2 (43%)
INSULATED PANELS WITH DIFFERING WIDTHS
300, 750, 900, 1200 AND 1500 MM
CLEAR GLASS BALCONY
300MM WIDTH ON NON-OPENABLE PANELS
01 TOWER A: TRUE ELEVATION. NORTH
G
J
7500
LIVING AREA
STAIR CORE
LIVING AREA
COLOUR SCHEME FOR OPAQUE PANELS
External :
T1 - Orange, RAL 2004
T2 - Red, RAL 3000
T3 - dark Red, RAL 3005
FIN- Red, RAL 3000
Parapet +70.000
Roof +69.800
22 FFL +66.450
21 FFL +63.500
BALCONY
BALCONY
BALCONY
BALCONY
02 TOWER A: ELEVATION. NORTH CTD
03 EAST/ WEST ELEVATION
DOUBLE GLAZED WINDOWS
13 FFL +45 800
CONCRETE SLAP
SLIDING DOORS TO BALCONY
HEAT STRENGTHEND FIXED DOUBLE GLAZING
13 FFL +42 850
CONCRETE SLAB
METALL PANEL
HEAT STRENGTHEND AND LAMINATED NON STRUCTURAL GLASS BALUSTRADE
13 FFL +39 900
CONCRETE SLAB
METALL PANEL RAL 9006 silver
METALL PANEL RAL 7037 Dust Grey
METALL PANEL RAL 9002 Creme White
HEAT STRENGTHEND AND LAMINATED NON STRUCTURAL GLASS BALUSTRADE
INSULATED METAL SANDWICH PANEL
B-C-D PART ELEVATION
B-C-D PART Section
BED ROOM
LIVING ROOM
BALCONY
B-C-D PART PLAN
CONCRETE SLAB
insulated metal sandwich panel
panell junction and fixing
steel bracket for facade panels
double glazed window
Detail 1:5

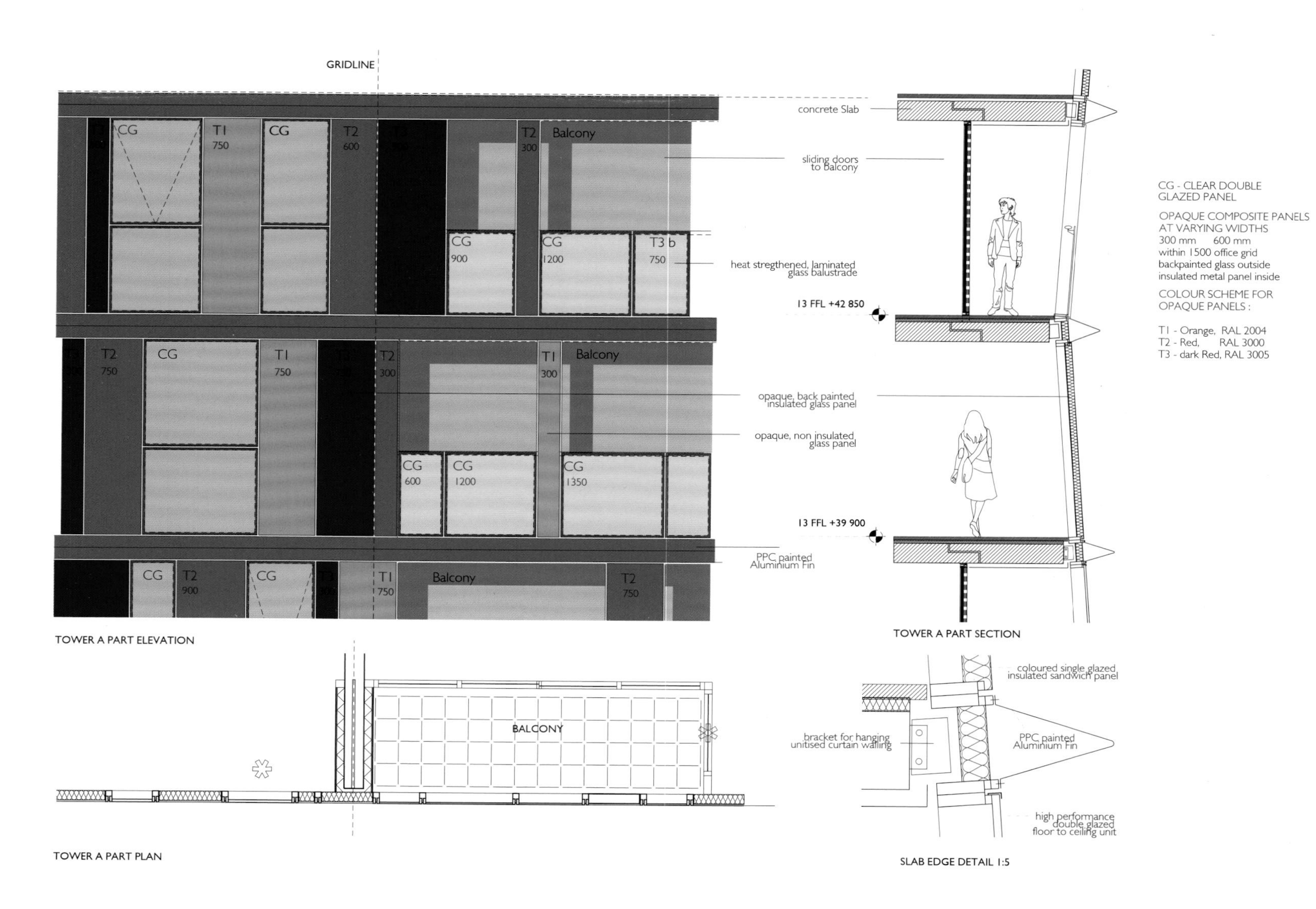
GRIDLINE
CG
T1
750
CG
T2
600
T2
300
Balcony
CG
900
CG
1200
T3 b
750
T2
750
CG
T1
750
T2
300
T1
300
Balcony
CG
600
CG
1200
CG
1350
CG
T2
900
CG
T1
750
Balcony
T2
750
TOWER A PART ELEVATION
concrete Slab
sliding doors to balcony
heat stregthened, laminated glass balustrade
13 FFL +42 850
opaque, back painted insulated glass panel
opaque, non insulated glass panel
13 FFL +39 900
PPC painted Aluminium Fin
TOWER A PART SECTION
CG - CLEAR DOUBLE GLAZED PANEL
OPAQUE COMPOSITE PANELS AT VARYING WIDTHS
300 mm 600 mm
within 1500 office grid
backpainted glass outside
insulated metal panel inside
COLOUR SCHEME FOR OPAQUE PANELS :
T1 - Orange, RAL 2004
T2 - Red, RAL 3000
T3 - dark Red, RAL 3005
BALCONY
TOWER A PART PLAN
coloured single glazed insulated sandwich panel
bracket for hanging unitised curtain walling
PPC painted Aluminium Fin
high performance double glazed floor to ceiling unit
SLAB EDGE DETAIL 1:5

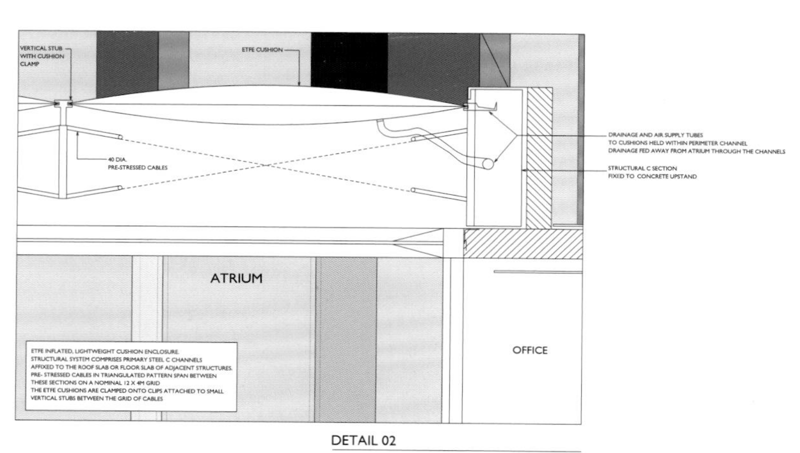
ATRIUM
OFFICE
DETAIL 02

APARTMENTS
OFFICE
OFFICE
ATRIUM
OFFICE
OFFICE
RETAIL
RECEPTION AREA
DETAIL 03

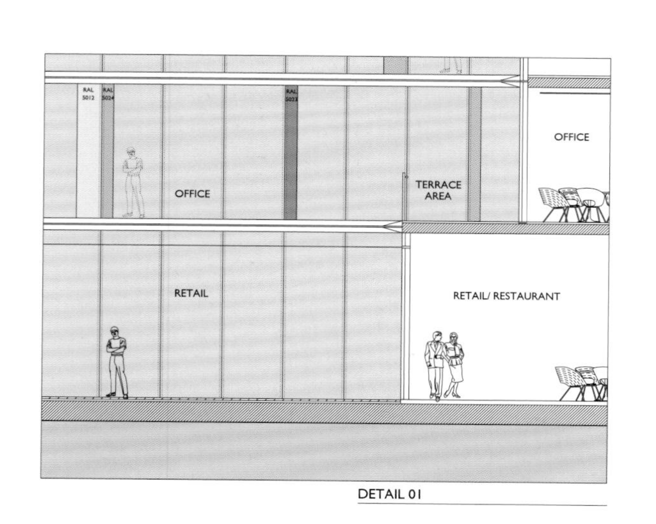
OFFICE
OFFICE
TERRACE AREA
RETAIL
RETAIL/ RESTAURANT
DETAIL 01

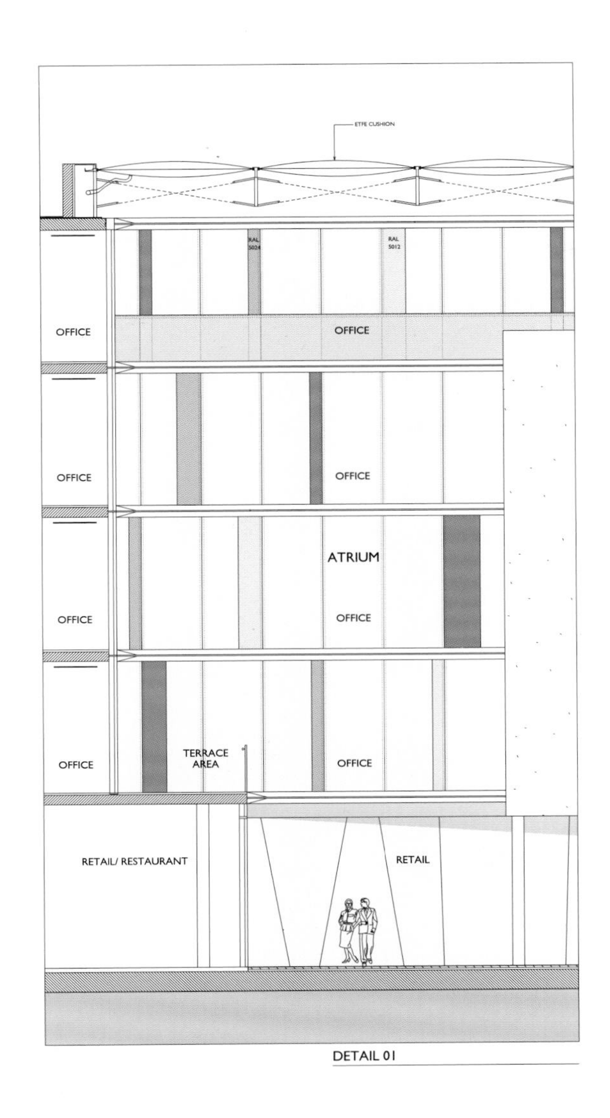
ETFE CUSHION
OFFICE
OFFICE
OFFICE
OFFICE
ATRIUM
OFFICE
OFFICE
OFFICE
OFFICE
TERRACE AREA
OFFICE
RETAIL/ RESTAURANT
RETAIL
DETAIL 01

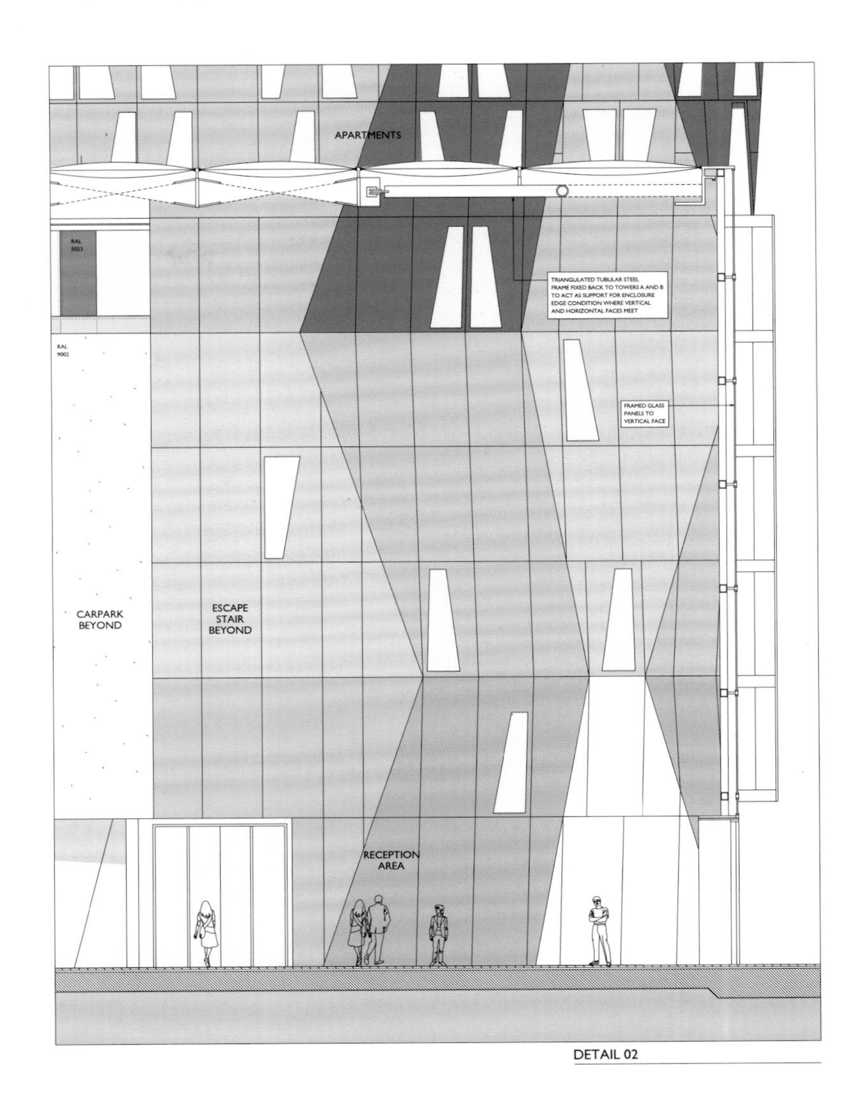
APARTMENTS
TRIANGULATED TUBULAR STEEL FRAME FIXED BACK TO TOWERS A AND B TO ACT AS SUPPORT FOR ENCLOSURE EDGE CONDITION WHERE VERTICAL AND HORIZONTAL FACES MEET
FRAMED GLASS PANELS TO VERTICAL FACE
CARPARK BEYOND
ESCAPE STAIR BEYOND
RECEPTION AREA
DETAIL 02

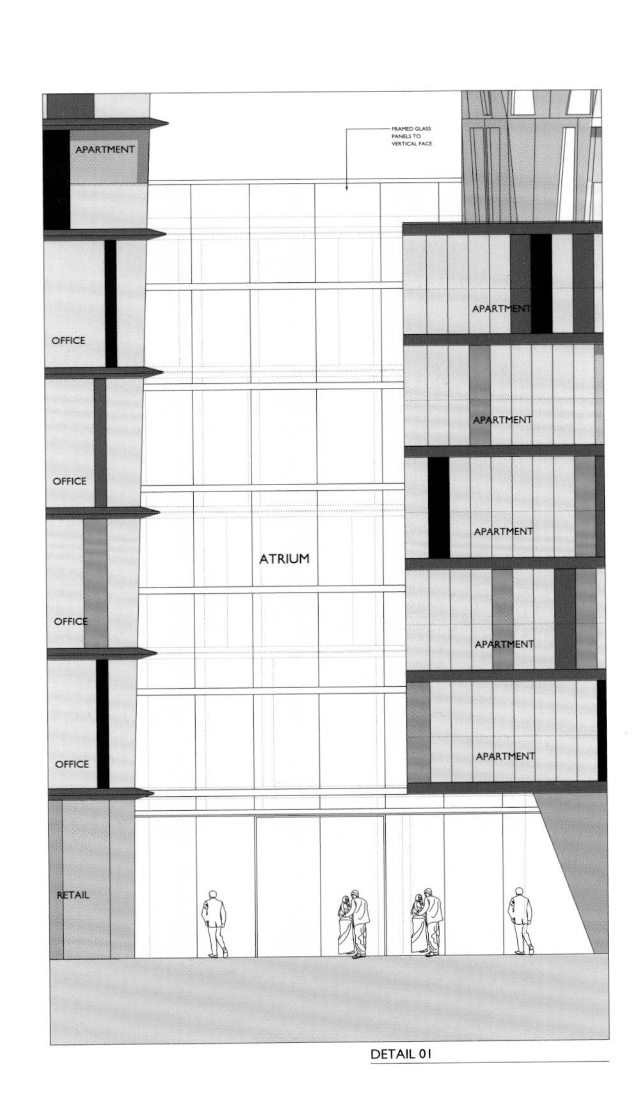
APARTMENT
FRAMED GLASS PANELS TO VERTICAL FACE
APARTMENT
OFFICE
APARTMENT
OFFICE
APARTMENT
ATRIUM
OFFICE
APARTMENT
OFFICE
APARTMENT
RETAIL
DETAIL 01

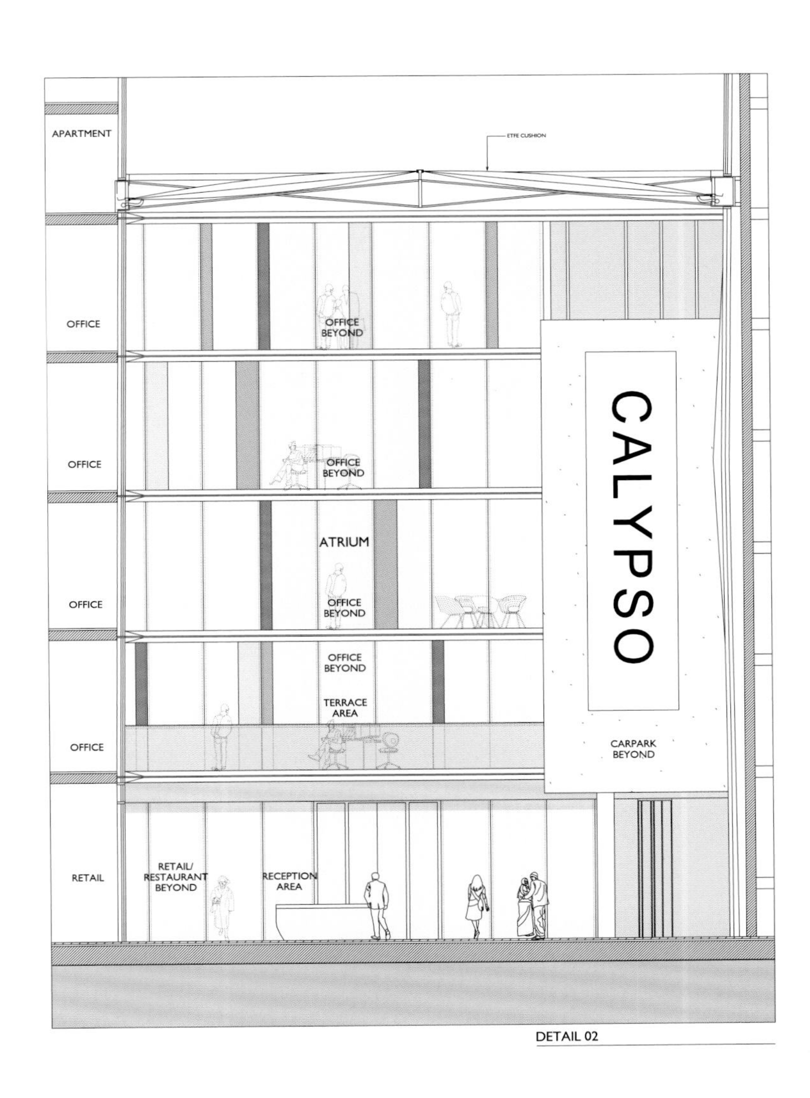
APARTMENT
ETFE CUSHION
OFFICE
OFFICE BEYOND
OFFICE
OFFICE BEYOND
ATRIUM
OFFICE
OFFICE BEYOND
OFFICE BEYOND
TERRACE AREA
OFFICE
CALYPSO
CARPARK BEYOND
RETAIL
RETAIL/ RESTAURANT BEYOND
RECEPTION AREA
DETAIL 02

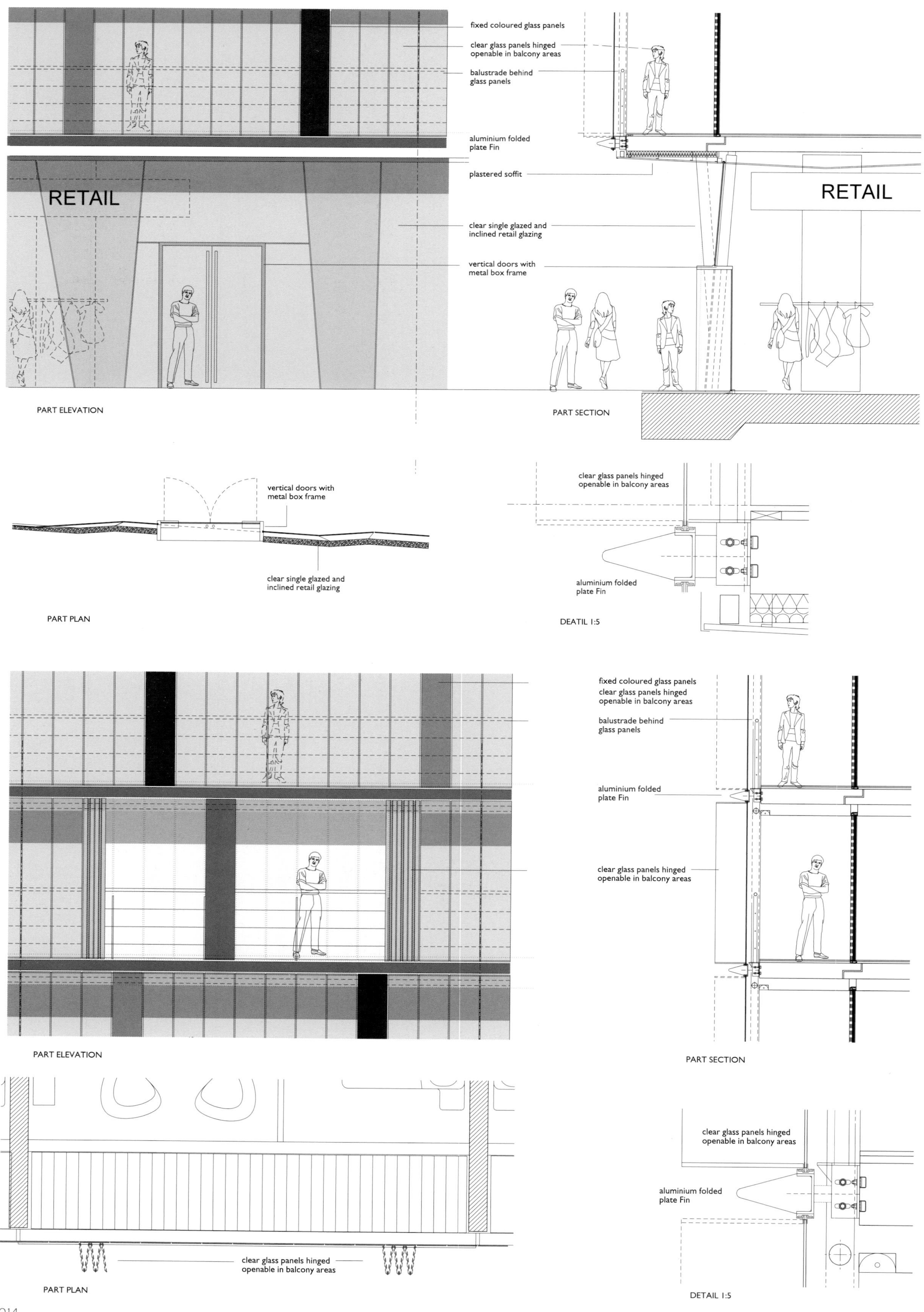
fixed coloured glass panels
clear glass panels hinged openable in balcony areas
balustrade behind glass panels
aluminium folded plate Fin
plastered soffit
clear single glazed and inclined retail glazing
vertical doors with metal box frame
RETAIL
RETAIL
PART ELEVATION
PART SECTION
vertical doors with metal box frame
clear single glazed and inclined retail glazing
PART PLAN
clear glass panels hinged openable in balcony areas
aluminium folded plate Fin
DEATIL 1:5
fixed coloured glass panels
clear glass panels hinged openable in balcony areas
balustrade behind glass panels
aluminium folded plate Fin
clear glass panels hinged openable in balcony areas
PART ELEVATION
PART SECTION
clear glass panels hinged openable in balcony areas
PART PLAN
clear glass panels hinged openable in balcony areas
aluminium folded plate Fin
DETAIL 1:5

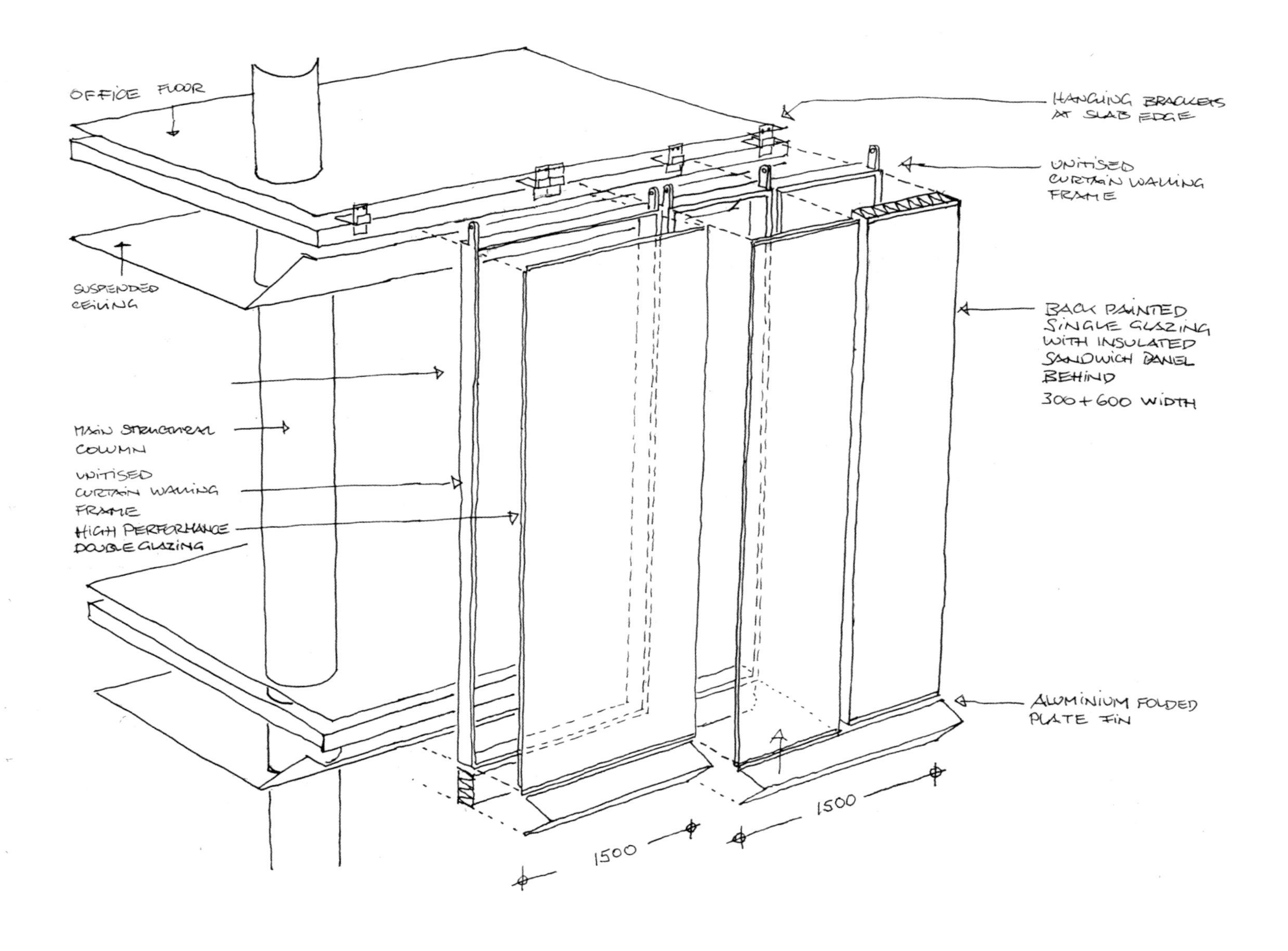
OFFICE FLOOR
HANGING BRACKETS AT SLAB EDGE
UNITISED CURTAIN WALLING FRAME
SUSPENDED CEILING
BACK PAINTED SINGLE GLAZING WITH INSULATED SANDWICH PANEL BEHIND
300 + 600 WIDTH
MAIN STRUCTURAL COLUMN
UNITISED CURTAIN WALLING FRAME
HIGH PERFORMANCE DOUBLE GLAZING
ALUMINIUM FOLDED PLATE FIN
1500
1500

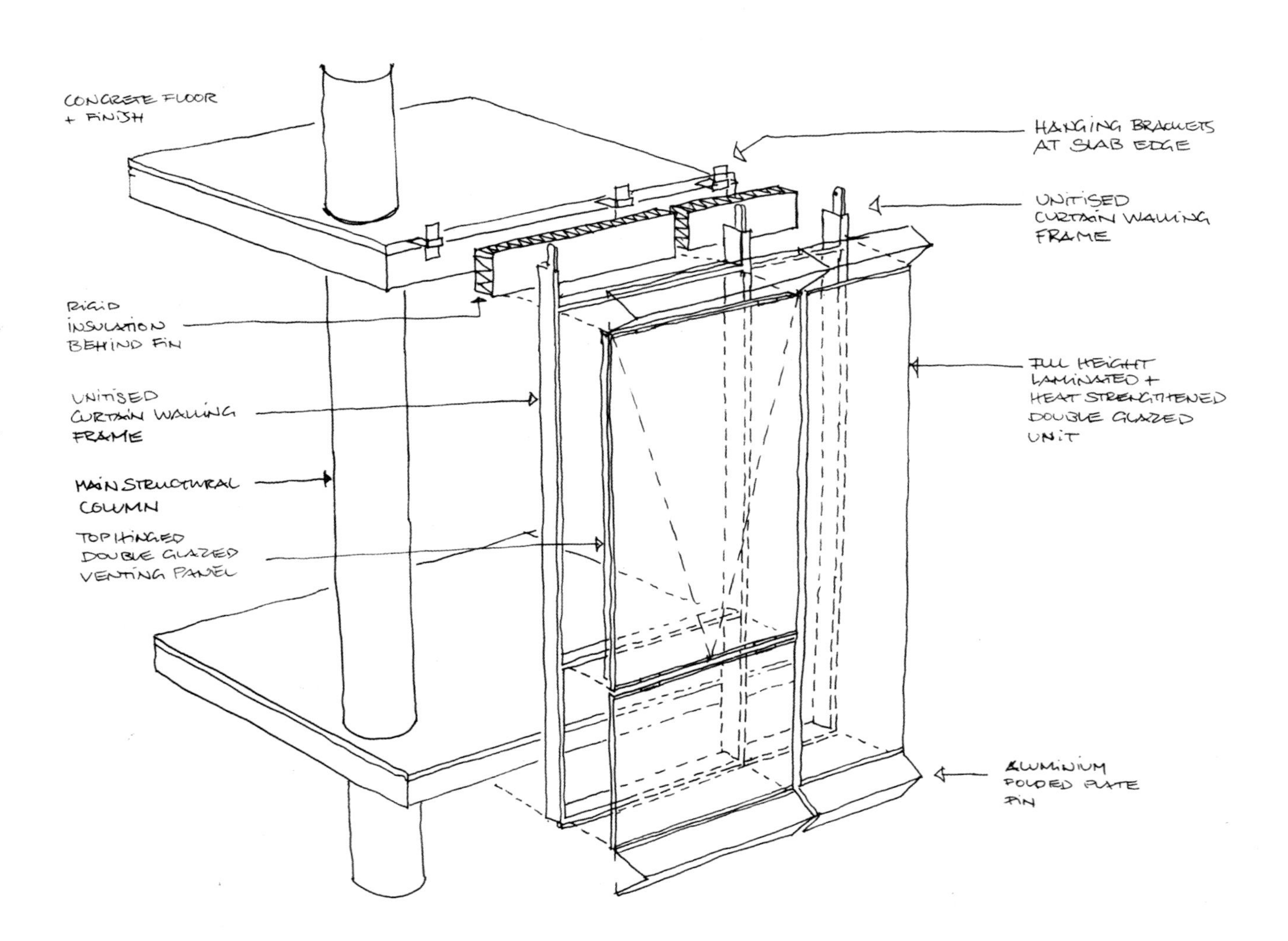
CONCRETE FLOOR + FINISH
HANGING BRACKETS AT SLAB EDGE
UNITISED CURTAIN WALLING FRAME
RIGID INSULATION BEHIND FIN
FULL HEIGHT LAMINATED + HEAT STRENGTHENED DOUBLE GLAZED UNIT
UNITISED CURTAIN WALLING FRAME
MAIN STRUCTURAL COLUMN
TOP HINGED DOUBLE GLAZED VENTING PANEL
ALUMINIUM FOLDED PLATE FIN

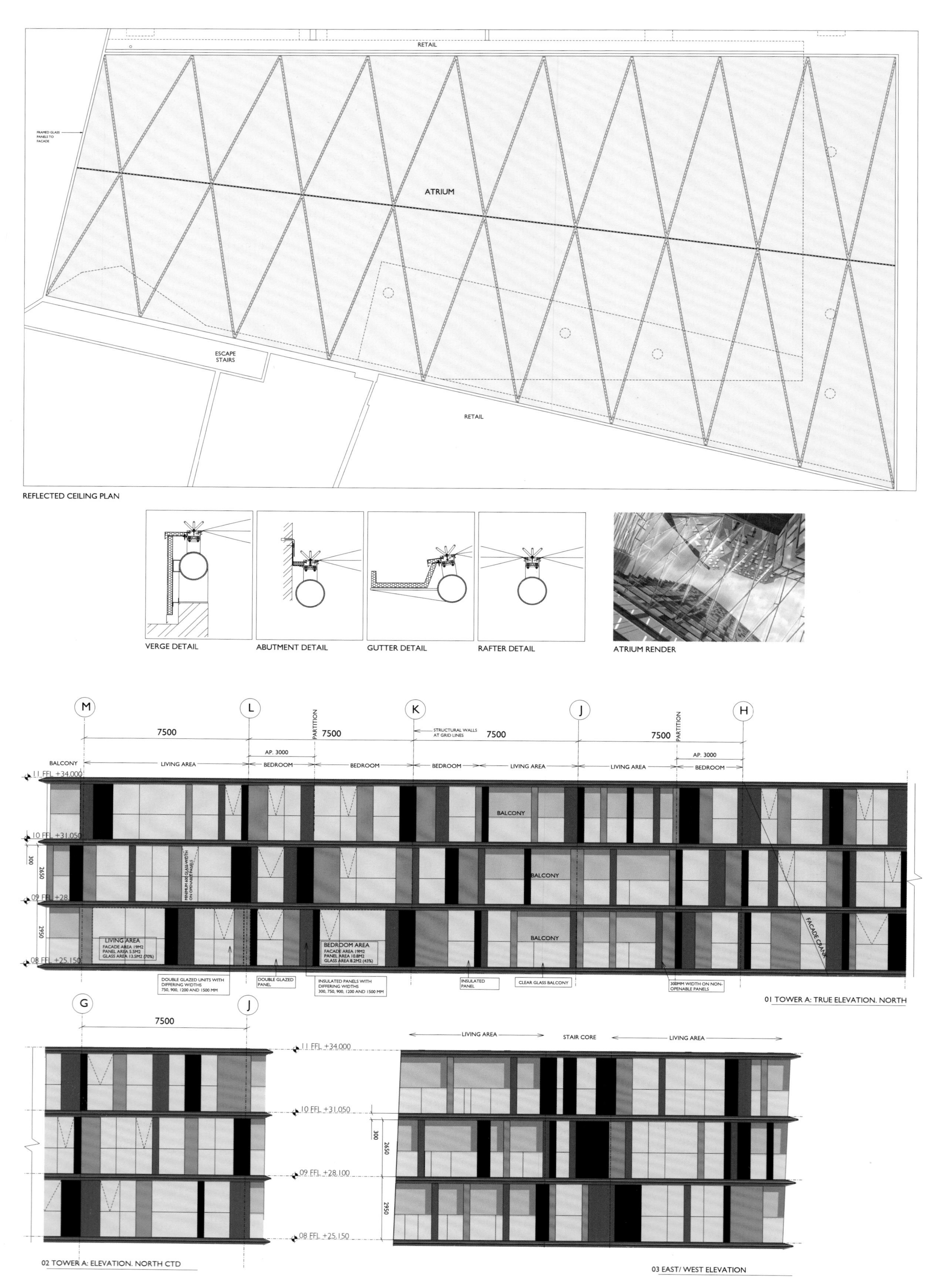

RETAIL
FRAMED GLASS PANELS TO FACADE
ATRIUM
ESCAPE STAIRS
RETAIL
REFLECTED CEILING PLAN
VERGE DETAIL
ABUTMENT DETAIL
GUTTER DETAIL
RAFTER DETAIL
ATRIUM RENDER
M
L
K
J
H
7500
7500
7500
7500
PARTITION
STRUCTURAL WALLS AT GRID LINES
AP. 3000
BALCONY
LIVING AREA
BEDROOM
BEDROOM
BEDROOM
LIVING AREA
LIVING AREA
BEDROOM
11 FFL +34.000
10 FFL +31.050
09 FFL +28
08 FFL +25.150
300
2650
2950
BALCONY
MINIMUM 600 GLASS WIDTH ON OPENABLE PANELS
LIVING AREA
FACADE AREA 19M2
PANEL AREA 5.5M2
GLASS AREA 13.5M2 (70%)
BEDROOM AREA
FACADE AREA 19M2
PANEL AREA 10.8M2
GLASS AREA 8.2M2 (43%)
FACADE CRANK
DOUBLE GLAZED UNITS WITH DIFFERING WIDTHS 750, 900, 1200 AND 1500 MM
DOUBLE GLAZED PANEL
INSULATED PANELS WITH DIFFERING WIDTHS 300, 750, 900, 1200 AND 1500 MM
INSULATED PANEL
CLEAR GLASS BALCONY
300MM WIDTH ON NON-OPENABLE PANELS
01 TOWER A: TRUE ELEVATION. NORTH
G
J
7500
LIVING AREA
STAIR CORE
LIVING AREA
11 FFL +34.000
10 FFL +31.050
09 FFL +28.100
08 FFL +25.150
02 TOWER A: ELEVATION. NORTH CTD
03 EAST/ WEST ELEVATION

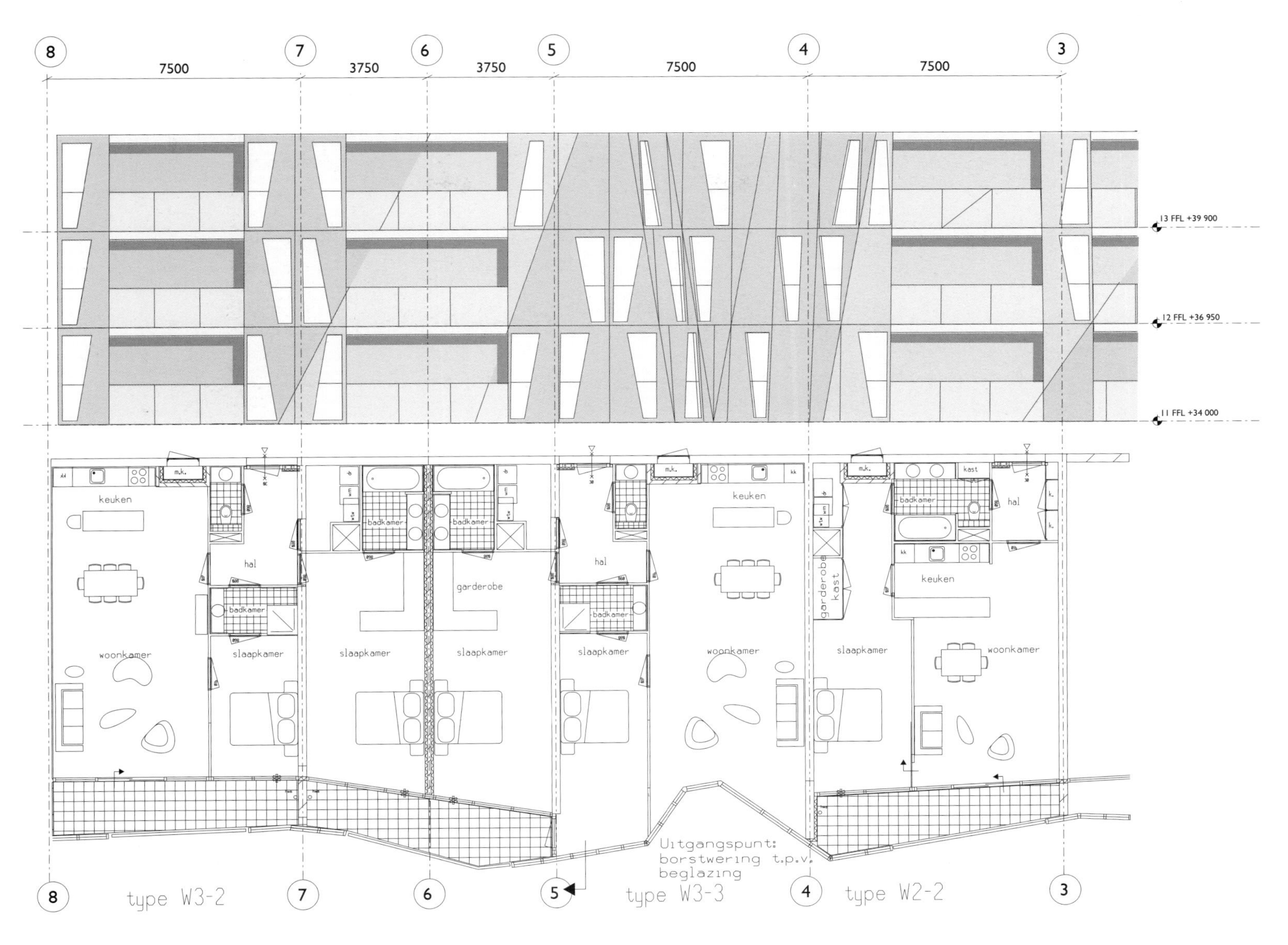

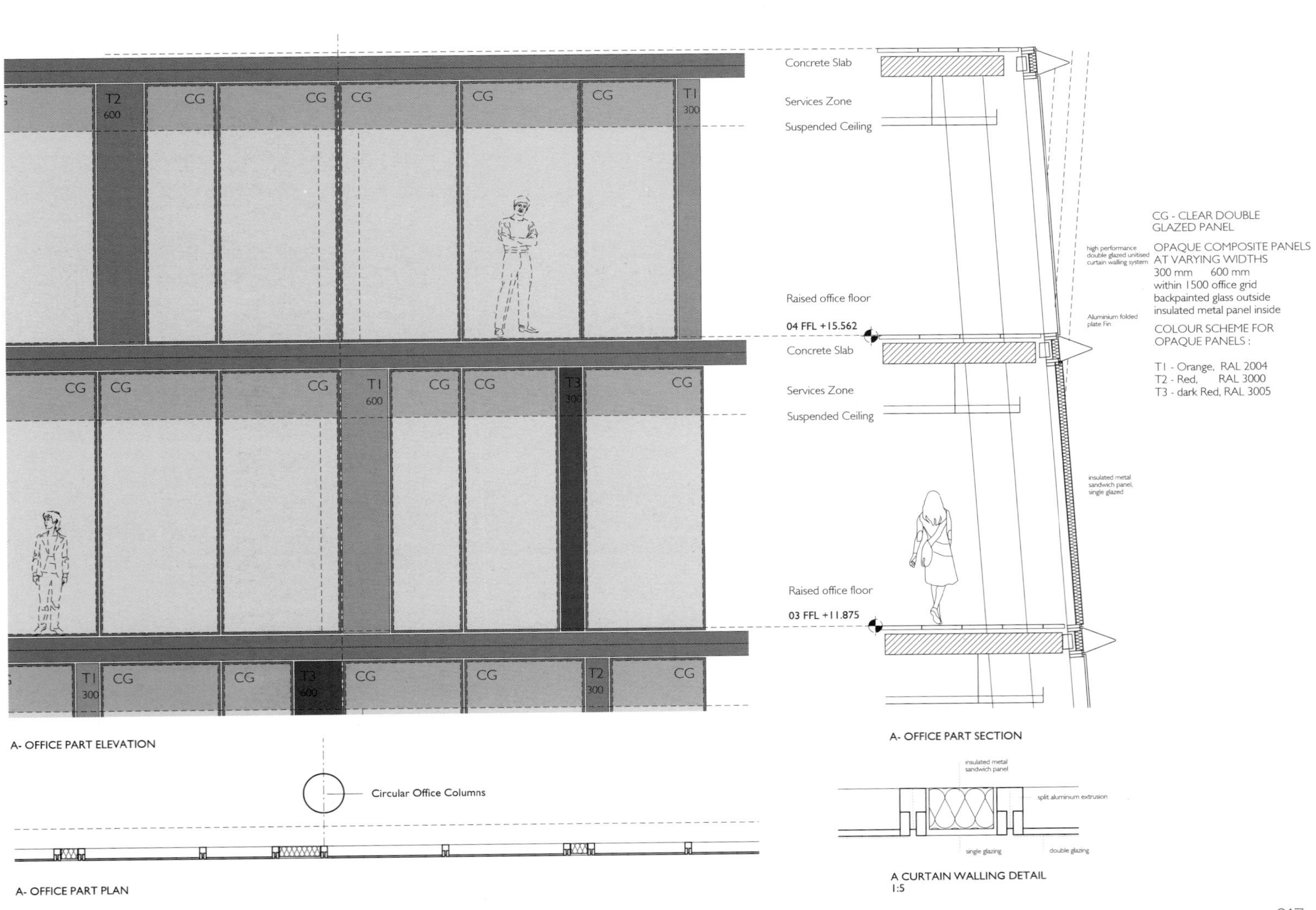

A- OFFICE PART ELEVATION

A- OFFICE PART SECTION

A- OFFICE PART PLAN

A CURTAIN WALLING DETAIL
1:5

Quattro Corti

Location: Saint Petersburg, Russia
Architect: Studio Piuarch
Client: Galaxy.LLC
Built Area: 23,500 m^2
Completion Date: 2006-2010
Photographer: Andrea Martiradonna

The Quattro Corti business center is located in the historic area of Saint Petersburg, in the immediate vicinity of St. Isaac's cathedral. The lots on which the building was erected are typical of this part of the city: about 60 meters deep with a single side facing the street and the remaining three bounded by walls.

The project called for the realization of a contemporary building while maintaining the historic facade of the two buildings that previously occupied the site. These were incorporated into the new structure by a metal roof whose continuous form reconnects the varying slopes of the existing roofs, the materials and geometry of which are integrated harmoniously into the city skyline.

Within the volume defined by these limits, four courts were created to illuminate the interior spaces and to serve as gathering places capable of hosting art installations, exhibitions and other public activities. The facades of these courts are composed of reflective glass panels set at different angles. The result is the fracturing of the overall reflection, which generates a kaleidoscopic effect that is enlivened by the natural variations in the quantity and intensity of sunlight. Each court uses a different colored glass, which in turn creates different atmospheres. The four colors—gold, green, azure and white—were inspired by the chromatic richness of St. Petersburg's historic architecture.

The building has an area of 23500 square meters distributed over six above-ground floors, one basement floor and two more below ground equipped with automated parking. The complex is mainly dedicated to Gazprom Neft offices, but also hosts complementary activities such as a restaurant. The Mansarda restaurant is on the top floor and features a terrace that offers spectacular panoramic views of the courtyards of the complex and the city center, particularly of the facade and cupola of St. Isaac just across the way.

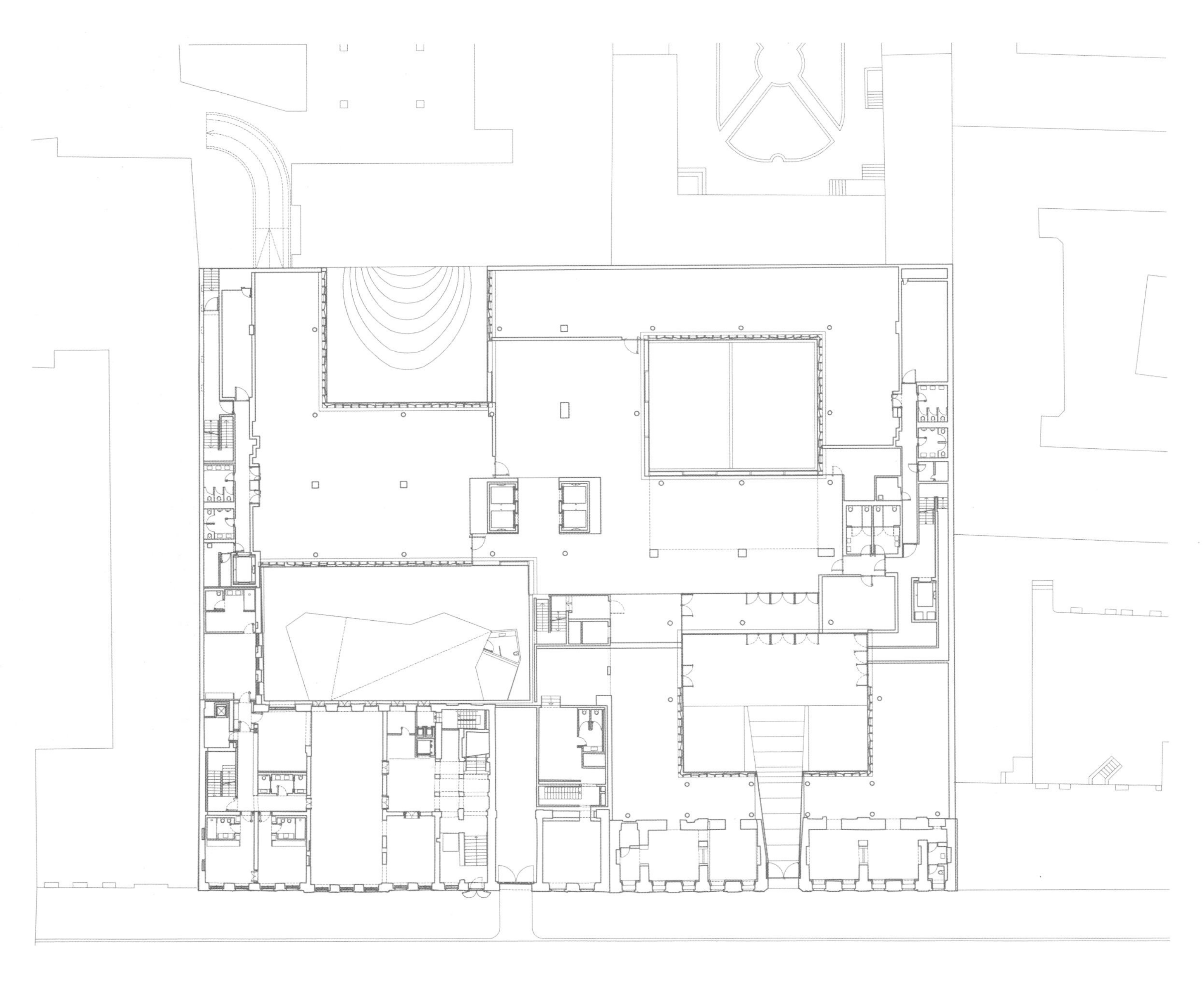

Ground floor plan

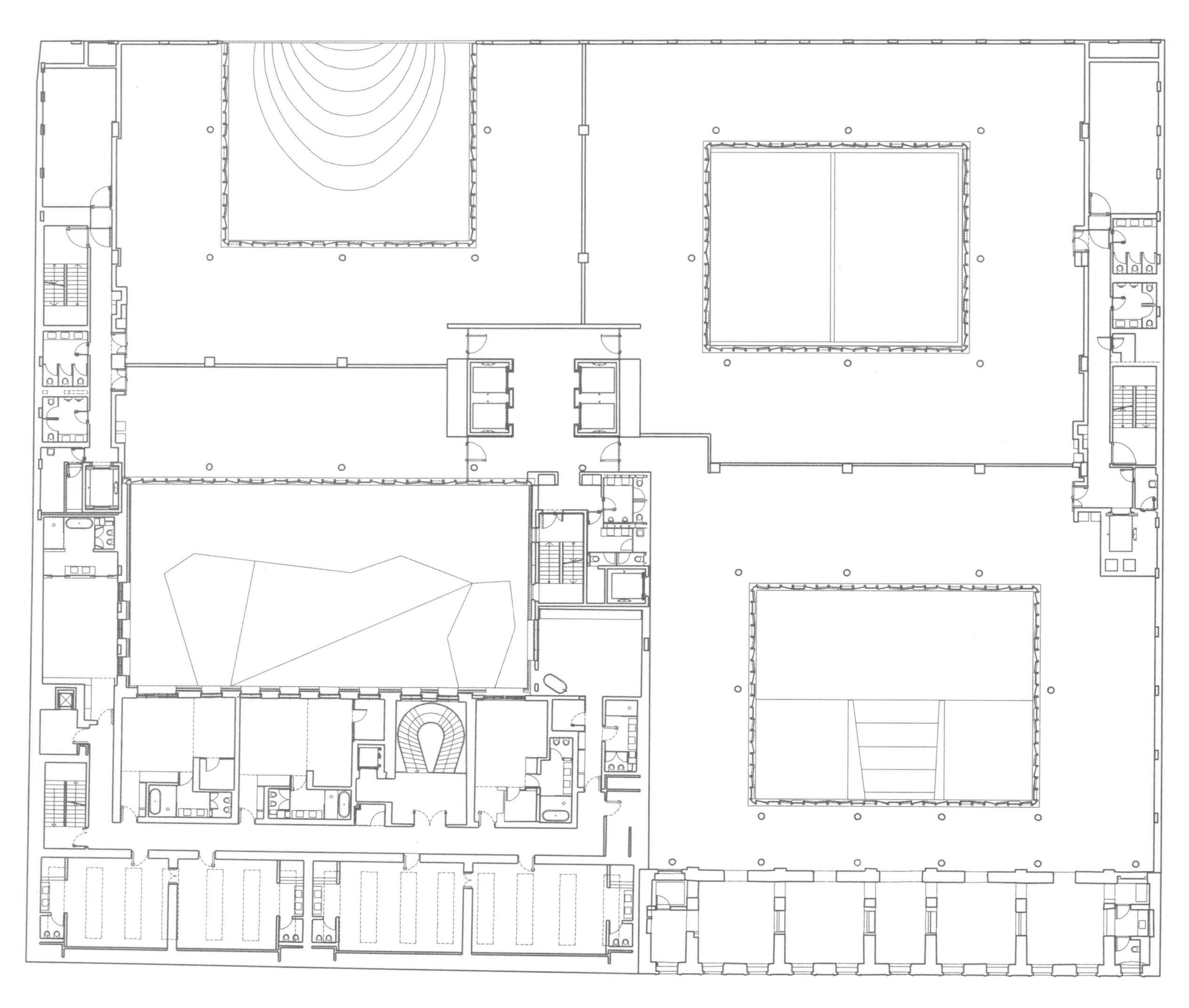

Typical floor

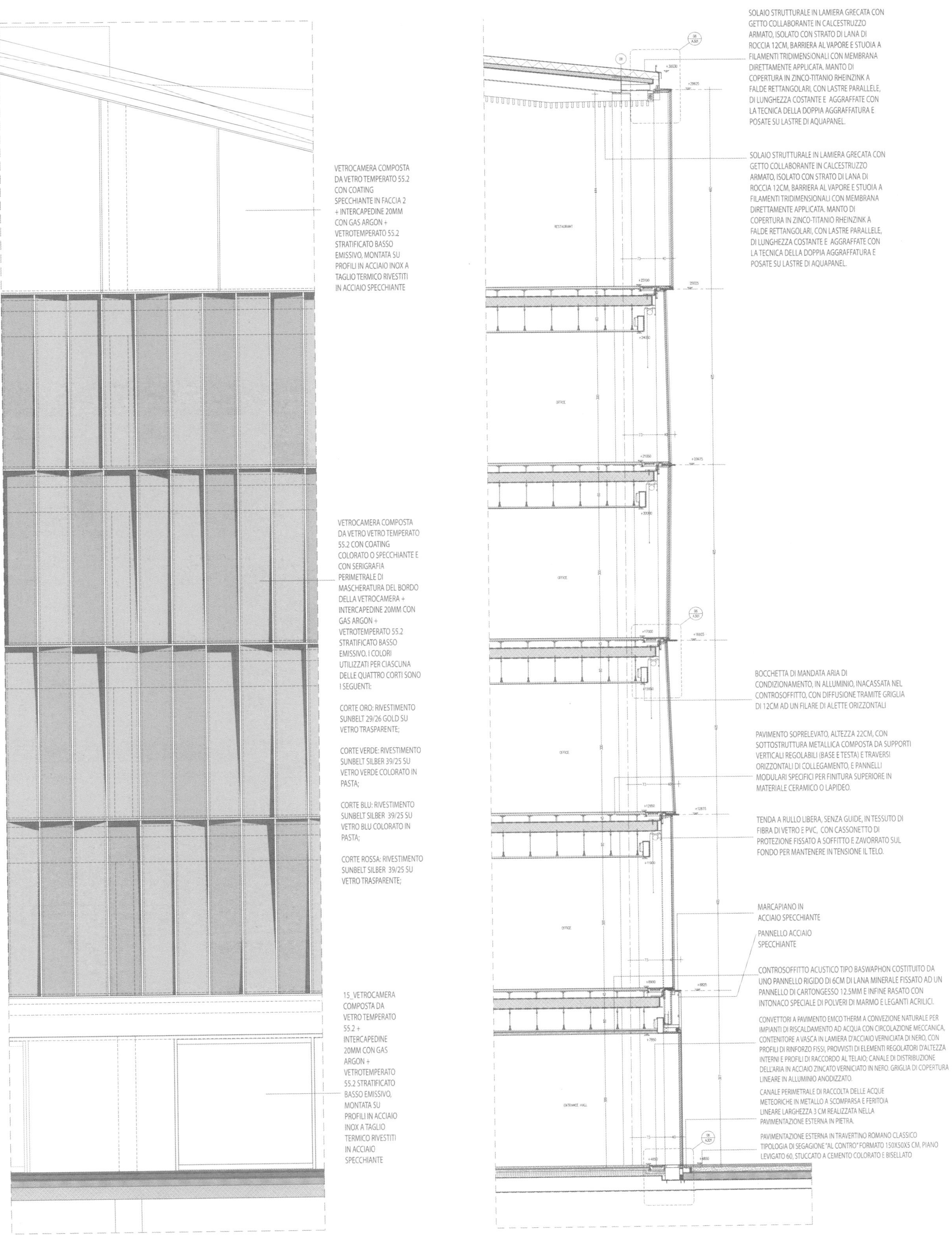

Dett facciata

Financial and Commercial Department of Voest Alpine Stahl GmbH

Location: Linz, Austria
Architect: Dietmar Feichtinger Architectes
Client: voestalpine Stahl GmbH
Site Area: 36,700 m²
Completion Date: 2009

The Voest steelworks in Linz has erected its new, representative sales and finances head office around an extensive open area. The urban planning concept and the design of the buildings on the square were carried out by Austria's most important architect living abroad, following a competition between several top-rank designers.

Dietmar Feichtinger Architectes rigorously cleared the entire competition site of motor vehicles. Built of reinforced concrete of remarkable quality, flanked on the open edges by planted embankments, naturally cross-ventilated, lit from planted atriums and broken up into almost domestic proportions by glass walls, some transparent, some satin finish, this garage with its light- coloured polished floor continues the business of representation begun above ground level in a highly skilled way. Whoever does not exploit the privilege of driving under the new steel and glass canopy (also designed by Dietmar Feichtinger Architectes) that projects far in front of the BG41 building and instead, when the weather is inclement, wishes to stay dry by walking through the garage, will not be disappointed by the underground setting.

However the main entrance to the sales and finance office lies, impossible to overlook below a daring cantilever that exploits steel's structural possibilities at the north-eastern end of the building. Here you can see an entire a glazed, seemingly unbounded, foyer that focuses the attention of visitors from the steel reception desk on the one side to the impressive landscape of cranes, waste heaps, and chimneys on the other. From here a panorama lift takes you to the meeting rooms, accessible to a wider public, that are located in this part of the building. In terms of number, size and fittings these rooms respond to the needs of the location Linz as a whole and are augmented by a terrace on the top floor, surrounded by glass walls and open to the sky, which offers exceptional views. All the rooms on the ground floor, which is clearly shaped as a plinth and has a delicately profiled glass facade, are also reserved for "public" functions such as the company's own travel agency, an archive, advertising media department and similar.

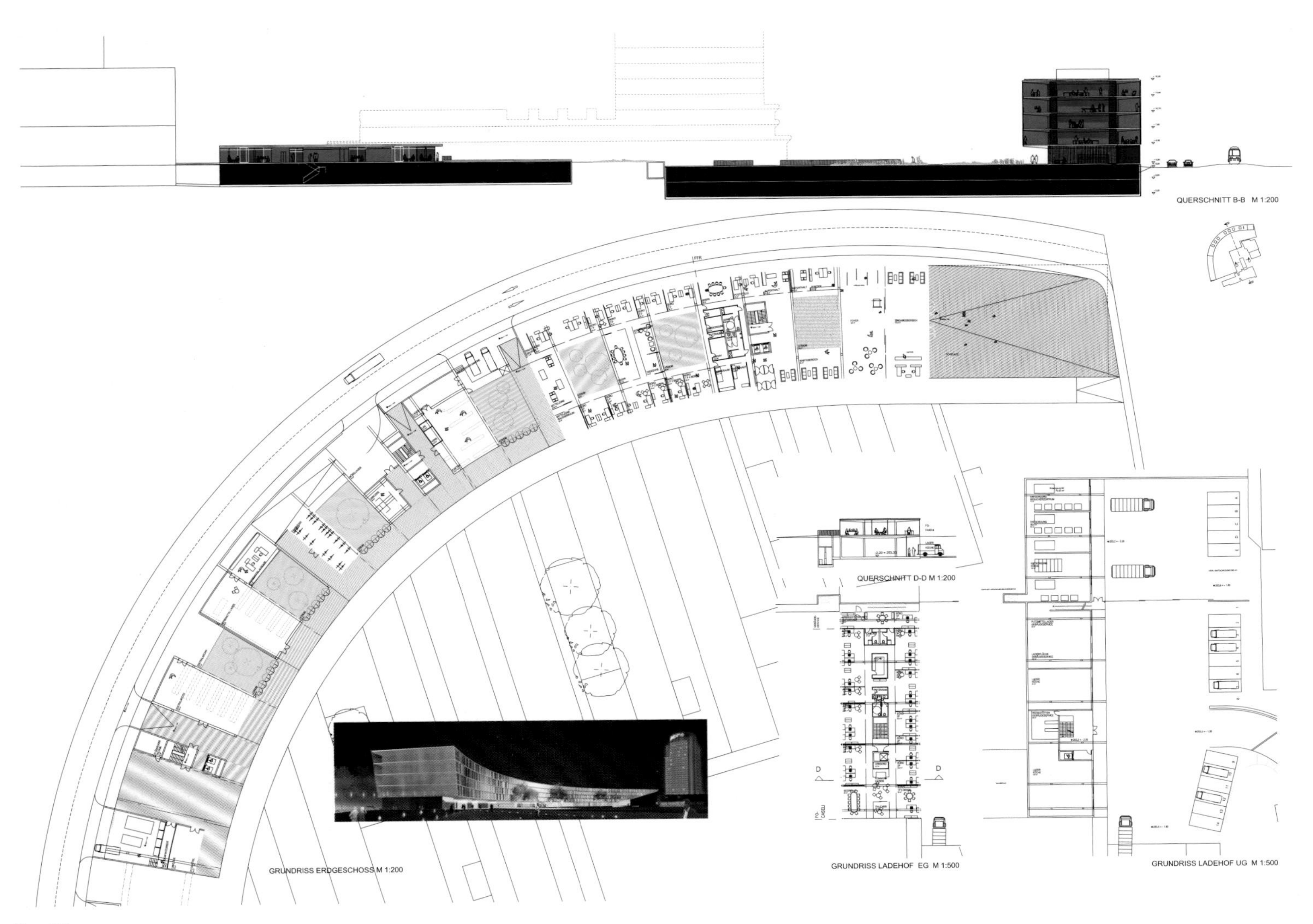

Plan 04

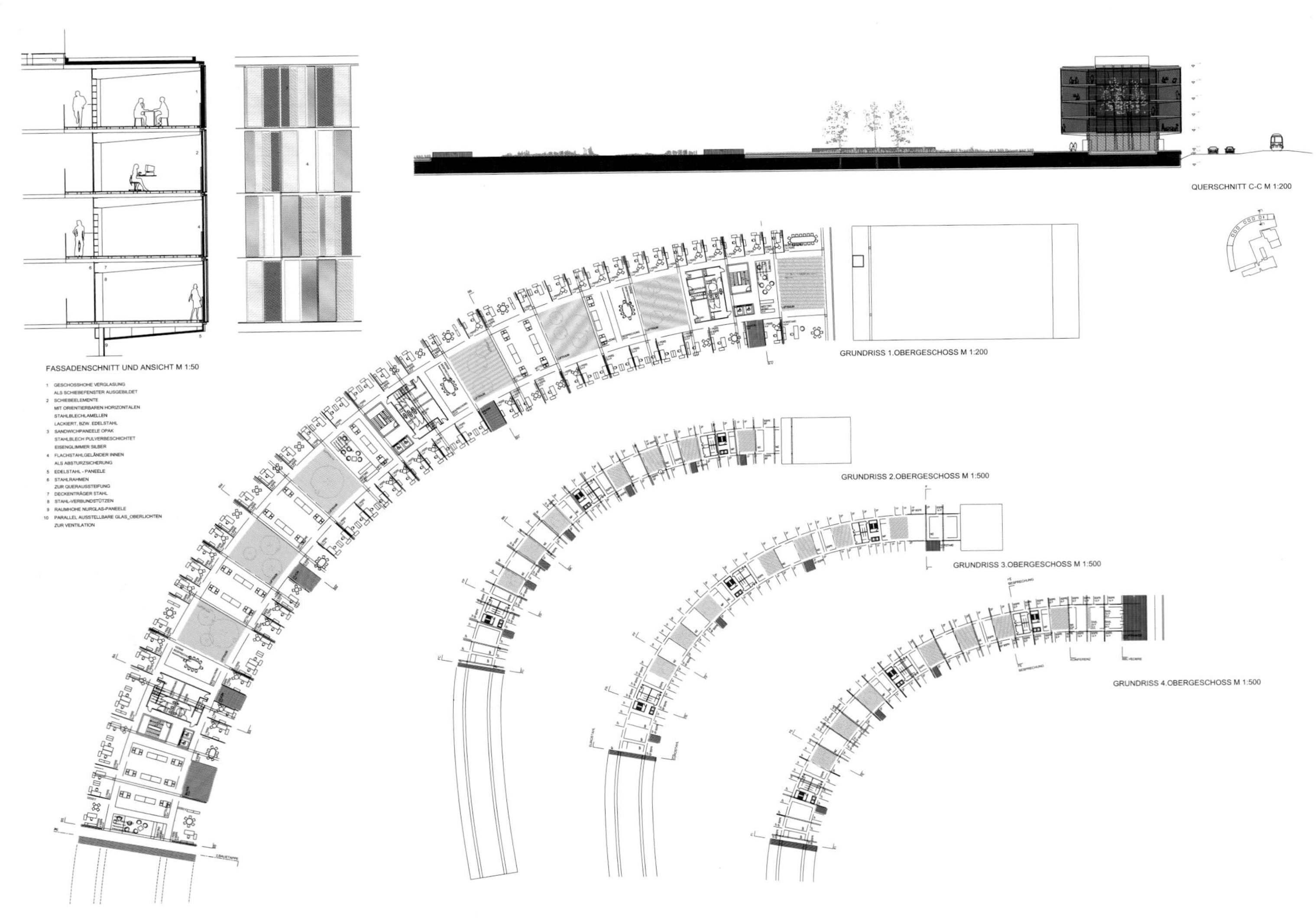

Plan 05

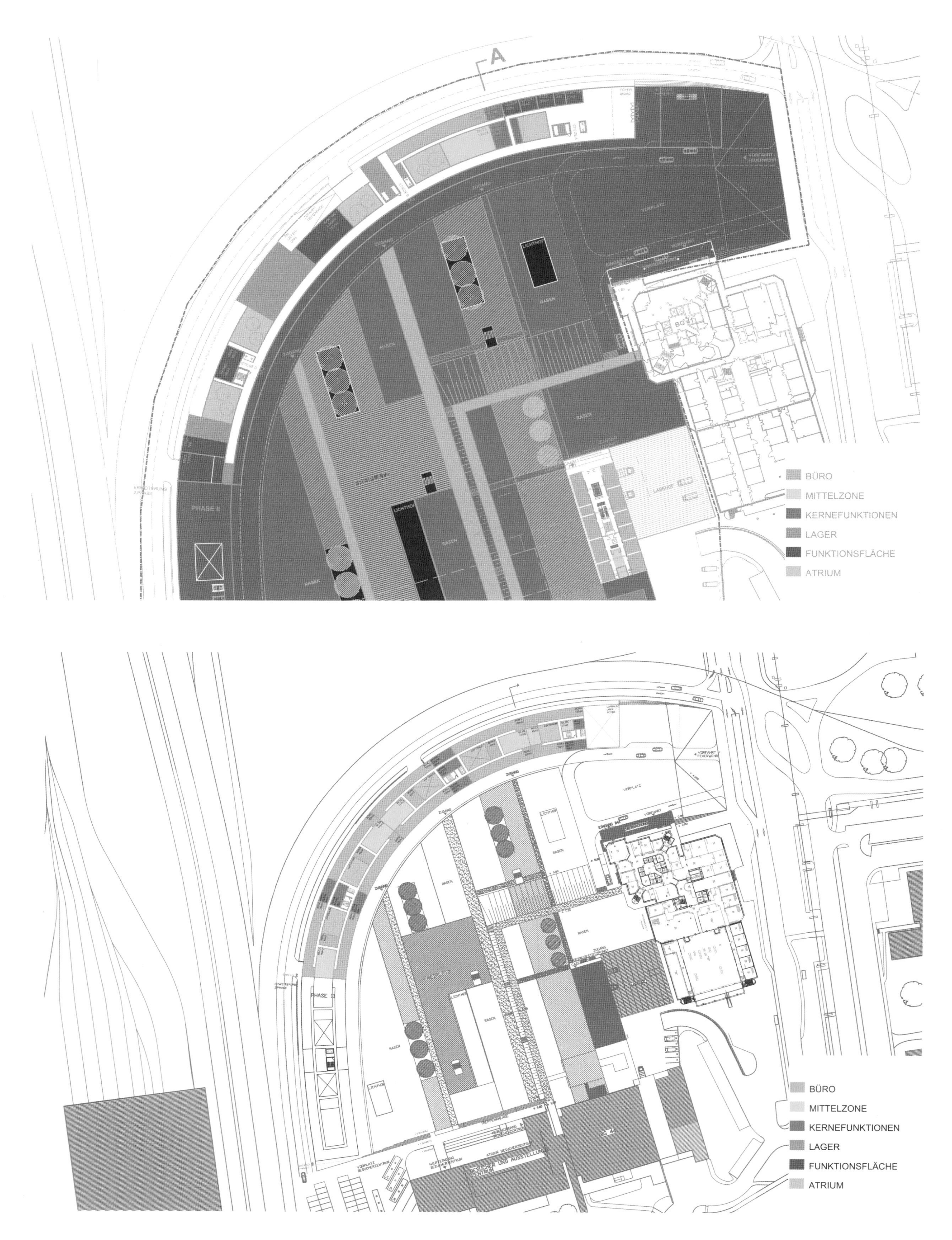
BÜRO
MITTELZONE
KERNEFUNKTIONEN
LAGER
FUNKTIONSFLÄCHE
ATRIUM
LICHTHOF
RASEN
FREIPLATZ
VORPLATZ
PHASE II
LADEHOF
BÜRO
MITTELZONE
KERNEFUNKTIONEN
LAGER
FUNKTIONSFLÄCHE
ATRIUM

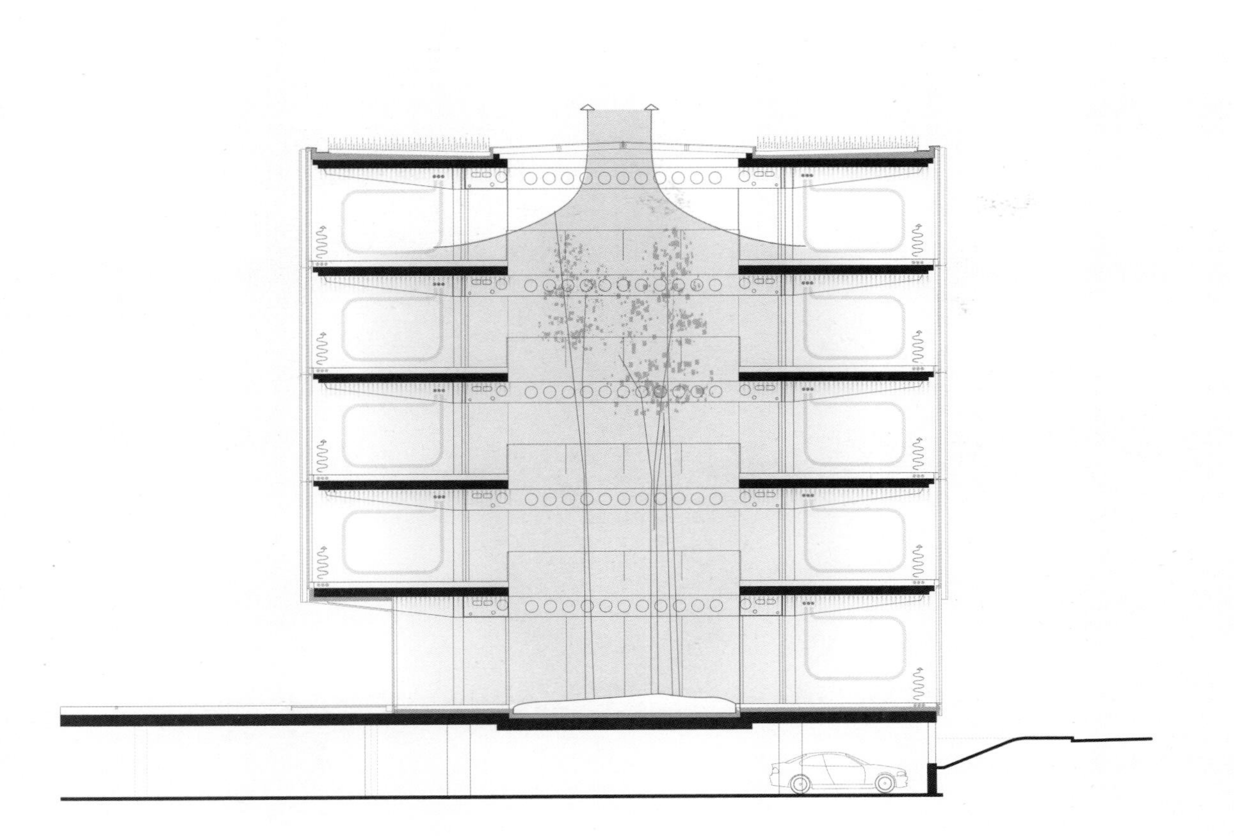

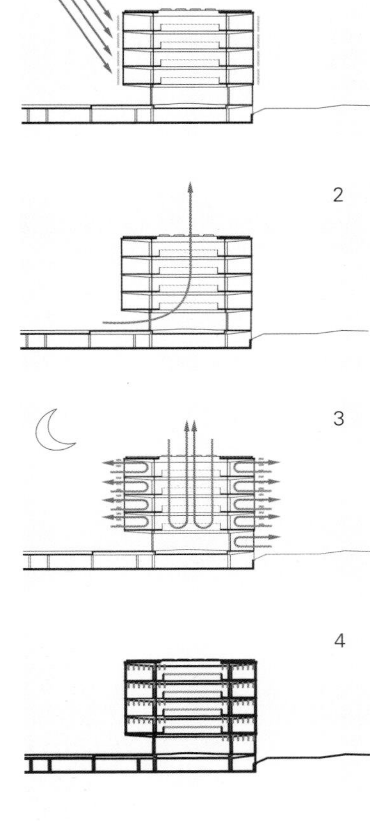

Description
1 Sunscreen
2 Cross-ventilation
3 Night cooling through gap ventilation
4 Maximizing the effective storage mass by exposed concrete ceilings
5 Treatment of waste heat from industrial plant

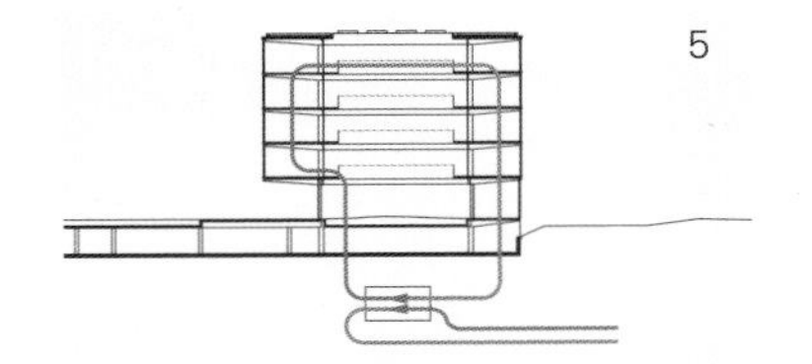

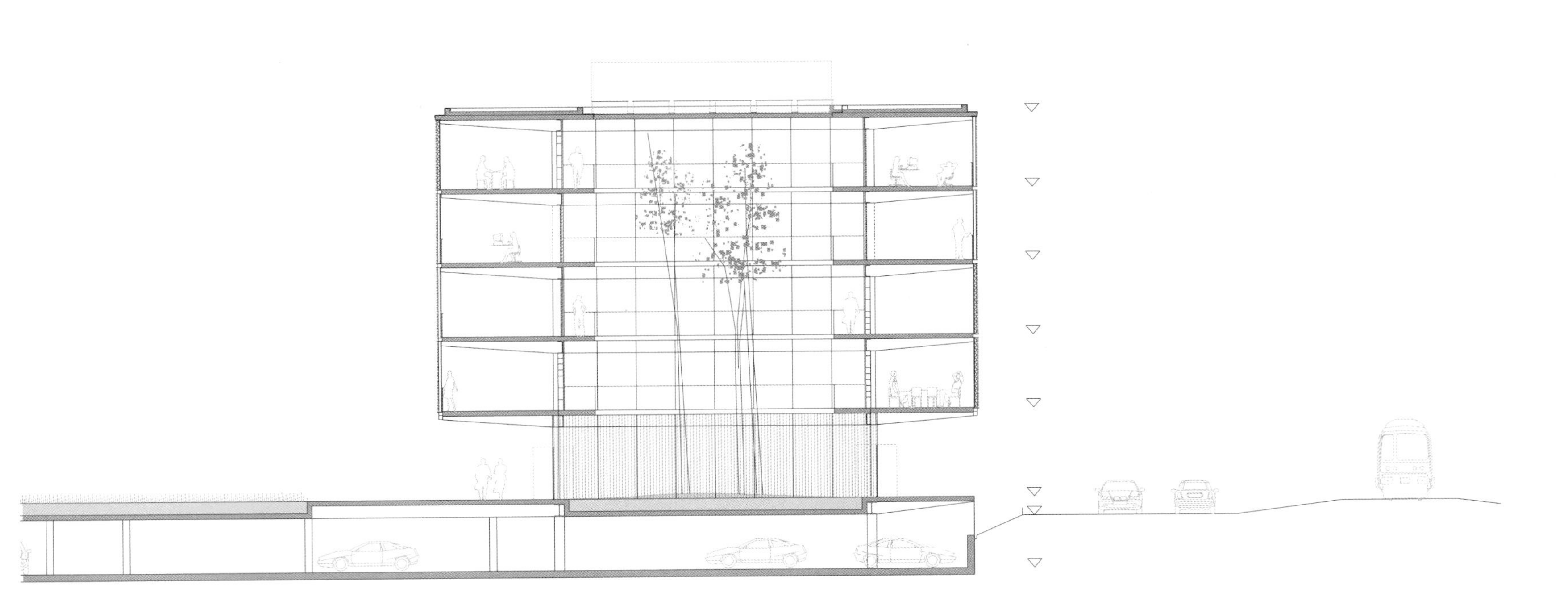

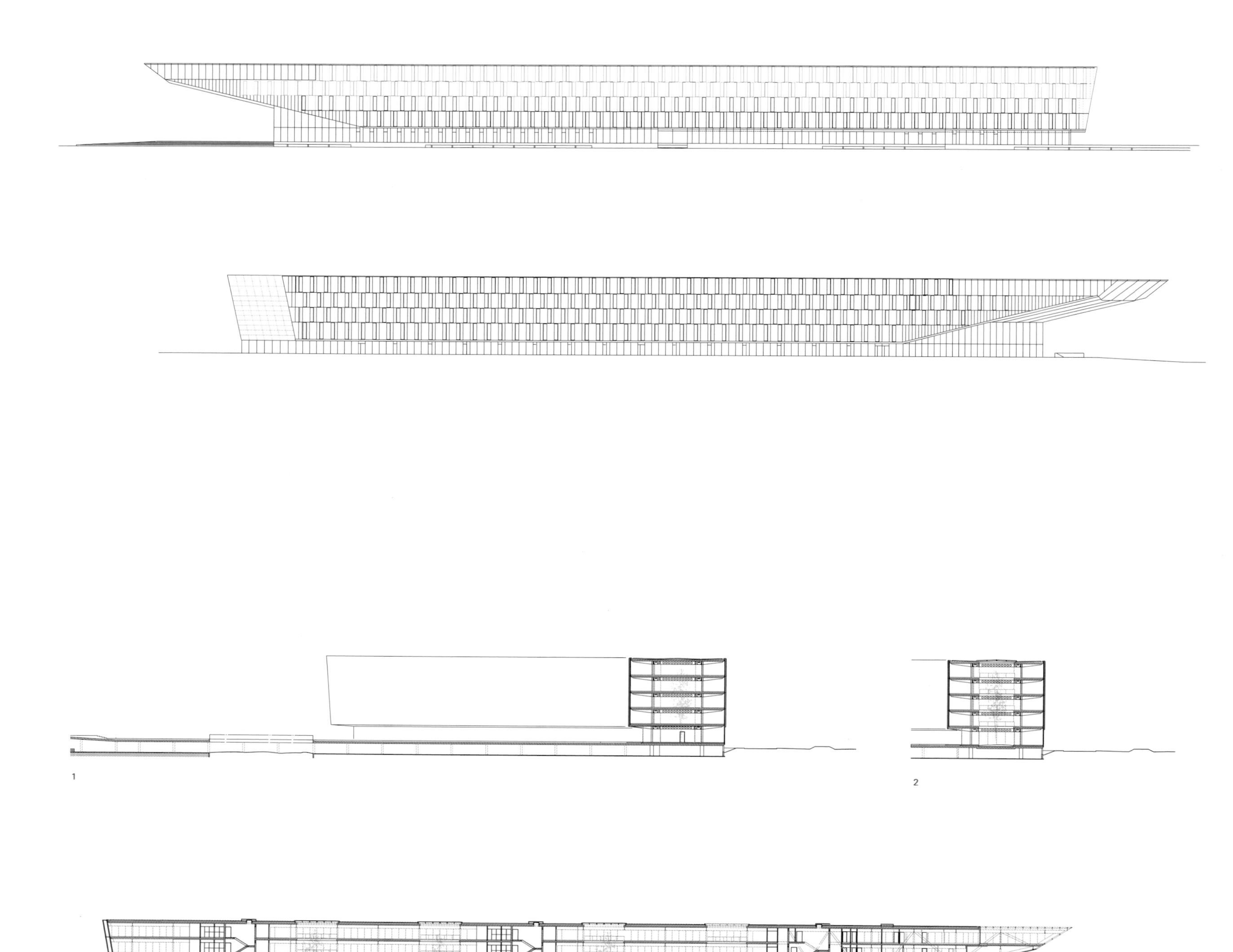
1
2
3

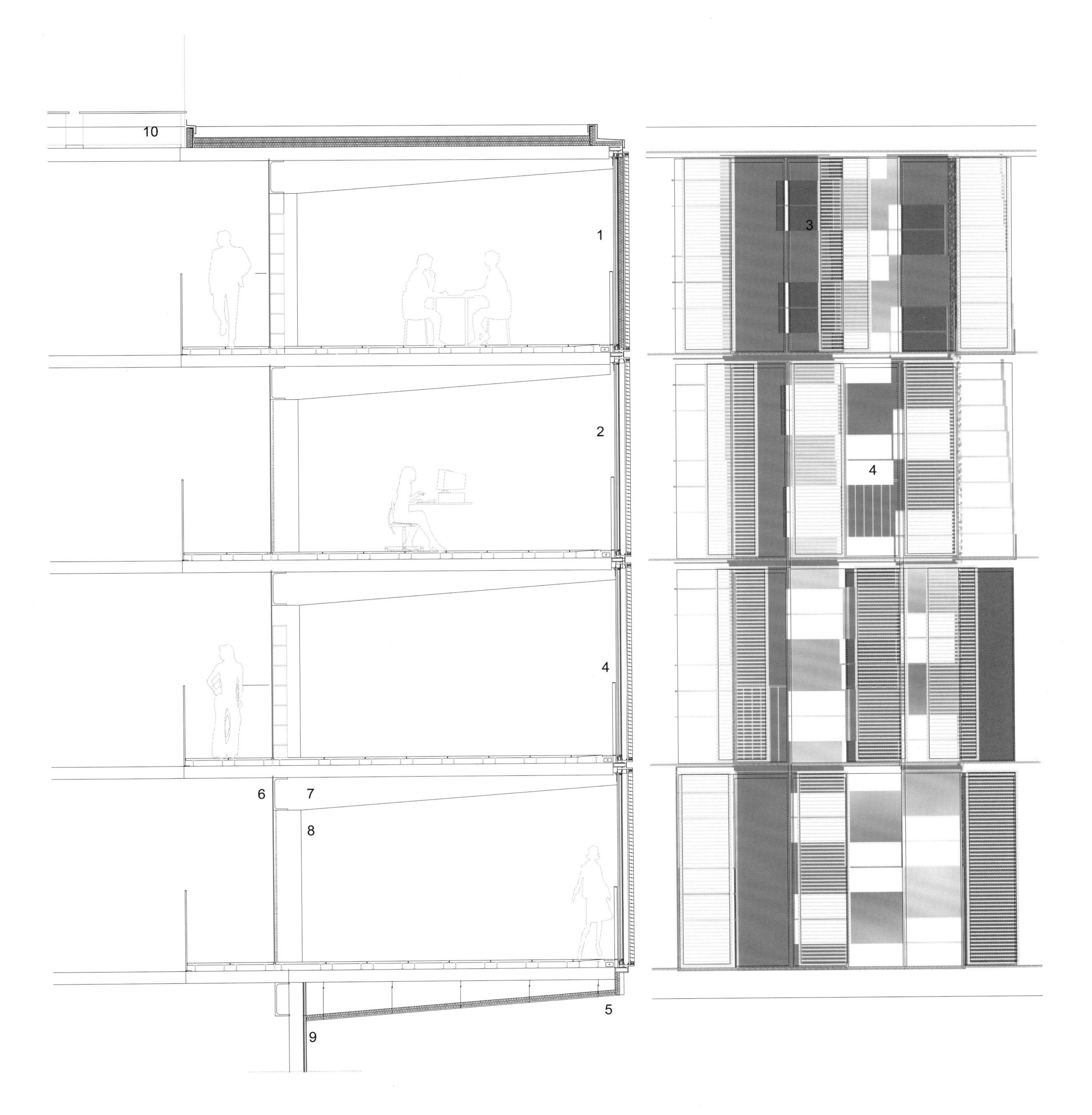

FACADE AND SECTION WITH VIEW
1. Basement ceiling glazing designed as a sliding window
2. Slide elements with mobile horizontal steel laminations painted or stainless steel
3. Sandwich panels opaque sheet steel powder coated silver Eisenglimmer
4. Flat steel railings internally absturzsicherung
5. Stainless steel panels
6. Steel frame for transverse reinforcement
7. Steel joists
8. Steel composite columns
9. Room height only glass panels
10. Parallel opening front glass skylights for ventilation

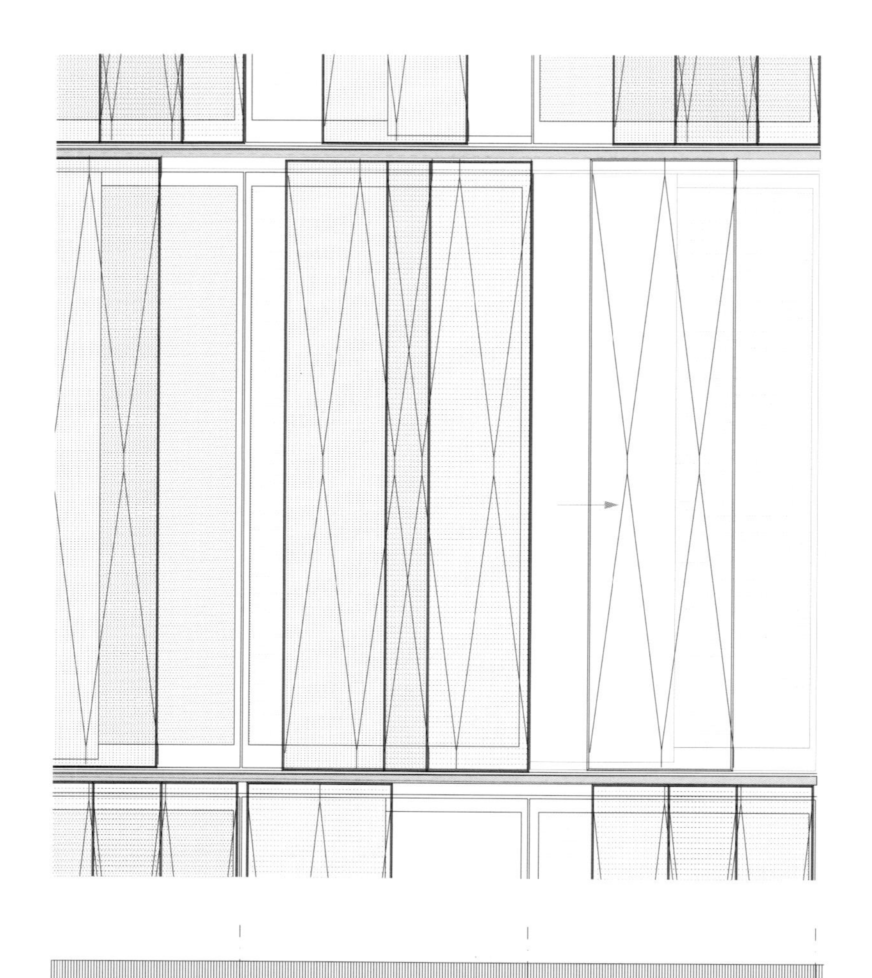

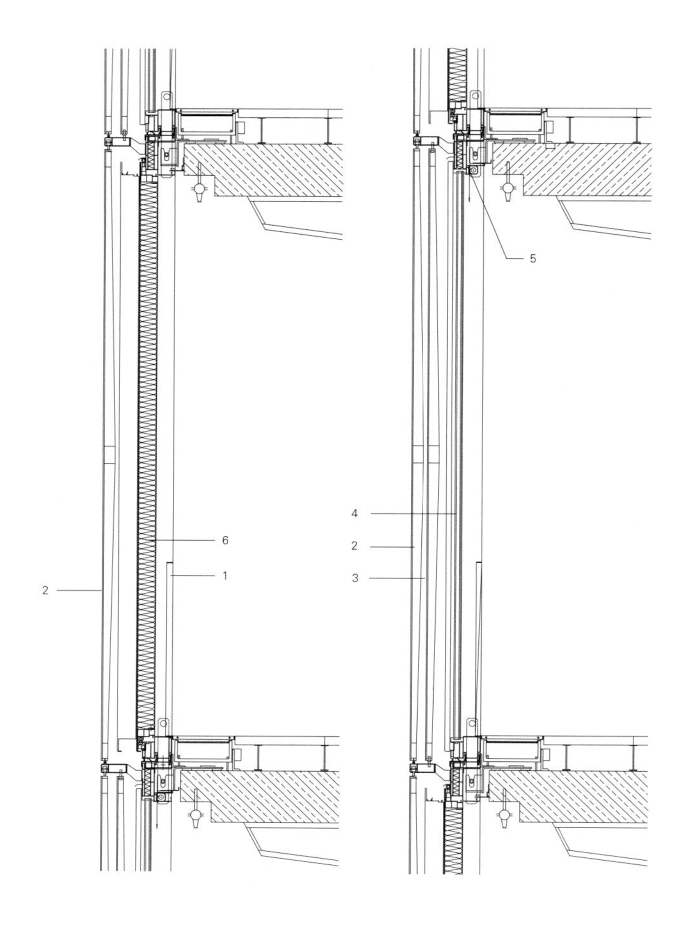

1. Fall protection flachstahl powder coated ral 7036
2. Sun protection sliding element made of expanded metal flat steel construction
3. Sun protection fixed element metallic yellow powdercoated
4. Storey height fixed glazing
5. Antiglare
6. Parallel opening sash window als sandwich powder coated ral 7036

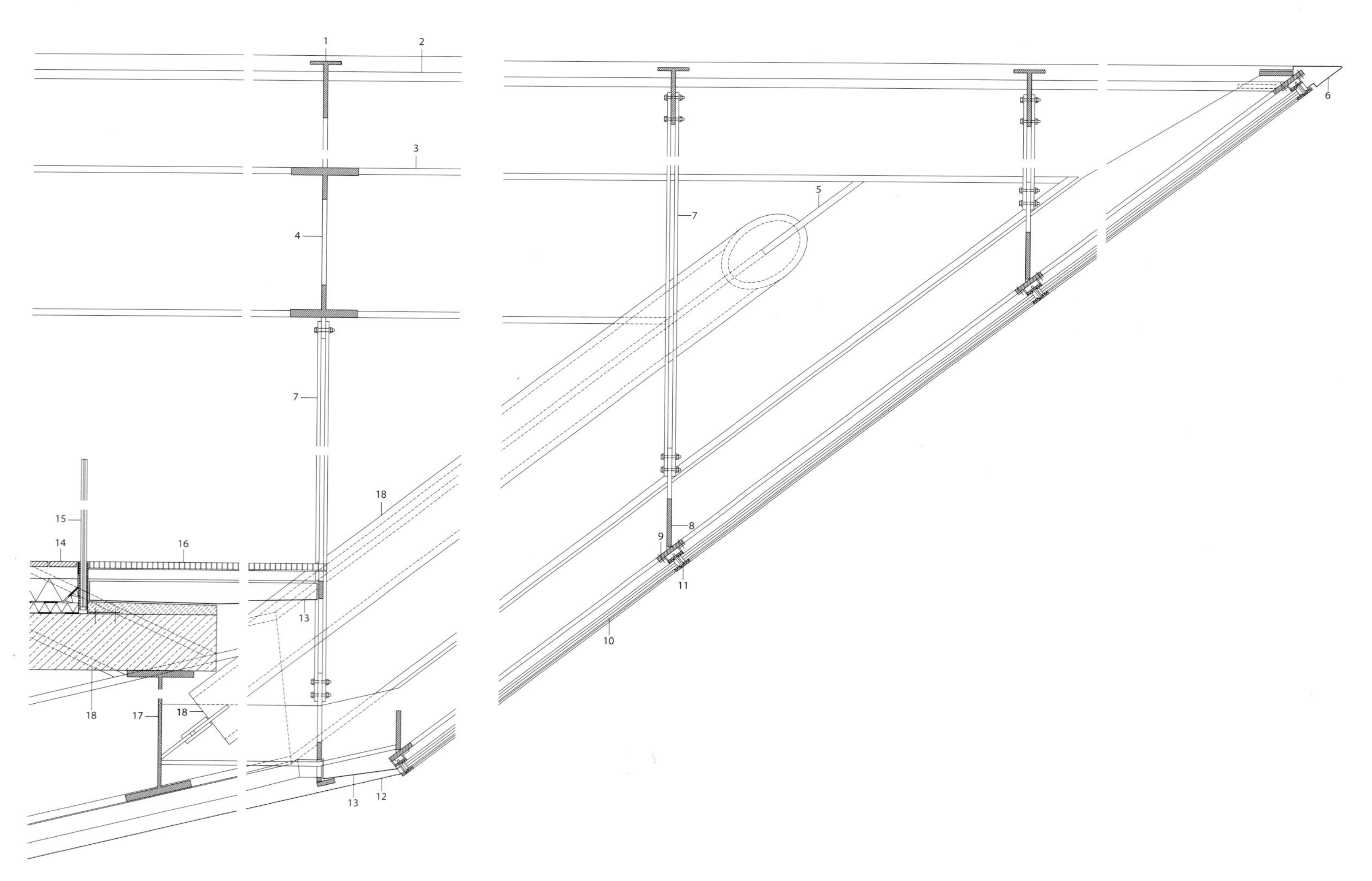

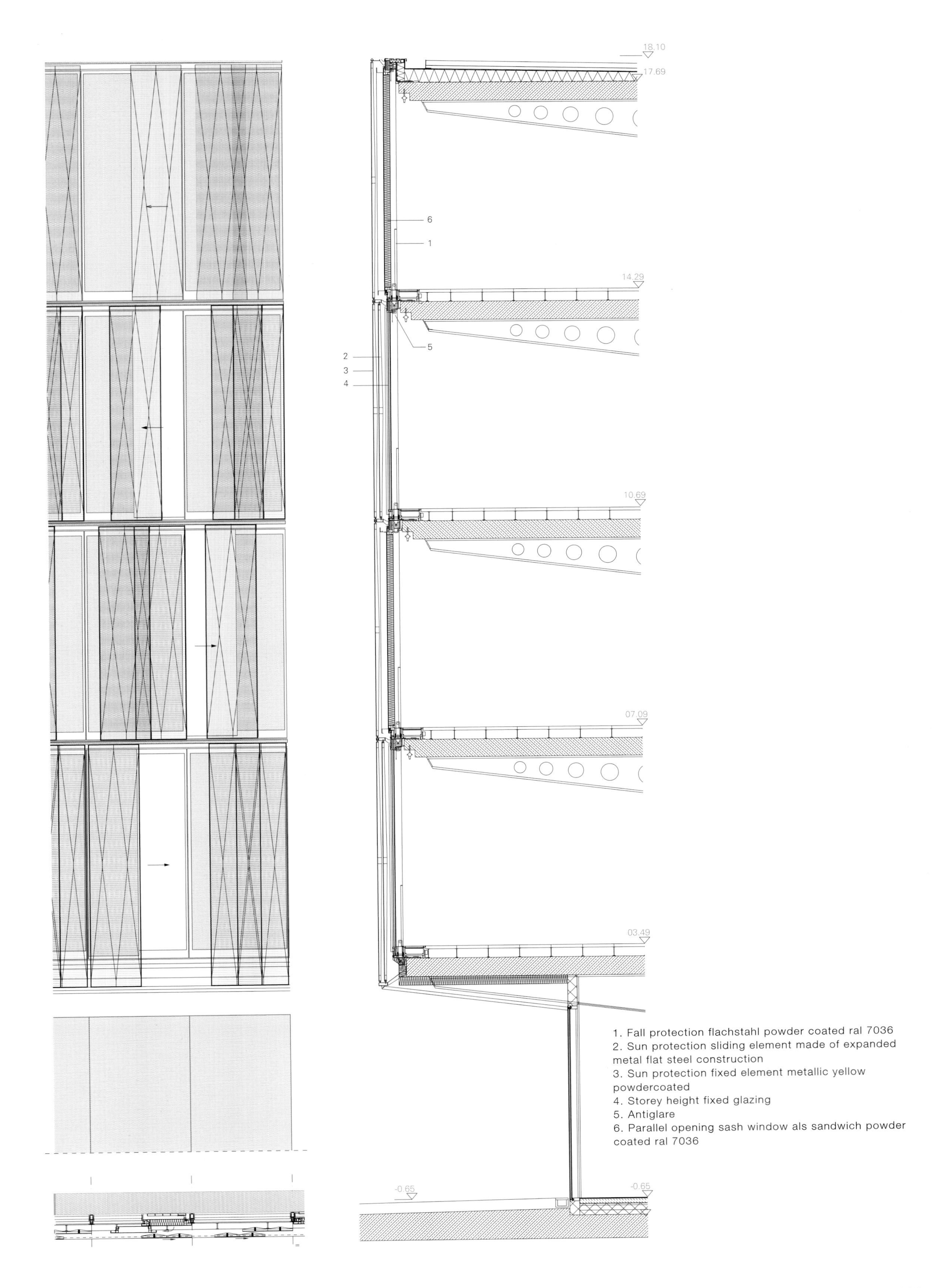

1. Fall protection flachstahl powder coated ral 7036
2. Sun protection sliding element made of expanded metal flat steel construction
3. Sun protection fixed element metallic yellow powdercoated
4. Storey height fixed glazing
5. Antiglare
6. Parallel opening sash window als sandwich powder coated ral 7036

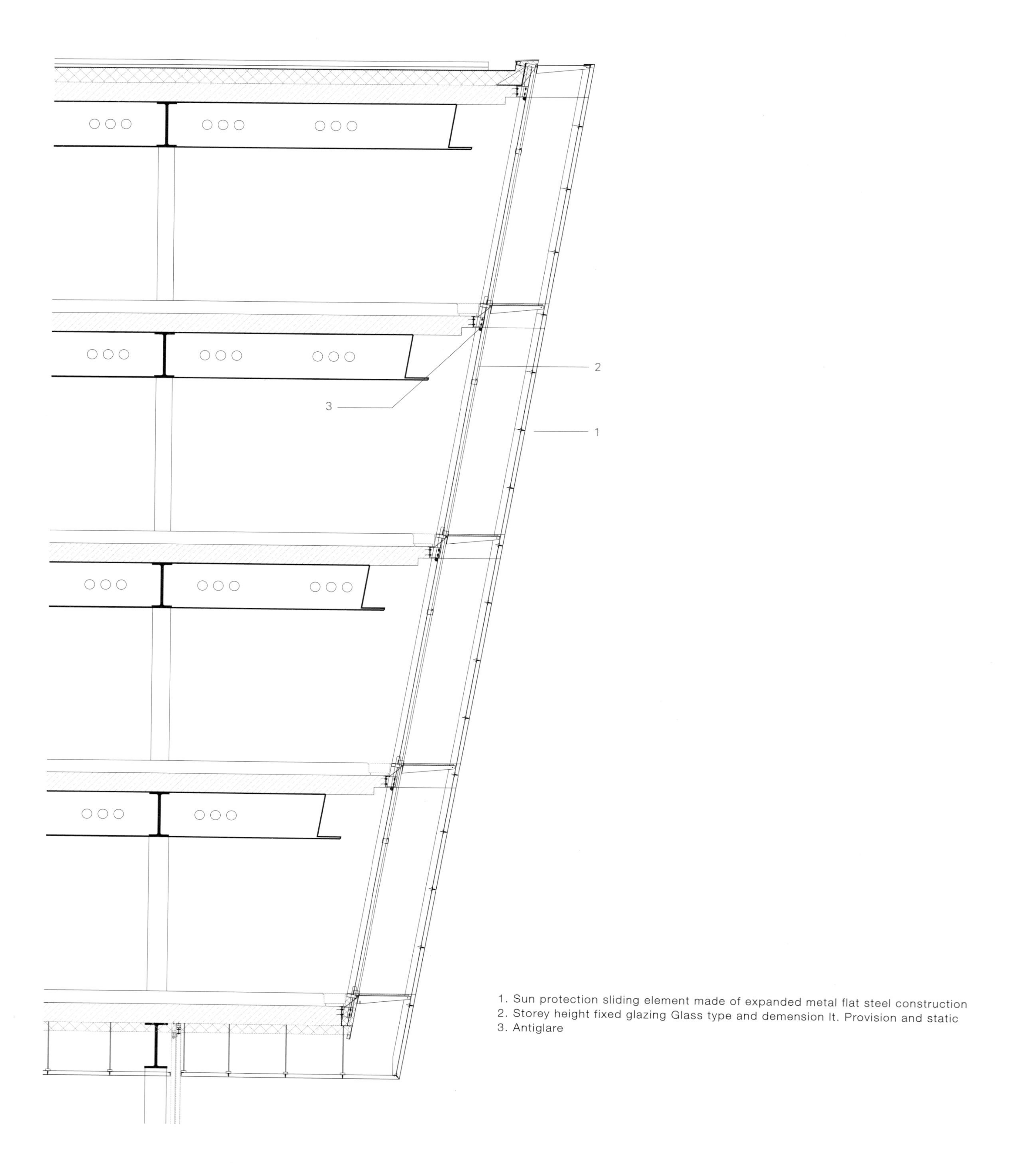

1. Sun protection sliding element made of expanded metal flat steel construction
2. Storey height fixed glazing Glass type and demension lt. Provision and static
3. Antiglare

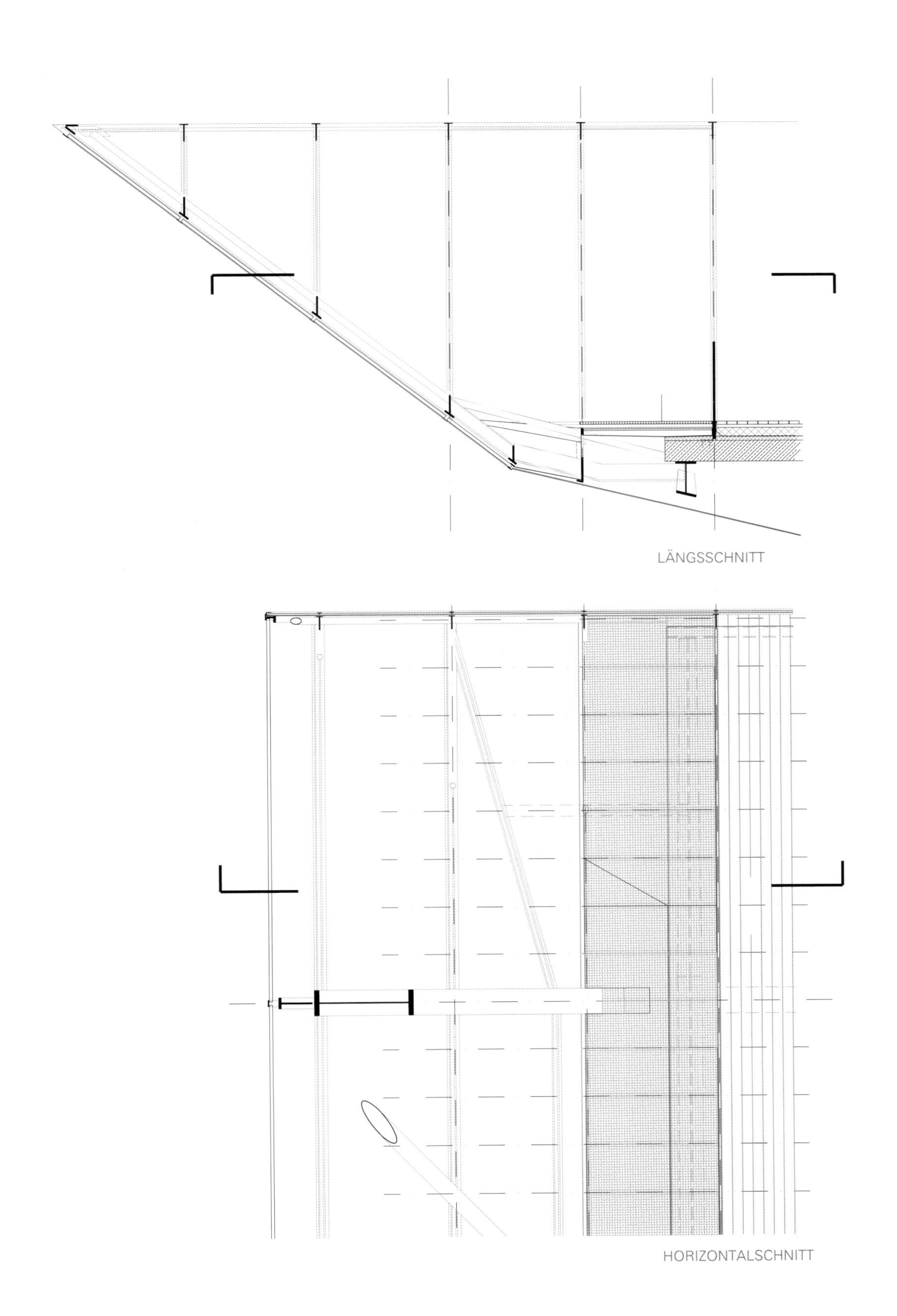
LÄNGSSCHNITT
HORIZONTALSCHNITT

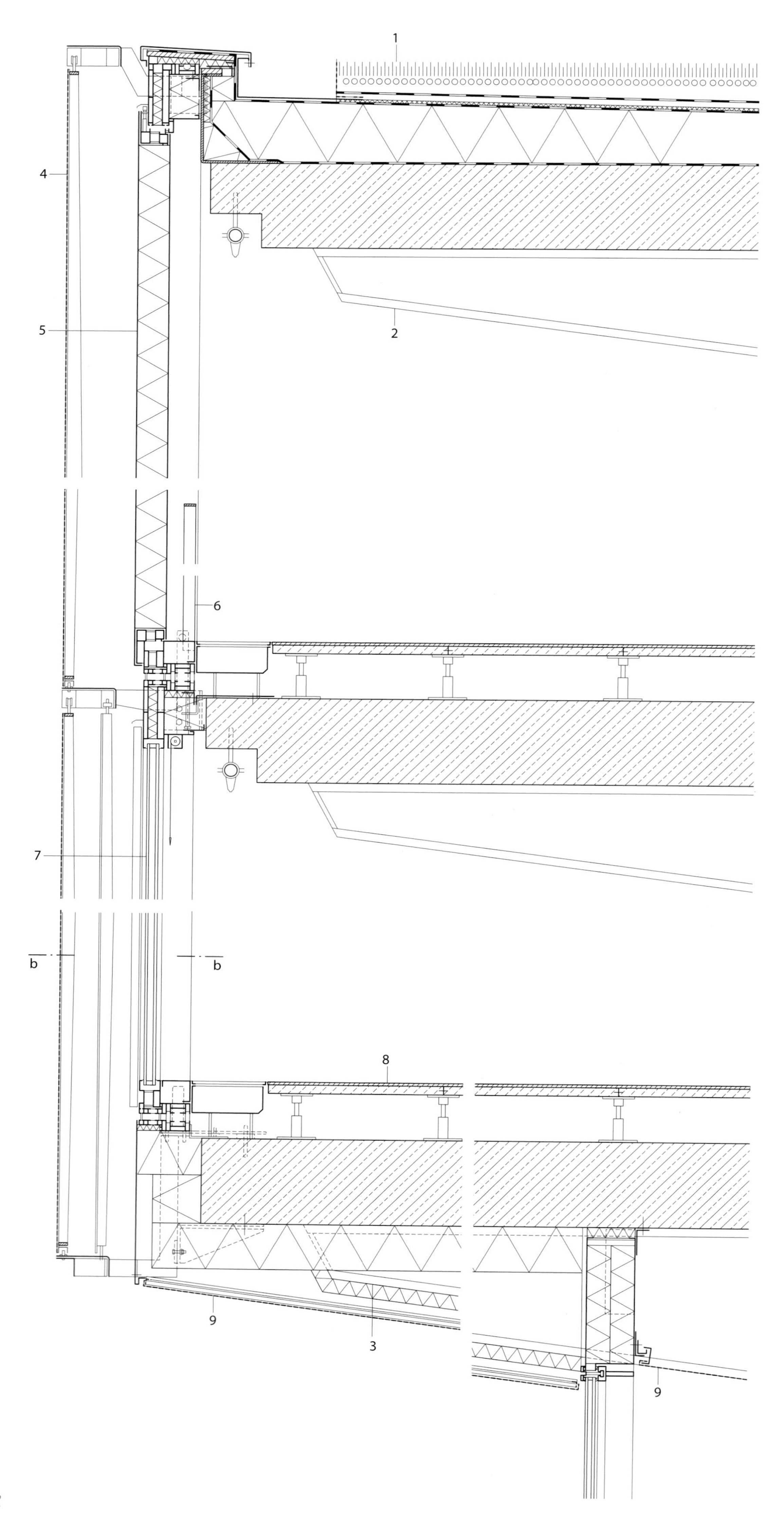
1
4
5
2
6
7
b
b
8
9
3
9

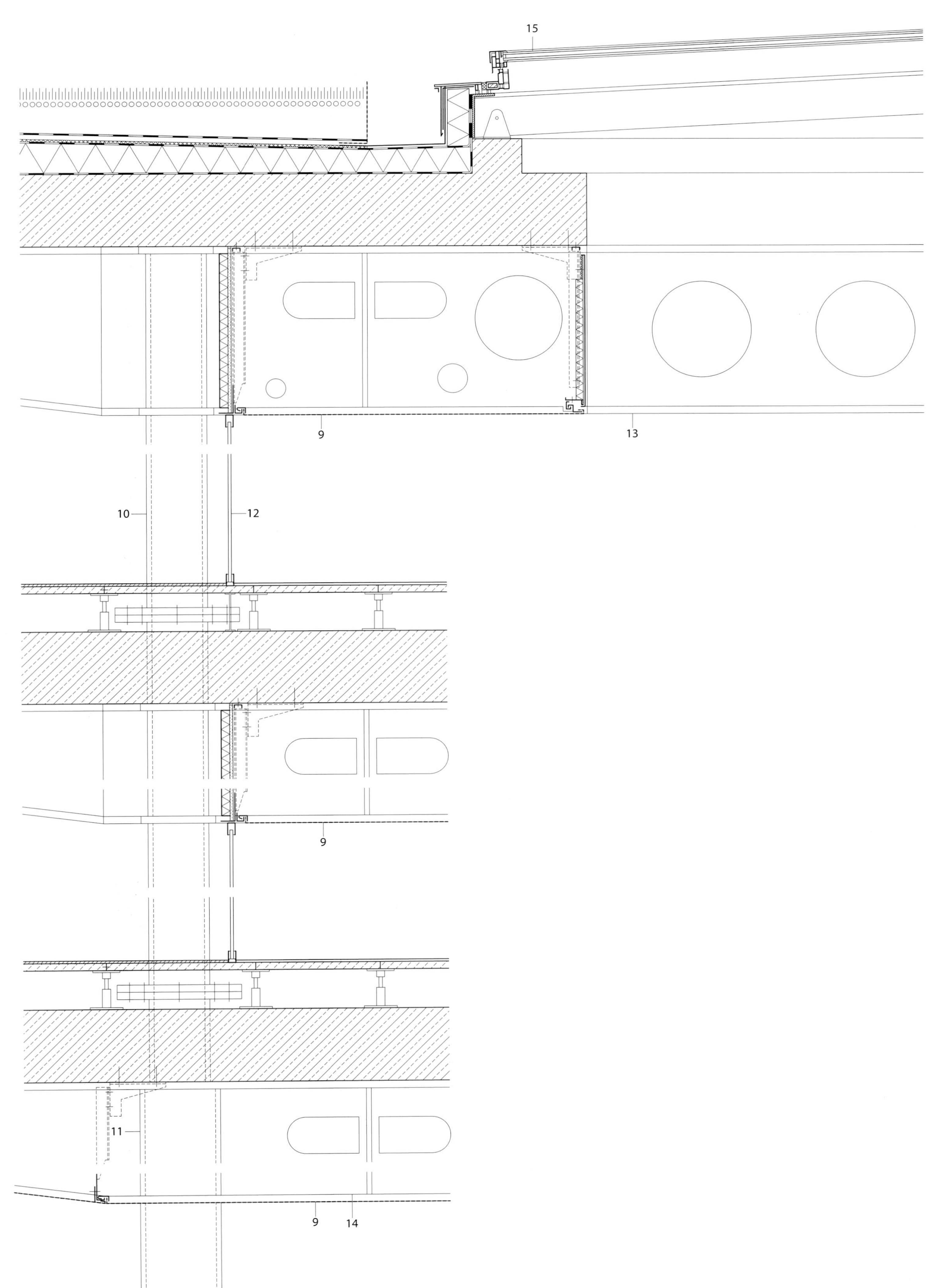
15
9
13
10
12
9
11
9
14

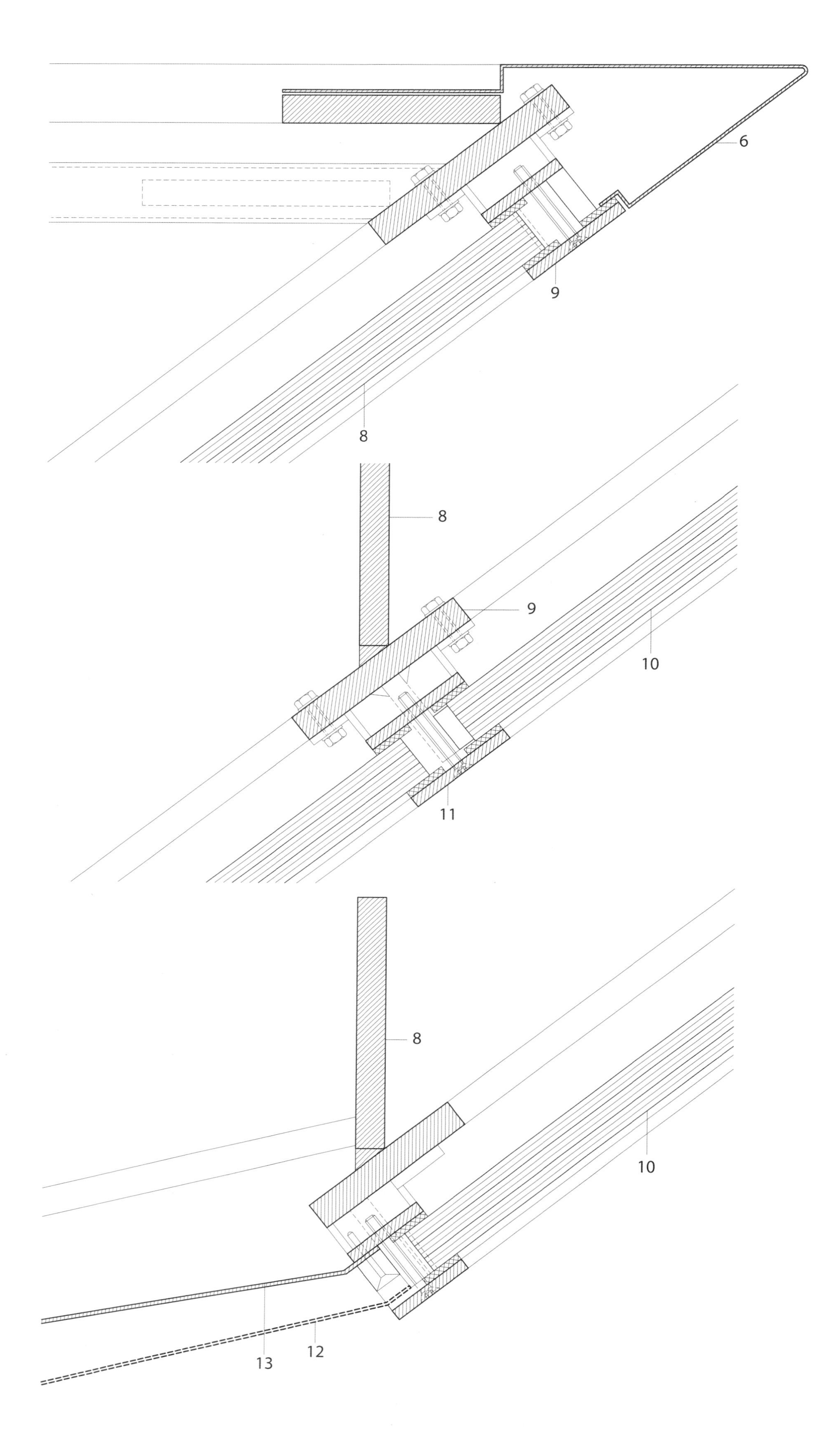
6
9
8
8
9
10
11
8
10
13
12

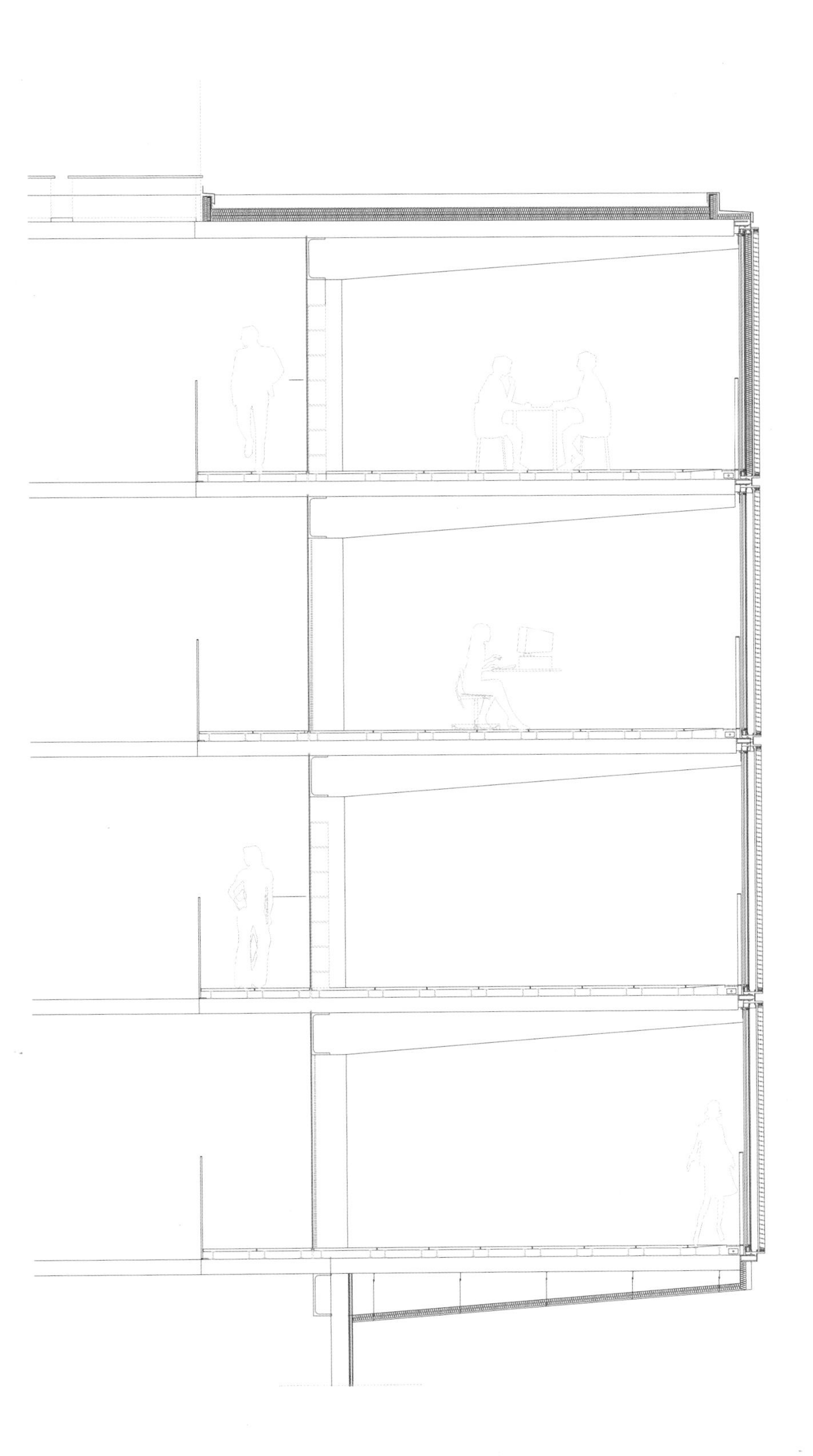

3

Trade Fair Centre

Location: La Spezia, Italy
Architects: MMAA
Client: Centro Fieristico della Spezia
Site Area: 9,200 m²
Floor Area: 5,500 m²
Completion Date: 2007
Photographer: Roberto Buratta

The new expo pavilion has been built in a former industrial area previously occupied by a pasta factory. The building, which features a coated iron structure and envelope made of titanium-zinc panels and vast glazed surfaces, takes the form of a long parallelepiped, folded over on itself that soars up from the ground, jutting out sharply at the end to shelter the entrance facing the pre-existing multiplex. The shape of the building is based on the idea of creating an architectural promenade that unfurls from the highest point of the building, through an exhibition space, down to the bottom.

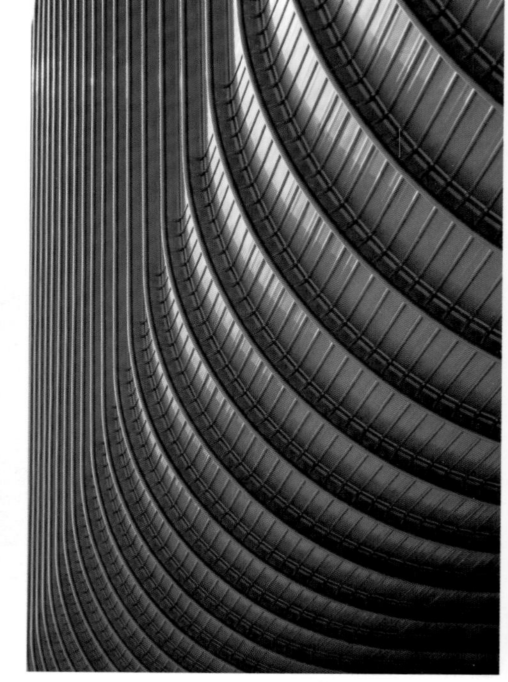

The shape of the ramp that leads through the interior makes the outside of the building look like a giant, twisted metal tube. This striking form also becomes an element that structures the interior layout, creating the right hierarchy of space without sacrificing overall harmony: the unified space can house different functions at the same time, or be used as single venue.

On the ground floor are public and service entrances, and many of the exhibition areas, which then seamlessly continue through the space of the intermediate and upper floor.

This top floor can be used separately from the rest of the building, for exhibitions that are thematic or reserved for a select public, since it has its own entrance and a section with stairs and elevators that can be dedicated to this specific use. Due to its location, the fair becomes an ideal gateway to the city, looking onto Via Carducci through a glazed wall parallel to the street, topped by a single metal roof that also becomes a unifying element in the architectural whole.

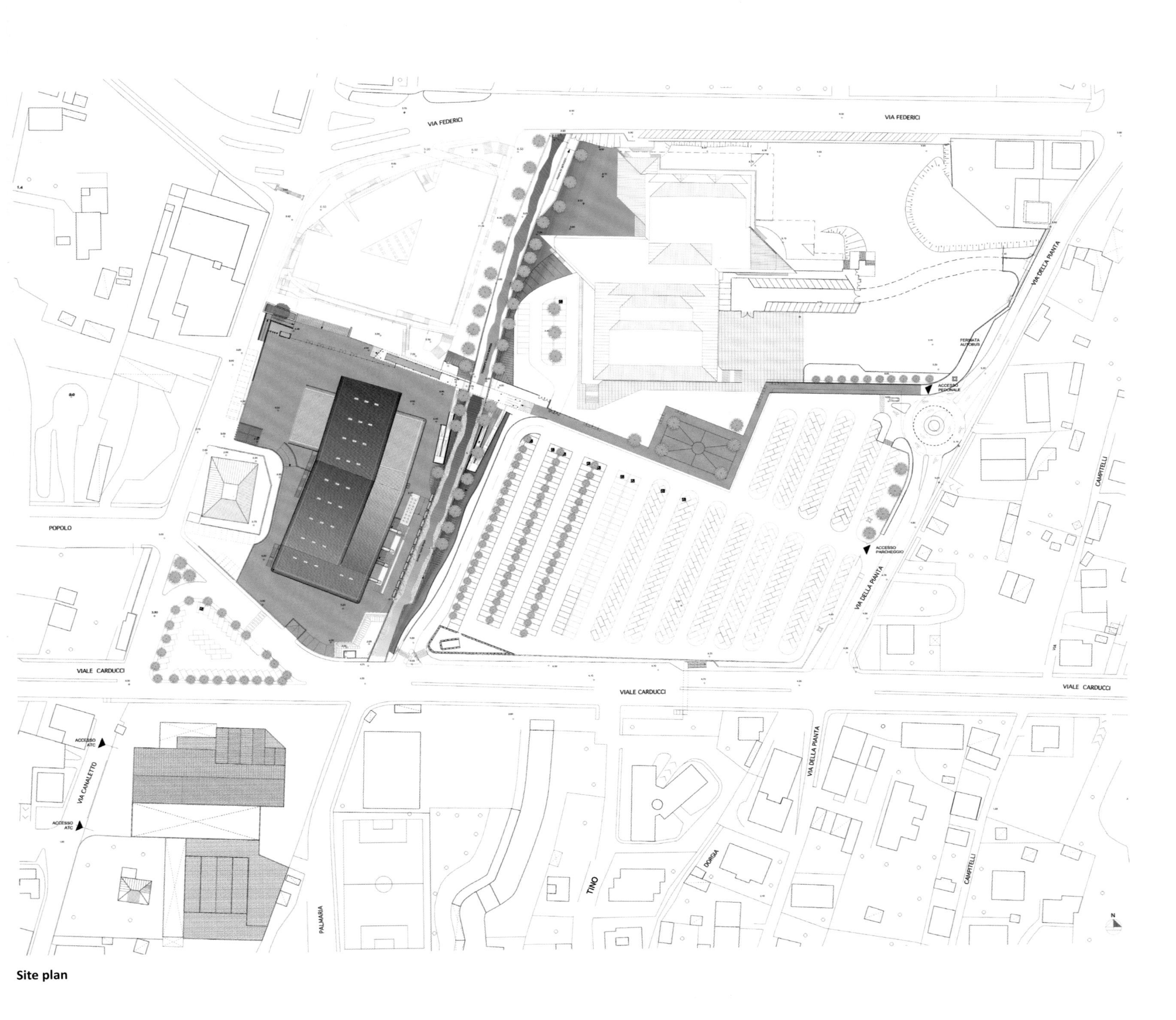
VIA FEDERICI
VIA FEDERICI
VIA DELLA PIANTA
FERMATA AUTOBUS
ACCESSO PEDONALE
CAMPITELLI
POPOLO
ACCESSO PARCHEGGIO
VIA DELLA PIANTA
VIALE CARDUCCI
VIALE CARDUCCI
VIALE CARDUCCI
ACCESSO ATC
VIA CANALETTO
ACCESSO ATC
VIA DELLA PIANTA
DORGIA
CAMPITELLI
TINO
PALMARIA
N

Site plan

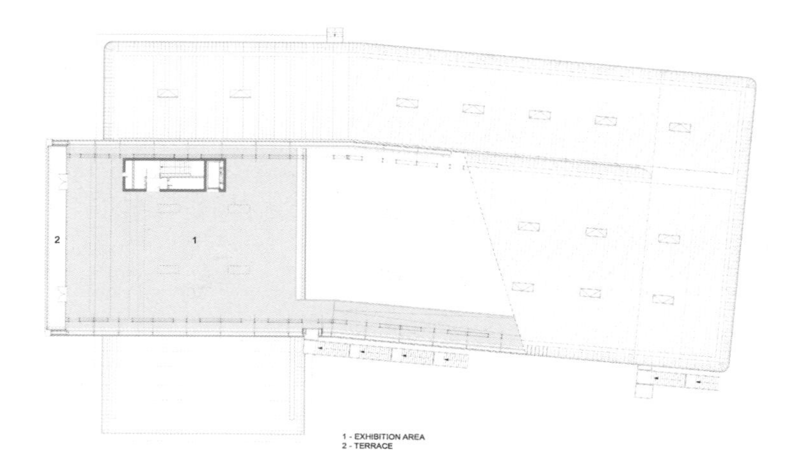

First floor plan

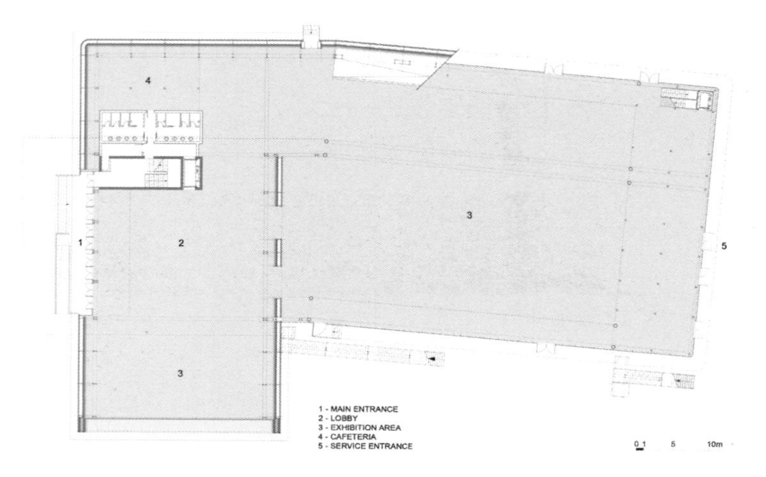

Ground floor plan

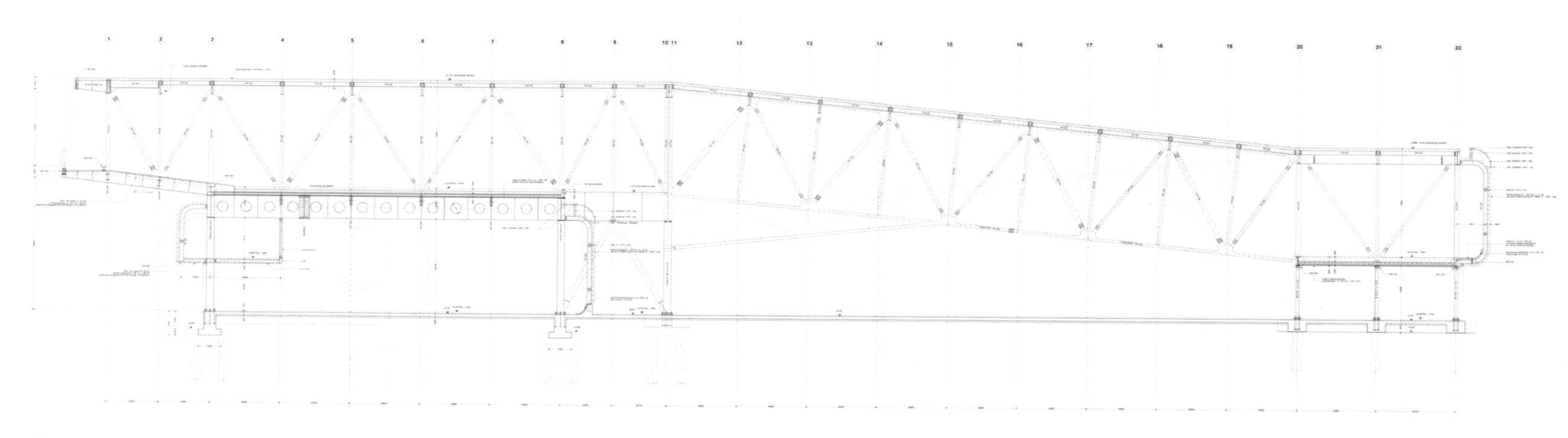

Structural section AA

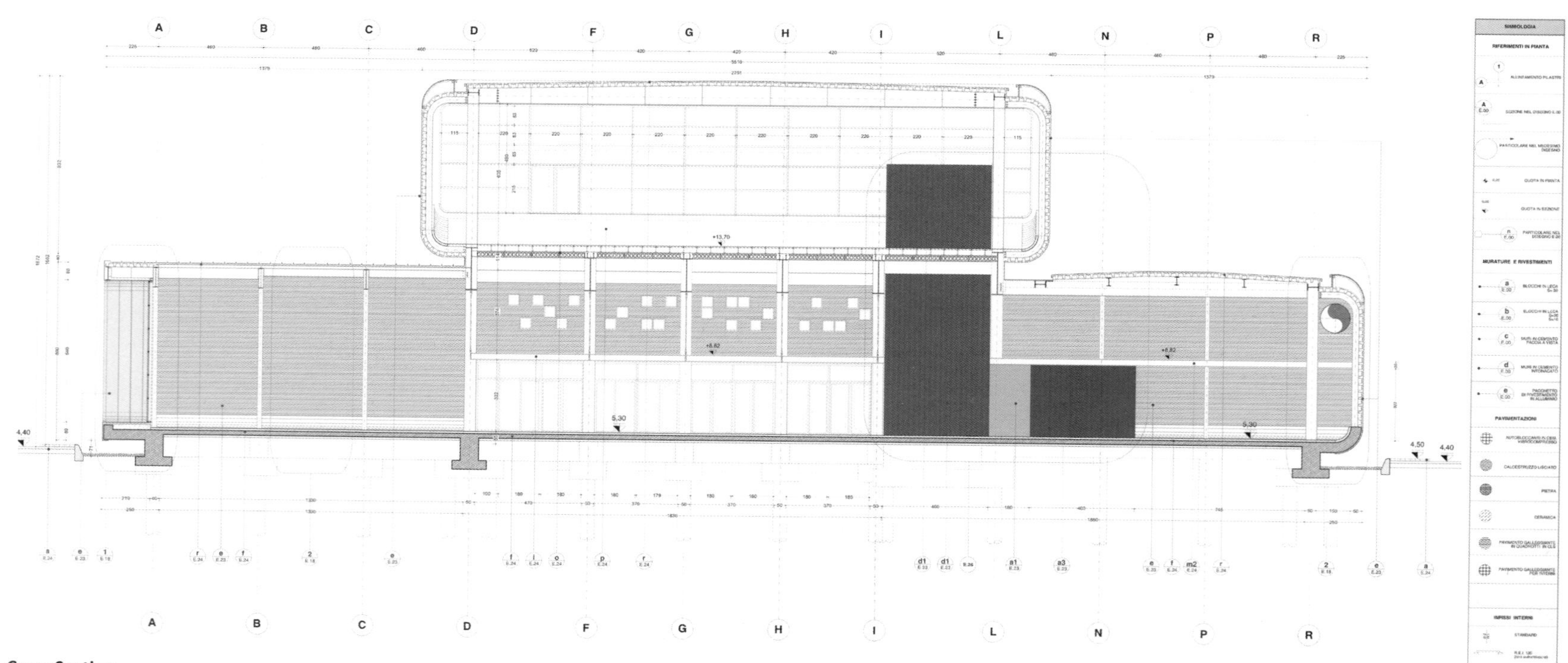

Cross Section

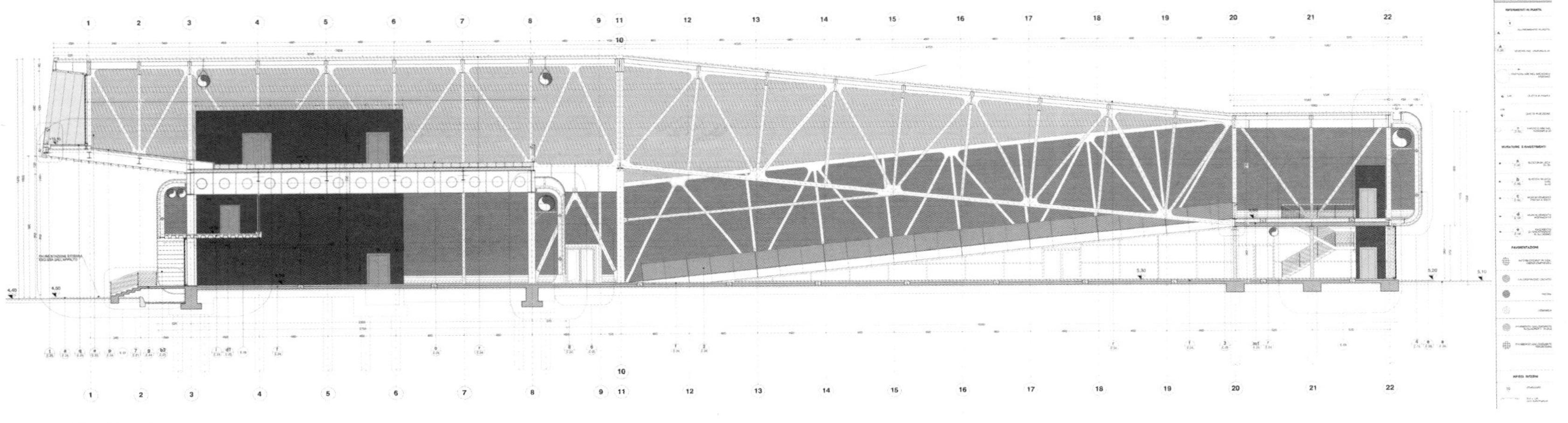

Longitudinal Section

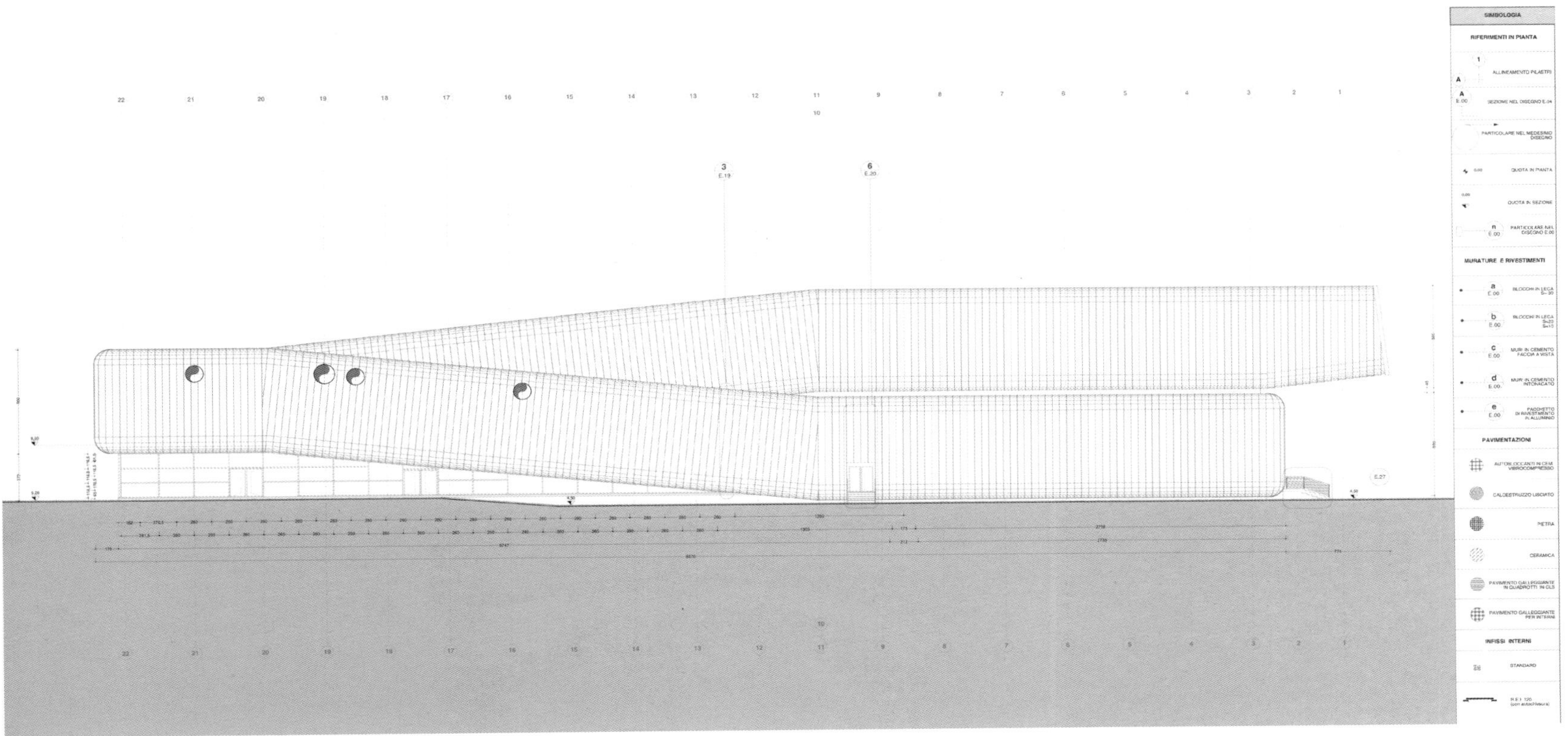

East Elevation

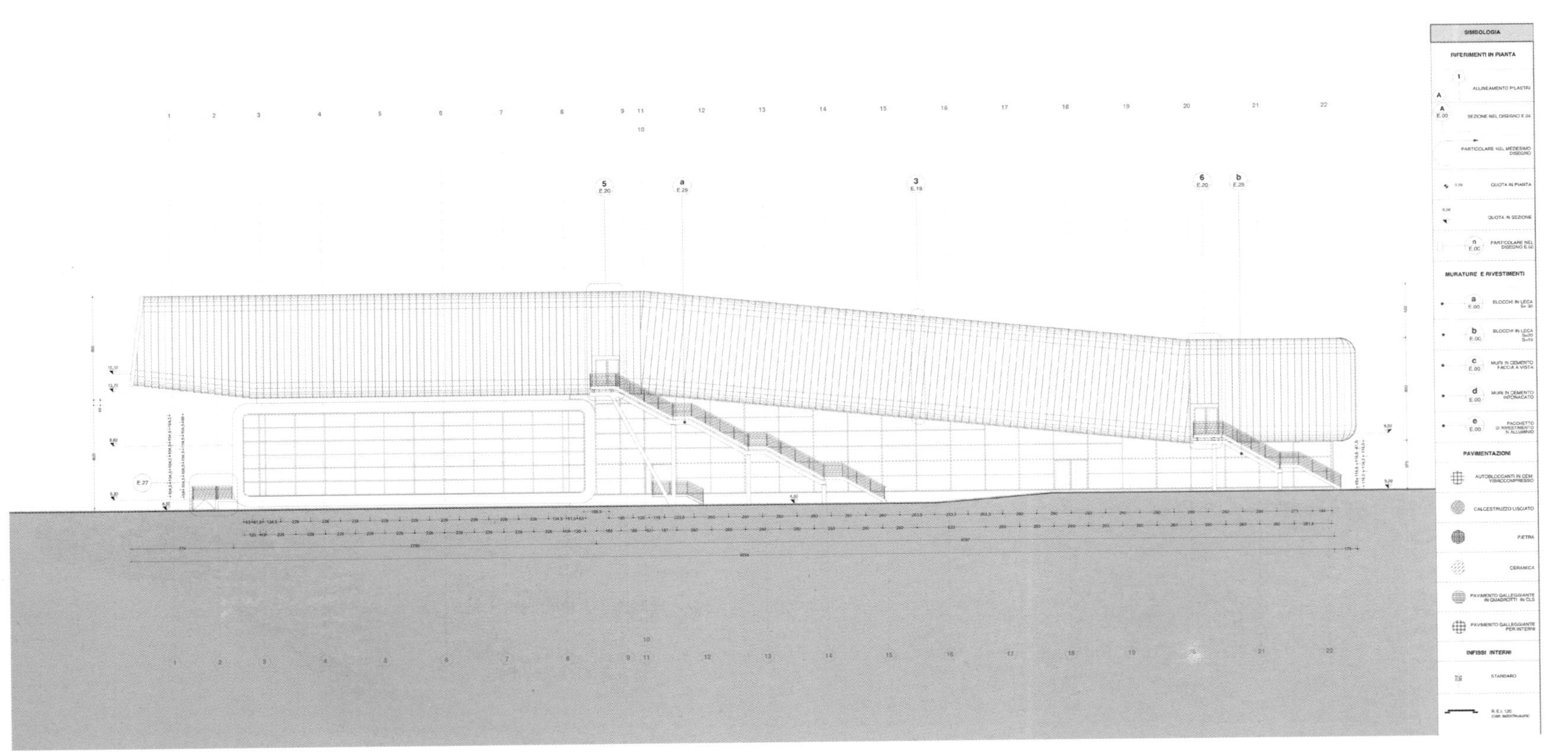

West Elevation

Hotel Lone

Location: Rovinj, Croatia
Architect: 3LHD
Client: Maistra d.d.
Site Area: 22,157 m^2
Completion Date: 2011
Photography: Cat Vinton, Damir Fabijanić, 3LHD

Hotel Lone, the first design hotel in Croatia, is situated in the Zlatni Rt (Golden Cape) forest park, Rovinj's most attractive tourist zone, located in the immediate vicinity of the legendary Eden Hotel and the new Monte Mulini hotel. The surrounding grounds and parkland is a unique and protected region of the Zlatni Rt forest in the Lone Bay.

The term design hotel is meant to illustrate this as a space that nurtures the concept of an interesting and functional design. It was created by a team of renowned Croatian creatives comprised of a new generation of architects, conceptual artists, product, fashion and graphic designers. The architects from studio 3LHD were responsible for the design and construction of the hotel building. In addition to the overall architecture, the interiors and the furniture were designed and chosen especially for the hotel in order to achieve a distinct and recognizable identity.

The key to the concept was the awareness of the necessity to avoid the sterility of most hotel facilities; that is why the designers used textiles with rich textures and quality oak veneer treated with eco-friendly lacquers, which give the visual and tactile impression of untreated massive wood. This material is usually too rustic but in this execution it manages to integrate walls and spaces into a harmonious composition through contemporary design and purified, spatially articulated shapes of panelling that overflow into furniture.

The conceptual assumptions used in the design of the hotel and its interior show evidence of a deep respect towards the achievements of hotel architecture on the Adriatic Coast from the previous century, combining it with a strong modernity expressed primarily in materials, functions and typologies and consequently in architectural forms.

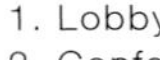

1. Lobby
2. Conference bar
3. Restaurant
4. Staff room
5. Transformer station
6. Conference hall
7. Business club
8. Sanitary facilities
9. Rooms
10. Service/evacuation staircase
11. Terrace
12. Shallow decorative pool
13. Congress entrance
14. Maid room
15. Garage
16. Service entrance

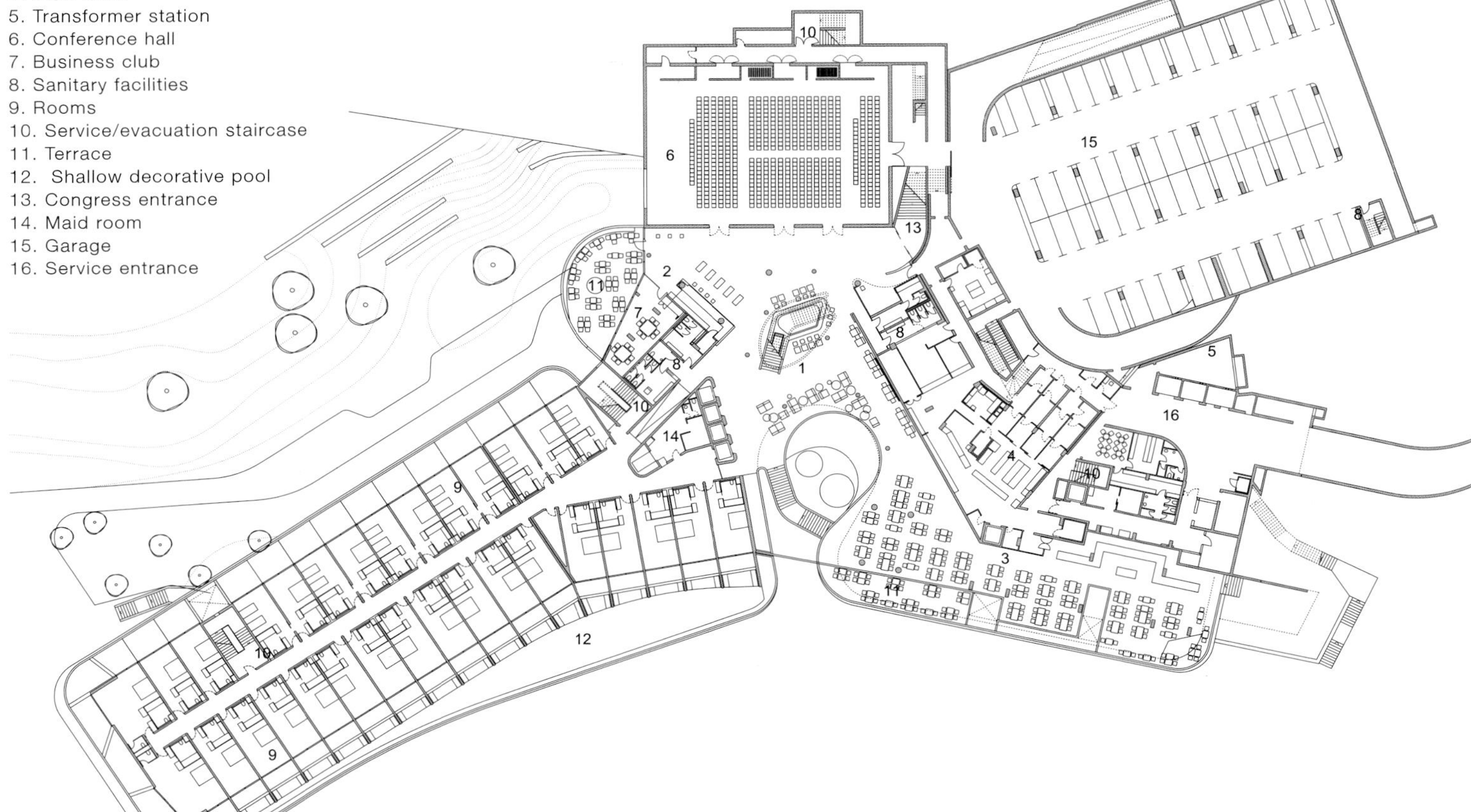

Floor plan -1

1. Lobby
2. Staircase
3. Restaurant
4. Boiler room
5. Kitchen
6. Sprinkler station
7. Jazz club
8. Children's club
9. Wardrobe-staff
10. Wardrobe-guests
11. Pool 1
12. Fitness
13. Sauna
14. Relax zone
15. Vital bar
16. Sinked rooms
17. Massage
18. Staircase
19. Entrance jazz club
20. Terrace restaurant
21. Entrance lobby
22. Terasa bazen
23. Entrance fitness
24. Terrace relax
25. Terrace vital bar

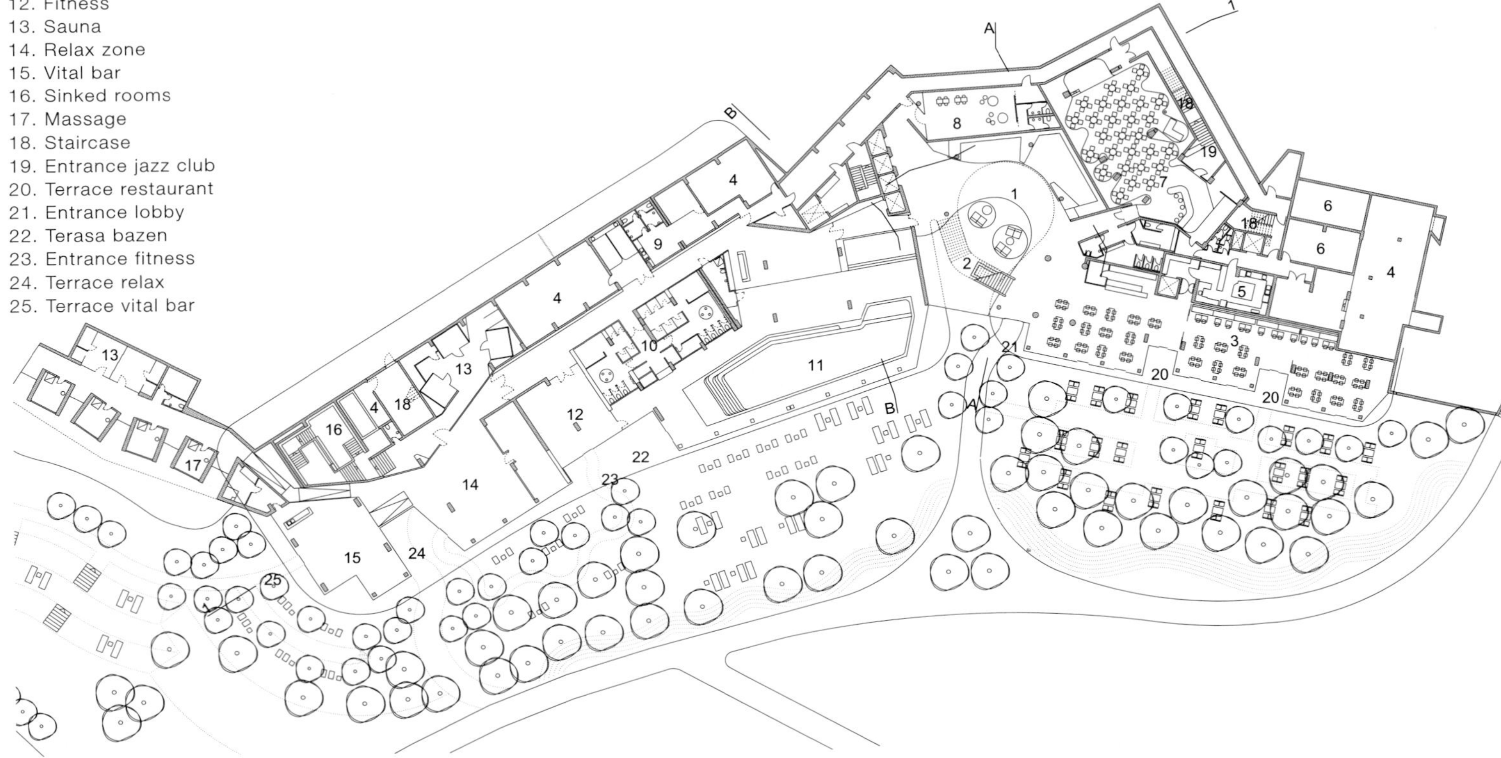

Floor plan -2

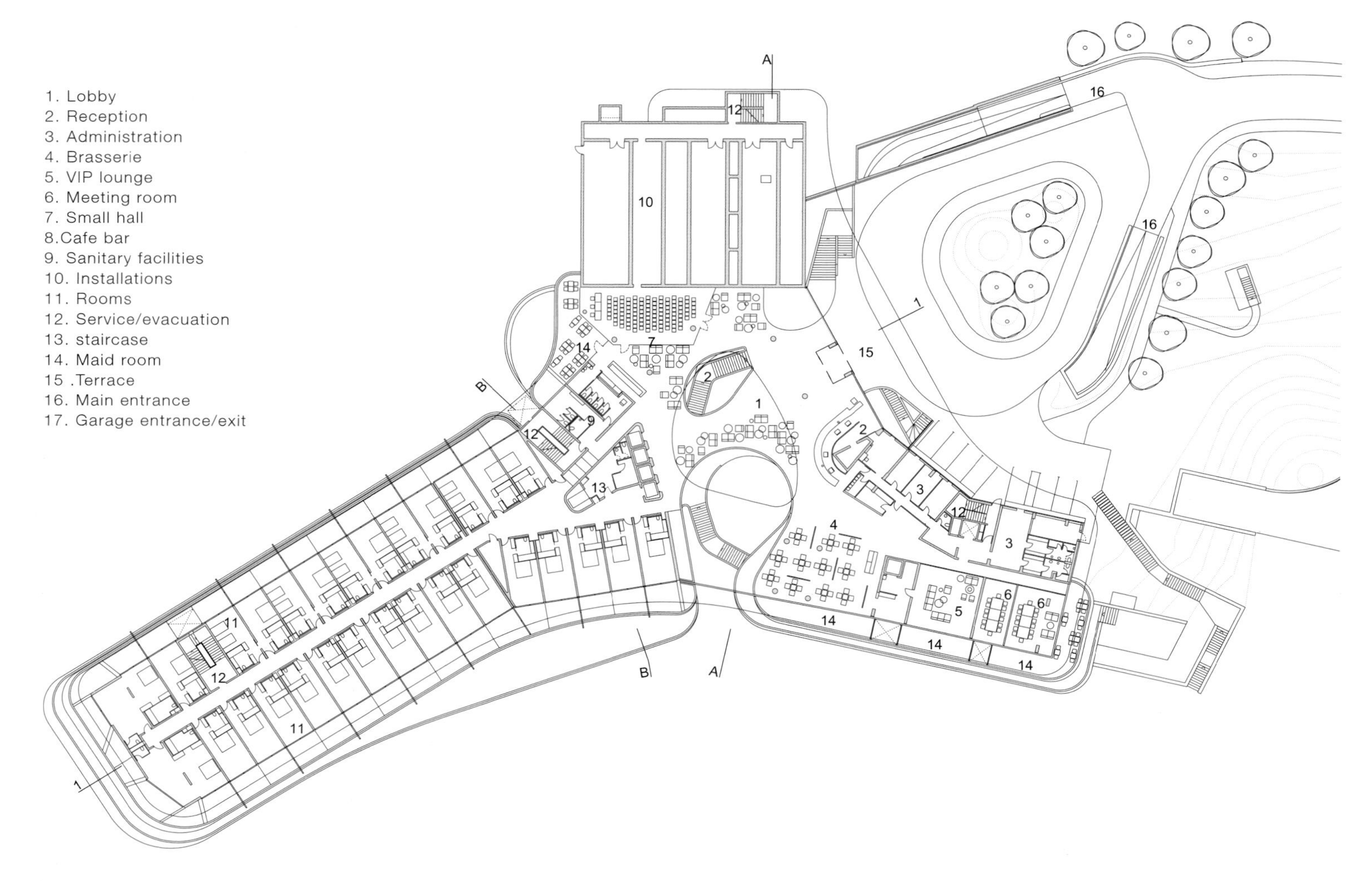

Floor plan 1

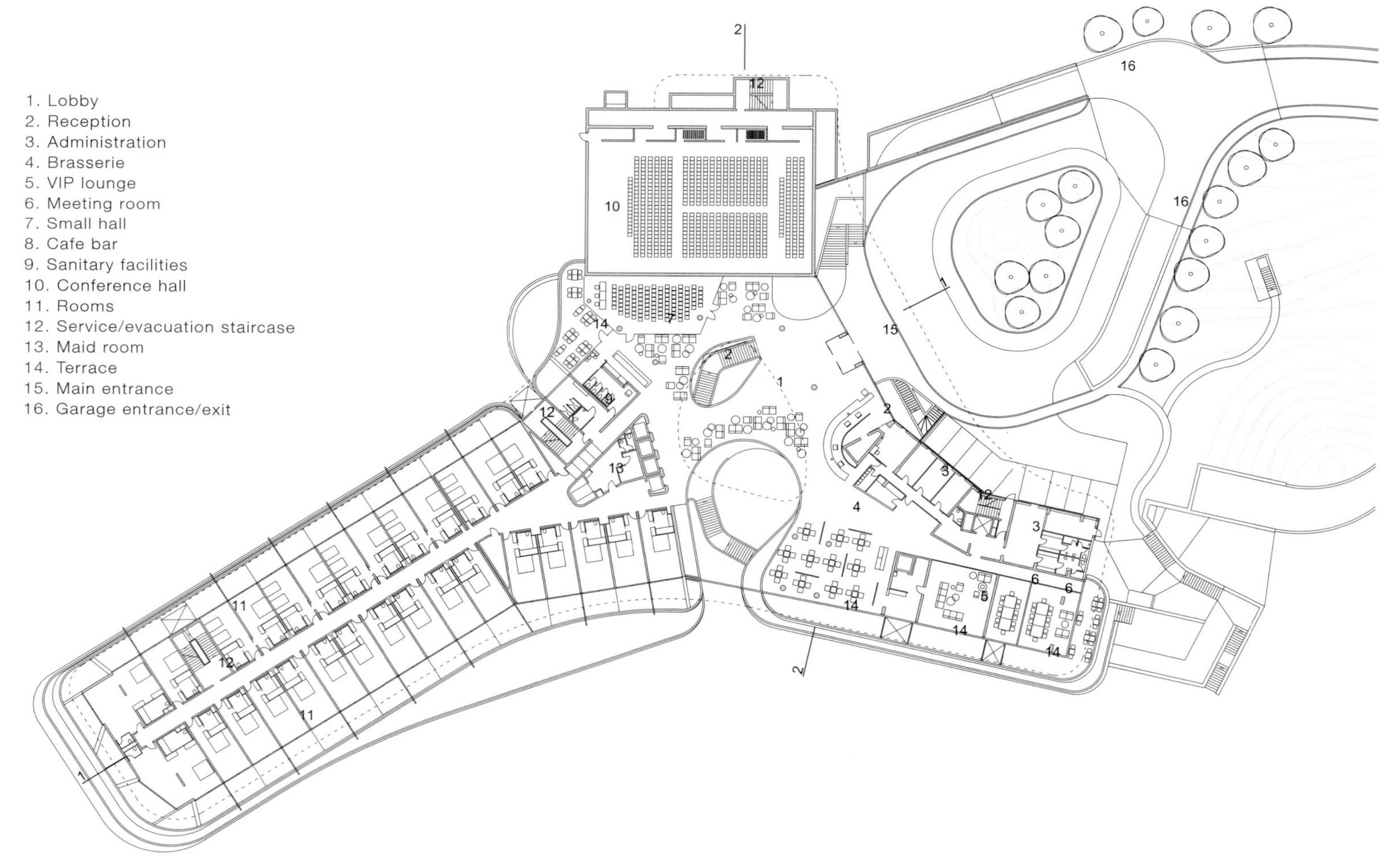

Floor plan congress

Elevation north

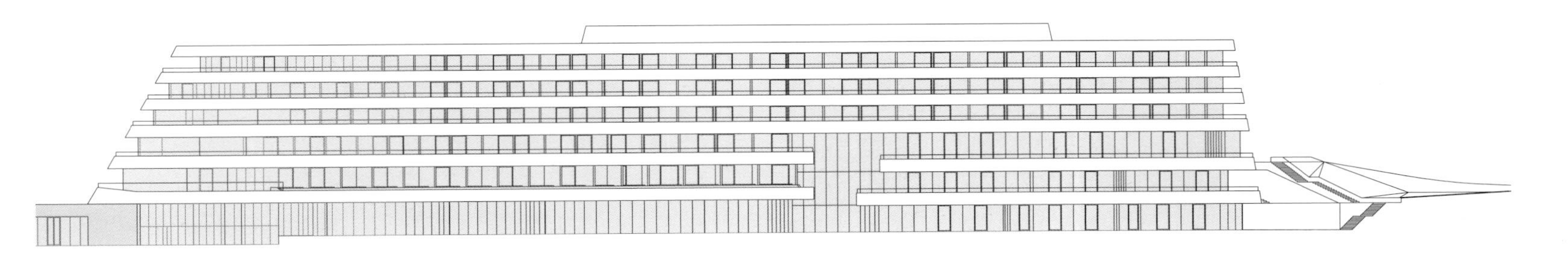

Elevation south

Elevation west

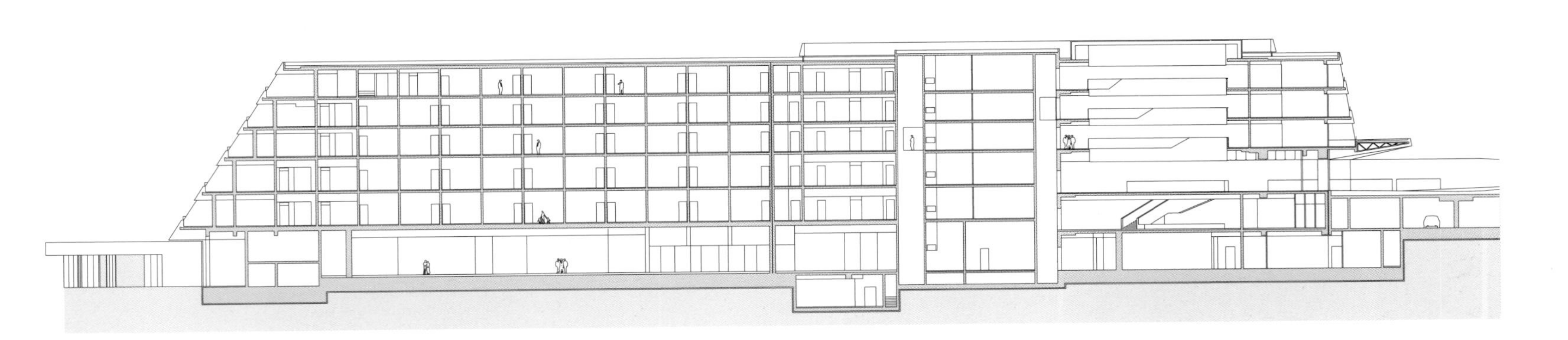

Section 1-1

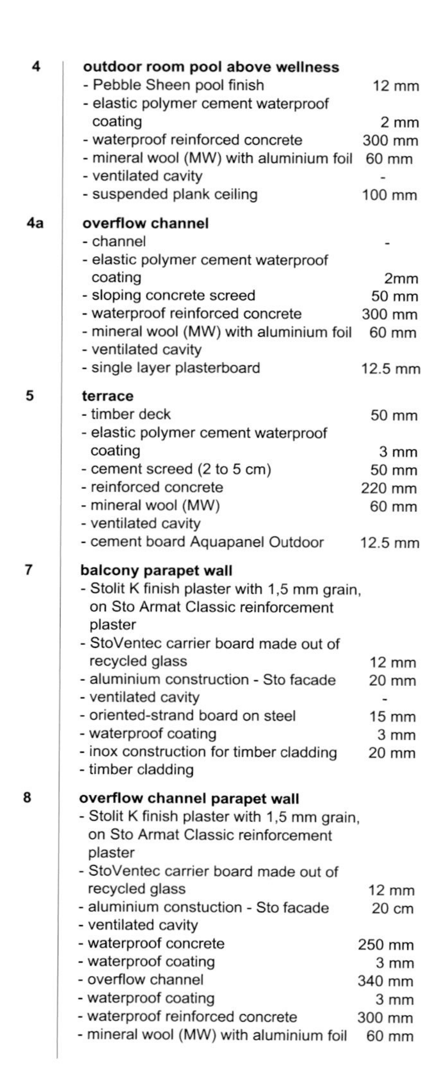

4 **outdoor room pool above wellness**
- Pebble Sheen pool finish 12 mm
- elastic polymer cement waterproof coating 2 mm
- waterproof reinforced concrete 300 mm
- mineral wool (MW) with aluminium foil 60 mm
- ventilated cavity -
- suspended plank ceiling 100 mm

4a **overflow channel**
- channel -
- elastic polymer cement waterproof coating 2mm
- sloping concrete screed 50 mm
- waterproof reinforced concrete 300 mm
- mineral wool (MW) with aluminium foil 60 mm
- ventilated cavity
- single layer plasterboard 12.5 mm

5 **terrace**
- timber deck 50 mm
- elastic polymer cement waterproof coating 3 mm
- cement screed (2 to 5 cm) 50 mm
- reinforced concrete 220 mm
- mineral wool (MW) 60 mm
- ventilated cavity
- cement board Aquapanel Outdoor 12.5 mm

7 **balcony parapet wall**
- Stolit K finish plaster with 1,5 mm grain, on Sto Armat Classic reinforcement plaster
- StoVentec carrier board made out of recycled glass 12 mm
- aluminium construction - Sto facade 20 mm
- ventilated cavity -
- oriented-strand board on steel 15 mm
- waterproof coating 3 mm
- inox construction for timber cladding 20 mm
- timber cladding

8 **overflow channel parapet wall**
- Stolit K finish plaster with 1,5 mm grain, on Sto Armat Classic reinforcement plaster
- StoVentec carrier board made out of recycled glass 12 mm
- aluminium constuction - Sto facade 20 cm
- ventilated cavity
- waterproof concrete 250 mm
- waterproof coating 3 mm
- overflow channel 340 mm
- waterproof coating 3 mm
- waterproof reinforced concrete 300 mm
- mineral wool (MW) with aluminium foil 60 mm

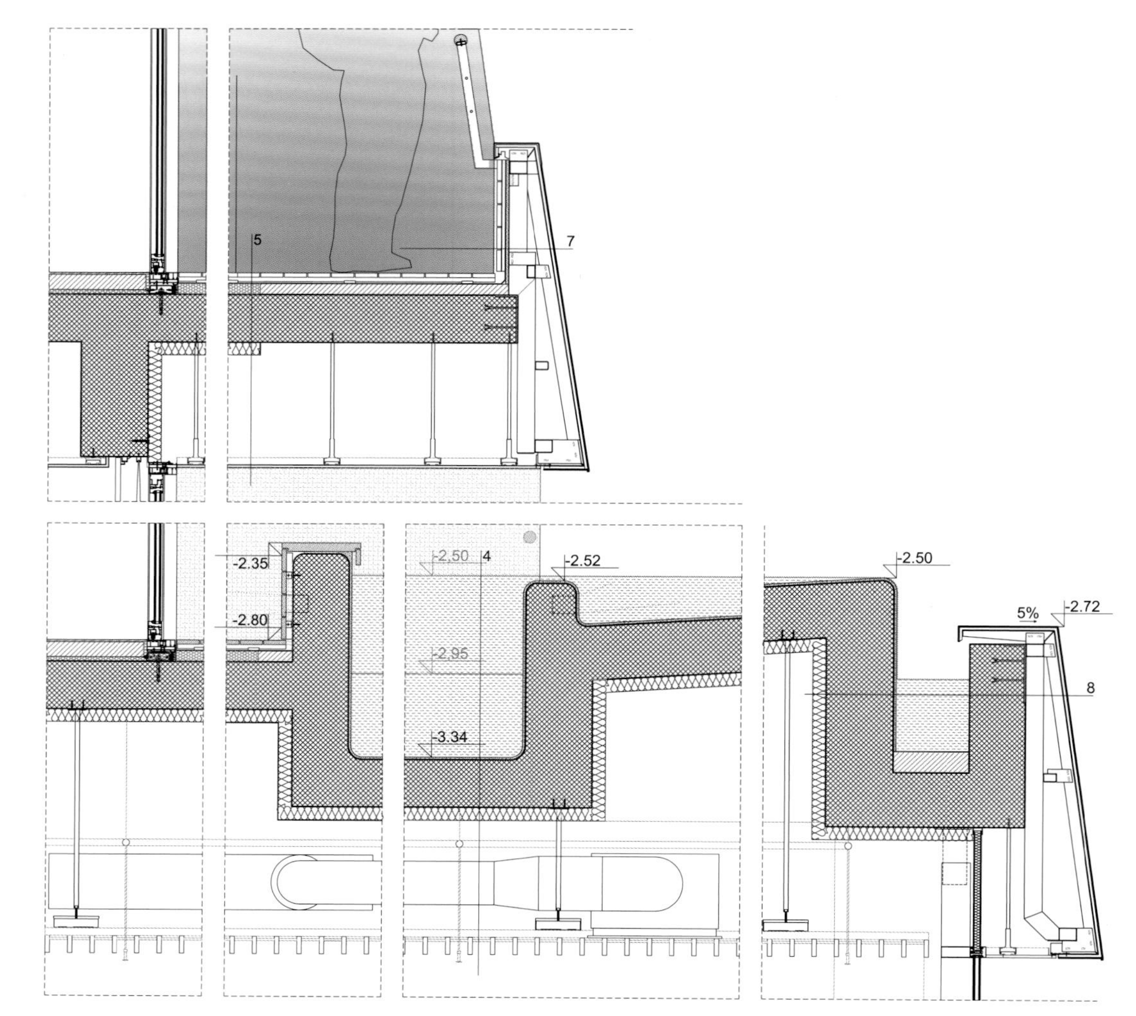

Detail of room pool

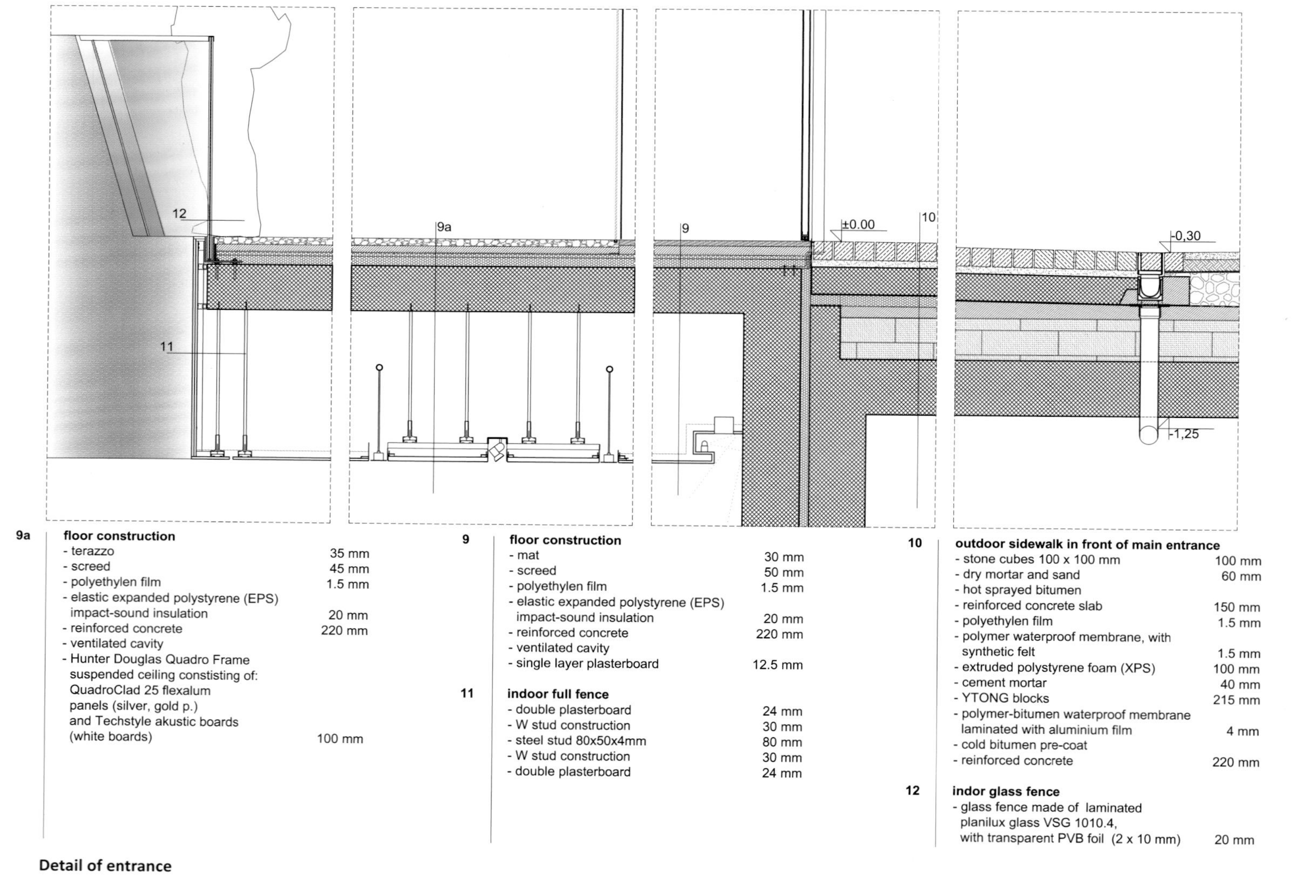

9a **floor construction**
- terazzo 35 mm
- screed 45 mm
- polyethylen film 1.5 mm
- elastic expanded polystyrene (EPS) impact-sound insulation 20 mm
- reinforced concrete 220 mm
- ventilated cavity
- Hunter Douglas Quadro Frame suspended ceiling constisting of: QuadroClad 25 flexalum panels (silver, gold p.) and Techstyle akustic boards (white boards) 100 mm

9 **floor construction**
- mat 30 mm
- screed 50 mm
- polyethylen film 1.5 mm
- elastic expanded polystyrene (EPS) impact-sound insulation 20 mm
- reinforced concrete 220 mm
- ventilated cavity
- single layer plasterboard 12.5 mm

11 **indoor full fence**
- double plasterboard 24 mm
- W stud construction 30 mm
- steel stud 80x50x4mm 80 mm
- W stud construction 30 mm
- double plasterboard 24 mm

10 **outdoor sidewalk in front of main entrance**
- stone cubes 100 x 100 mm 100 mm
- dry mortar and sand 60 mm
- hot sprayed bitumen
- reinforced concrete slab 150 mm
- polyethylen film 1.5 mm
- polymer waterproof membrane, with synthetic felt 1.5 mm
- extruded polystyrene foam (XPS) 100 mm
- cement mortar 40 mm
- YTONG blocks 215 mm
- polymer-bitumen waterproof membrane laminated with aluminium film 4 mm
- cold bitumen pre-coat
- reinforced concrete 220 mm

12 **indor glass fence**
- glass fence made of laminated planilux glass VSG 1010.4, with transparent PVB foil (2 x 10 mm) 20 mm

Detail of entrance

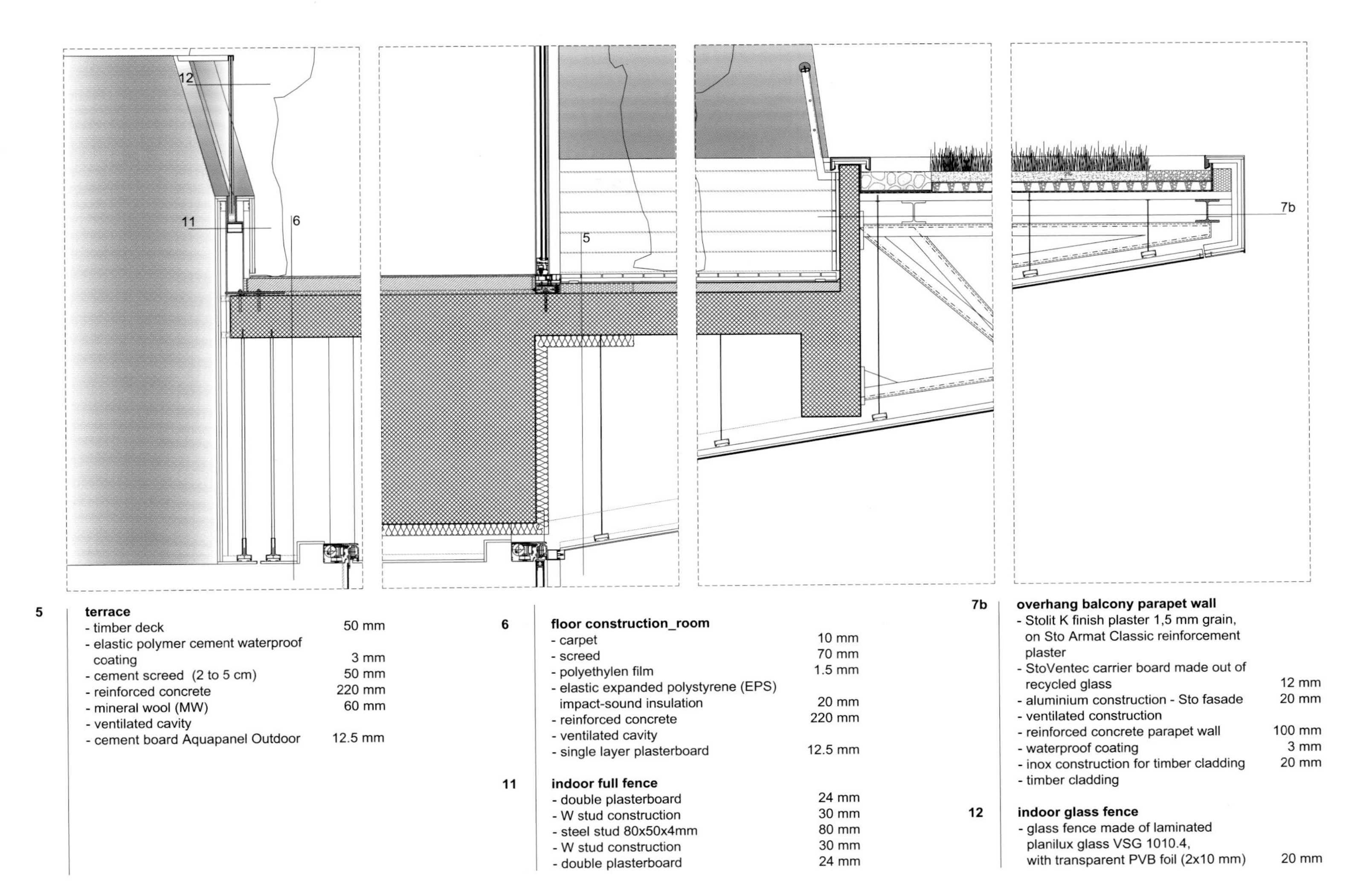

5 terrace

- timber deck 50 mm
- elastic polymer cement waterproof coating 3 mm
- cement screed (2 to 5 cm) 50 mm
- reinforced concrete 220 mm
- mineral wool (MW) 60 mm
- ventilated cavity
- cement board Aquapanel Outdoor 12.5 mm

6 floor construction_room

- carpet 10 mm
- screed 70 mm
- polyethylen film 1.5 mm
- elastic expanded polystyrene (EPS) impact-sound insulation 20 mm
- reinforced concrete 220 mm
- ventilated cavity
- single layer plasterboard 12.5 mm

11 indoor full fence

- double plasterboard 24 mm
- W stud construction 30 mm
- steel stud 80x50x4mm 80 mm
- W stud construction 30 mm
- double plasterboard 24 mm

7b overhang balcony parapet wall

- Stolit K finish plaster 1,5 mm grain, on Sto Armat Classic reinforcement plaster
- StoVentec carrier board made out of recycled glass 12 mm
- aluminium construction - Sto fasade 20 mm
- ventilated construction
- reinforced concrete parapet wall 100 mm
- waterproof coating 3 mm
- inox construction for timber cladding 20 mm
- timber cladding

12 indoor glass fence

- glass fence made of laminated planilux glass VSG 1010.4, with transparent PVB foil (2x10 mm) 20 mm

Detail of overhang

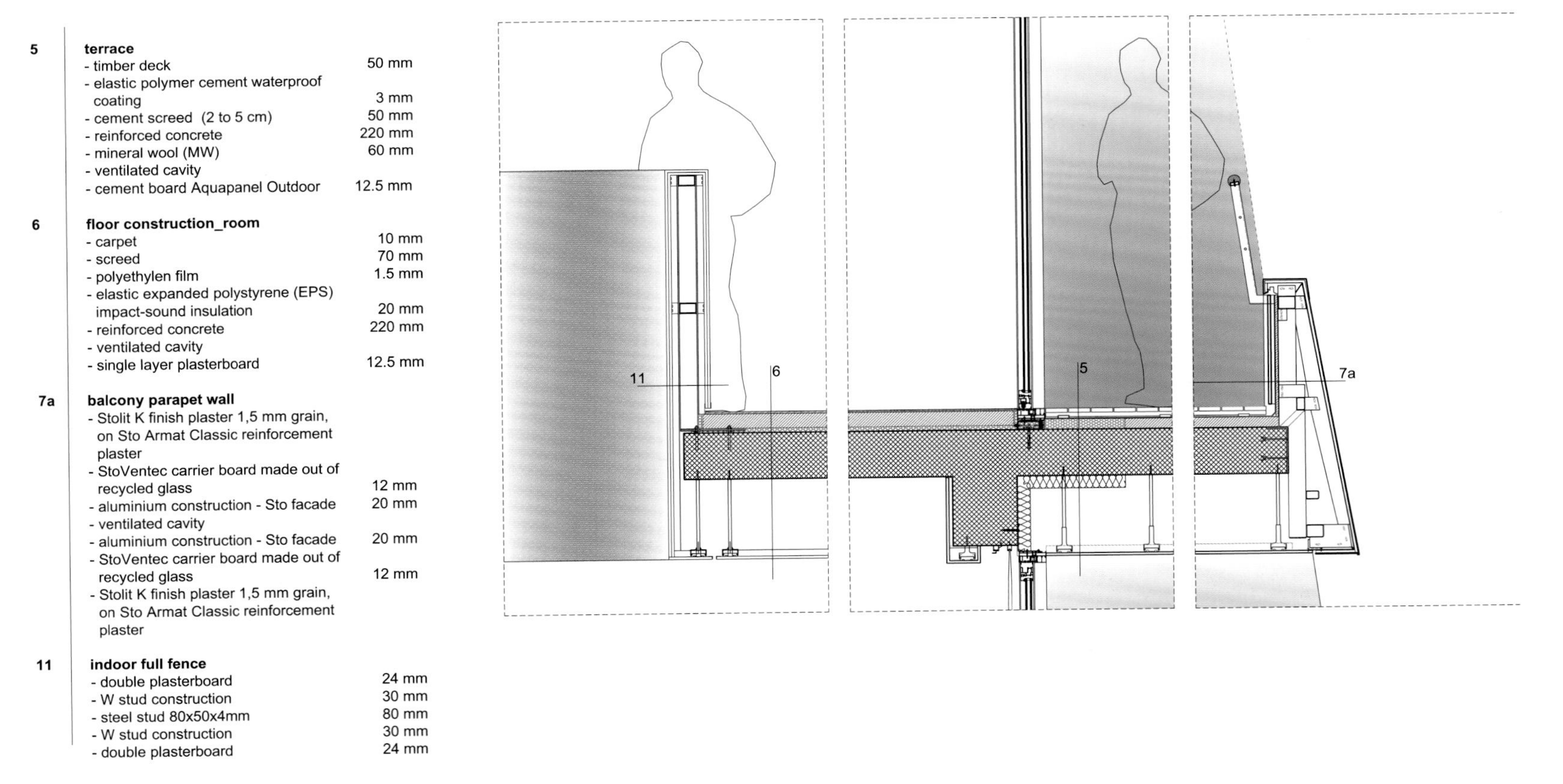

5 terrace

- timber deck 50 mm
- elastic polymer cement waterproof coating 3 mm
- cement screed (2 to 5 cm) 50 mm
- reinforced concrete 220 mm
- mineral wool (MW) 60 mm
- ventilated cavity
- cement board Aquapanel Outdoor 12.5 mm

6 floor construction_room

- carpet 10 mm
- screed 70 mm
- polyethylen film 1.5 mm
- elastic expanded polystyrene (EPS) impact-sound insulation 20 mm
- reinforced concrete 220 mm
- ventilated cavity
- single layer plasterboard 12.5 mm

7a balcony parapet wall

- Stolit K finish plaster 1,5 mm grain, on Sto Armat Classic reinforcement plaster
- StoVentec carrier board made out of recycled glass 12 mm
- aluminium construction - Sto facade 20 mm
- ventilated cavity
- aluminium construction - Sto facade 20 mm
- StoVentec carrier board made out of recycled glass 12 mm
- Stolit K finish plaster 1,5 mm grain, on Sto Armat Classic reinforcement plaster

11 indoor full fence

- double plasterboard 24 mm
- W stud construction 30 mm
- steel stud 80x50x4mm 80 mm
- W stud construction 30 mm
- double plasterboard 24 mm

Detail of outdoor parapet

1	**wellness pool**	
	- Pebble Sheen pool finish	12 mm
	- elastic polymer cement waterproof coating	2 mm
	- cement screed	50 mm
	- polyethylen film	1.5 mm
	- extruded polystyrene (XPS)	60 mm
	- single-layer polymer-bitumen waterproof membrane (reinforcment in fibreglass), with heat-bonded overlaps	4 mm
	- cold bitumen pre-coat	
	- concrete screed	80 mm
	- bed of gravel	
2	**pool deck - overflow channel**	
	- ceramic tiles, glued	20 mm
	- inox grating	5 mm
	- channel	
	- elastic polymer cement waterproof coating	2 mm
	- sloping concrete screed	50 mm
	- waterproof reinforced concrete	300 mm
	- concrete fill	220 mm
	- polyethylen film	1.5 mm
	- extruded polystyrene (XPS)	60 mm
	- single-layer polymer-bitumen waterproof membrane (reinforcment in fibreglass), with heat-bonded overlaps	4 mm
	- cold bitumen pre-coat	
	- concrete screed	80 mm
	- bed of gravel	
3	**pool deck**	
	- ceramic tiles, glued	20 mm
	- sloping cement screed	50 mm
	- expanded polystyrene (EPS)	60 mm
	- precast reinforced concrete slab	100 mm
	- ventilated cavity	-
	- cement screed	50 mm
	- polyethylen film	1.5 mm
	- extruded polystyrene (XPS)	60 mm
	- single-layer polymer-bitumen waterproof membrane (reinforcment in fibreglass), with heat-bonded overlaps	4 mm
	- cold bitumen pre-coat	
	- concrete screed	80 mm
	- bed of gravel	
13	**glass wall**	
	- double glazed insulating glass (IZO), with one low-emissivity glass (Low-E)	
	- aluminium construction	

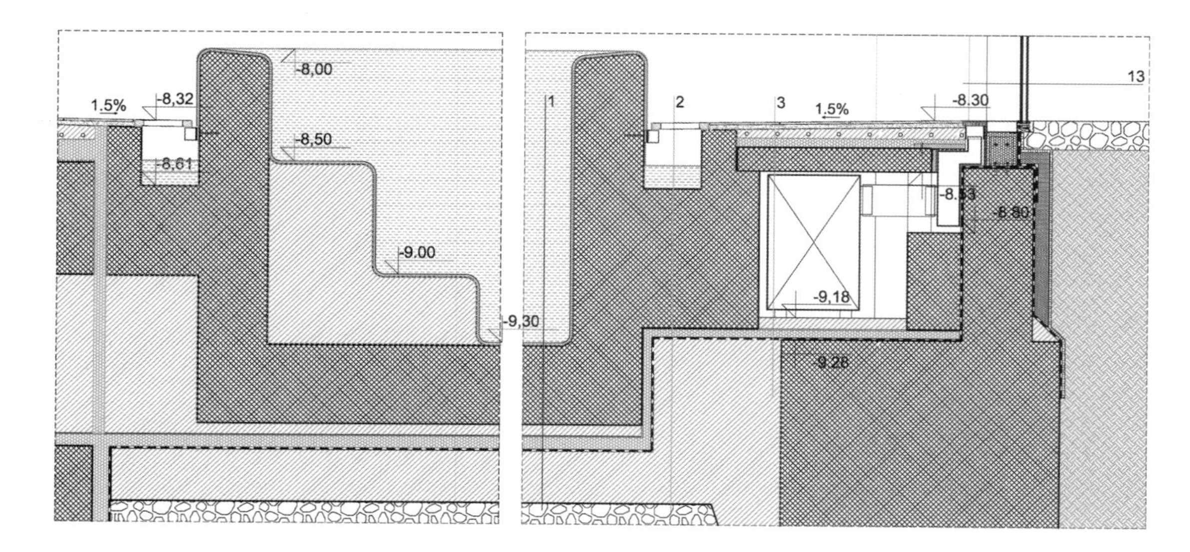

Detail of recreational pool

5	**terrace**	
	- timber deck	50 mm
	- elastic polymer cement waterproof coating	3 mm
	- cement screed (2 to 5 cm)	50 mm
	- reinforced concrete	220 mm
	- mineral wool (MW)	60 mm
	- ventilated cavity	
	- cement board Aquapanel Outdoor	12.5 mm
6	**floor construction_room**	
	- carpet	1.0 mm
	- screed	70 mm
	- polyethylen film	1.5 mm
	- elastic expanded polystyrene (EPS) impact-sound insulation	20 mm
	- reinforced concrete	220 mm
	- ventilated cavity	
	- single layer plasterboard	12.5 mm
7	**balcony parapet wall**	
	- Stolit K finish plaster 1,5 mm grain, on Sto Armat Classic reinforcement plaster	
	- StoVentec carrier board made out of recycled glass	12 mm
	- aluminium construction - Sto facade	20 mm
	- ventilated cavity	
	- oriented-strand board on steel	15 mm
	- waterproof coating	3 mm
	- inox construction for timber cladding	20 mm
	- timber cladding	
7a	**balcony parapet wall**	
	- Stolit K finish plaster 1,5 mm grain, on Sto Armat Classic reinforcement plaster	
	- StoVentec carrier board made out of recycled glass	12 mm
	- aluminium construction - Sto facade	20 mm
	- ventilated cavity	
	- aluminium construction - Sto facade	20 mm
	- StoVentec carrier board made out of recycled glass	12 mm
	- Stolit K finish plaster 1,5 mm grain, on Sto Armat Classic reinforcement plaster	

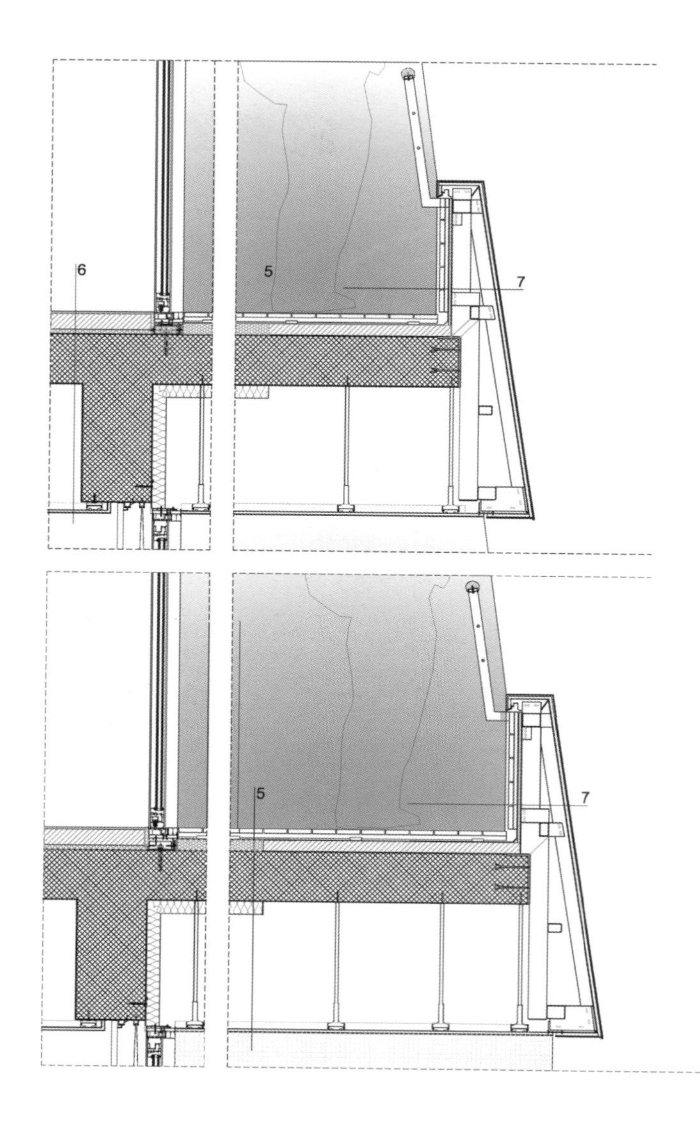

Detail of parapet wall on terraces

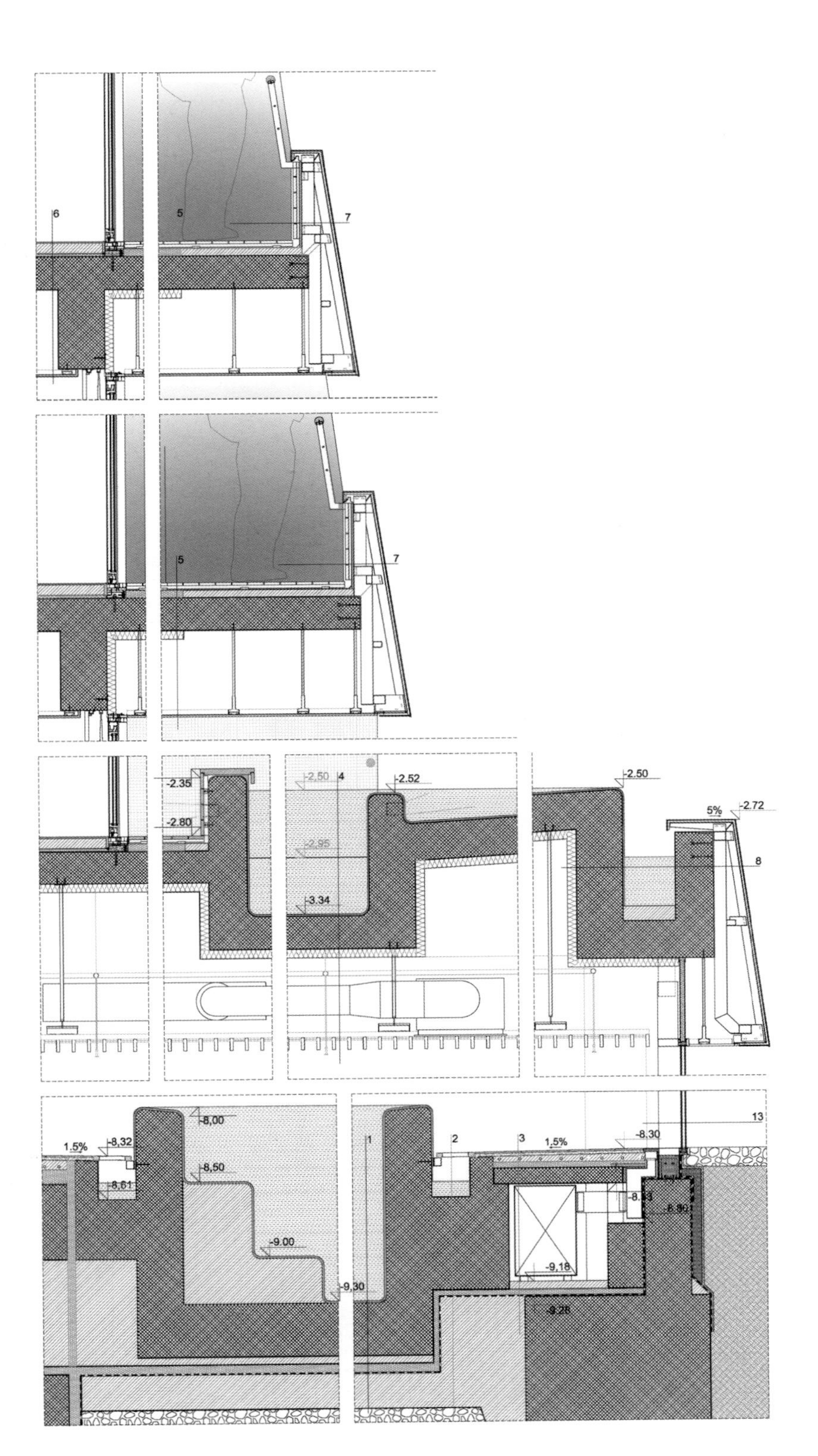

Details of recreational pool

1 wellness pool
- Pebble Sheen pool finish 12 mm
- elastic polymer cement waterproof coating 2 mm
- cement screed 50 mm
- polyethylen film 1.5 mm
- extruded polystyrene (XPS) 60 mm
- single-layer polymer-bitumen waterproof membrane (reinforcment in fibreglass), with heat-bonded overlaps 4 mm
- cold bitumen pre-coat
- concrete screed 80 mm
- bed of gravel

2 pool deck - overflow channel
- ceramic tiles, glued 20 mm
- inox grating 5 mm
- channel
- elastic polymer cement waterproof coating 2 mm
- sloping concrete screed 50 mm
- waterproof reinforced concrete 300 mm
- concrete fill 220 mm
- polyethylen film 1.5 mm
- extruded polystyrene (XPS) 60 mm
- single-layer polymer-bitumen waterproof membrane (reinforcment in fibreglass), with heat-bonded overlaps 4 mm
- cold bitumen pre-coat
- concrete screed 80 mm
- bed of gravel

3 pool deck
- ceramic tiles, glued 20 mm
- sloping cement screed 50 mm
- expanded polystyrene (EPS) 60 mm
- precast reinforced concrete slab 100 mm
- ventilated cavity -
- cement screed 50 mm
- polyethylen film 1.5 mm
- extruded polystyrene (XPS) 60 mm
- single-layer polymer-bitumen waterproof membrane (reinforcment in fibreglass), with heat-bonded overlaps 4 mm
- cold bitumen pre-coat
- concrete screed 80 mm
- bed of gravel

4 outdoor room pool above wellness
- Pebble Sheen pool finish 12 mm
- elastic polymer cement waterproof coating 2 mm
- waterproof reinforced concrete 300 mm
- mineral wool (MW) with aluminium foil 60 mm
- ventilated cavity -
- suspended plank ceiling 100 mm

5 terrace
- timber deck 50 mm
- elastic polymer cement waterproof coating 3 mm
- cement screed (2 to 5 cm) 50 mm
- reinforced concrete 220 mm
- mineral wool (MW) 60 mm
- ventilated cavity
- cement board Aquapanel Outdoor 12.5 mm

6 floor construction_room
- carpet 10 mm
- screed 70 mm
- polyethylen film 1.5 mm
- elastic expanded polystyrene (EPS) impact-sound insulation 20 mm
- reinforced concrete 220 mm
- ventilated cavity
- single layer plasterboard 12.5 mm

7 balcony parapet wall
- Stolit K finish plaster 1,5 mm grain, on Sto Armat Classic reinforcement plaster
- StoVentec carrier board made out of recycled glass 12 mm
- aluminium construction - Sto facade 20 mm
- ventilated cavity
- oriented-strand board on steel 15 mm
- waterproof coating 3 mm
- inox construction for timber cladding 20 mm
- timber cladding

8 overflow channel parapet wall
- Stolit K finish plaster with 1,5 mm grain, on Sto Armat Classic reinforcement plaster
- StoVentec carrier board made out of recycled glass 12 mm
- aluminium constuction - Sto facade 20 cm
- ventilated cavity
- waterproof concrete 250 mm
- waterproof coating 3 mm
- overflow channel 340 mm
- waterproof coating 3 mm
- waterproof reinforced concrete 300 mm
- mineral wool (MW) with aluminium foil 60 mm

13 glass wall
- double glazed insulating glass (IZO), with one low-emissivity glass (Low-E)
- aluminium construction

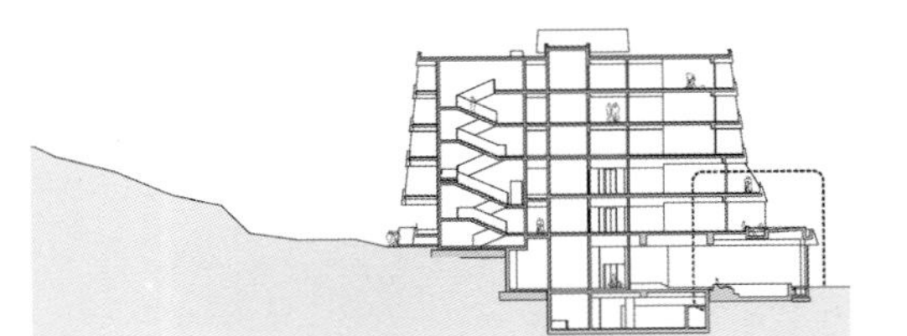

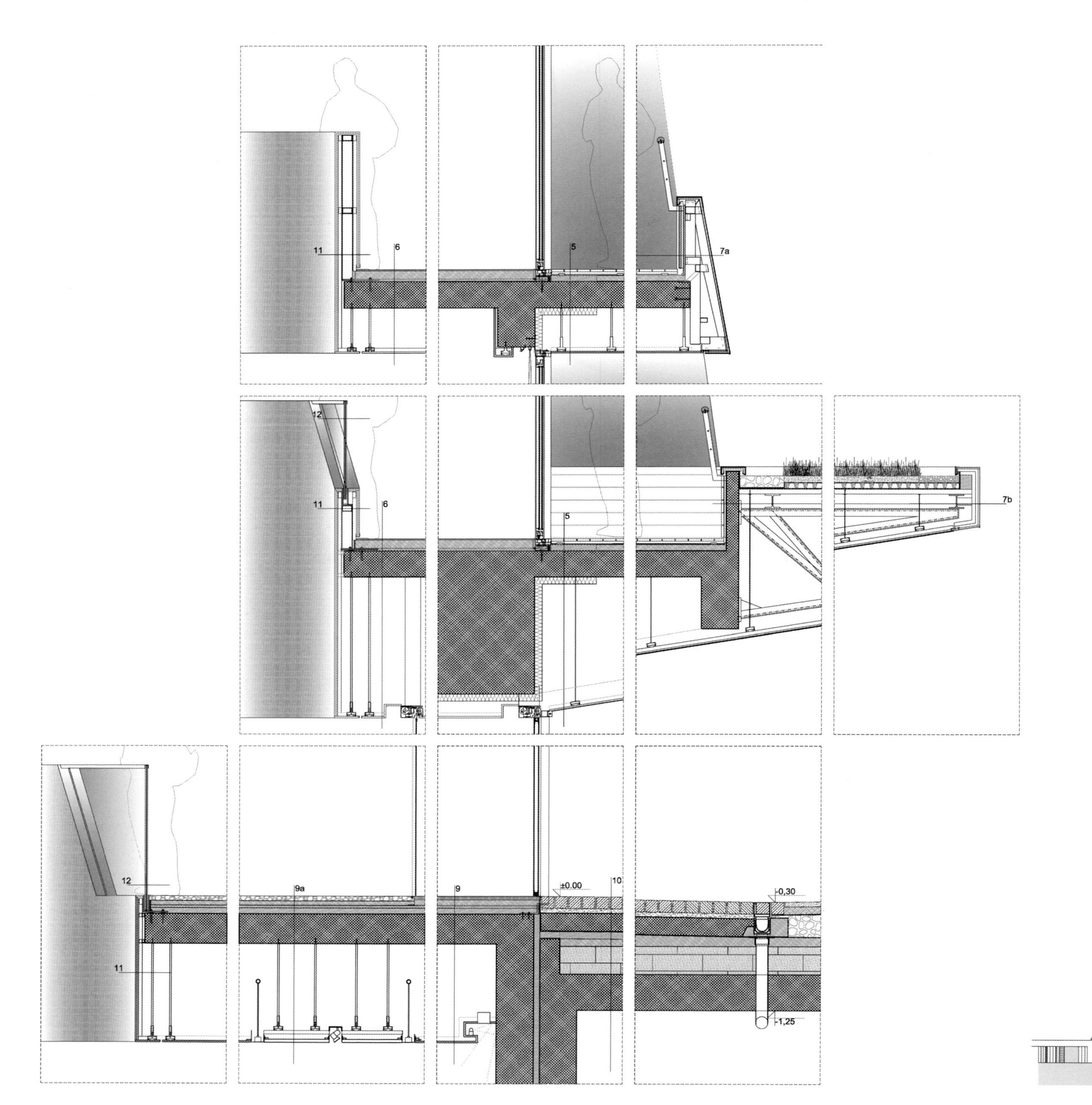

Detail of overhang and parapets

5 terrace
- timber deck 50 mm
- elastic polymer cement waterproof coating 3 mm
- cement screed (2 to 5 cm) 50 mm
- reinforced concrete 220 mm
- mineral wool (MW) 60 mm
- ventilated cavity
- cement board Aquapanel Outdoor 12.5 mm

6 floor construction_room
- carpet 10 mm
- screed 70 mm
- polyethylen film 1.5 mm
- elastic expanded polystyrene (EPS) impact-sound insulation 20 mm
- reinforced concrete 220 mm
- ventilated cavity
- single layer plasterboard 12.5 mm

7 balcony parapet wall
- Stolit K finish plaster 1,5 mm grain, on Sto Armat Classic reinforcement plaster
- StoVentec carrier board made out of recycled glass 12 mm
- aluminium construction - Sto facade 20 mm
- ventilated cavity
- oriented-strand board on steel 15 mm
- waterproof coating 3 mm
- inox construction for timber cladding 20 mm
- timber cladding

7a balcony parapet wall
- Stolit K finish plaster 1,5 mm grain, on Sto Armat Classic reinforcement plaster
- StoVentec carrier board made out of recycled glass 12 mm
- aluminium construction - Sto facade 20 mm
- ventilated cavity
- aluminium construction - Sto facade 20 mm
- StoVentec carrier board made out of recycled glass 12 mm
- Stolit K finish plaster 1,5 mm grain, on Sto Armat Classic reinforcement plaster

7b overhang balcony parapet wall
- Stolit K finish plaster 1,5 mm grain, on Sto Armat Classic reinforcement plaster
- StoVentec carrier board made out of recycled glass 12 mm
- aluminium construction - Sto fasade 20 mm
- ventilated construction
- reinforced concrete parapet wall 100 mm
- waterproof coating 3 mm
- inox construction for timber cladding 20 mm
- timber cladding

9 floor construction
- mat 30 mm
- screed 50 mm
- polyethylen film 1.5 mm
- elastic expanded polystyrene (EPS) impact-sound insulation 20 mm
- reinforced concrete 220 mm
- ventilated cavity
- single layer plasterboard 12.5 mm

9a floor construction
- terazzo 35 mm
- screed 45 mm
- polyethylen film 1.5 mm
- elastic expanded polystyrene (EPS) impact-sound insulation 20 mm
- reinforced concrete 220 mm
- ventilated cavity
- Hunter Douglas Quadro Frame suspended ceiling constisting of: QuadroClad 25 flexalum panels (silver, gold p.) and Techstyle akustic boards (white boards) 100 mm

10 outdoor sidewalk in front of main entrance
- stone cubes 100 x 100 mm 100 mm
- dry mortar and sand 60 mm
- hot sprayed bitumen
- reinforced concrete slab 150 mm
- polyethylen film 1.5 mm
- polymer waterproof membrane, with synthetic felt 1.5 mm
- extruded polystyrene foam (XPS) 100 mm
- cement mortar 40 mm
- YTONG blocks 215 mm
- polymer-bitumen waterproof membrane laminated with aluminium film 4 mm
- cold bitumen pre-coat
- reinforced concrete 220 mm

11 indoor full fence
- double plasterboard 24 mm
- W stud construction 30 mm
- steel stud 80x50x4mm 80 mm
- W stud construction 30 mm
- double plasterboard 24 mm

12 indor glass fence
- glass fence made of laminated planilux glass VSG 1010.4, with transparent PVB foil (2 x 10 mm) 20 mm

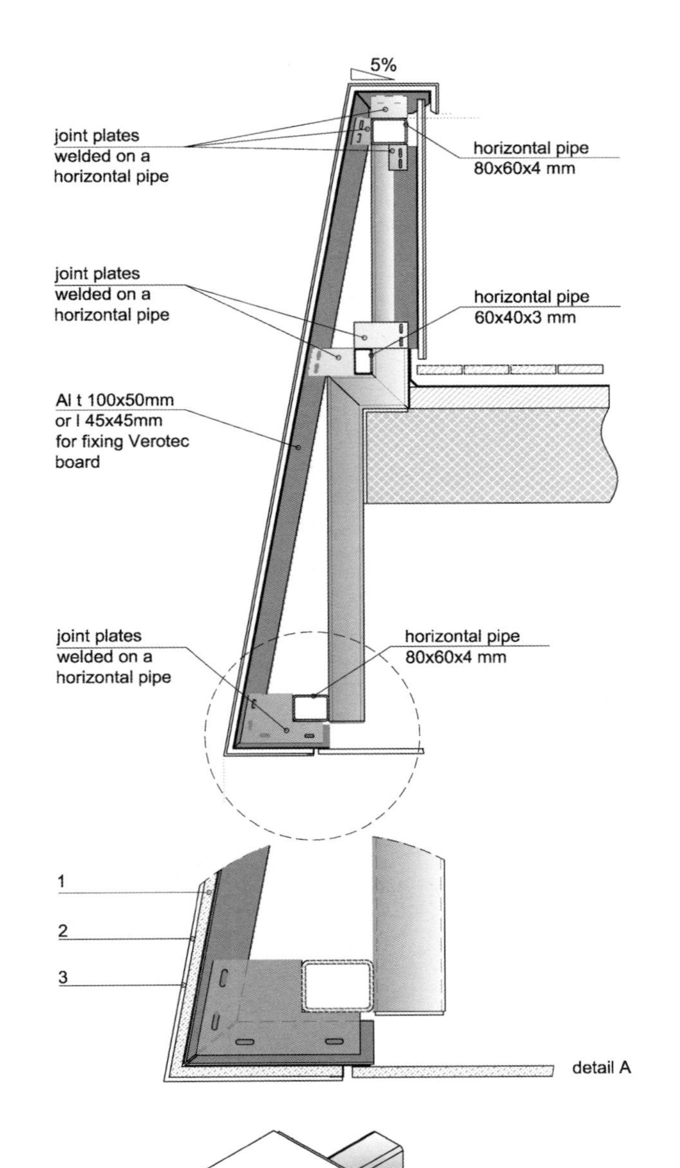

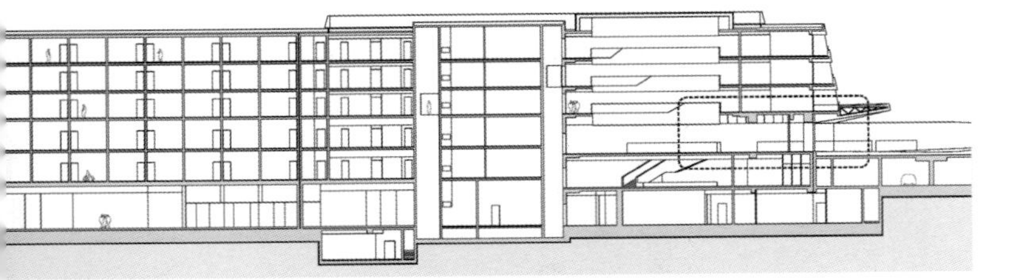

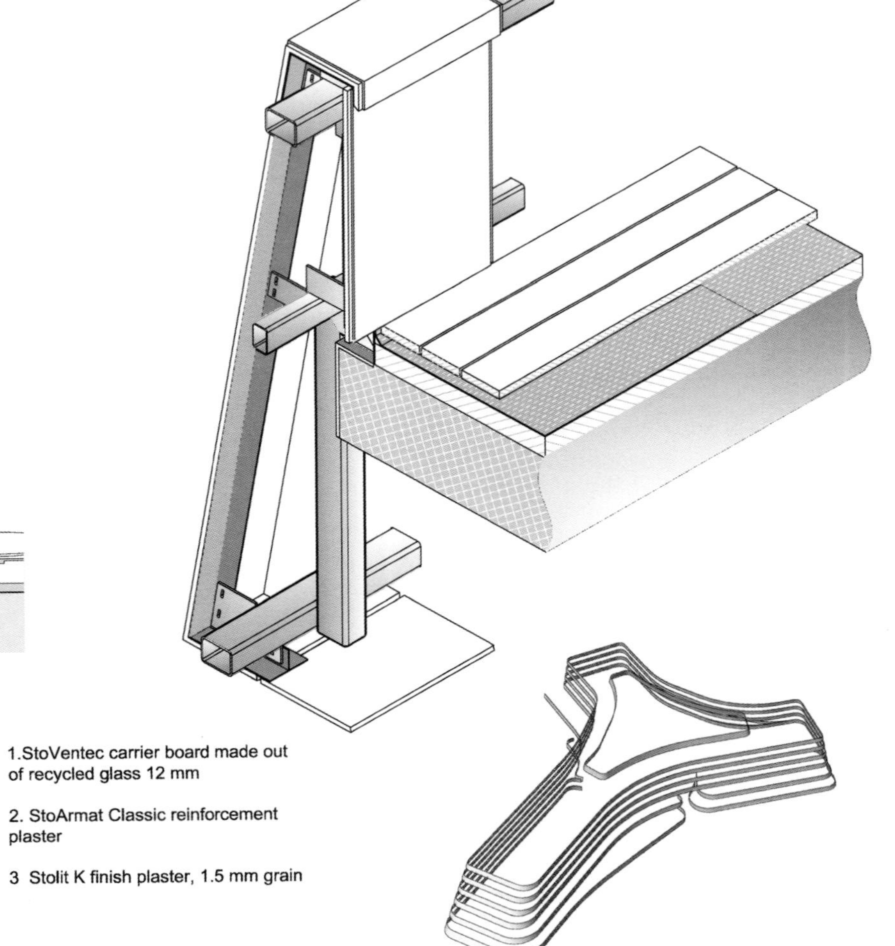

1.StoVentec carrier board made out of recycled glass 12 mm

2. StoArmat Classic reinforcement plaster

3 Stolit K finish plaster, 1.5 mm grain

Detail of parapet wall

Zamet Centre

Location: Rijeka, Croatia
Architect: 3LHD
Client: Rijeka Sport d.o.o.
Site Area: 12,289 m²
Gross Floor Area: 16,830 m²
Completion Date: 2009
Photography: Domagoj Blažević, Miljenko Bernfest, Damir Fabijanić , 3LHD

Situated in Rijeka's quarter Zamet, the new Centre Zamet in complete size of 16830 m² hosts various facilities: sports hall with a maximum of 2380 seats, local community office, city library, 13 commercial and service facilities and a garage with 250 parking spaces.

The ribbon-like stripes were inspired by "gromača", a type of rocks specific to Rijeka, which the centre artificially reinterprets by colour and shape. Stripes are covered with 51.000 ceramic tiles designed by 3LHD and manufactured specially for the centre. Steel girders of 55 meters span and different heights enable the natural light illumination of the sports hall.

The basic characteristic of the design is the integration of a big project task into the urban structure of Zamet, with the objective of minimizing disruption and to evaluate its given urban conditions-unlevelling the terrain, the pedestrian link in a north-south direction, the quality plateau in front of the primary school, the park zone, placing the programme in the centre of Zamet at the intersection of communications. The joint conceptual and design element of the handball hall and the Zamet centre are "ribbons" stretching in a north-south direction, functioning at the same time as an architectural design element of the objects and as a zoning element which forms a public square and a link between the north - park-school and the south - the street. One third of the hall's volume is built into the terrain, and the building with its public and service facilities has been completely integrated into the terrain.

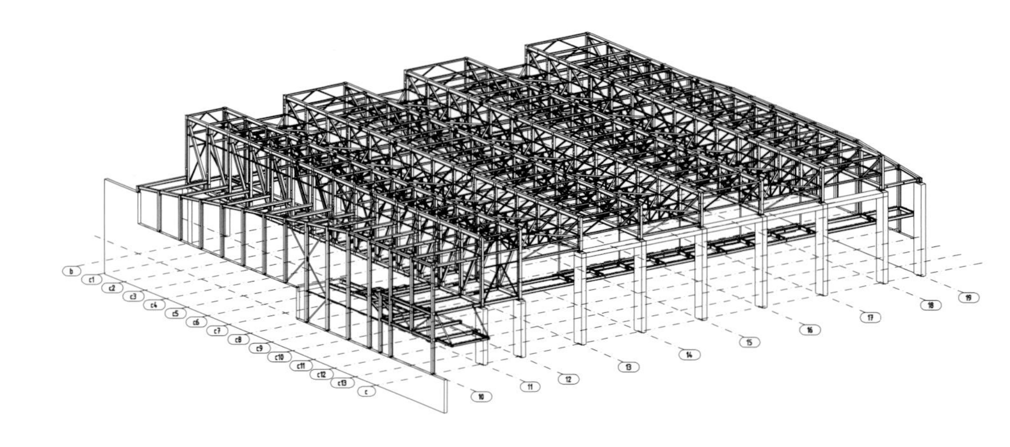

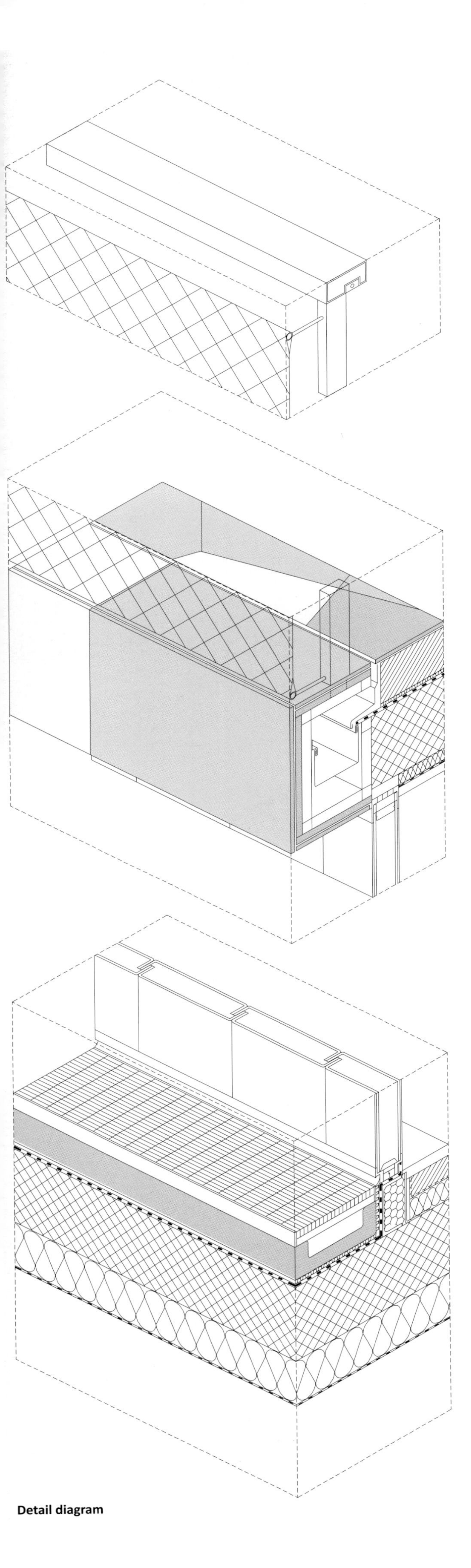

Detail diagram

1. Square
2. Utility court
3. Entrance for visitors
4. Entrance players
5. Press entrance
6. Vip entrance
7. Local comunity
8. Library
9. Entrance shops
10.Shop
11. Staircase
12. Entrance
13. Entrance control
14. Lobby
15. Superintendent
16. Storage
17. Sauna
18. Massage
19. Locker room
20. Hall
21. Hall
22. WC
23. Referee
24. Utility room
25. Coach
26. Dopping control
27. Office
28. Conference room
29. Auditorium
30. Coat check
31. Spectators
32. Caffe

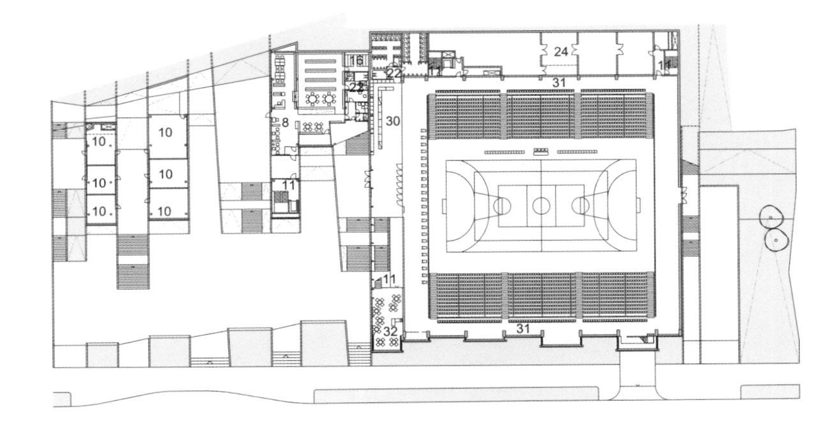

First floor Plan

1. Square
2. Utility court
3. Entrance for visitors
4. Entrance players
5. Press entrance
6. VIp entrance
7. Local community
8. Library
9. Entrance shops
10. Shop
11. Staircase
12. Entrance
13. Entrance control
14. Lobby
15. Superintendent
16. Storage
17. Sauna
18. Massage
19. Locker room
20. Hall
21. Hall
22. WC
23. Referee
24. Utility room
25. Coach
26. Dopping control
27. Office
28. Conference room
29. Auditorium

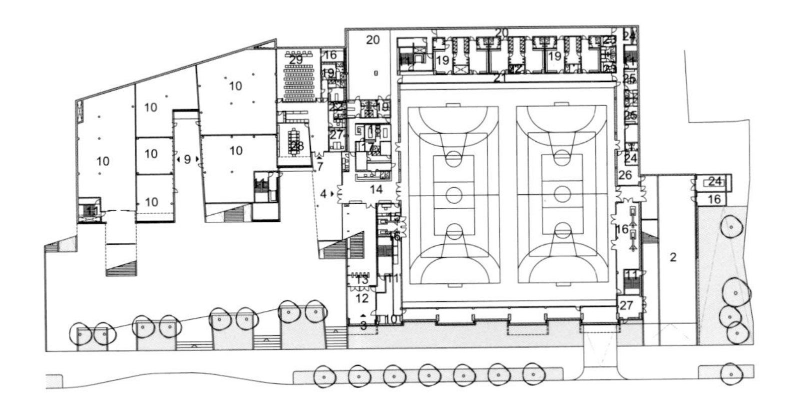

Ground floor Plan

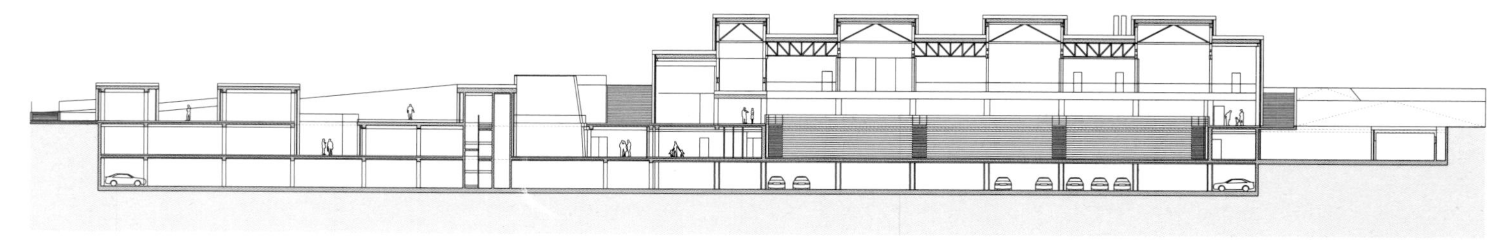
Section 1

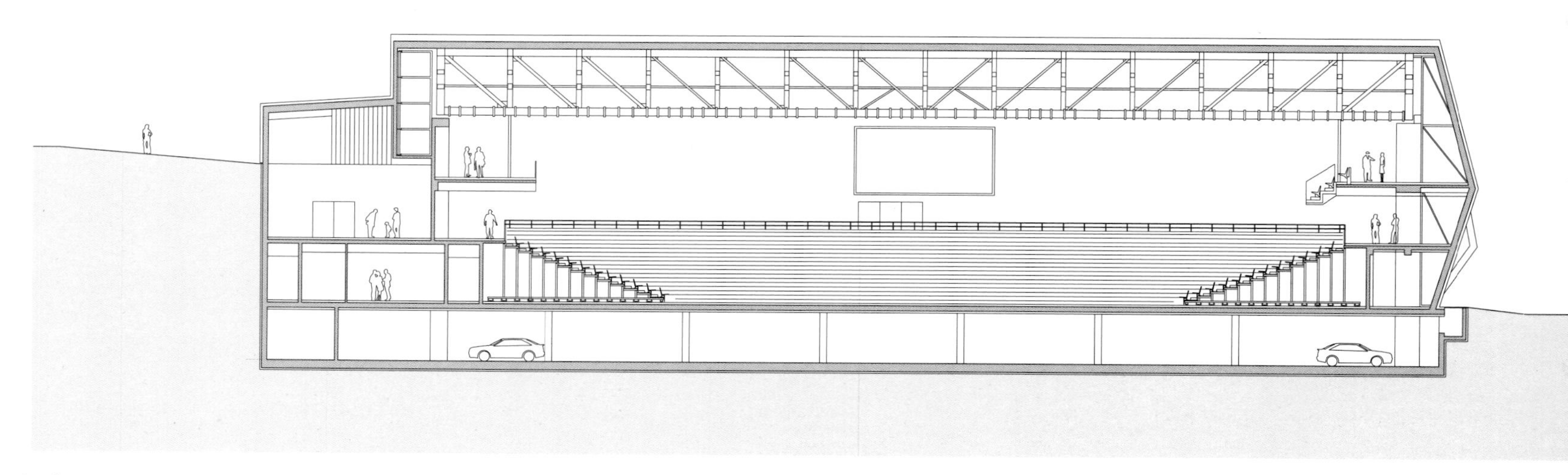
Section a

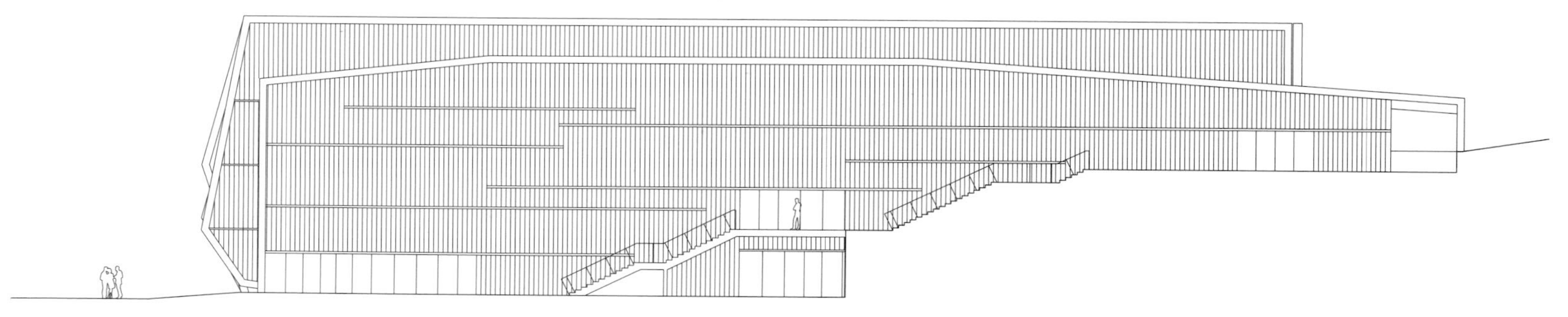
Elevation east

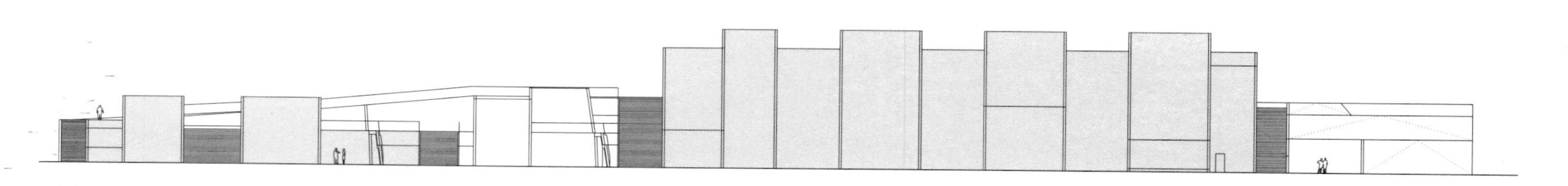
Elevation south

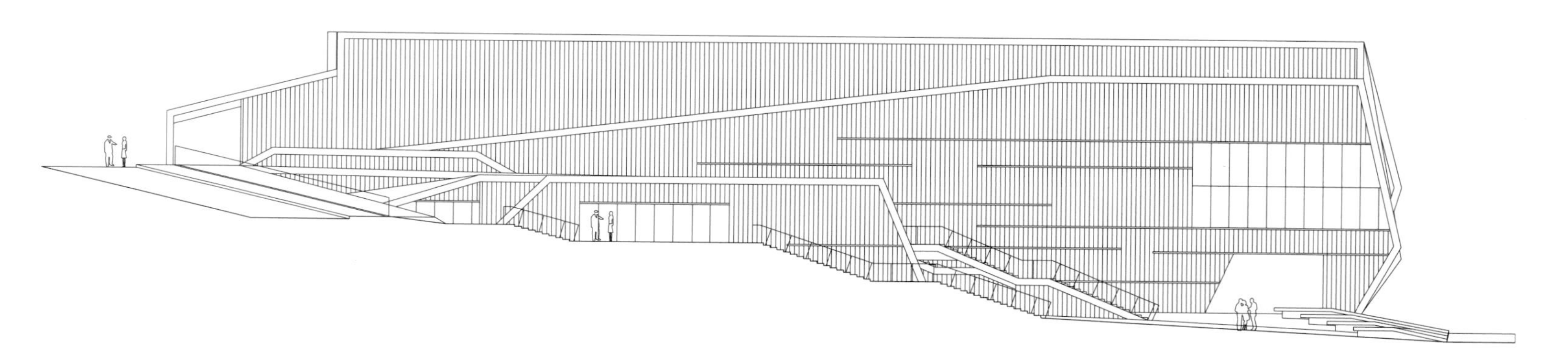
Elevation west

1	**roof construction**	
	ceramic tiles on aluminium construction	11 mm
	ventilated cavity	-
	two-layer TPO roof seal	4 mm
	sloping polystyrene rigid-foam thermal insulation	200 mm
	vapour barrier	5 mm
	corrugated steel sheeting	137 mm
	steel structure	
	wood wool acoustic panels	
2	**wall construction - ceramic tiles**	
	wood wool acoustic panels	25 mm
	oriented-strand board on wood construction	20 mm
	air cavity	-
	vapour barrier	0.2 mm
	mineral wool thermal insulation	120 mm
	oriented-strand board	20 mm
	sealing layer	4 mm
	ventilated cavity	-
	ceramic tiles on aluminium construction	11 mm
3	**wall construction - ceramic tiles**	
	fibre-cement board	25 mm
	oriented-strand board on wood construction	20 mm
	air cavity	-
	oriented-strand board on steel construction	20 mm
	vapour-permeable and water-resistant membrane	1 mm
	air cavity	-
	ceramic tiles on aluminium construction	11 mm
10	**floor construction**	
	cast floor - polyurethane	3 mm
	screed	65 mm
	separating layer	0.2 mm
	expanded polystyrene impact-sound insulation EPS	40 mm
	elastic expanded polystyrene EPS-T	20 mm
	reinforced concrete slab	300 mm
	mineral wool thermal insulation	120 mm
	plaster	2 mm

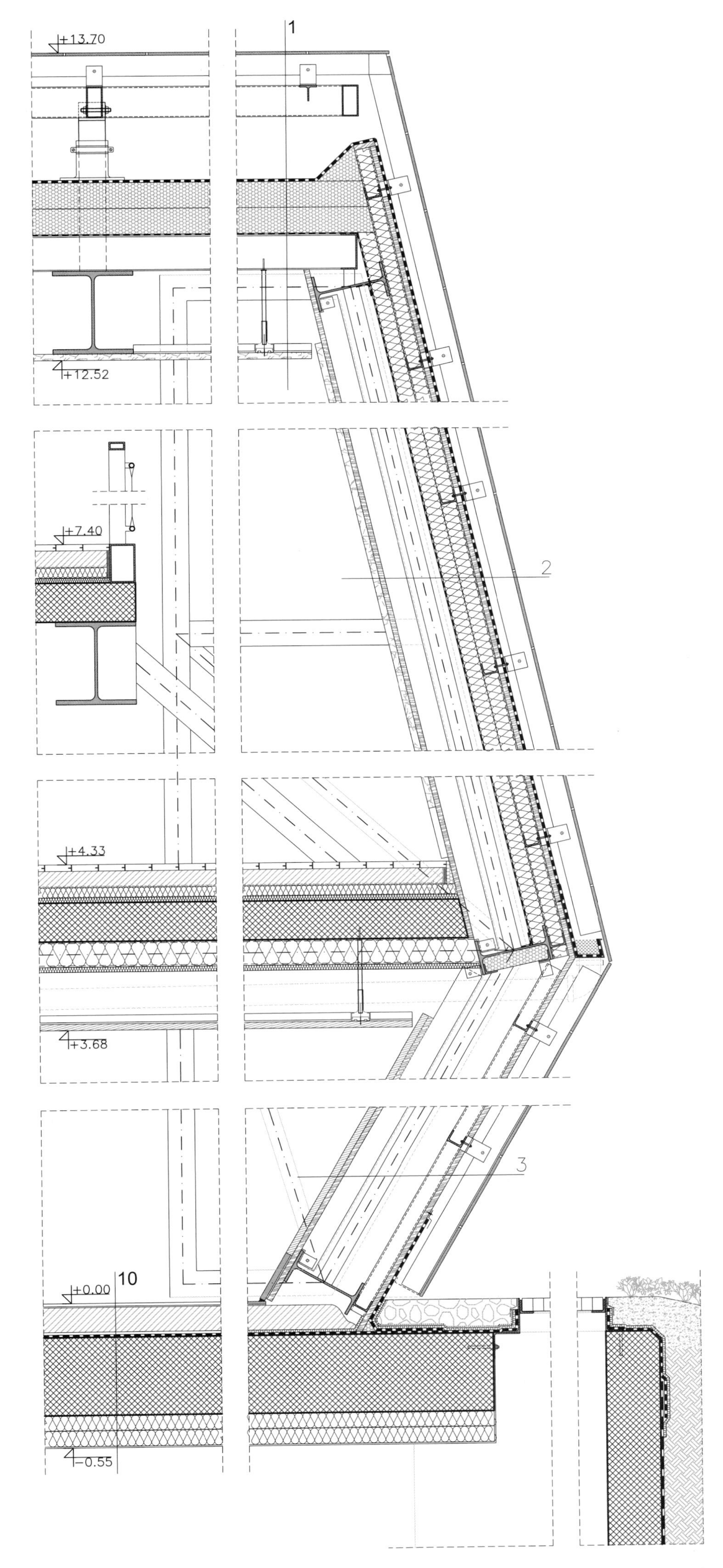

1	**roof construction**	
	ceramic tiles on aluminium construction	11 mm
	ventilated cavity	-
	two-layer TPO roof seal	4 mm
	sloping polystyrene rigid-foam thermal insulation	200 mm
	vapour barrier	5 mm
	corrugated steel sheeting	137 mm
	steel structure	
	wood wool acoustic panels	
4	**wall construction - U profiled glass**	
	2x12.5mm plasterboard	25 mm
	oriented-strand board on steel construction	20 mm
	air cavity	-
	vapour barrier	0.2 mm
	oriented-strand board on steel construction	20 mm
	rock wool thermal insulation	120 mm
	ventilated cavity	40 mm
	U-profiled glass in aluminium frame	83 mm
5	**wall construction - U profiled glass**	
	reinforced concrete wall	300 mm
	rock wool thermal insulation	120 mm
	ventilated cavity	40 mm
	U-profiled glass in aluminium frame	83 mm
6	**wall construction in ground**	
	reinforced concrete wall	300 mm
	bentonite-geotextile waterproofing membrane	6.4 mm
	sloping polystyrene rigid-foam thermal insulation	80 mm
	soil layer	100 mm
	drainage layer of gravel	
7	**wall construction in ground**	
	reinforced concrete wall	300 mm
	bentonite-geotextile waterproofing membrane	6.4 mm
	soil layer	100 mm
	drainage layer of gravel	
10	**floor construction**	
	cast floor - polyurethane	3 mm
	screed	65 mm
	separating layer	0.2 mm
	expanded polystyrene impact-sound insulation EPS	40 mm
	elastic expanded polystyrene EPS-T	20 mm
	reinforced concrete slab	300 mm
	mineral wool thermal insulation	120 mm
	plaster	2 mm
11	**ground floor construction**	
	epoxy coating	3 mm
	reinforced concrete foundation slab	500 mm
	bentonite-geotextile waterproofing membrane	6.4 mm
	soil layer	100 mm
	bed of gravel	

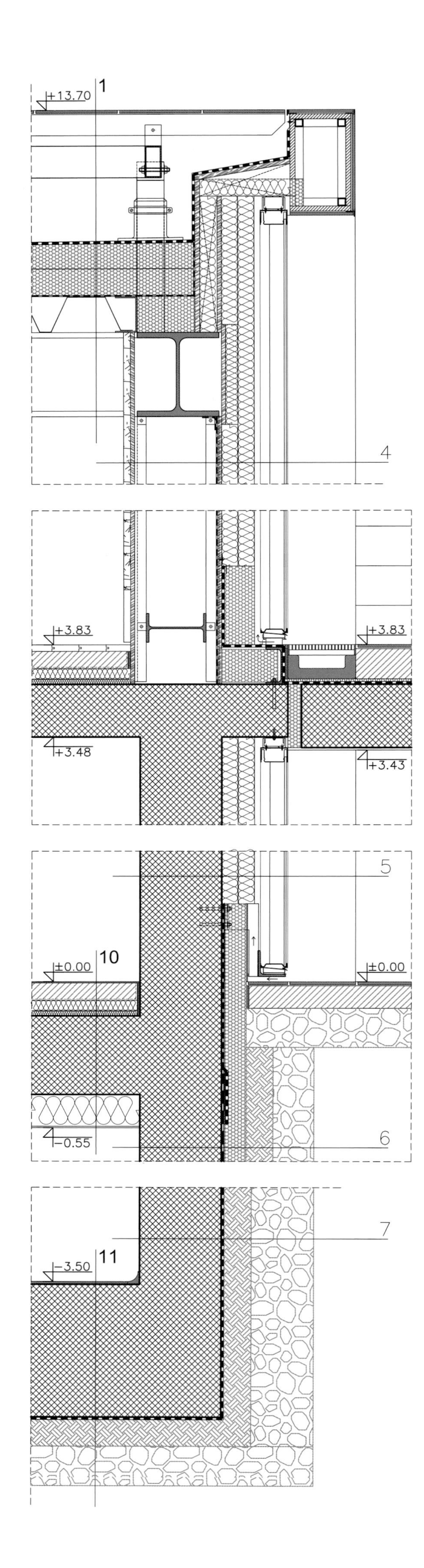

1	**roof construction**	
	ceramic tiles on aluminium construction	11 mm
	ventilated cavity	-
	two-layer TPO roof seal	4 mm
	sloping polystyrene rigid-foam thermal insulation	200 mm
	vapour barrier	5 mm
	corrugated steel sheeting	137 mm
	steel structure	
	wood wool acoustic panels	
8	**floor construction**	
	parquet	15 mm
	screed	65 mm
	separating layer	0.2 mm
	expanded polystyrene impact-sound insulation EPS	40 mm
	elastic expanded polystyrene EPS-T	20 mm
	reinforced concrete slab	150 mm
	air cavity	-
	wood wool acoustic panels	25 mm
9	**floor construction**	
	parquet	15 mm
	screed	65 mm
	separating layer	0.2 mm
	expanded polystyrene impact-sound insulation EPS	40 mm
	elastic expanded polystyrene EPS-T	20 mm
	reinforced concrete slab	200 mm
	rock wool thermal insulation	120 mm
10	**floor construction**	
	cast floor - polyurethane	3 mm
	screed	65 mm
	separating layer	0.2 mm
	expanded polystyrene impact-sound insulation EPS	40 mm
	elastic expanded polystyrene EPS-T	20 mm
	reinforced concrete slab	300 mm
	mineral wool thermal insulation	120 mm
	plaster	2 mm
12	U-profiled glass - double glazing in aluminium frame	83mm

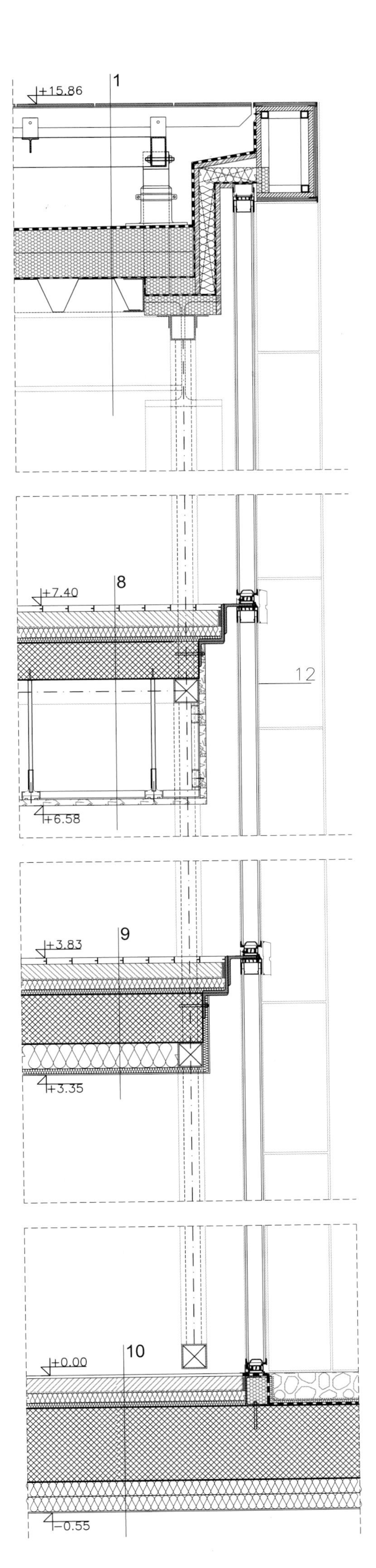

Culture & Arts

Messe Frankfurt – Tor Nord

Location: Frankfurt, German
Architect: Ingo Schrader Architekt BDA, Berlin
Client: Messe Frankfurt Venue Gmbh/ Marc Legg
Completion Date: 2013
Photographers: Ingo Schrader, Christian Richters, Messe Frankfurt

On October 30th, 2014 the Tor Nord on the premises of the Messe Frankfurt has been awarded the Special Award for Sustainable Steel Architecture of the Federal Ministry of the Environment, Nature Conservation, Building and Nuclear Safety (BMUB). The project had already been declared a winning project of the 2014 Iconic Awards by the German Design Council.

A highly visible white oval roof covers the control area of the main gate of the Frankfurt trade fair, creating a new landmark at the city entrance. The oval form stands out against the orthogonal buildings of the fairground and the diversity of directions focussing at the main gate, enabling easy orientation.

The supporting structure of the roof recalls natural, grown forms. This effect is not due to intentional formal mimicry. It results from a parametric design strategy and its mathematical and geometrical principles.

Like a tree, the roof's structure resonates with the specific conditions of the site, thus creating a formal expression with striking logic. The roof displays the intrinsic beauty of the laws of nature, the construction's delicacy and its elegant details. Close cooperation of planners and builders made this possible.

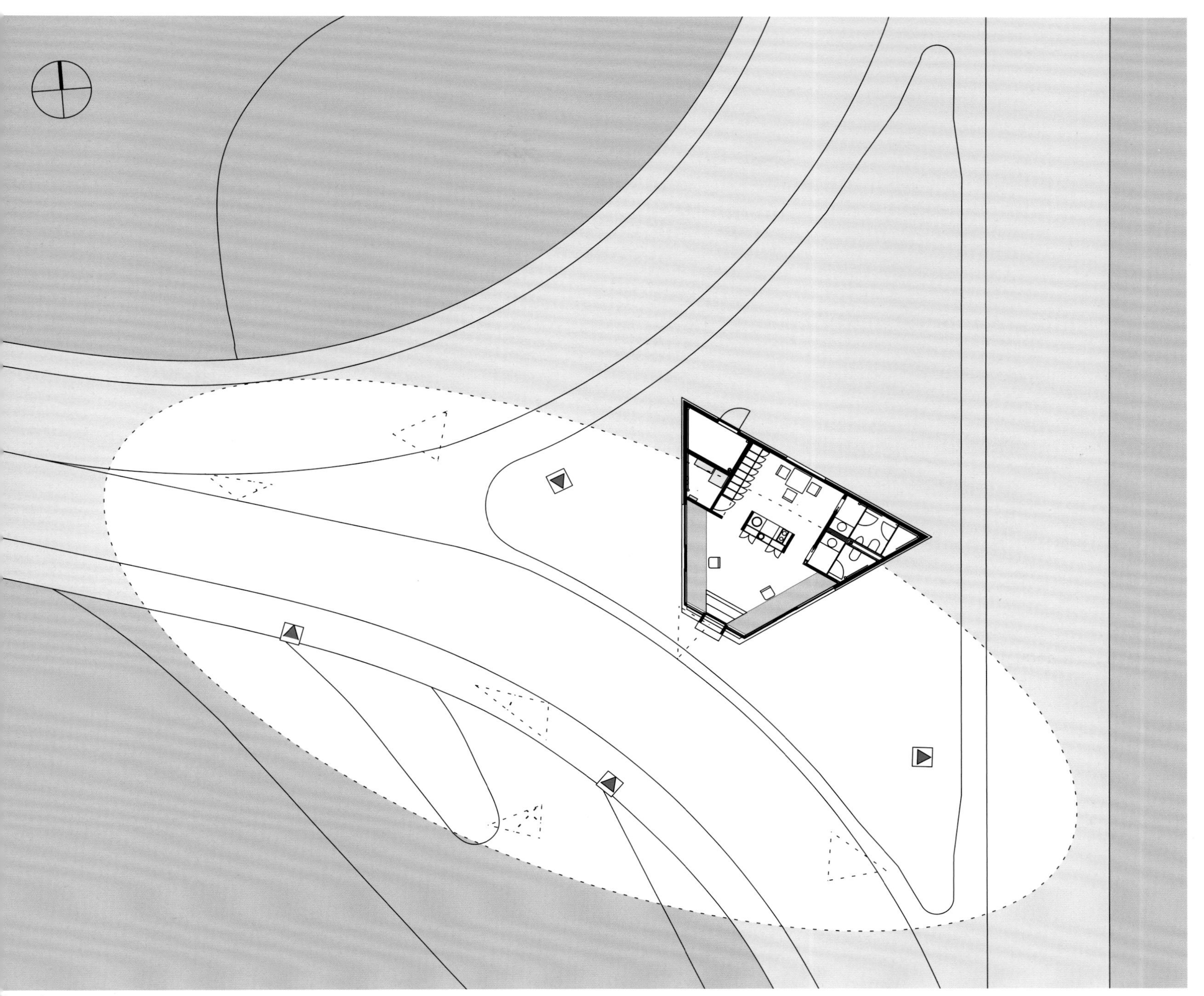

loor-Plan

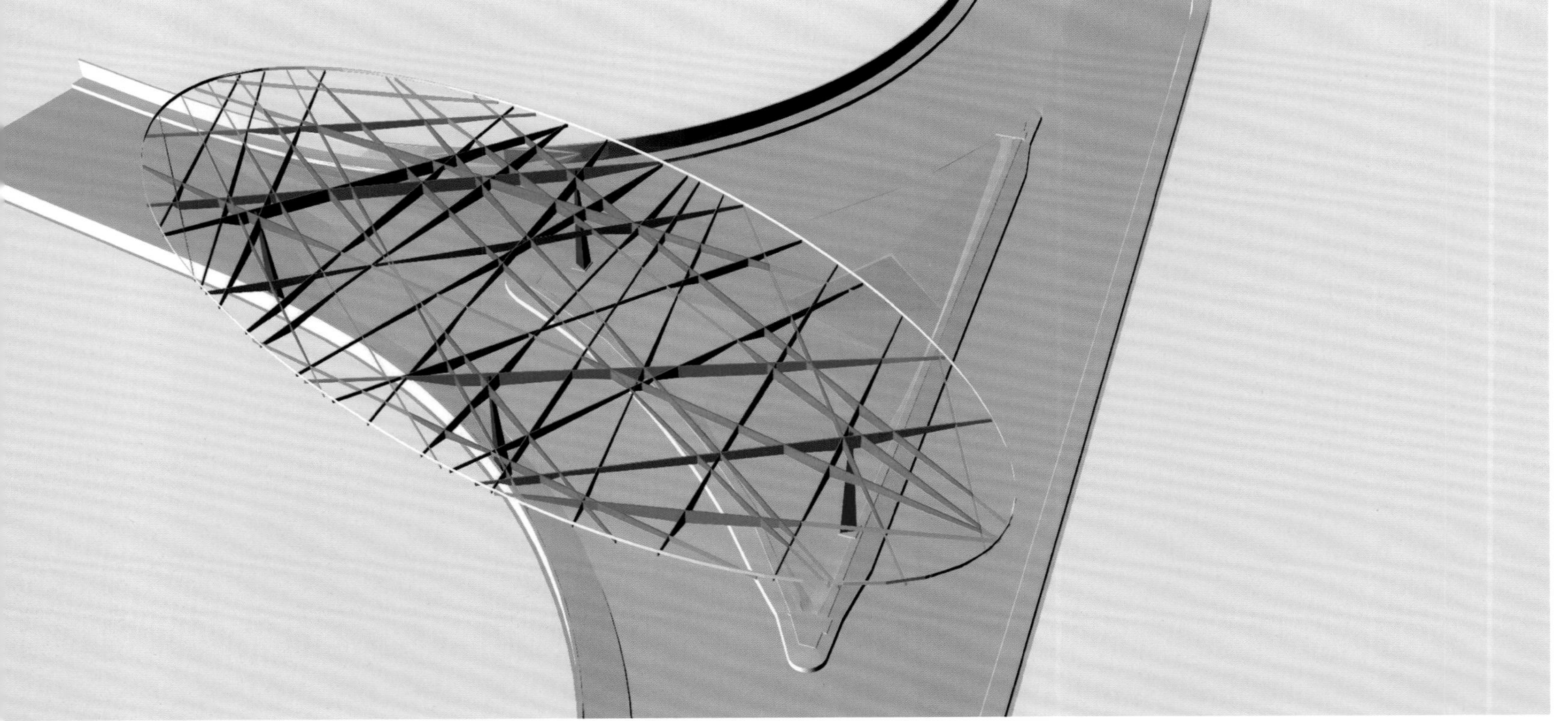

lord Structure

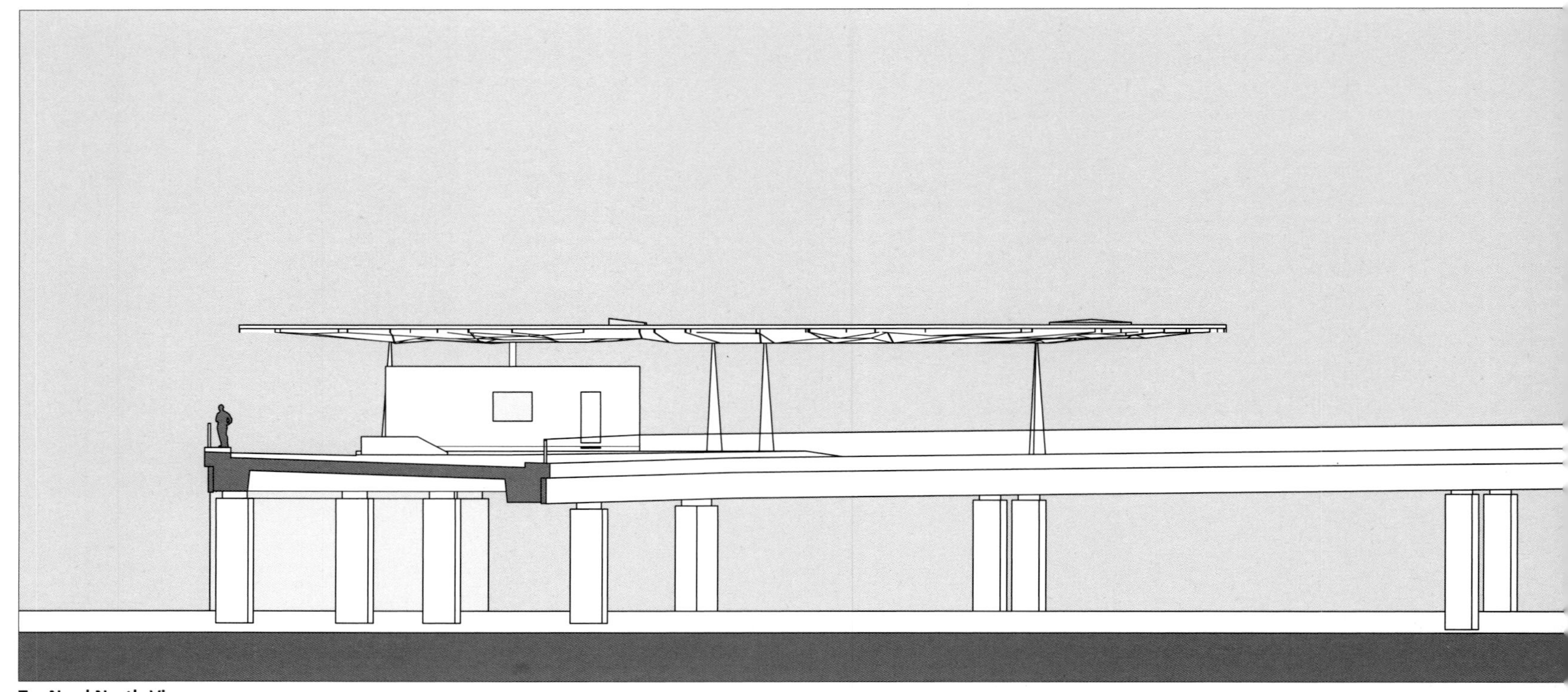

Tor Nord North-View

Tor Nord West-View

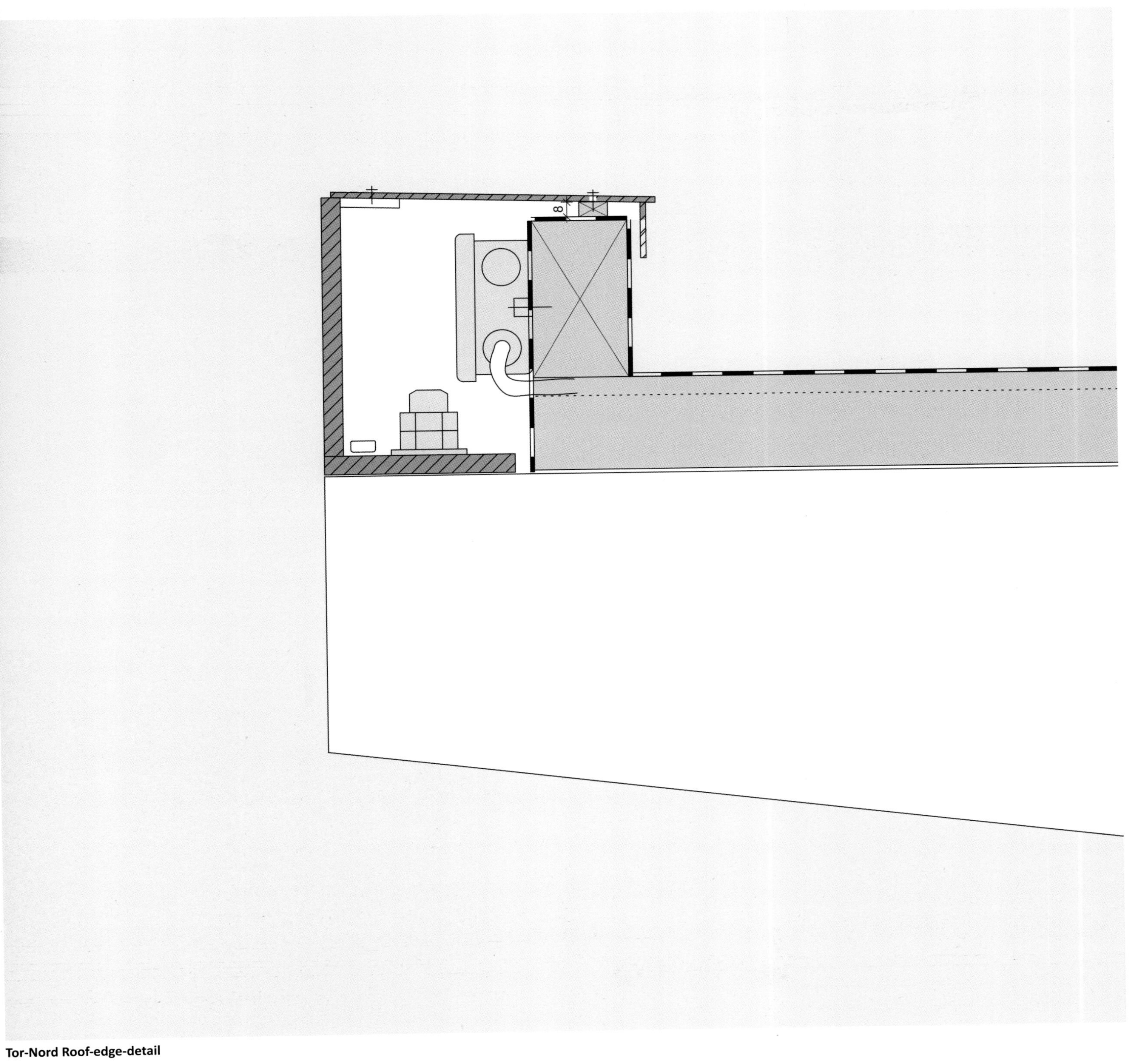

Tor-Nord Roof-edge-detail

Heydar Aliyev Center

Location: Baku, Azerbaijan
Architect: Zaha Hadid Architects
Client: The Republic of Azerbaijan
Site Area: 111,292 m^2
Total Floor Area: 101,801 m^2
Completion Date: 2012
Photography: Iwan Baan; Hélène Binet; Hufton + Crow

The design of the Heydar Aliyev Center establishes a continuous, fluid relationship between its surrounding plaza and the building's interior. The plaza, as the ground surface; accessible to all as part of Baku's urban fabric, rises to envelop an equally public interior space and define a sequence of event spaces dedicated to the collective celebration of contemporary and traditional Azeri culture. Elaborate formations such as undulations, bifurcations, folds, and inflections modify this plaza surface into an architectural landscape that performs a multitude of functions: welcoming, embracing, and directing visitors through different levels of the interior. With this gesture, the building blurs the conventional differentiation between architectural object and urban landscape, building envelope and urban plaza, figure and ground, interior and exterior.

Fluidity in architecture is not new to this region. In historical Islamic architecture, rows, grids, or sequences of columns flow to infinity like trees in a forest, establishing non-hierarchical space. Continuous calligraphic and ornamental patterns flow from carpets to walls, walls to ceilings, ceilings to domes, establishing seamless relationships and blurring distinctions between architectural elements and the ground they inhabit. Our intention was to relate to that historical understanding of architecture, not through the use of mimicry or a limiting adherence to the iconography of the past, but rather by developing a firmly contemporary interpretation, reflecting a more nuanced understanding.

Responding to the topographic sheer drop that formerly split the site in two, the project introduces a precisely terraced landscape that establishes alternative connections and routes between public plaza, building, and underground parking. This solution avoids additional excavation and landfill, and successfully converts an initial disadvantage of the site into a key design feature.

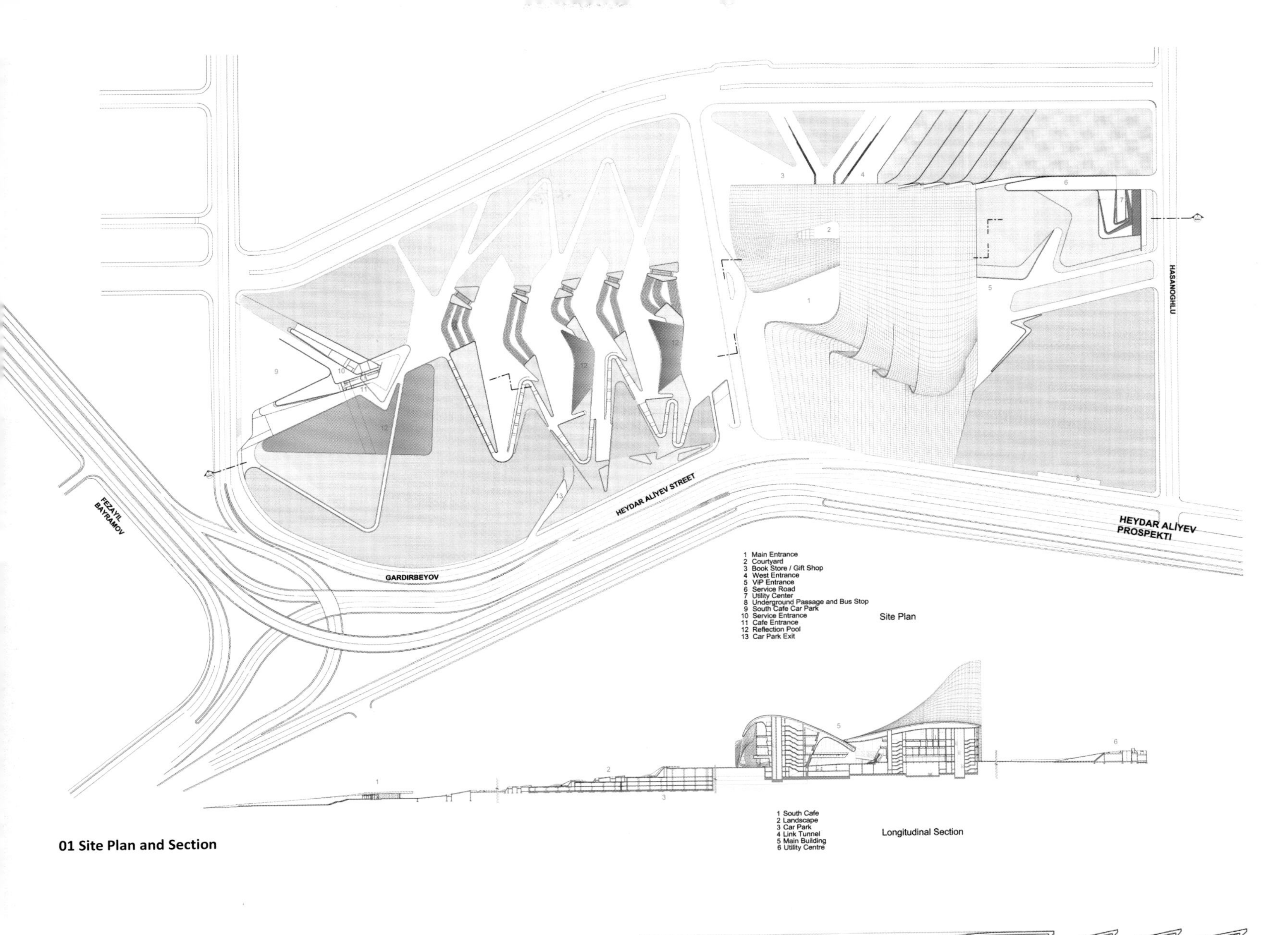

01 Site Plan and Section

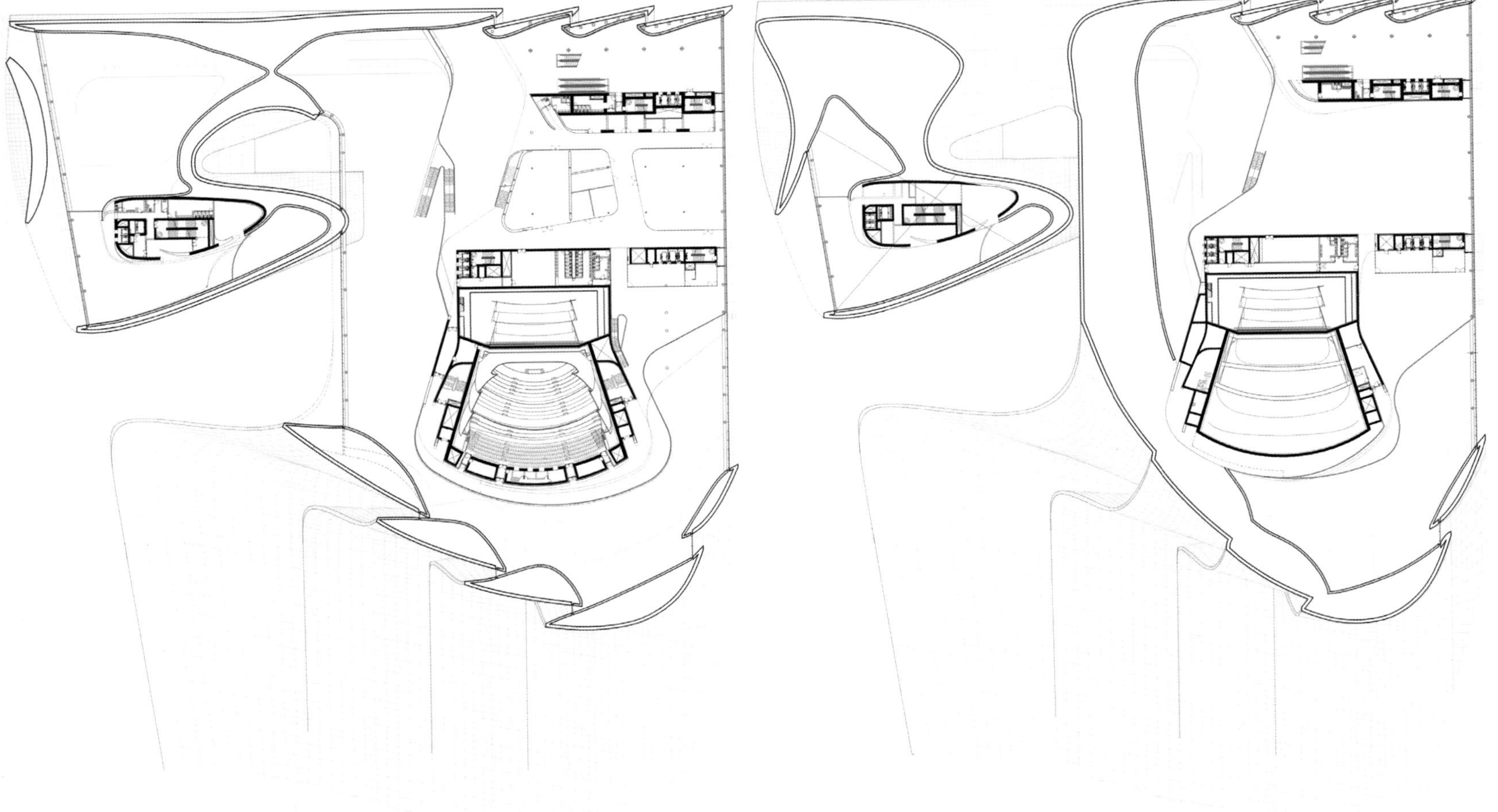

2nd Floor Plan

3rd Floor Plan

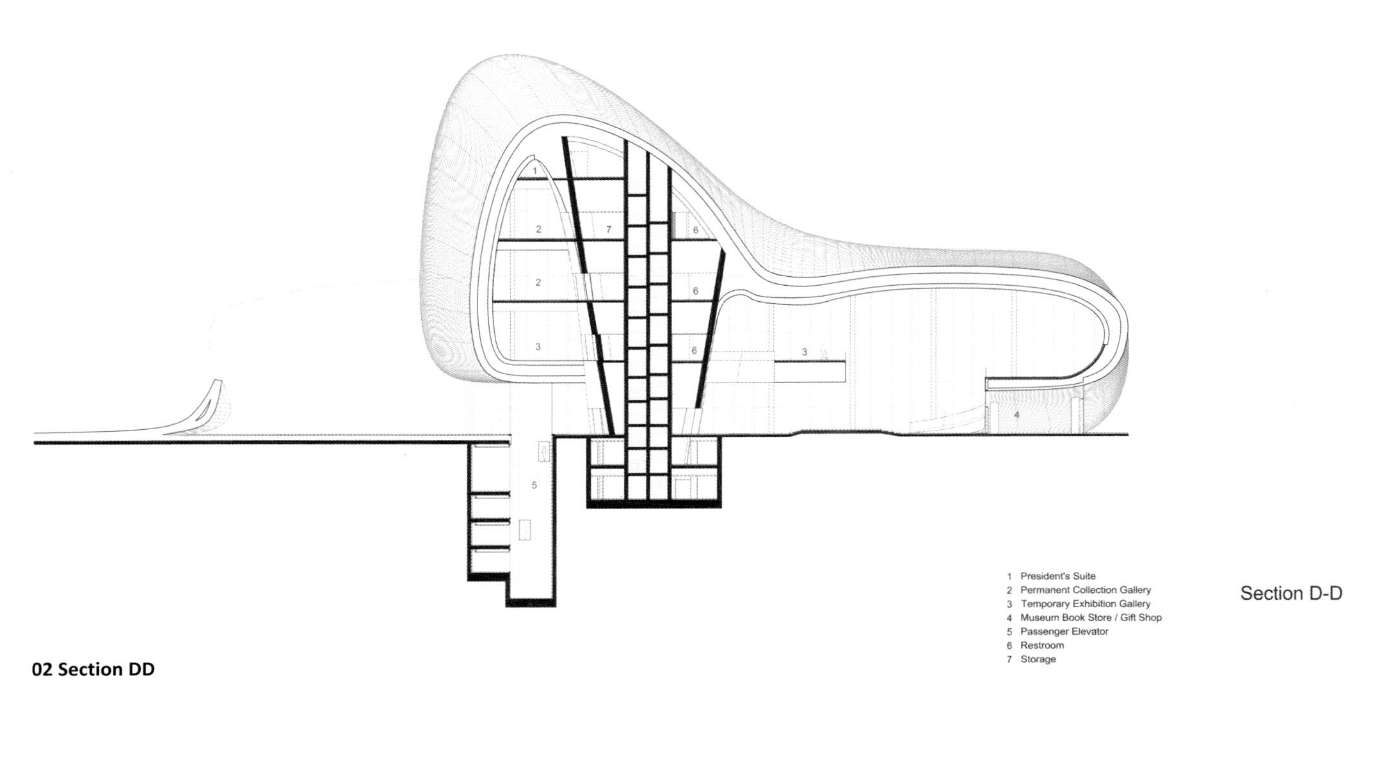

02 Section DD

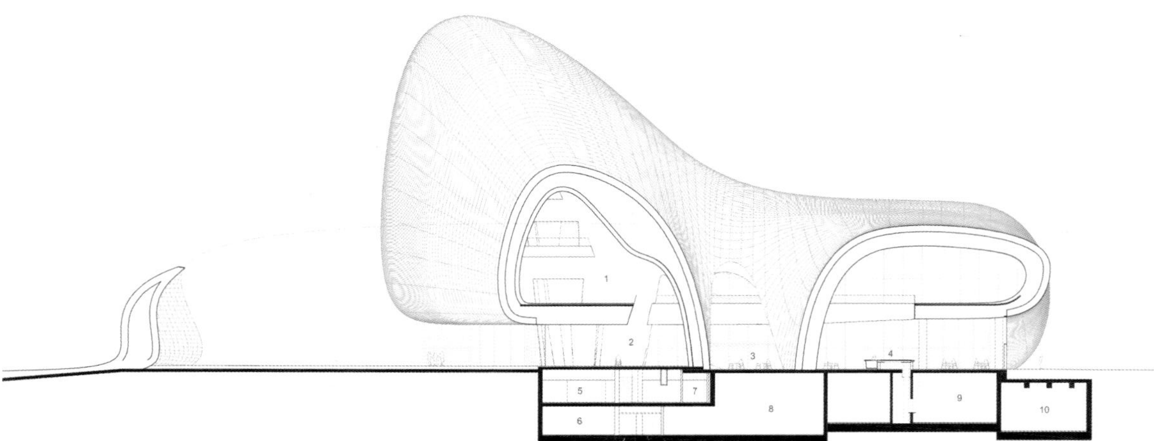

1 Temporary Collection Gallery
2 Museum Lobby
3 Courtyard
4 Museum Cafe
5 Cloakroom
6 Registration & Art Handling
7 Restroom
8 AHU Room
9 Cafe Kitchen
10 Loading Bay

Section E-E

03 Section EE

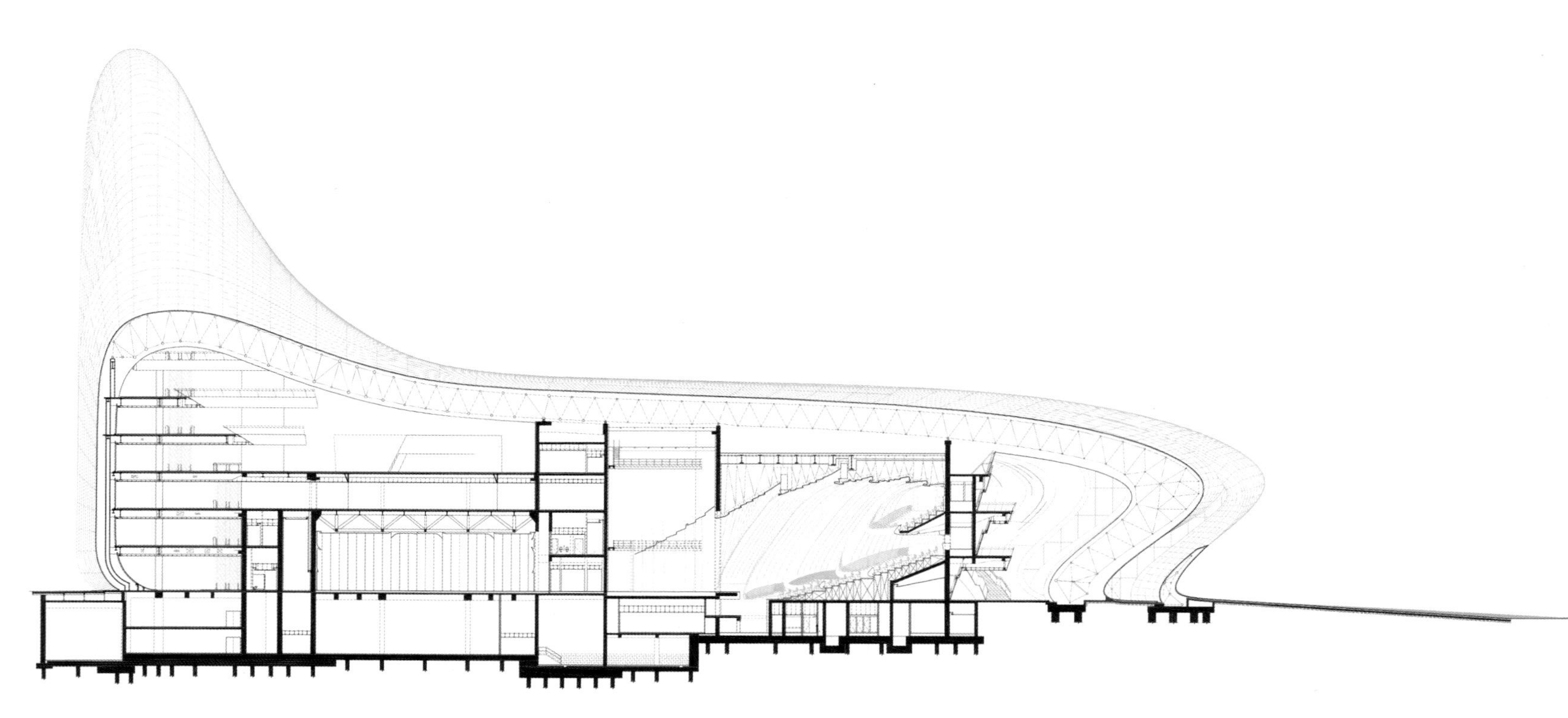

01 Section AA

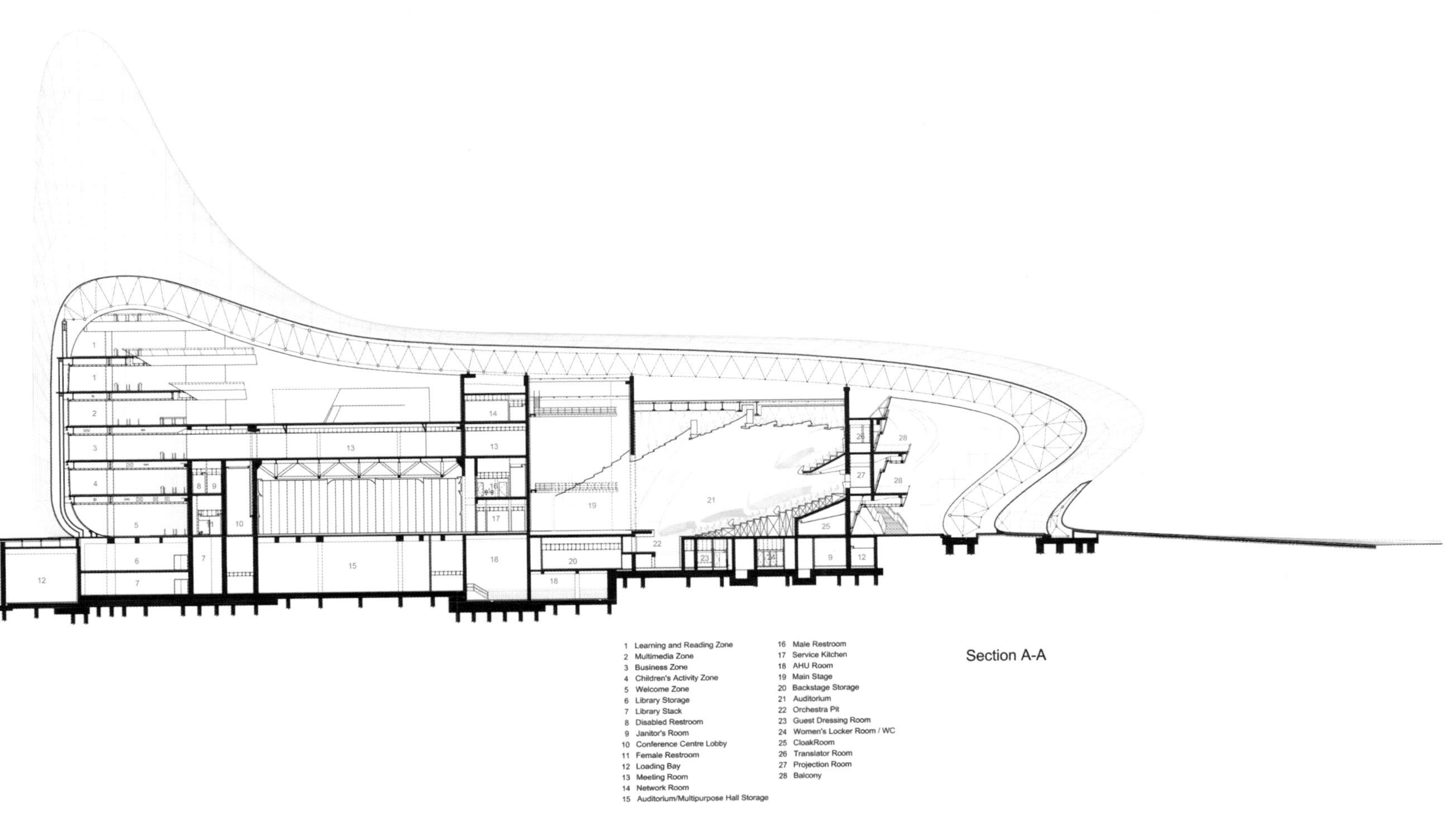

01 Section AA

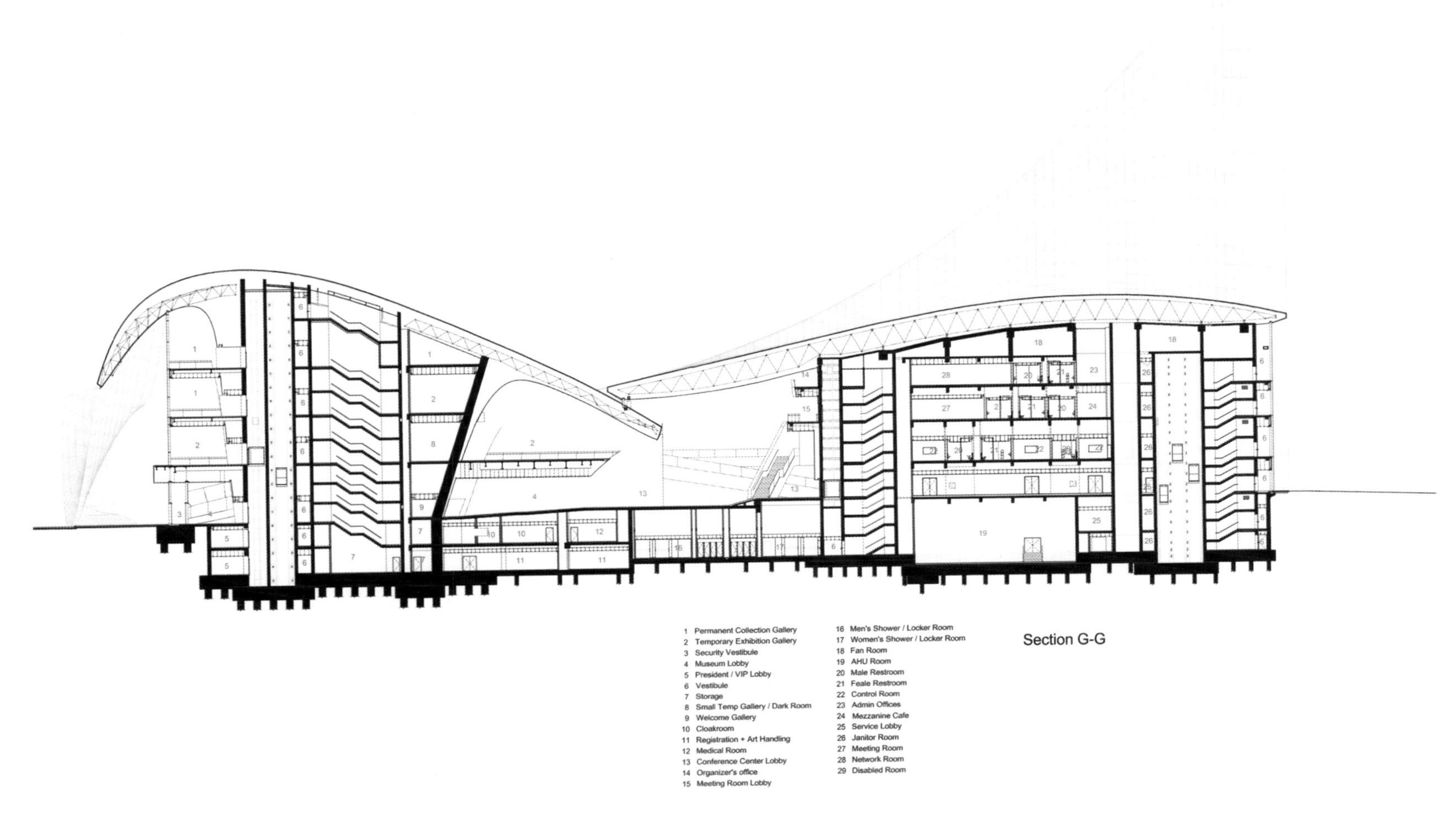

04 Section GG

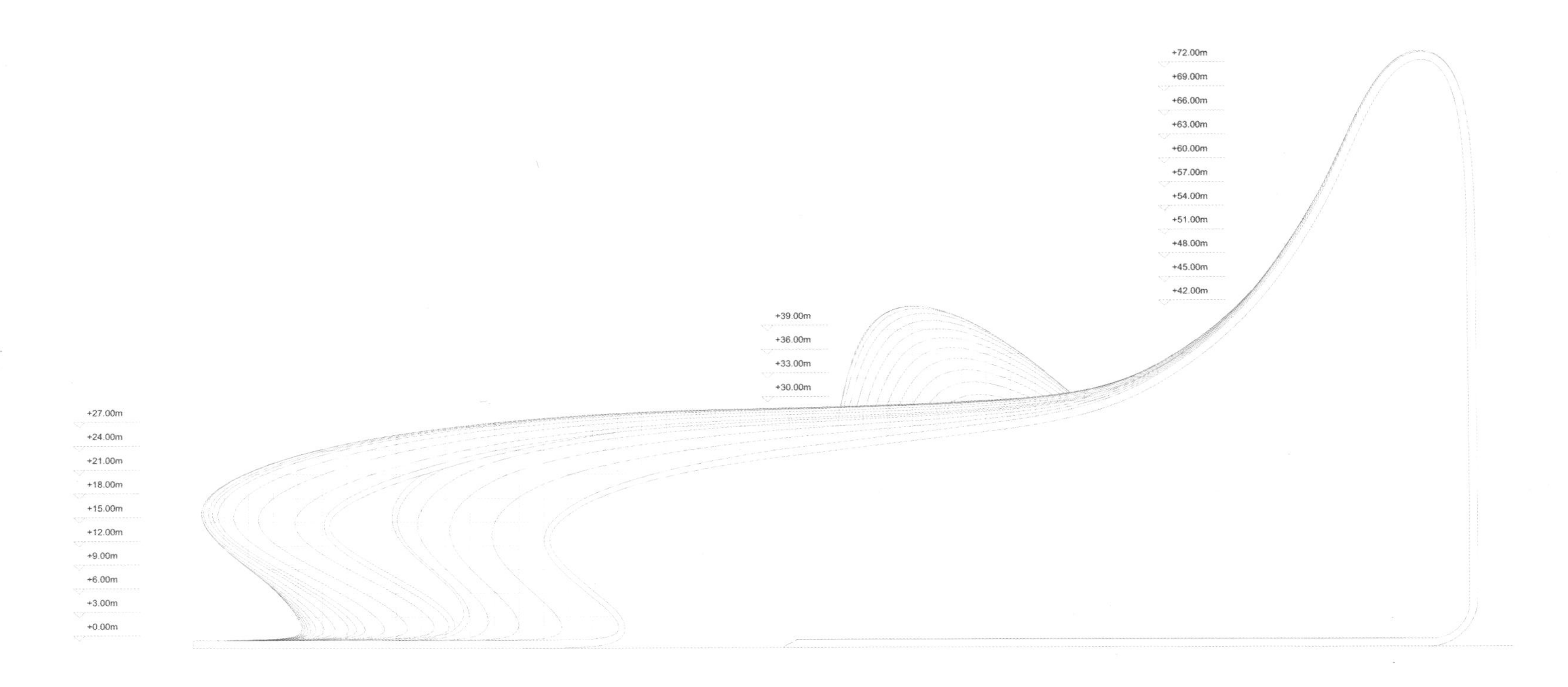

03 North East

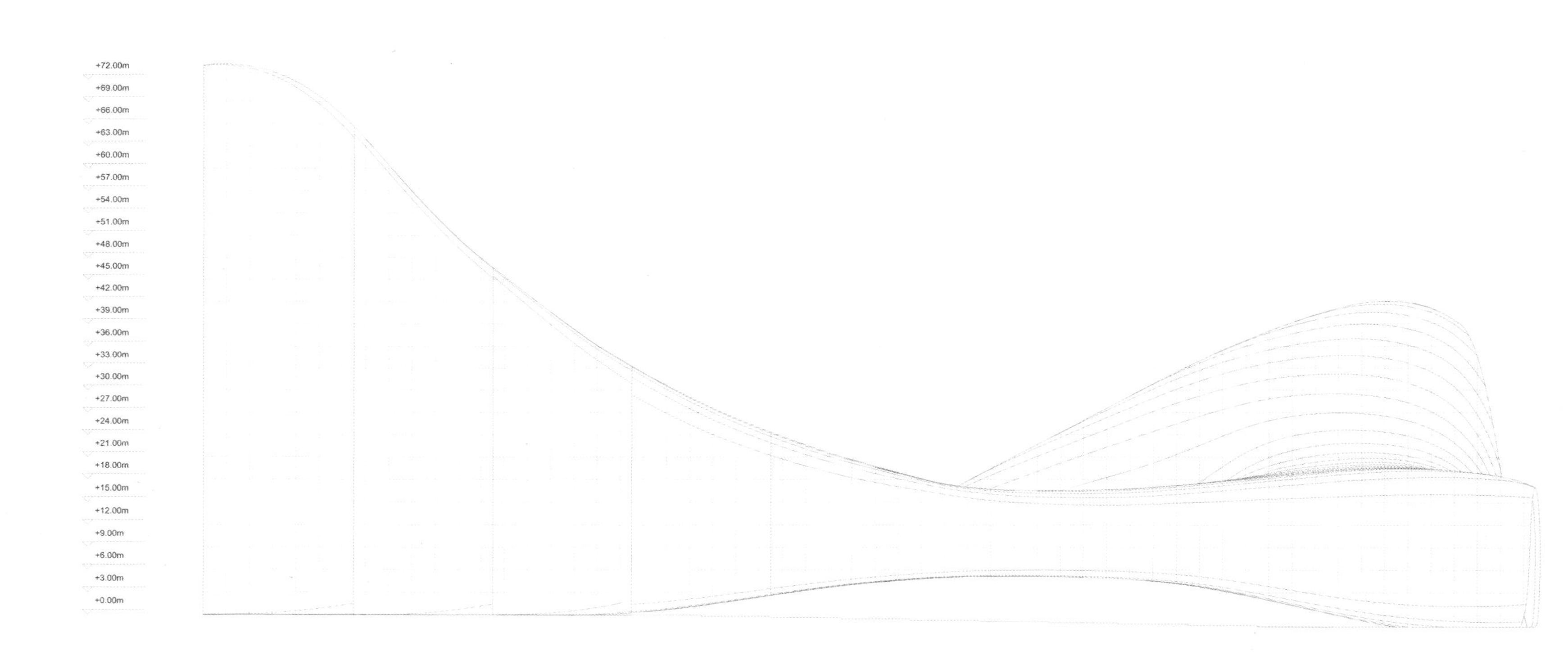

05 North West

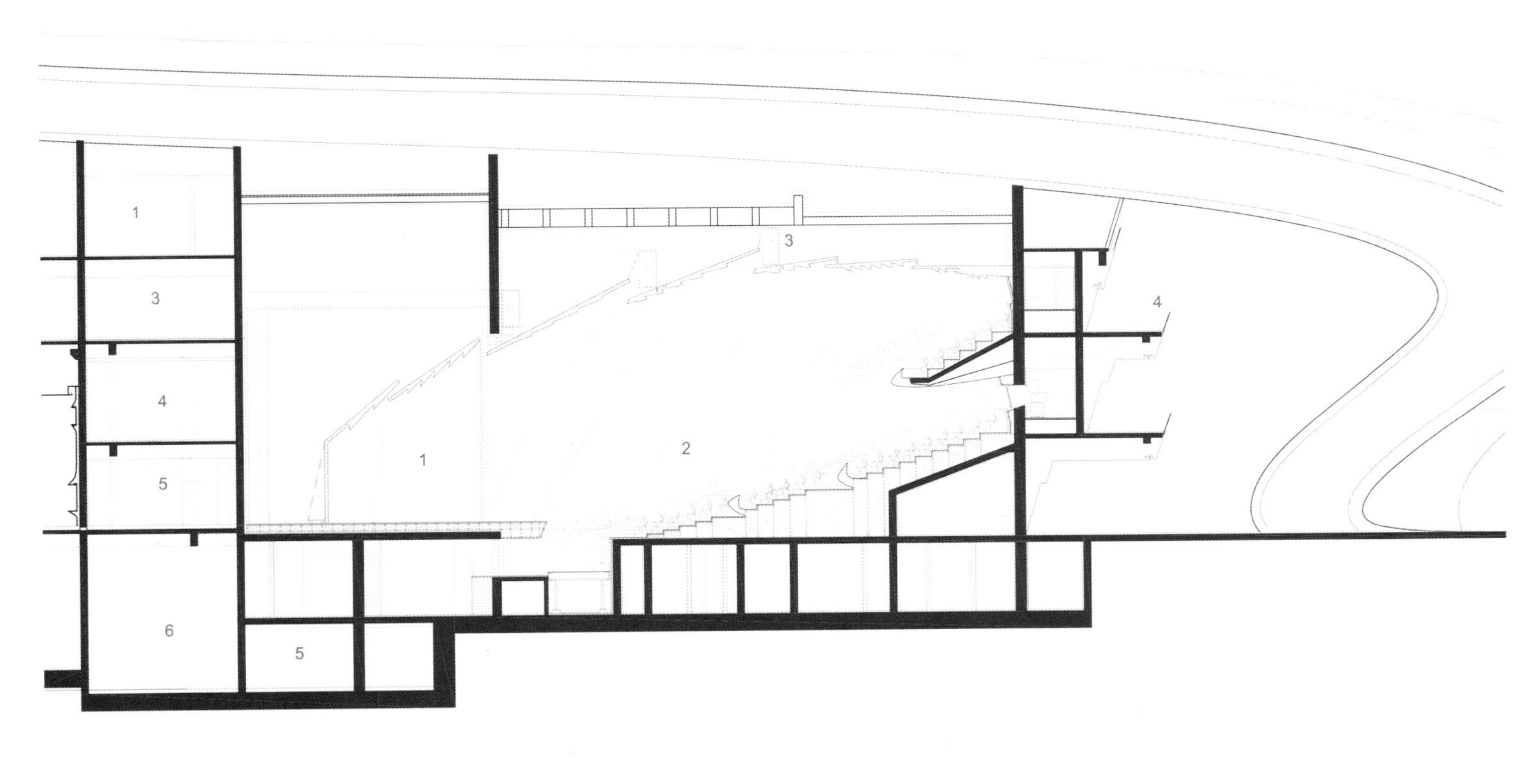

1 Network Room
2 Corridor
3 Meeting Room
4 Female Restroom
5 Service Kitchen
6 AHU Room
7 Main Stage
8 Multi-Purpose Auditorium
9 Balcony

Auditorium Section

Auditorium Section

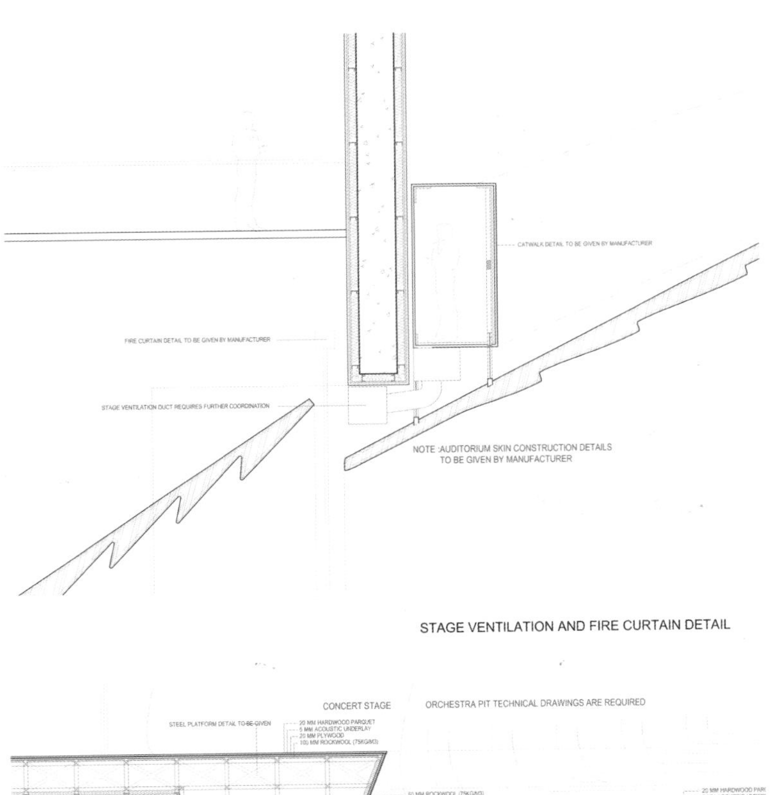

STAGE VENTILATION AND FIRE CURTAIN DETAIL

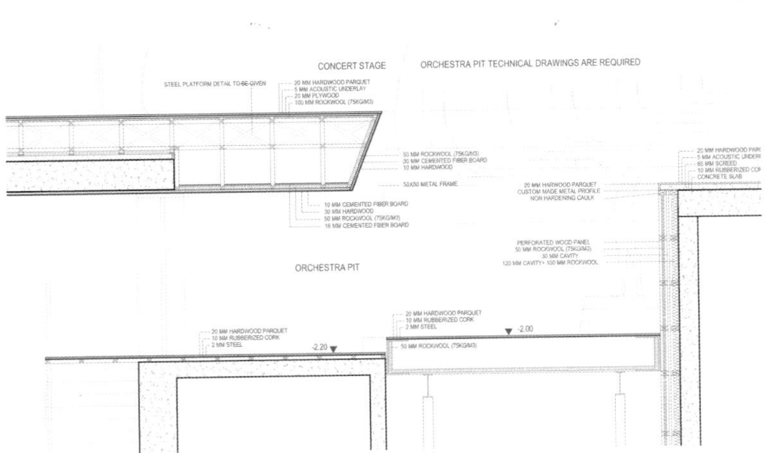

ORCHESTRA PIT DETAIL

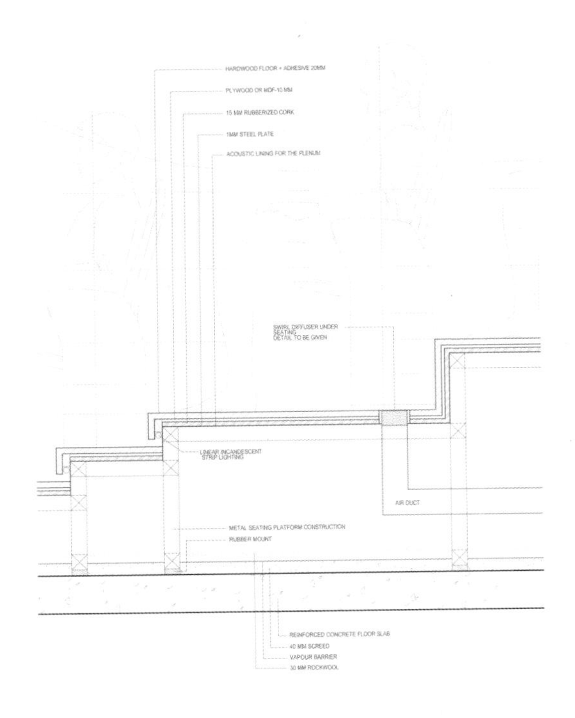

STEP LIGHTING AND VENTILATION DETAIL

Umwelt Arena

Location: Spreitenbach, Switzerland
Architect: Rene Schmid Architekten
Site Area: 5,400 m^2
Completion Date: 2013
Photography: Alex Buschor; Bruno Helbling; Michael Egloff

Aesthetics and ecology converge in the architecture of the Umwelt Arena. A striking architectural feature of the Umwelt Arena is the crystalline roof structure with the photovoltaic system integrated into the building. This is an intentional design. The large hall (arena) forms the centerpiece and conveys an open and inviting atmosphere. Solar heat is used to cool the Umwelt Arena in summer and heat it in winter. This is done by means of a distribution network (TABS) in the concrete floors and an earth tube system beneath the base plate of the bottom parking level. Cold water circulates through the network of pipes in summer, hot water in winter. The circulating water uses the difference in temperature in the earth collector and the large storage tanks to deposit excess heat or cold from the building interior in the storage media according to the season or requirements. Any excess heat or cold that cannot be accommodated by the water tanks is stored in the earth collector. The equipment used in this innovative heating and ventilation system runs on solar power and ultimately makes it possible to run a building with carbon-neutral services. The roof-integrated photovoltaic system produces more (CO2-free) electricity than is needed to run the building.

Umwelt Arena helps visitors to experience and understand sustainability and environmental technology. Visitors will find information and products on the theme of "modern life" under one roof. Experts and lay people alike can become acquainted with the subject over an area of 5,400 m^2 and compare technologies, products and services on the spot – in a clear and practical way. It's a unique setting for seminars, trade shows, conventions, events and banquets. Various multipurpose rooms are available for occasions ranging from small meetings to large-scale events.

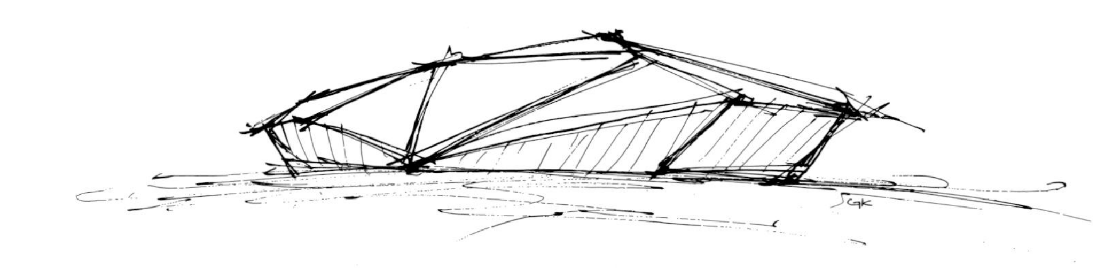

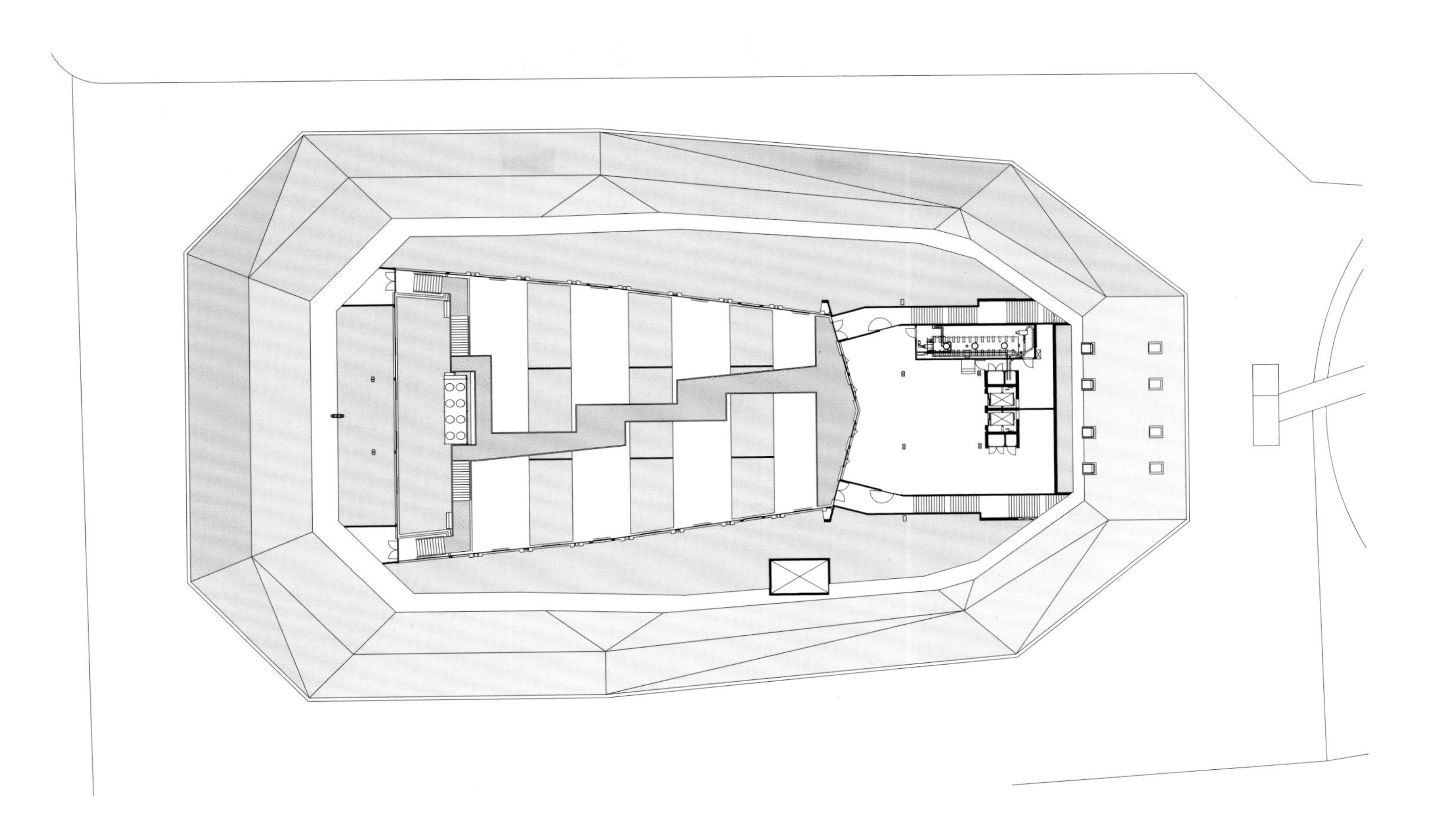

3rd floor plan roof

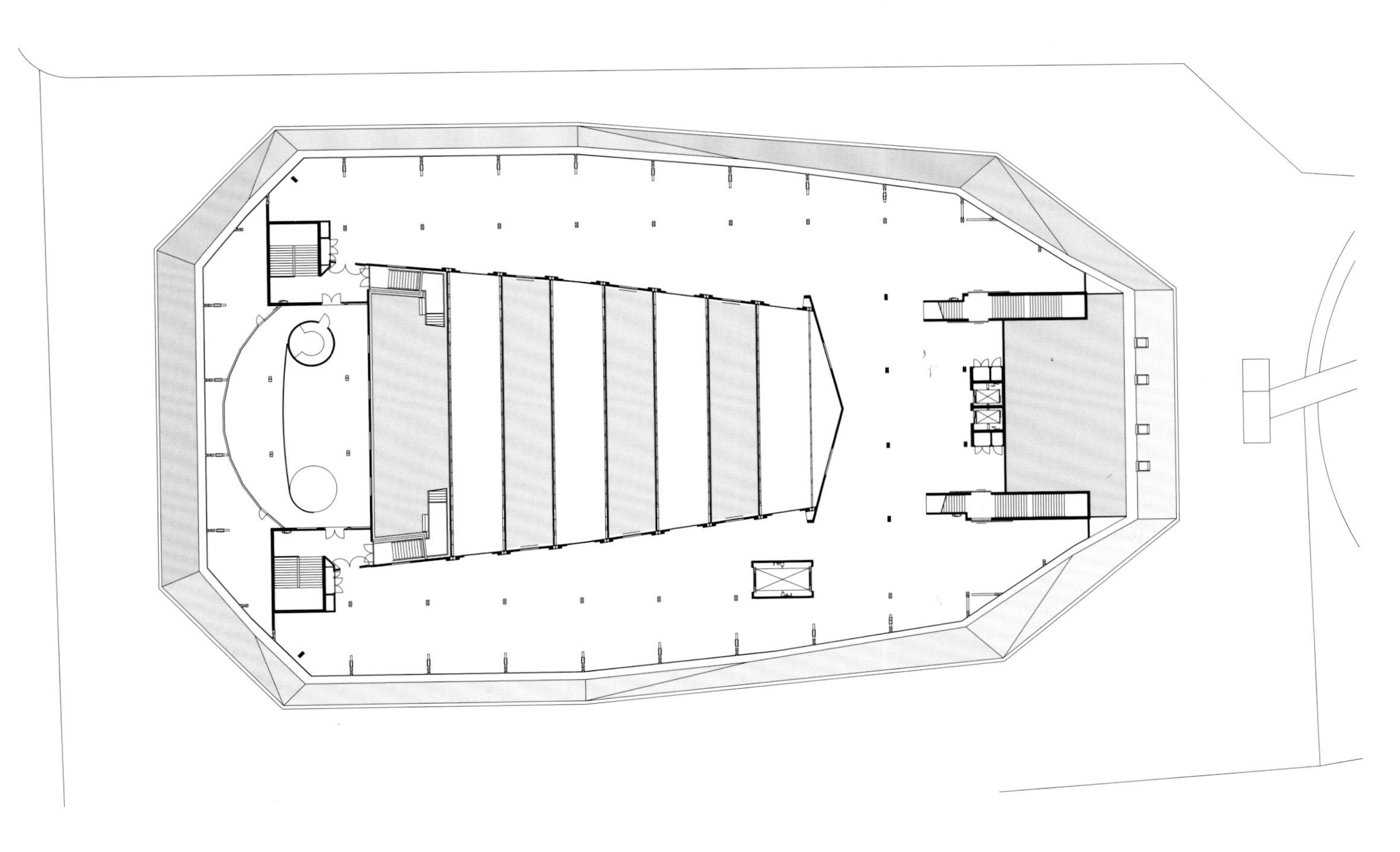

2nd floor plan

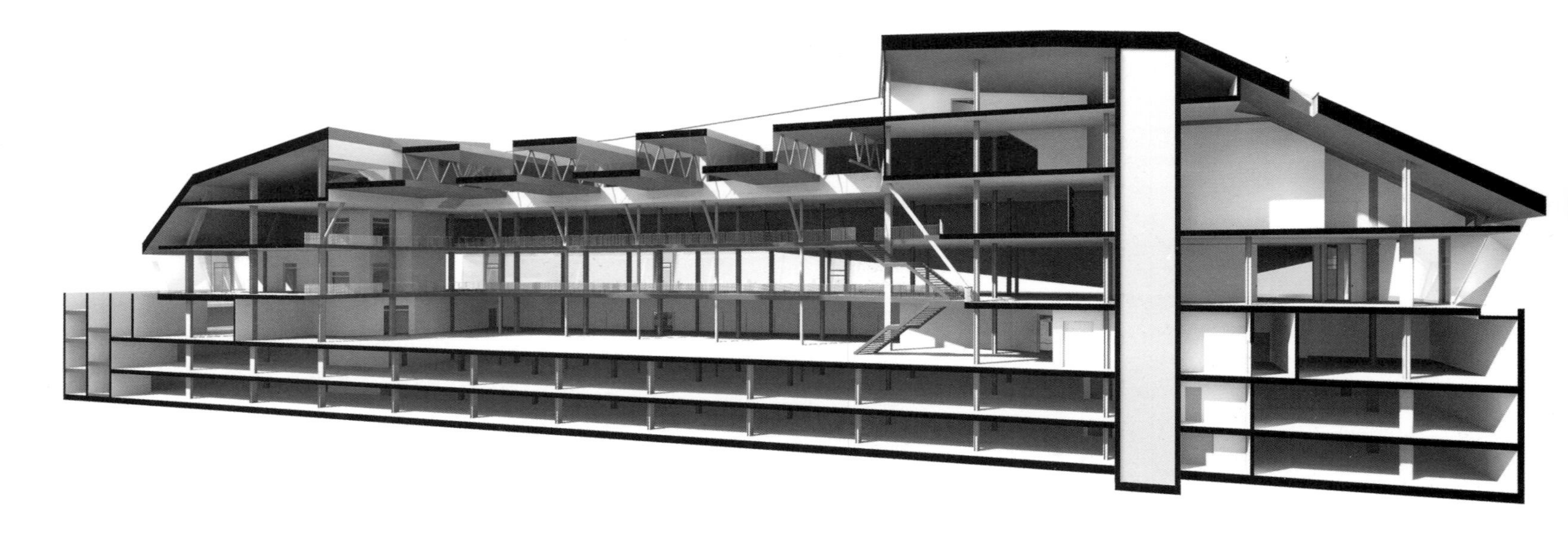

Longitudinal-Section

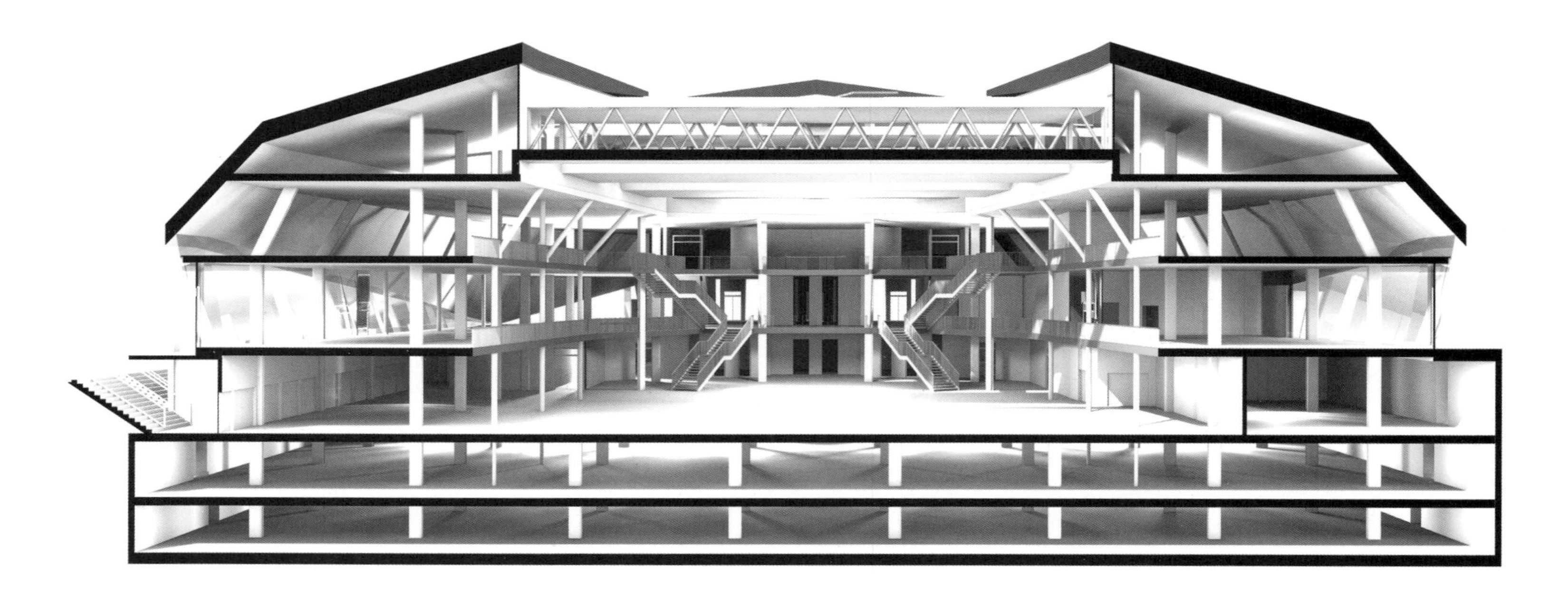

Cross-Section

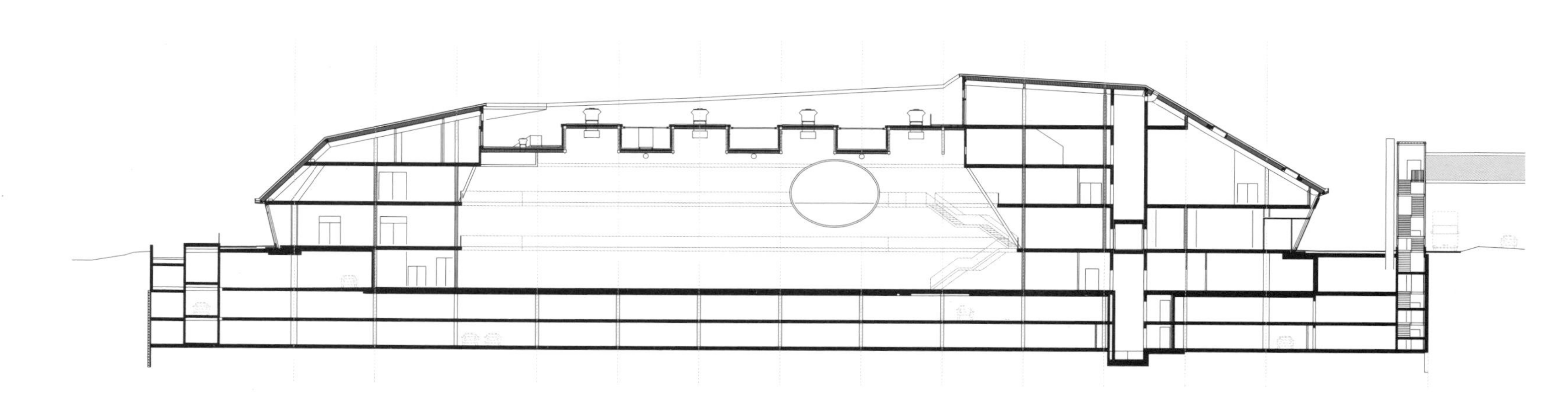

Longitudinal Section

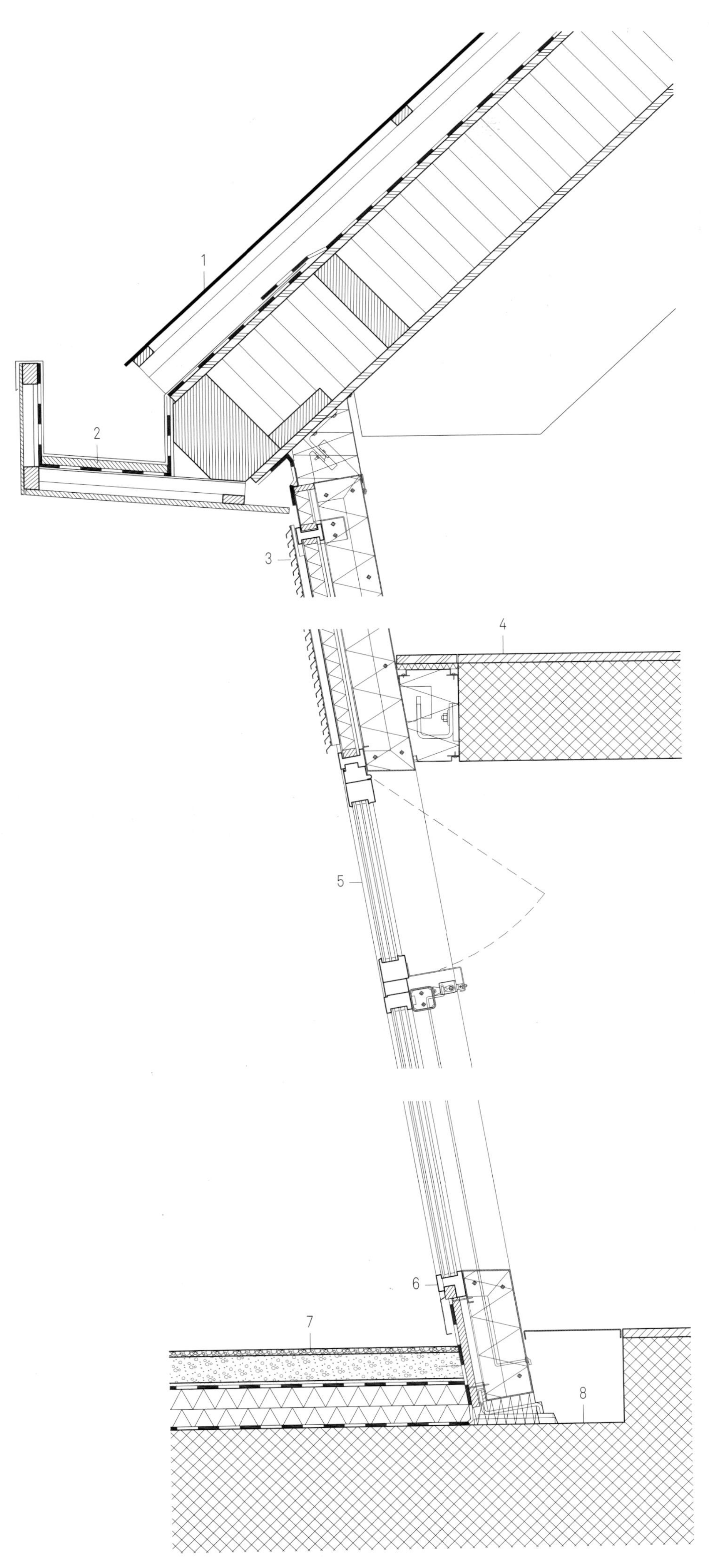

Detail Section Facade

1 Solar panels
Frying metal sheets 6.5 mm anthracite
Battens 3 cm
Battens variable
Roofing films
Wood elements 42 cm
Chrome sheet metal
Separating layer

2 Three-layer disc 4 cm
Seal / film
Substructure Steel 4mm
Fiber cement sheets 3 slats

3 Aluminum fins
Sandwich panel 8.5 cm
Steel section bent / welded 3.5 mm
Insulated

4 Hard concrete pigmented 3 cm
Concrete slab 34 cm
Sheet steel mounting facade 12 mm
Insulated

5 Opening wing with 3 triple glazing

6 Cover strip aluminum RAL 9005
CNS-sheet 2 mm
Water barrier
OSB 2.5 cm
Steel section bent / welded 3.5 mm
Insulated
Floor Console galvanized

7 Deck covering 2 cm
HMT 10cm
Balancing layer variable
Water barrier
Thermal insulation (XPS) 14 cm
Vapor barrier
Concrete slab 34 cm

8 Cable
Cover steel plate oiled 0.3 cm

Theatre de Stoep

Location: Spijkenisse, Netherland
Architect: UNStudio
Client: Municipality Of Spijkenisse
Floor Area: Gross: 7,000 m^2; Net: 5,700 m^2
Site Area: 3,600 m^2
Completion Date: 2014
Photographers: Peter Guenzel; Jan Paul Mioulet; Peter de Jong

Theatre de Stoep is part of the masterplan by Sjoerd Soeters for the town of Spijkenisse which aims to regenerate the town centre and to increase the appeal of the area. Located between the city of Rotterdam and the extension of the second Maasvlake (a land extension of 2000 ha), Spijkenisse has experienced exponential growth in the past 40 years (increasing from a population of 2500 to 70.000), leading the municipality to upgrade its urban fabric through a diversification of housing typologies, an upgrade of the retail heart of the city and an invigoration of its cultural infrastructure.

Theatre de Stoep responds to this revival by merging the archetypal function of a theatre - that of creating a world of illusion and enchantment - with the specific requirements of a regional theatre and its requisite to cater to the varied needs of the local community. The theatre is therefore designed with a dual emphasis on the chimeric nature of the world of the stage and the social aspects of the theatre experience.

In contrast to today's mediatised culture, theatre offers the participatory experience of the live event, often appropriately referred to as 'liveliness': the 'magic of live theatre', understood as the strange, elusive energy between audience and performer, the community forged together and the momentary collaboration necessitated by the live event. Theatre de Stoep is designed to fortify and inspire this liveliness, providing at one and the same time a place of performance, of social gathering and of experiencing contrasting realities: the world of the other, of fabrication, of expression and display, but simultaneously the very real sentient experience of ourselves as spectators within these worlds.

In the design of the 5,800 m^2 building a larger and a smaller theatrical space (with the main auditorium seating up to 650 guests and the smaller hall accommodating 200), several interlinked foyers, a grand café and a restaurant, an artist's cafe, a VIP lounge, numerous dressing rooms, multifunctional rooms and offices are all brought together within one volume. The placement of the various internal volumes results in a building in the form of a flower, with a large, column-free central foyer forming the heart of the structure.

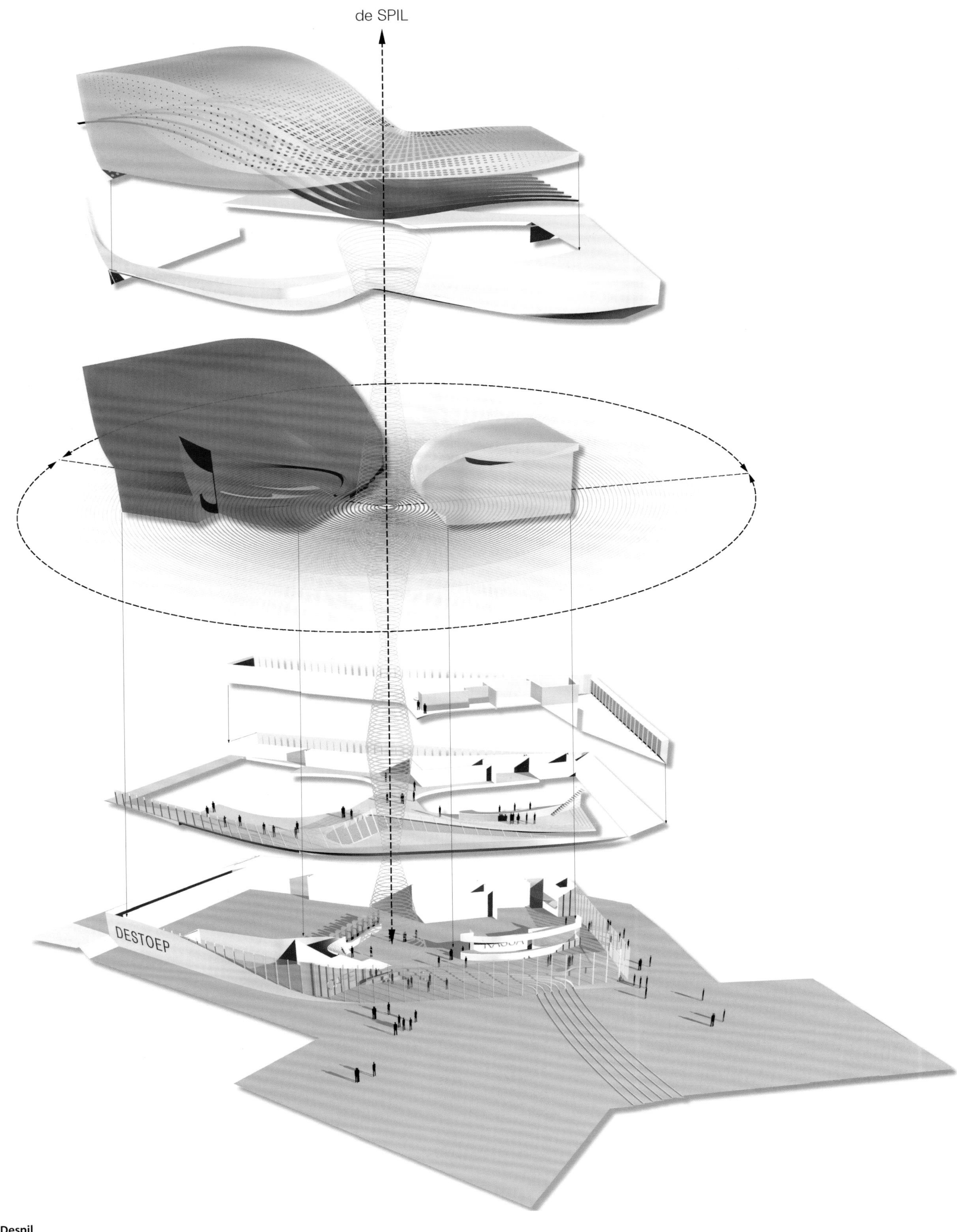

Spijk Despil

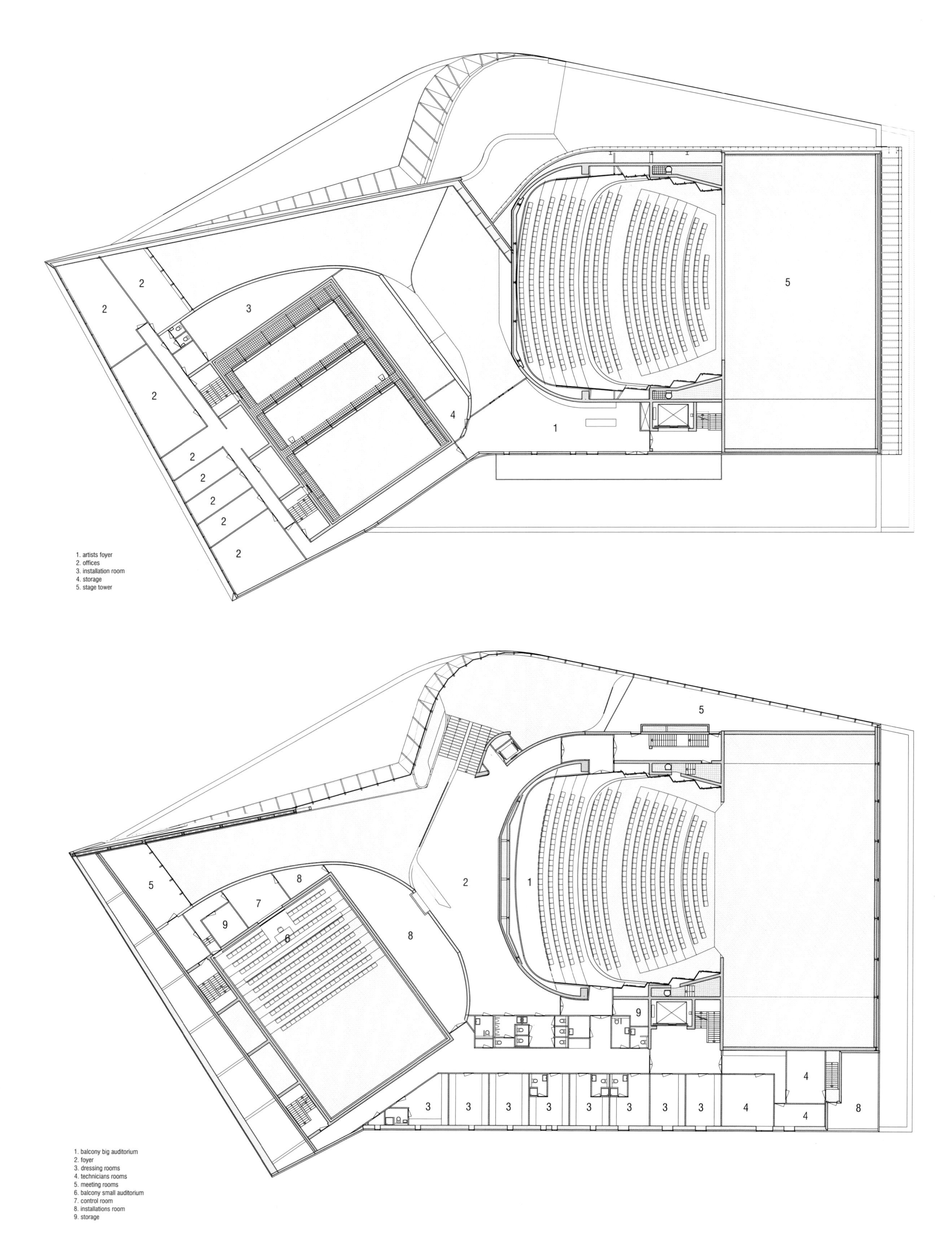
2
2
3
2
4
1
5
2
2
2
2
2
1. artists foyer
2. offices
3. installation room
4. storage
5. stage tower
5
5
8
7
9
6
8
2
1
9
3
3
3
3
3
3
3
3
4
4
4
8
1. balcony big auditorium
2. foyer
3. dressing rooms
4. technicians rooms
5. meeting rooms
6. balcony small auditorium
7. control room
8. installations room
9. storage

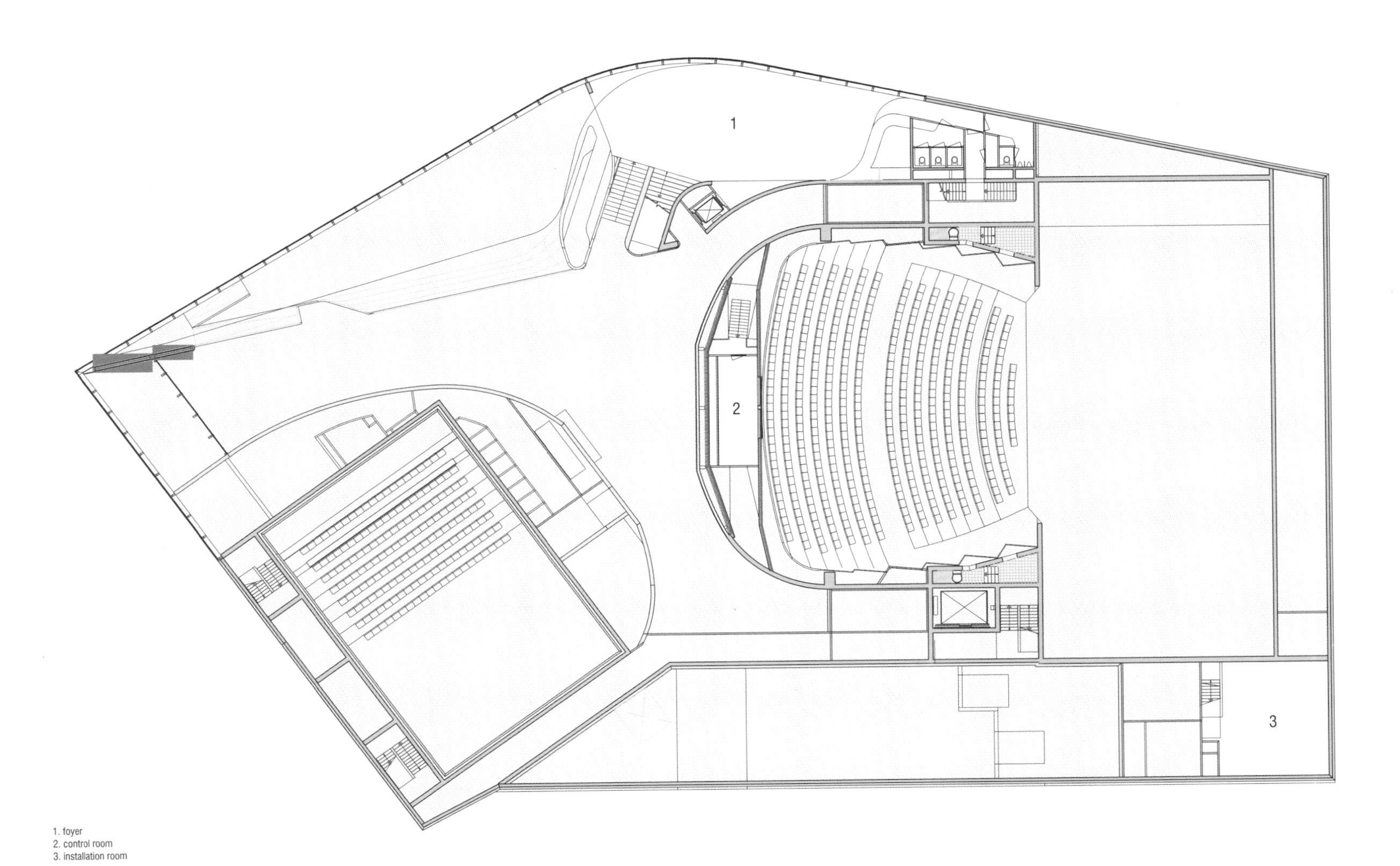

1. foyer
2. control room
3. installation room

1. big auditorium (656 seats)
2. small auditorium (200 seats)
3. foyer
4. theatre café
5. kitchen
6. expedition
7. ticket office
8. cloakroom
9. dressing room
10. installation room
11. atelier

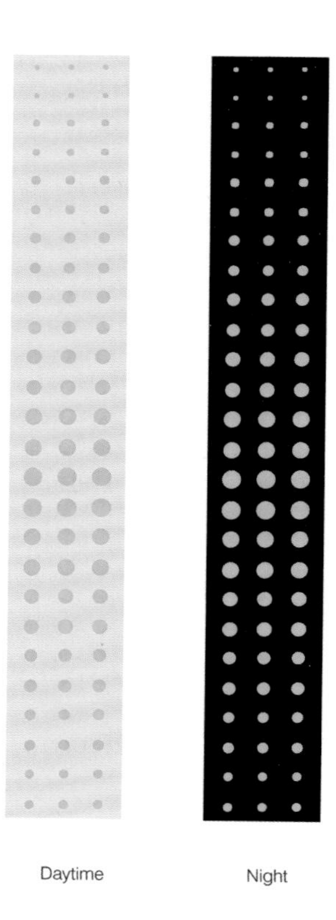

Daytime
Night

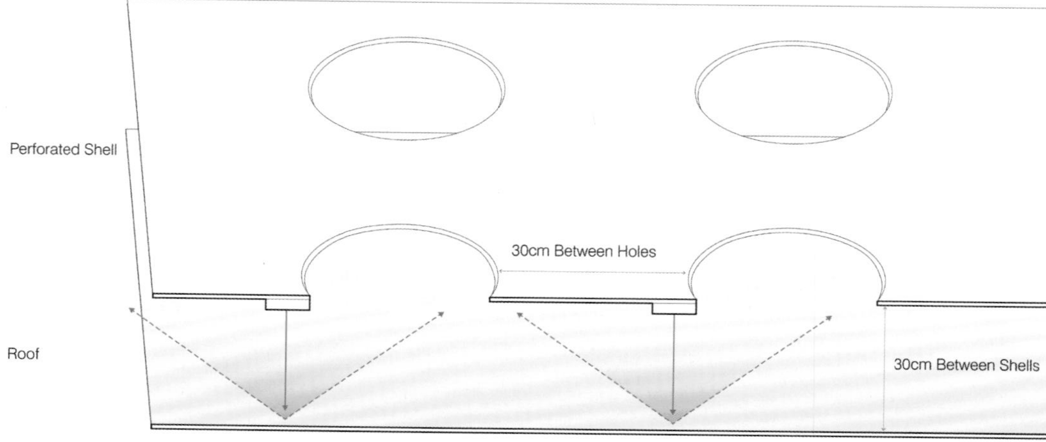

Perforated Shell
30cm Between Holes
Roof
30cm Between Shells

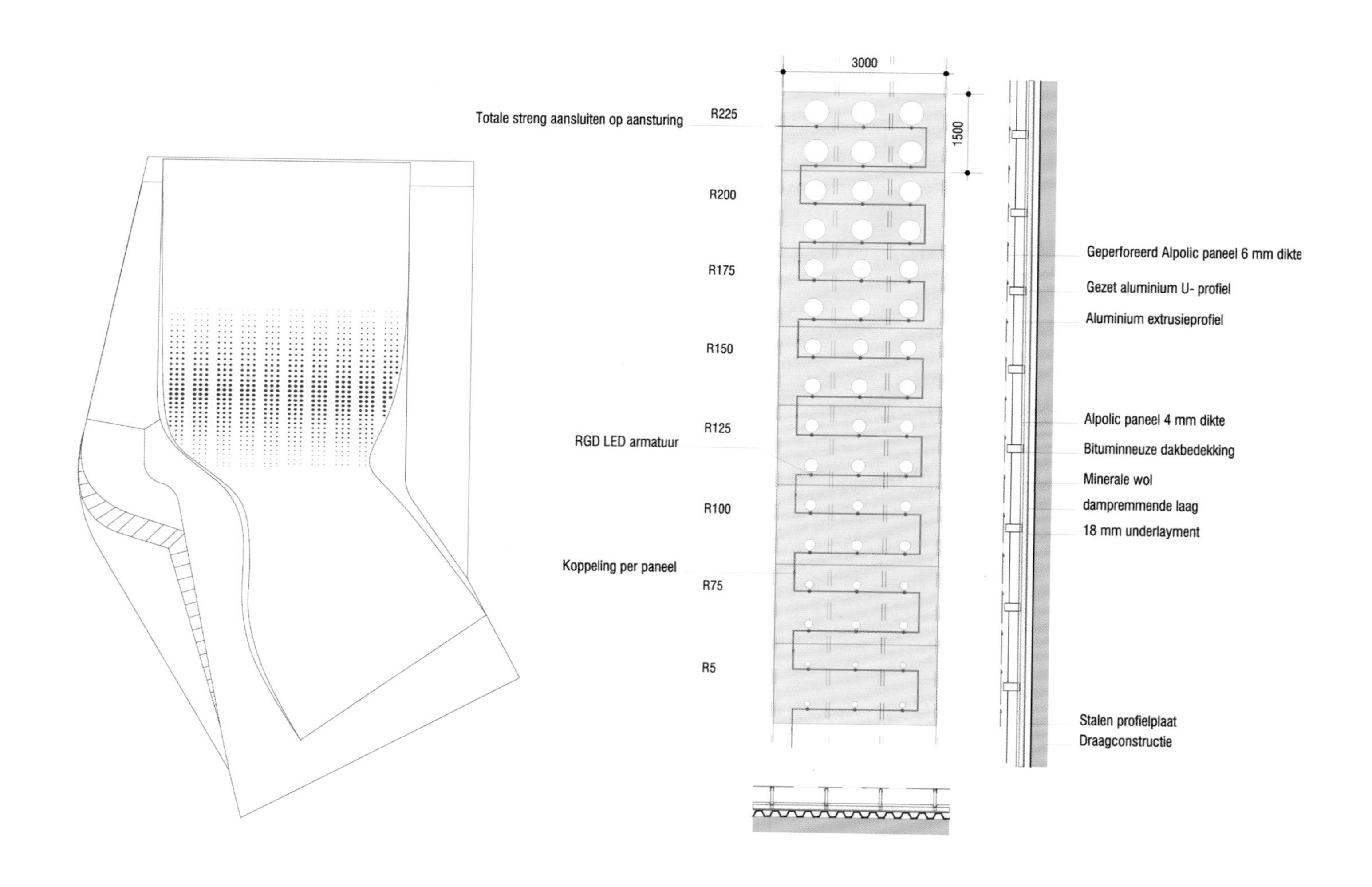

3000
1500
Totale streng aansluiten op aansturing
R225
R200
R175
R150
R125
RGD LED armatuur
R100
Koppeling per paneel
R75
R5
Geperforeerd Alpolic paneel 6 mm dikte
Gezet aluminium U- profiel
Aluminium extrusieprofiel
Alpolic paneel 4 mm dikte
Bituminneuze dakbedekking
Minerale wol
dampremmende laag
18 mm underlayment
Stalen profielplaat
Draagconstructie

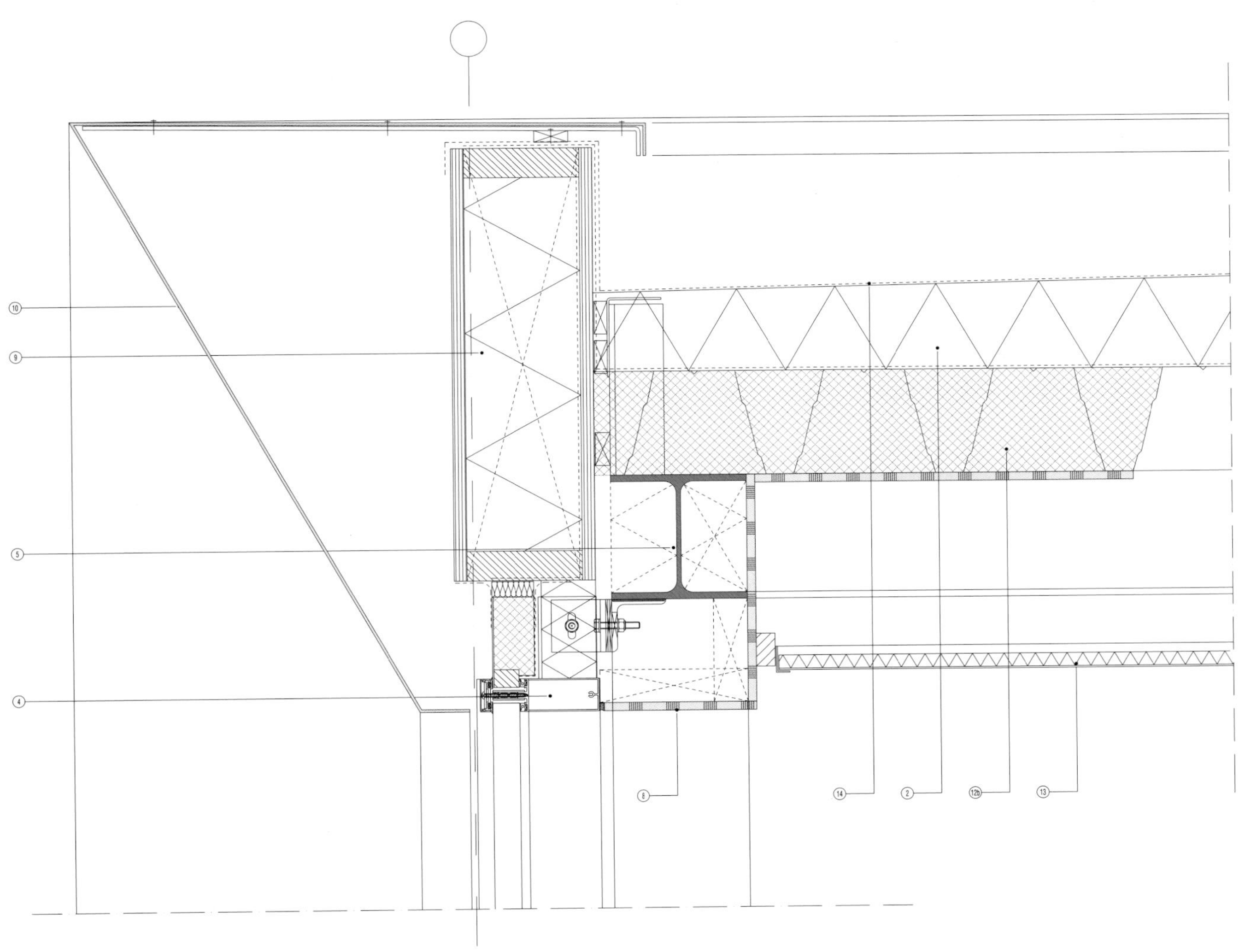

Detail dakrand kantoor

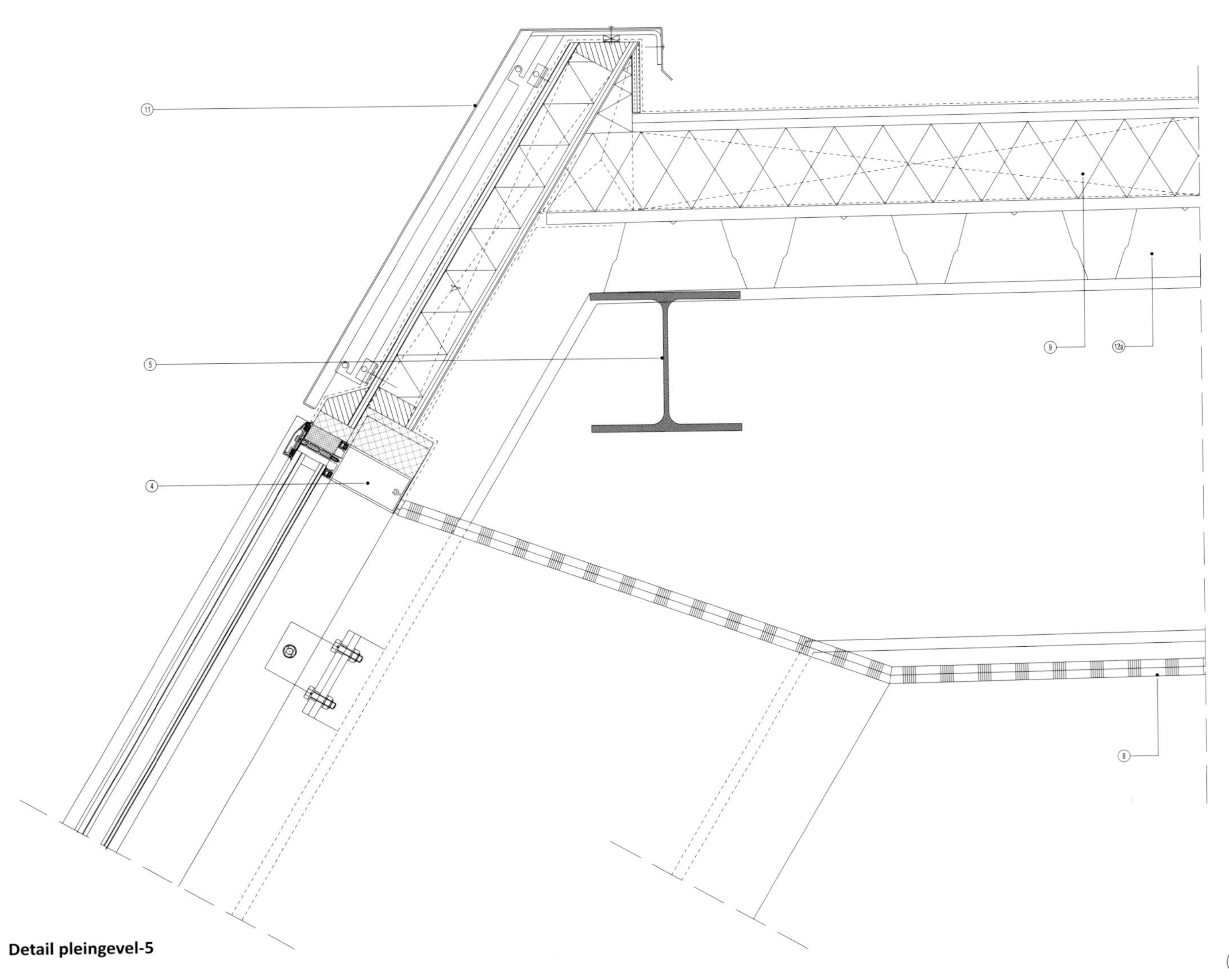

Detail pleingevel-5

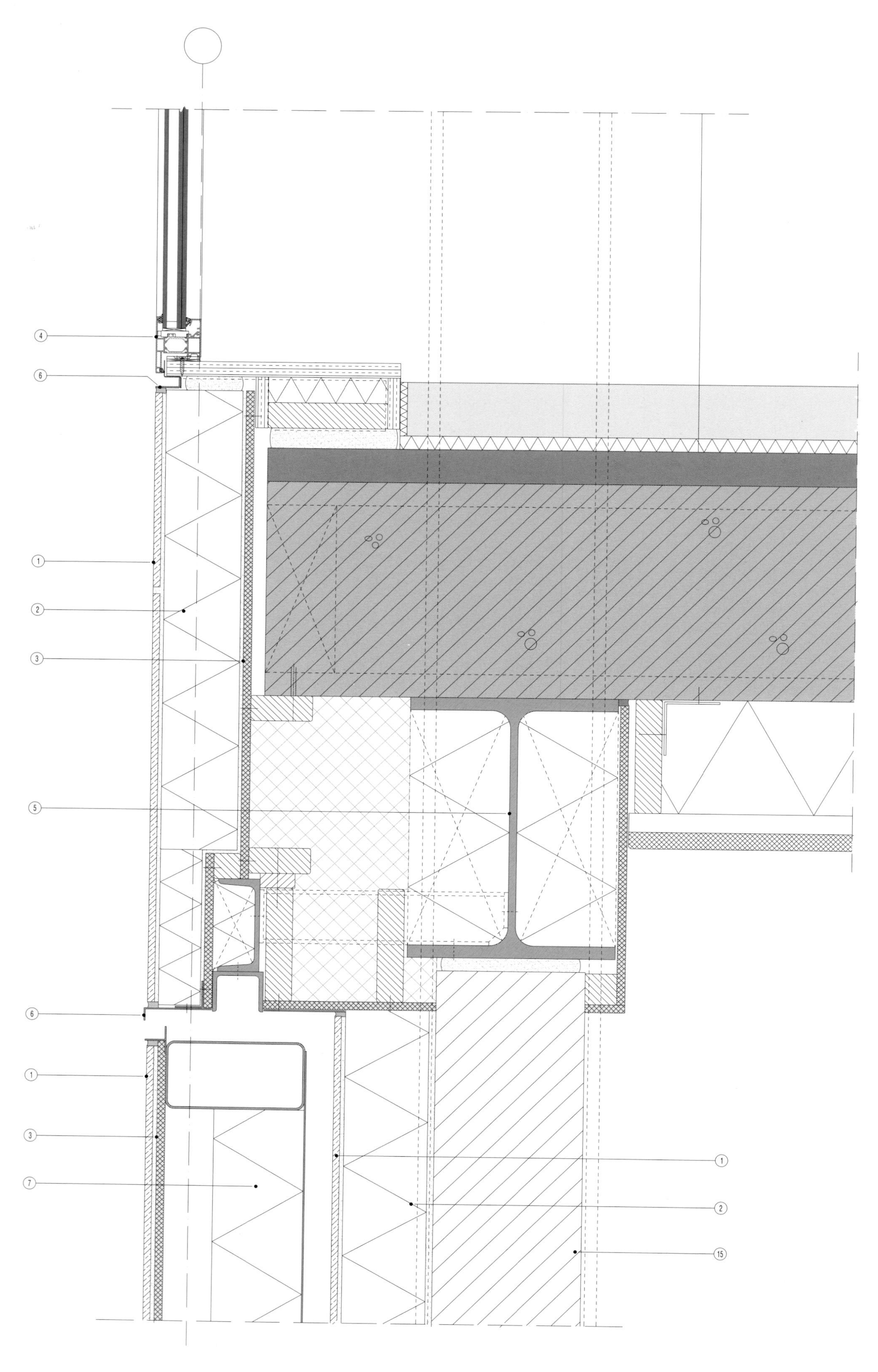

Detail expeditiedeur

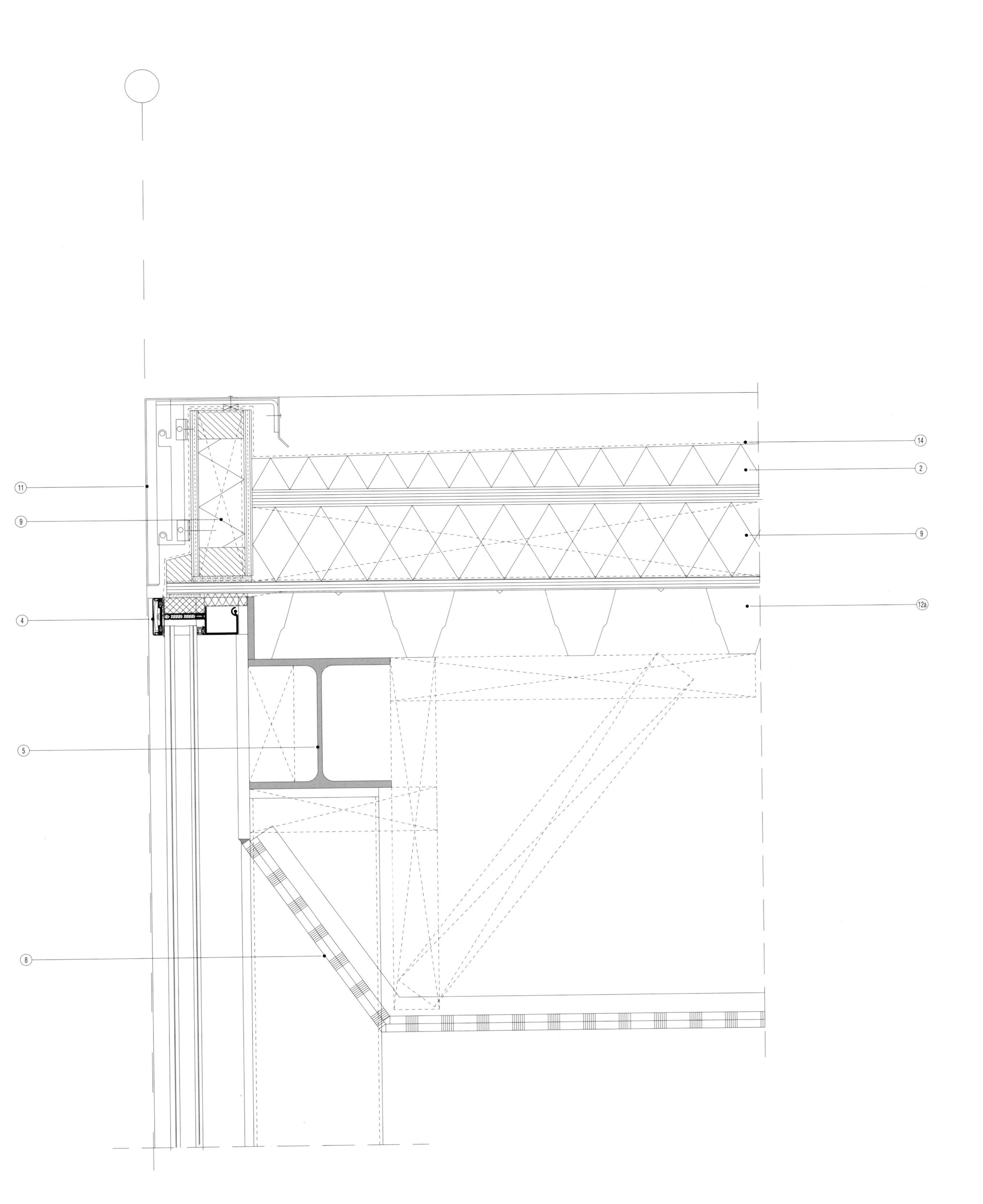

Detail pleingevel-1

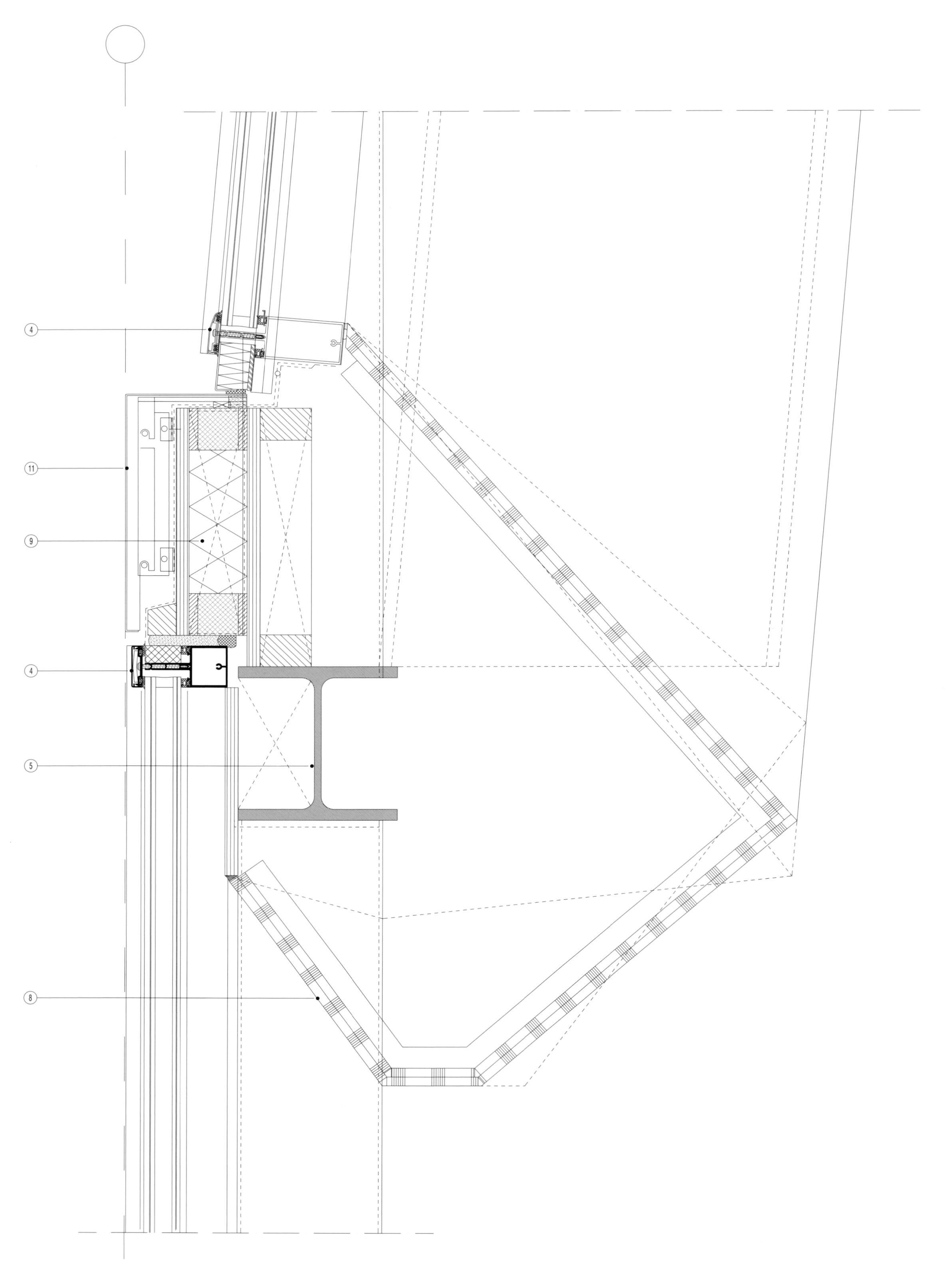

Detail pleingevel-2

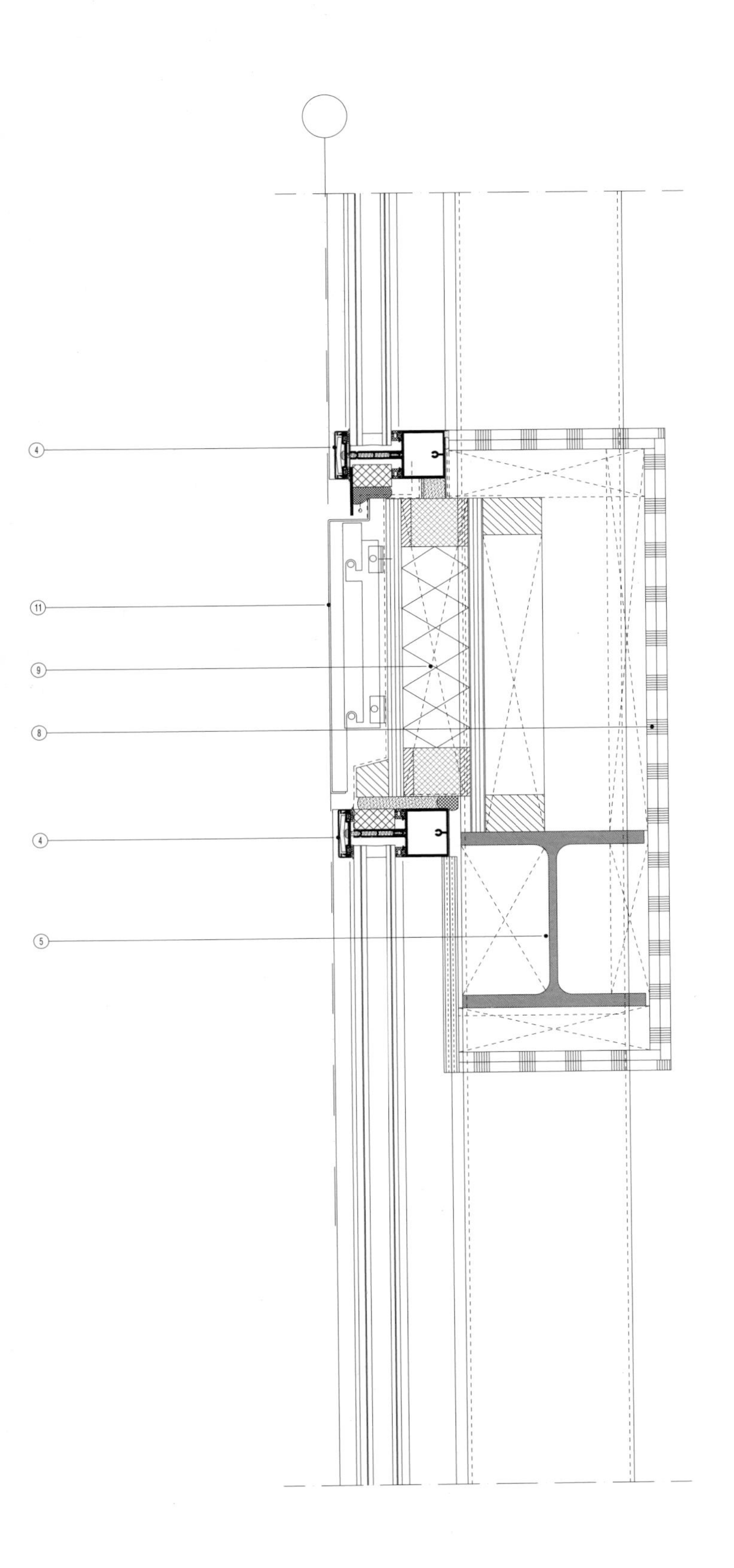

Detail pleingevel-3

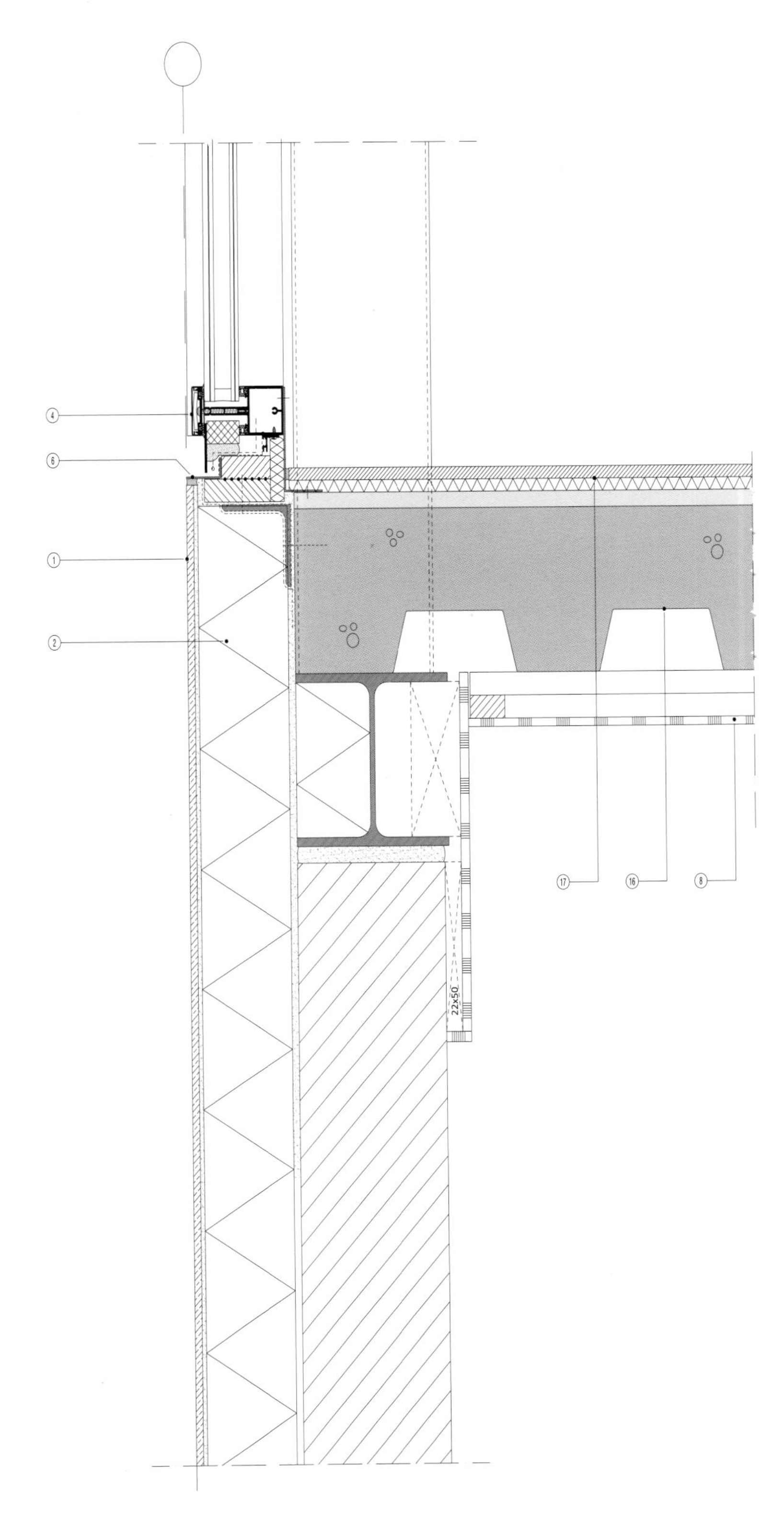

Detail pleingevel-4

Dongdaemun Design Plaza

Location: Seoul, Korea
Architect: Zaha Hadid Architects
Client: Seoul Metropolitan Government
Site Area: 62,692 m^2
Construction Floor Area: 86,574 m^2
Completion Date: 2014
Photographer: Virgile Simon Bertrand

The DDP has been designed as a cultural hub at the centre of Dongdaemun, an historic district of Seoul that is now renowned for its 24-hour shopping and cafes. DDP is a place for people of all ages; a catalyst for the instigation and exchange of ideas and for new technologies and media to be explored. The variety of public spaces within DDP include Art/Exhibition Halls, Conference Hall, Design Museum/Exhibition Hall/ Pathway, Design Labs & Academy Hall, Media Centre, Seminar Rooms and Designers Lounge, Design Market open 24 hours a day; enabling DDP to present the widest diversity of exhibitions and events that feed the cultural vitality of the city.

The DDP is an architectural landscape that revolves around the ancient city wall and cultural artefacts discovered during archaeological excavations preceding DDP's construction. These historic features form the central element of DDP's composition; linking the park, plaza and city together.

The design is the very specific result of how the context, local culture, programmatic requirements and innovative engineering come together - allowing the architecture, city and landscape to combine in both form and spatial experience – creating a whole new civic space for the city.

DDP's design and construction sets many new standards of innovation. DDP is the first public project in Korea to implement advanced 3-dimensional digital construction services that ensure the highest quality and cost controls. These include 3-dimensional Building Information Modelling (BIM) for construction management and engineering coordination, enabling the design process to adapt with the evolving client brief and integrate all engineering requirements. These innovations have enabled the team building DDP to control the construction with much greater precision than conventional processes and improve efficiencies. Implementing such construction technologies make DDP one of Korea's most innovative and technological advanced constructions to date.

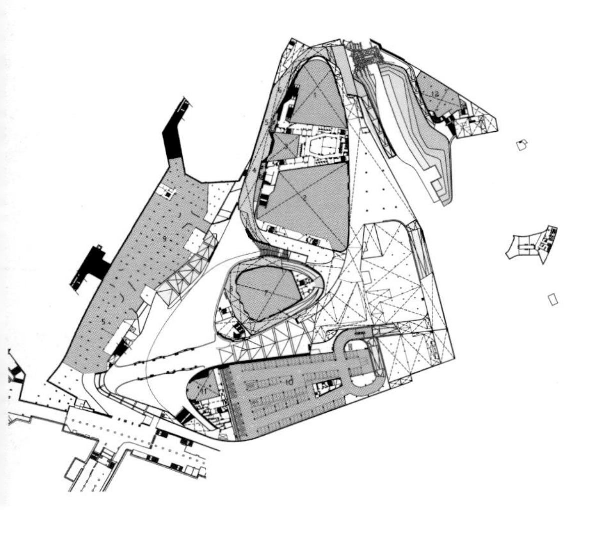

LEVEL B1

1 CONVENTION HALL #1
2 CONVENTION HALL #2
3 PRESS ROOM
4 VIP ROOM #1
5 VIP ROOM #2
6 BREAK-OUT SPACE
7 EXHIBITION HALL #1
8 EXHIBITION RAMP
9 RETAIL AREA
10 PARKING FACILITIES
11 INFORAMTION
12 EVENT HALL

B1 PLAN

0 5 10 20 40 80

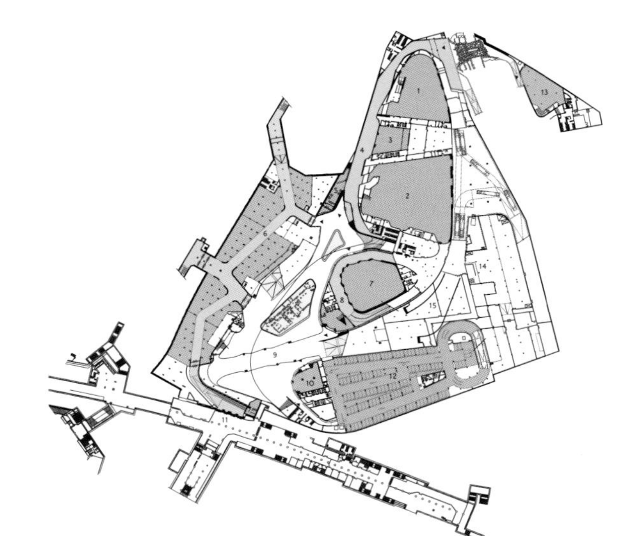

LEVEL B2

1 CONVENTION HALL #1
2 CONVENTION HALL #2
3 PRESS ROOM
4 BREAK-OUT SPACE
5 INFORMATION / TICKET CENTER
6 RETAIL AREA
7 EXHIBITION HALL #1
8 EXHIBITION RAMP
9 UNDERGROUND PLAZA
10 INFORMATION
11 SUBWAY ENTRANCE
12 PARKING FACILITIES
13 EVENT HALL
14 ENERGY CENTER
15 STORAGE

B2 PLAN

0 5 10 20 40 80

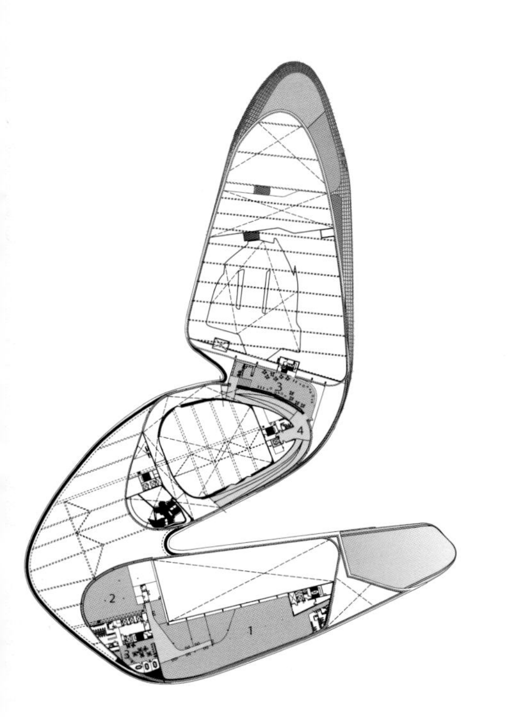

LEVEL 3F

1 DESIGN OFFICE
2 DIGITAL ARCHIVE
3 EXHIBITION LOUNGE
4 EXHIBITION RAMP

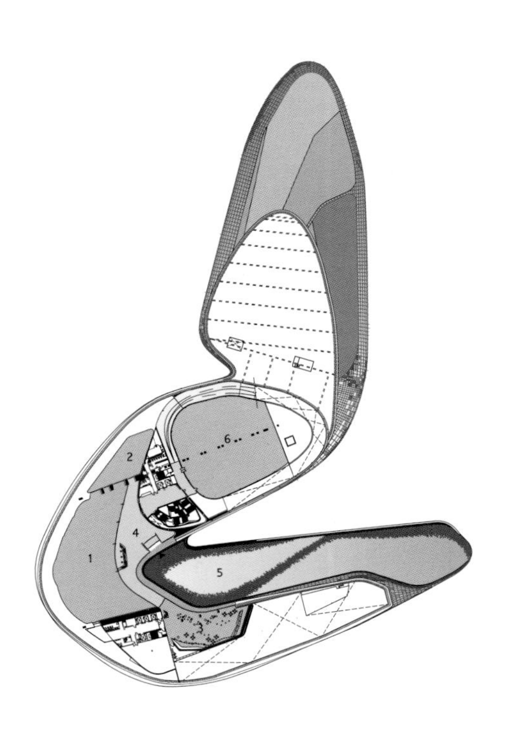

LEVEL 4F

1 MUSEUM OF MULTITUDES
2 MEDIA LAB
3 SKY LOUNGE
4 LOBBY
5 WALKABLE ROOF
6 MECHANICAL ROOM

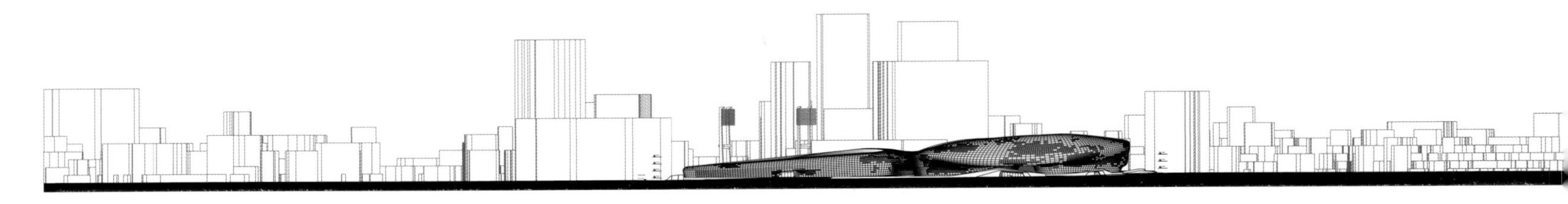

West Elevation

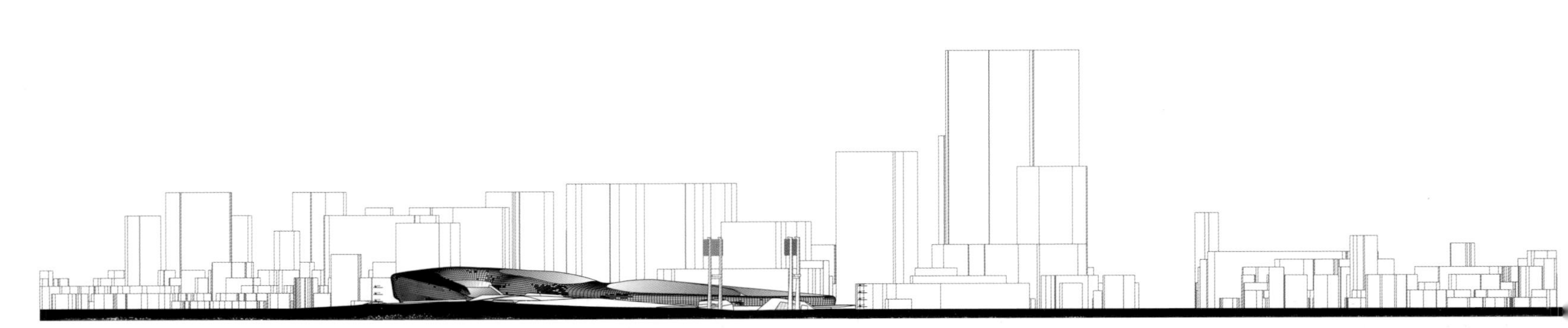

East Elevation

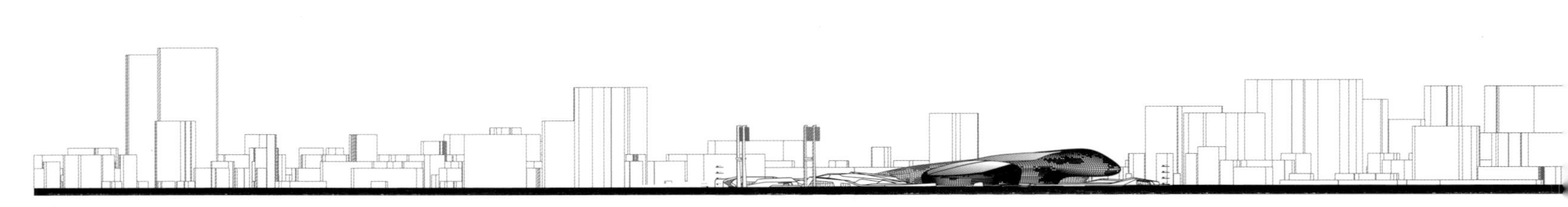

North Elevation

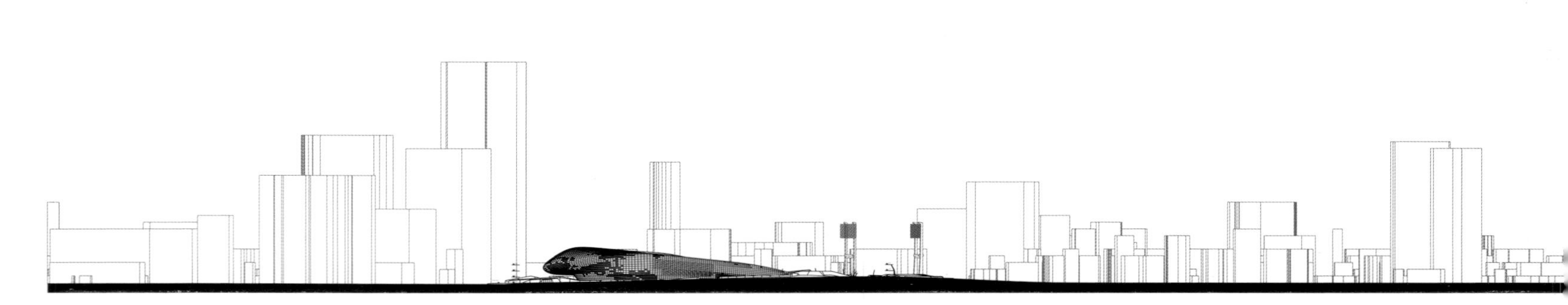

South Elevation

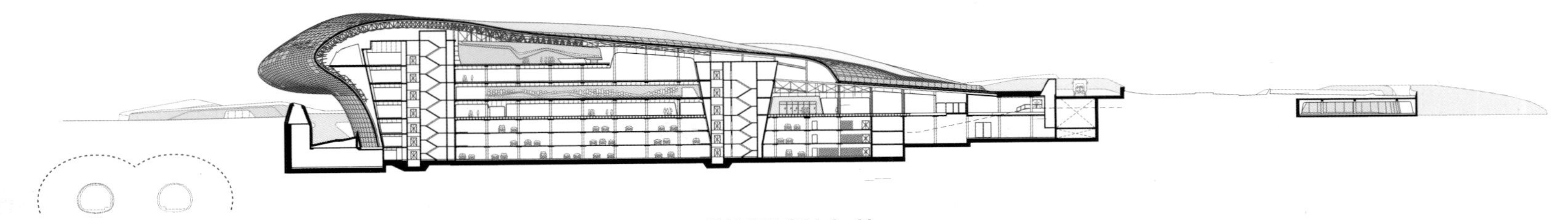

SECTION A-A'

SEC ABC

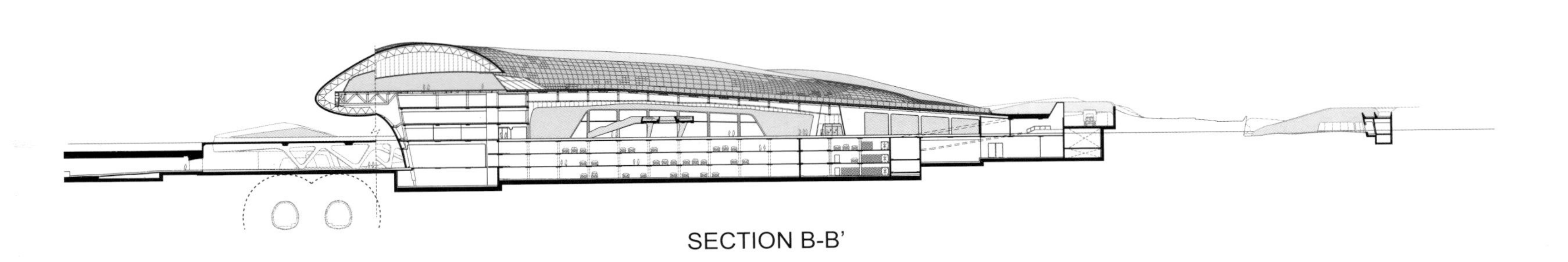

SECTION B-B'

SEC ABC

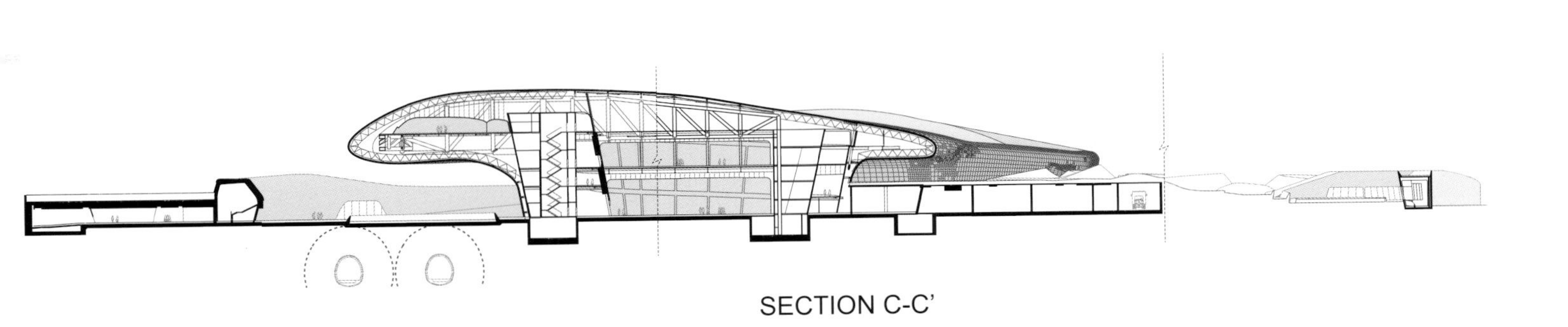

SECTION C-C'

SEC ABC

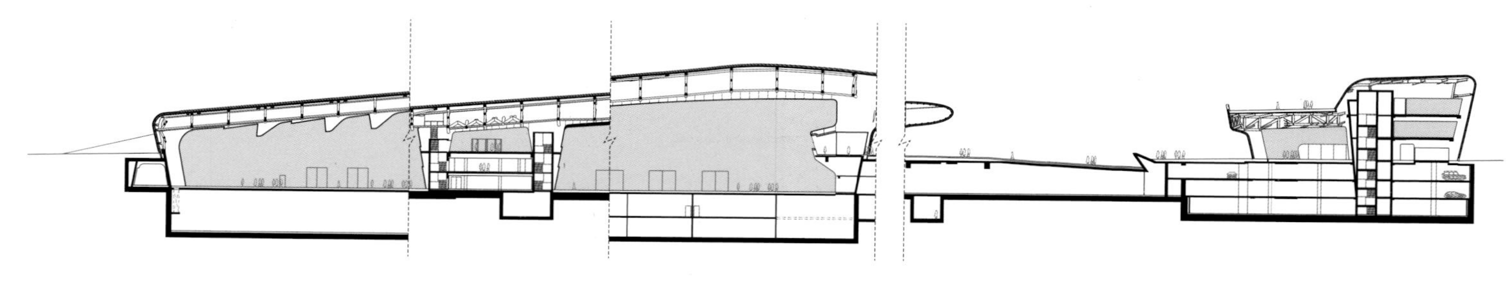

Section D-D

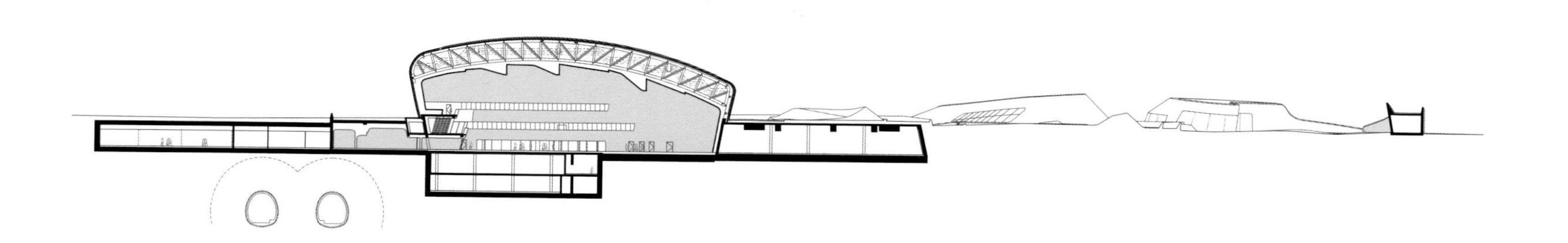

Section E-E

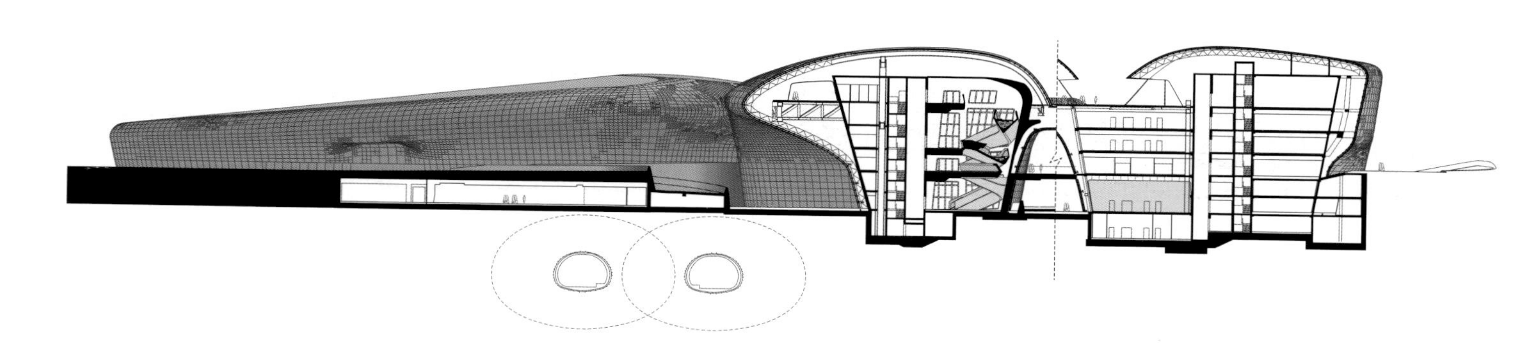

Section F-F

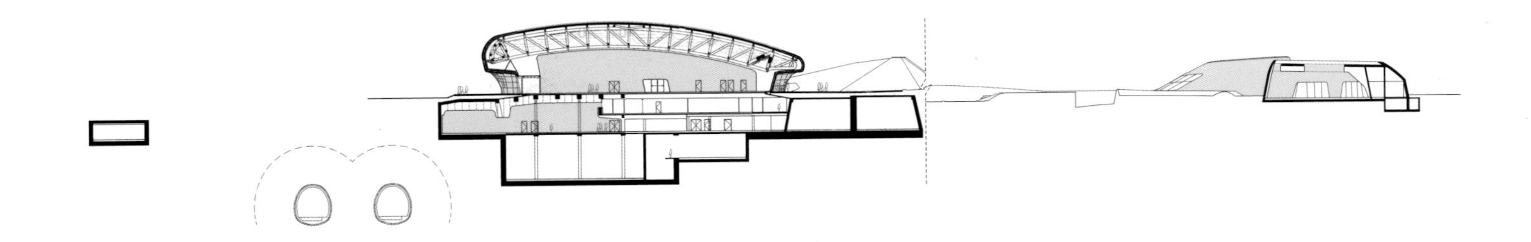

Section G-G

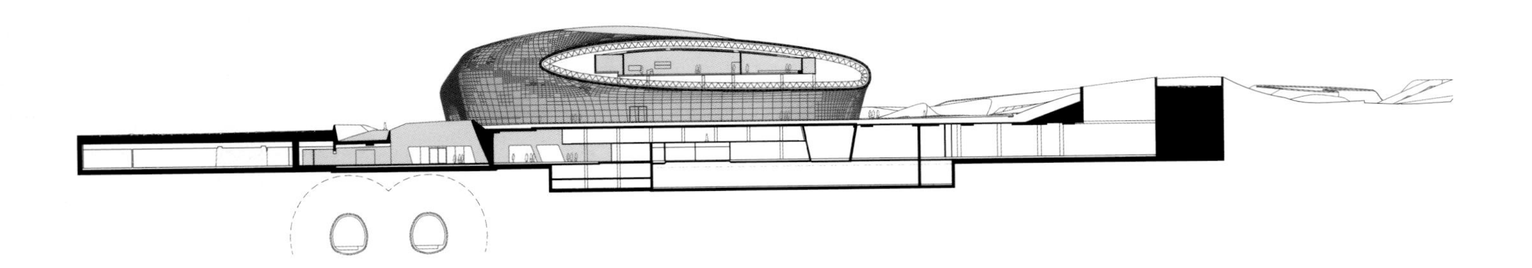

Section H-H

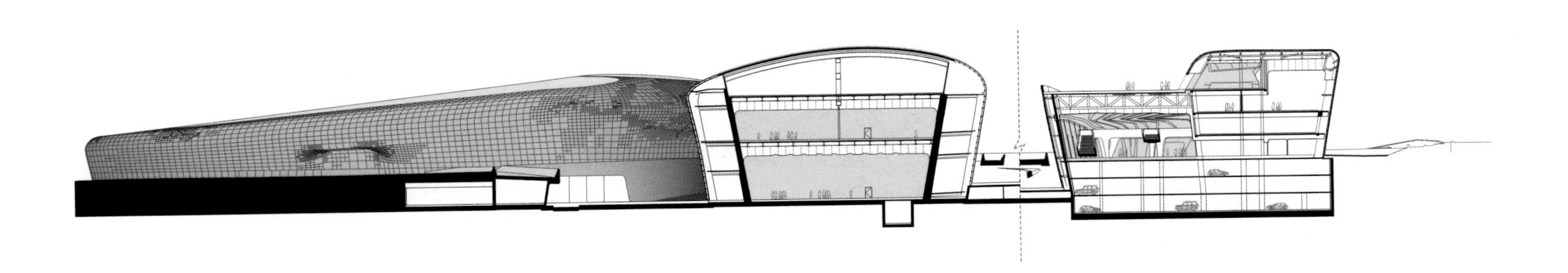

Section J-J

Kunsthaus Weiz

Location: Weiz, France
Architect: Dietmar Feichtinger Architectes
Client: BOOTES-Immorent GrundverwertungsgmbH
Site Area: 3,895 m^2
Floor Area: 7,150 m^2
Completion Date: 2005

The dimensions of the new building are integrated and translated into this historic urban mesh by modelling from this mass of "soft" elements. This is most strikingly shown at the side elevation of the building. Here the building lines a newly created small side street on the opposite side of which Feichtinger has erected an attractive office building that in turn absorbs the impetus of the impinging small town microstructure at its rear and channels this energy into the new side street through an opening in an otherwise hermetic façade. On the other side of the street, on the facade of the Kunsthaus, the dominant motif is that of a "wave" that formulates the change in height from the two-storey neighbouring buildings at the rear to the imposing three-storey ELIN building opposite the front of the Kunsthaus and makes it an aesthetic theme at cornice level. All these mediating elements of the building are made of glass, while the "massive" parts on the west side of the building are clad in copper.

The interior of this building with its continuous full-height glazing, its generously broad foyer, its glowing bar at first floor level running around the central events hall (capacity 645), and with its attached exhibition areas speaks a nonchalant language that employs precision. The entire ground floor is reserved for a supermarket that is intended to attract numbers of visitors who will also animate the old town centre. For this reason the large hall is positioned above the supermarket and is directly reached from the side street by a staircase (plus lift). On the basement levels there are car parking spaces.

The hall itself, equipped with a stage, a gallery and a high-tech booth, offers above all an aesthetic experience. The black walls are additionally covered with a metal mesh that can be illuminated in different colours by light diodes.

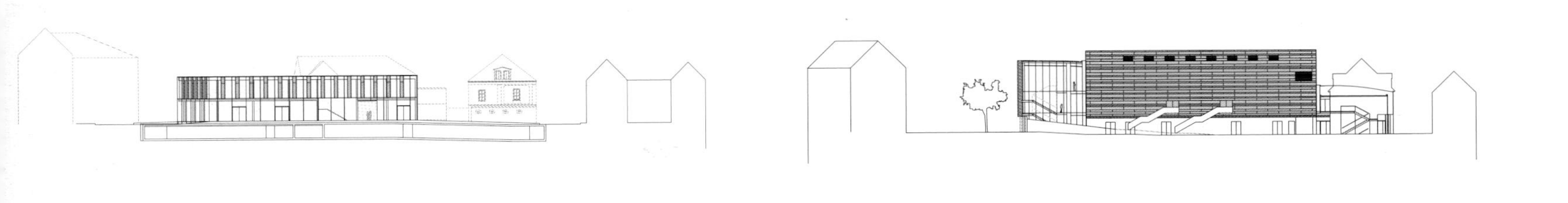

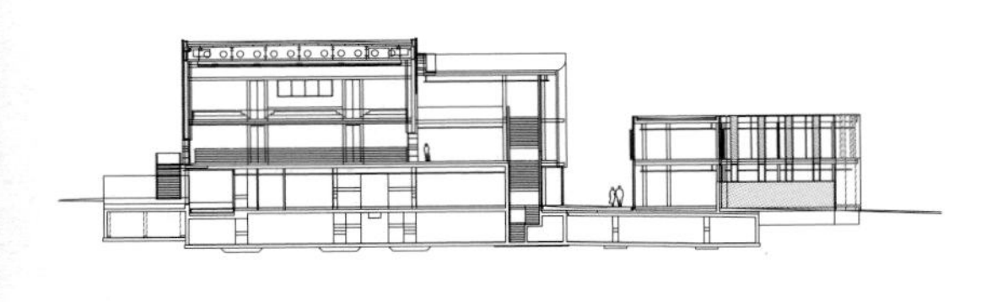

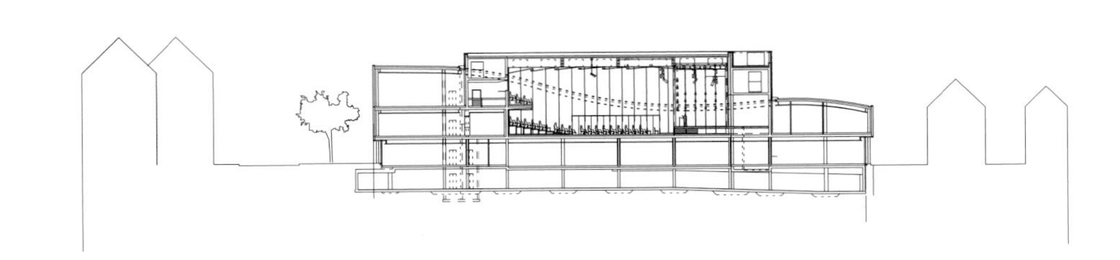

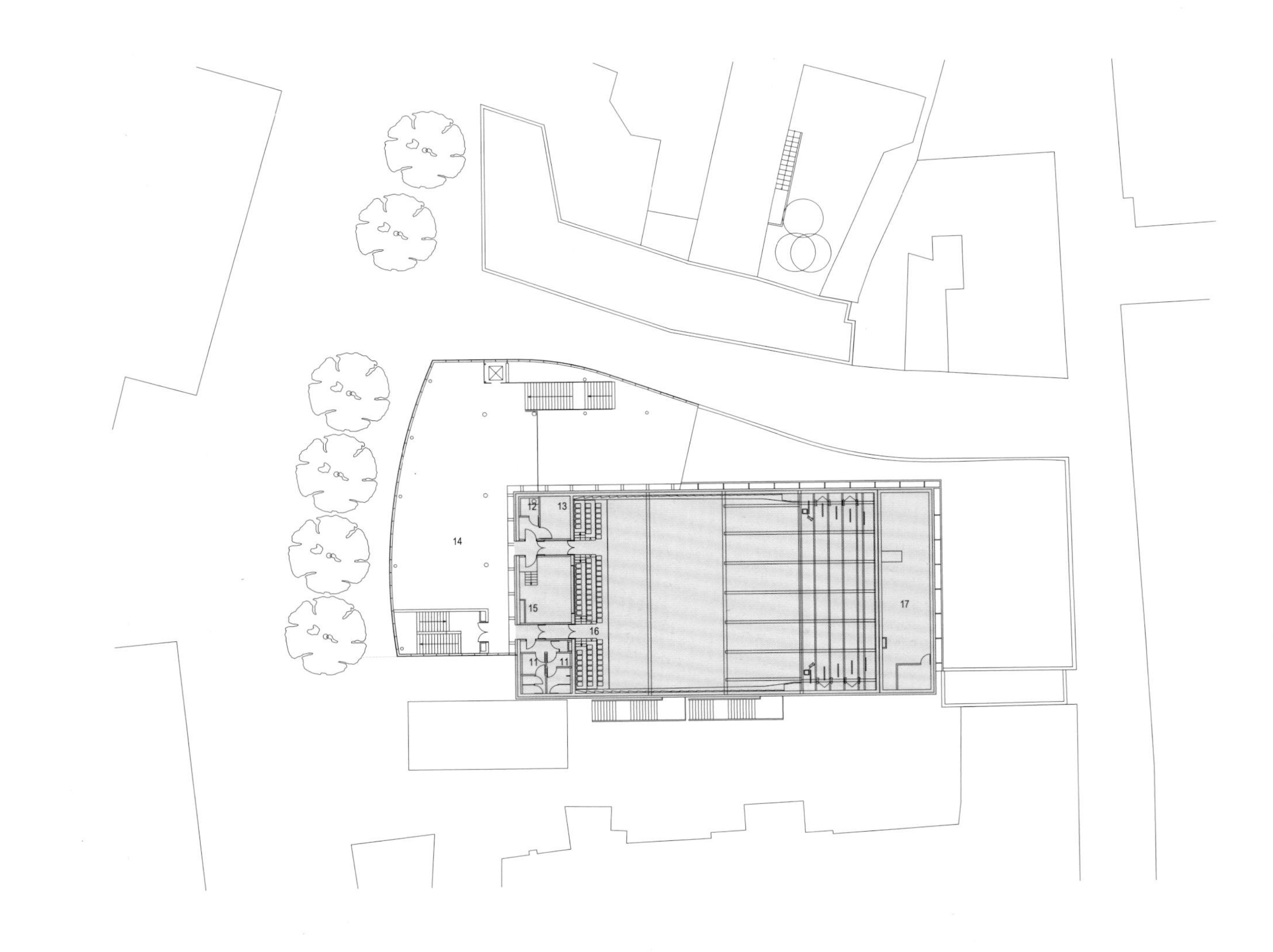

1 porch
2 box office
3 foyer
4 bar
5 kitchen
6 wardrobe
7 chair store
8 Frank Stronach Hall
9 stage
10 backstage
11 toilet
12 artist
13 storage
14 exhibition
15 stage control room
16 balcony
17 ventilation
18 offices
19 patio

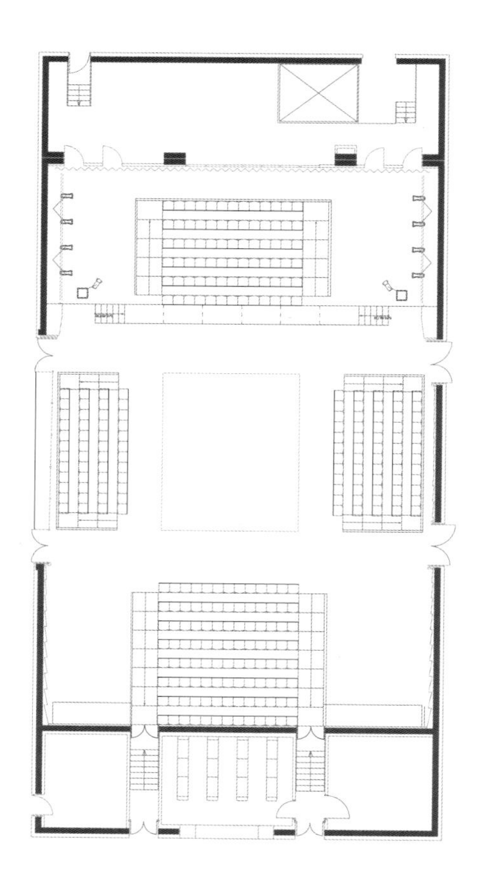

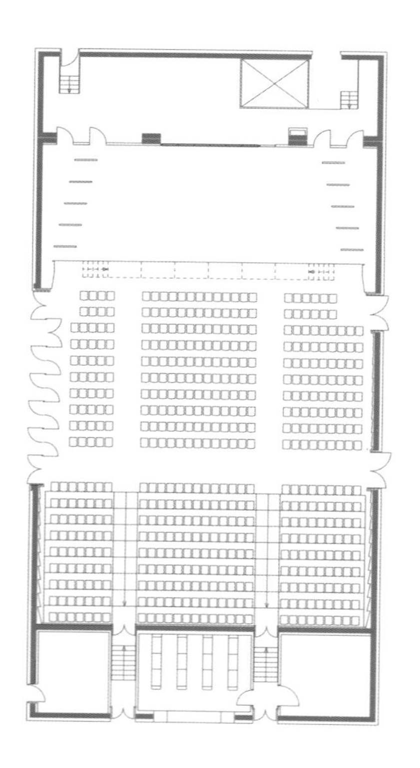

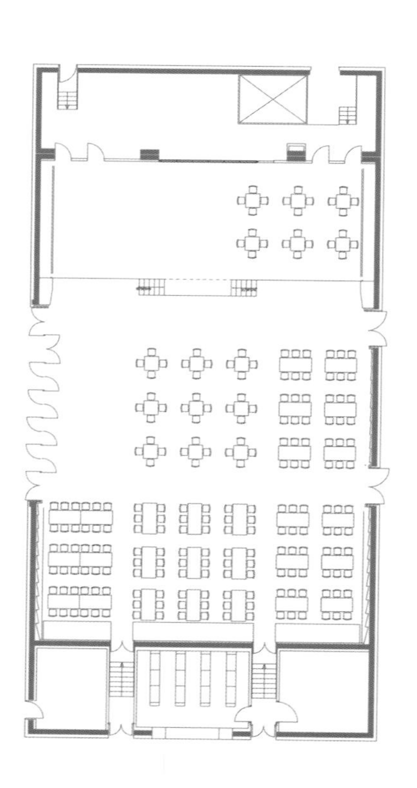

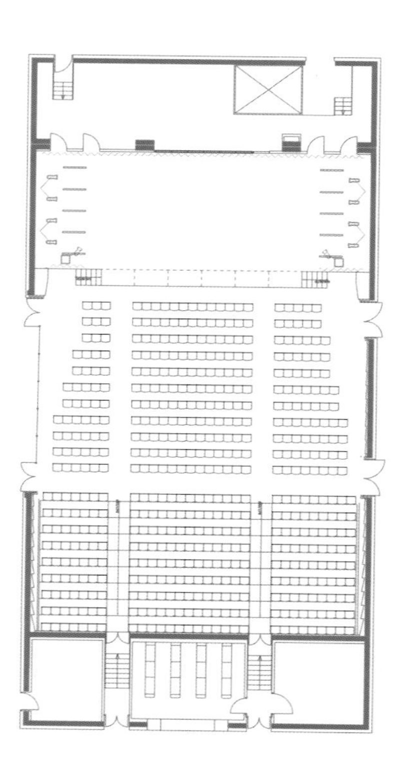

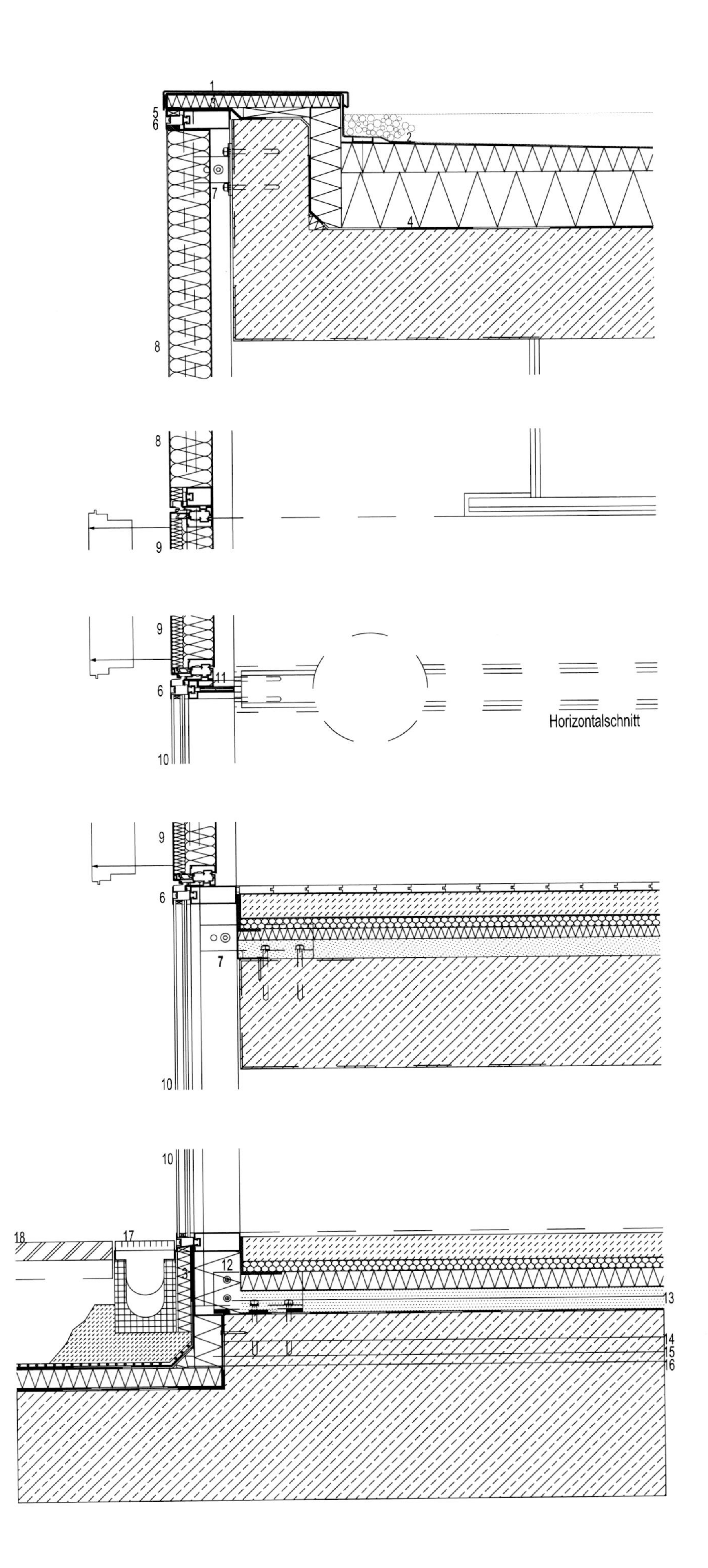

Fassadenschnitt Bürogebäude 1/20

1 Alu-Blech pulverbeschichtet
2 Kunststoffdachbahn 1,8mm, mechanisch befestigt
3 ALU-Paneel wärmegedämmt mit Gefälle
4 Dampfsperre Fassadenanschluss dampfdicht
5 Pfosten-Riegel Stossabdeckung regendicht
6 NIRO-Hutprofil gezogen, pulverbeschichtet mit Gummidichtung, geschraubt und silikonverfugt
7 Pfostenhalter verzinkt+pulverbeschichtet
8 oxidiertes Kupferpaneel, wärmegedämmt Fix-Paneel
9 oxidiertes Kupferpaneel, wärmegedämmt Parallel-Ausstellelement
10 ISO-Verglasung
11 ALU-Profil pulverbeschichtet als Distanzausgleich zum T-Profil
12 EPS 120mm
13 Pfostenmontageprofil, verzinkt
14 Blechhochzug, 2x gekantet, verzinkt
15 XPS-G-35 70mm
16 PS-Leiste 50/60
17 Rinne, Rinnengitter verzinkt, abnehmbar
18 Naturstein-Plattenbelag
Splittbett zementgebunden
Schutzschicht GGR-Matte
Bitumenbahn, zweilagig
Schaumglas Typ F, 50mm
Stahlbetonplatte mit Gefälle

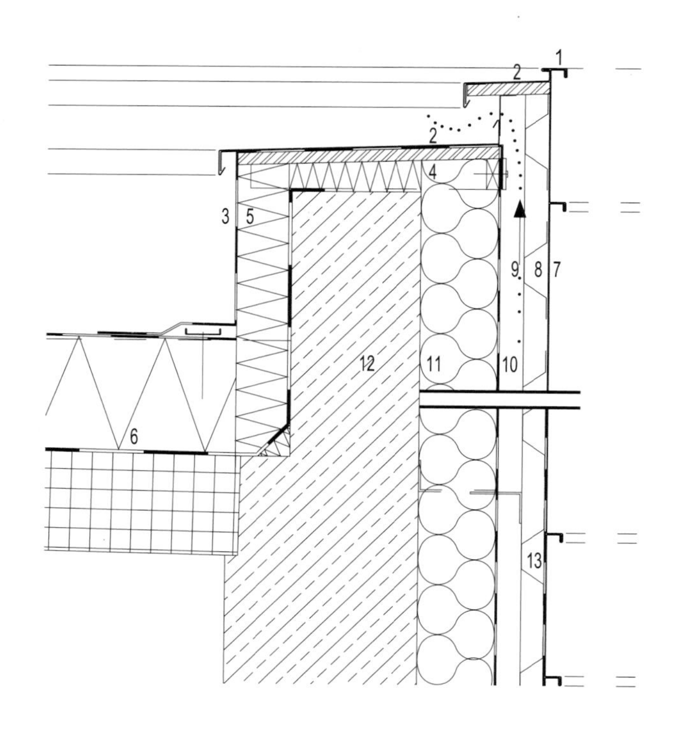

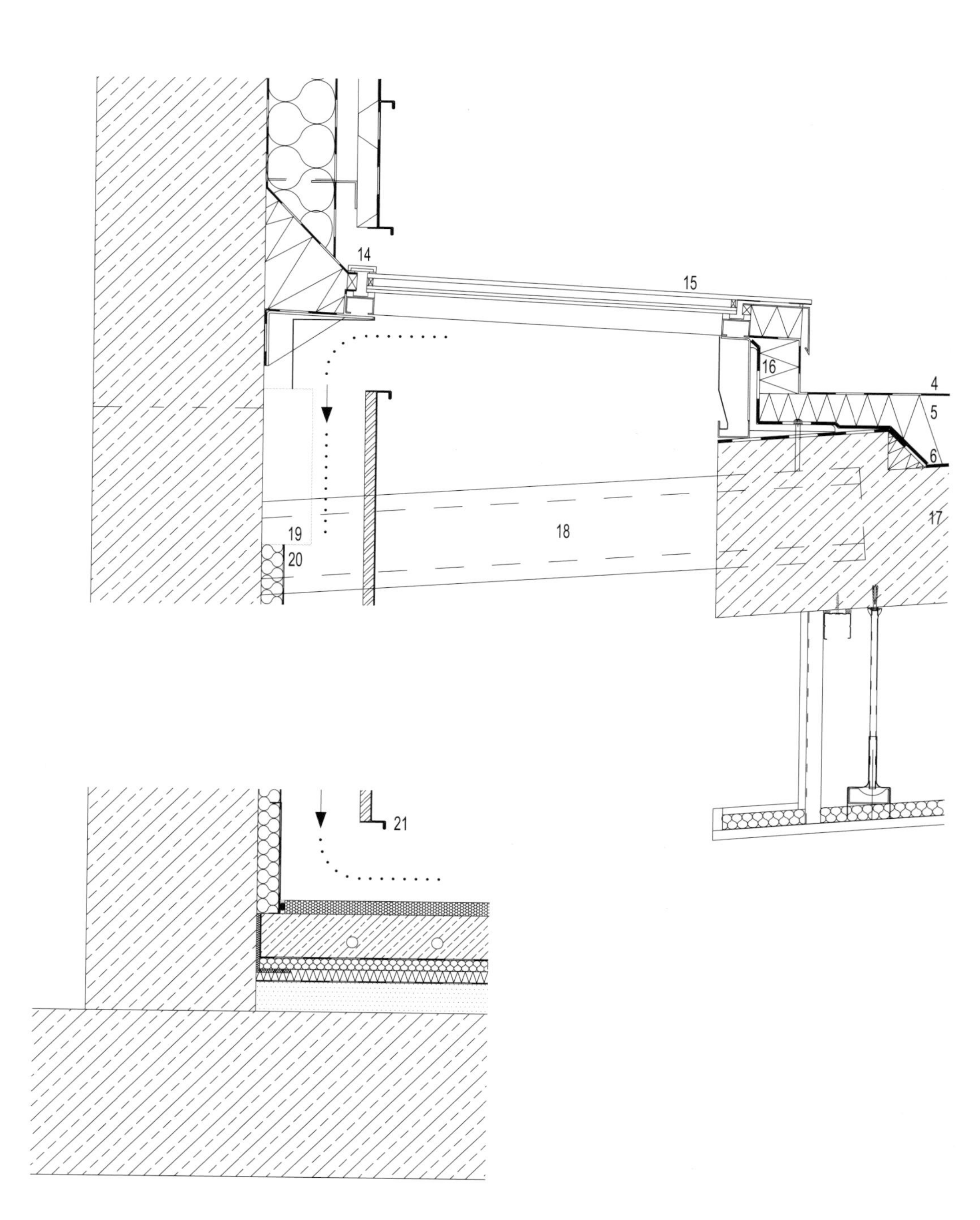

Fassadenschnitt Kunsthaus Kupferfassade 1/10

1 Attika mit Winkelflaz
2 Attika-Abdeckplatte 3% Gefälle
3 Kunststoffdachbahn 1,8mm
4 Gefällekeil / Wärmedämmplatten EPS-W20
5 Wärmedämmplatten EPS-W20
unkaschiert, trittfest
6 Dampfsperre
7 braun voroxidiertes Kupfer 0,7mm
waagr. Winkelstehfalzdeckung,
senkr. Querfalz als Einfachfalz
8 Trapezblech verzinkt
9 Unterkonstruktion / Hinterlüftung 60mm
10 hochdiffusionsoffene, regensichere und
winddichte Unterdeckbahn
11 formfeste Fassadendämmplatten 120mm
mechanisch befestigt
12 STB-Wand 300mm, im Attikabereich 200mm
13 Trennlage
14 Abdeckblech ALU 230mm 2x gekantet
15 Überstandsscheibe 130mm
16 ALU-Paneel wärmegedämmt
17 STB-Decke 250mm
18 Stahlschwert
19 Kühleinheit
20 Mineralwolle 40mm
21 Sockelabschluss
oxidiertes Kupferblech zweimal gekantet

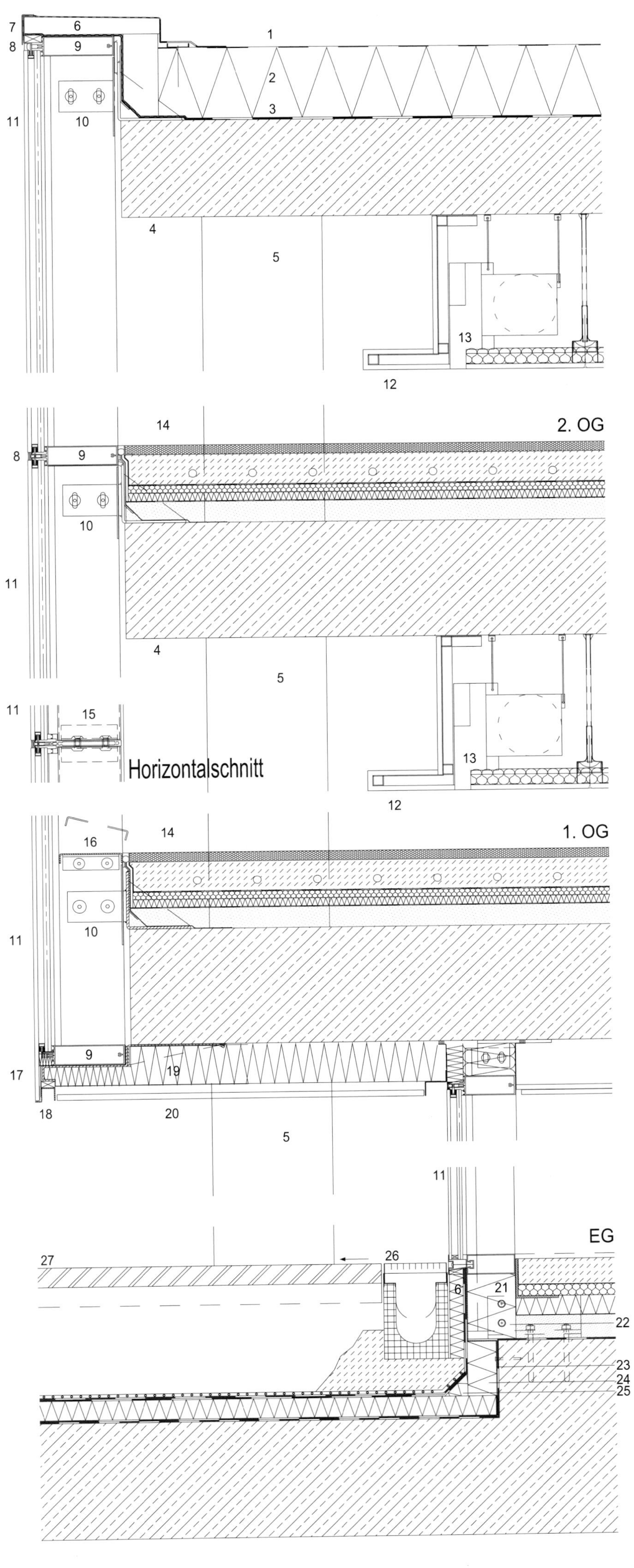

Fassadenschnitt Kunsthaus Glasfassade 1/10

1 Kunststoffdachbahn 1,8mm
2 Wärmedämmplatten EPS-20, 180mm
3 Dampfsperre
4 Stahlbetondecke mit Gefälle
Randbereich in Sichtbeton
5 Sichtbetonstütze Ø 30cm
6 ALU-Paneel wärmegedämmt mit Gefälle
7 NIRO-Blech, pulverbeschichtet
8 Silikonfuge, schwarz
9 Fassadenriegel, pulverbeschichtet
10 Pfostenhalter verzinkt und pulverbeschichtet
11 ISO-Verglasung
12 Gipskartondecke gelocht
13 Lüftungsanlage Induktivauslass
14 Terrazzo 20mm
Heizbetonestrich, bewehrt 80mm
PE-Folie einlagig 0,2mm
Trittschalldämmplatten TDPT 20/20 20mm
PE-Folie einlagig 0,2mm
Splittschuettung gebunden
15 Fassaden-T-Pfosten, pulverbeschichtet
16 Blech, verzinkt und pulverbeschichtet
17 Überstandsscheibe ca. 125mm, geschwärzt
18 U-Profil, NIRO oulverbeschichtet
19 Wärmedämmplatte FDP 100mm
20 HD-Bauplatte zementgebunden+glasfaserbewehrt
21 EPS 120mm
22 Pfostenmontageprofil, verzinkt
23 Blechhochzug, 2x gekantet, verzinkt
24 XPS-G-35 70mm
25 PS-Leiste 60/60
26 Rinne, Rinnengitter verzinkt, abnehmbar
27 Naturstein-Plattenbelag
Splittbett zementgebunden
Schutzschicht GGR-Matte
Bitumenbahn, zweilagig
Schaumglas Typ F, 50mm
Stahlbetonplatte mit Gefälle

Guangzhou Opera House

Location: Guangzhou, China
Architect: Zaha Hadid Architects
Client: Guangzhou Municipal Government
Size: 70,000m²
Completion Date: 2010
Photography: Christian Richters, Hufton + Crow, Iwan Baan, Virgile Simon Bertrand

Like pebbles in a stream smoothed by erosion, the Guangzhou Opera House sits in perfect harmony with its riverside location. The Opera House is at the heart of Guangzhou's cultural development. Its unique twin-boulder design enhances the city by opening it to the Pearl River, unifying the adjacent cultural buildings with the towers of international finance in Guangzhou's Zhujiang new town.

The 1,800-seat auditorium of the Opera House houses the very latest acoustic technology, and the smaller 400-seat multifunction hall is designed for performance art, opera and concerts in the round.

The design evolved from the concepts of a natural landscape and the fascinating interplay between architecture and nature; engaging with the principles of erosion, geology and topography. The Guangzhou Opera House design has been particularly influenced by river valleys – and the way in which they are transformed by erosion. Fold lines in this landscape define territories and zones within the Opera House, cutting dramatic interior and exterior canyons for circulation, lobbies and cafes, and allowing natural light to penetrate deep into the building. Smooth transitions between disparate elements and different levels continue this landscape analogy. Custom molded glass-fiber reinforced gypsum (GFRC) units have been used for the interior of the auditorium to continue the architectural language of fluidity and seamlessness.

The Guangzhou Opera House has been the catalyst for the development of cultural facilities in the city including new museums, library and archive. The Opera House design is the latest realization of Zaha Hadid Architects' unique exploration of contextual urban relationships, combining the cultural traditions that have shaped Guangzhou's history, with the ambition and optimism that will create its future.

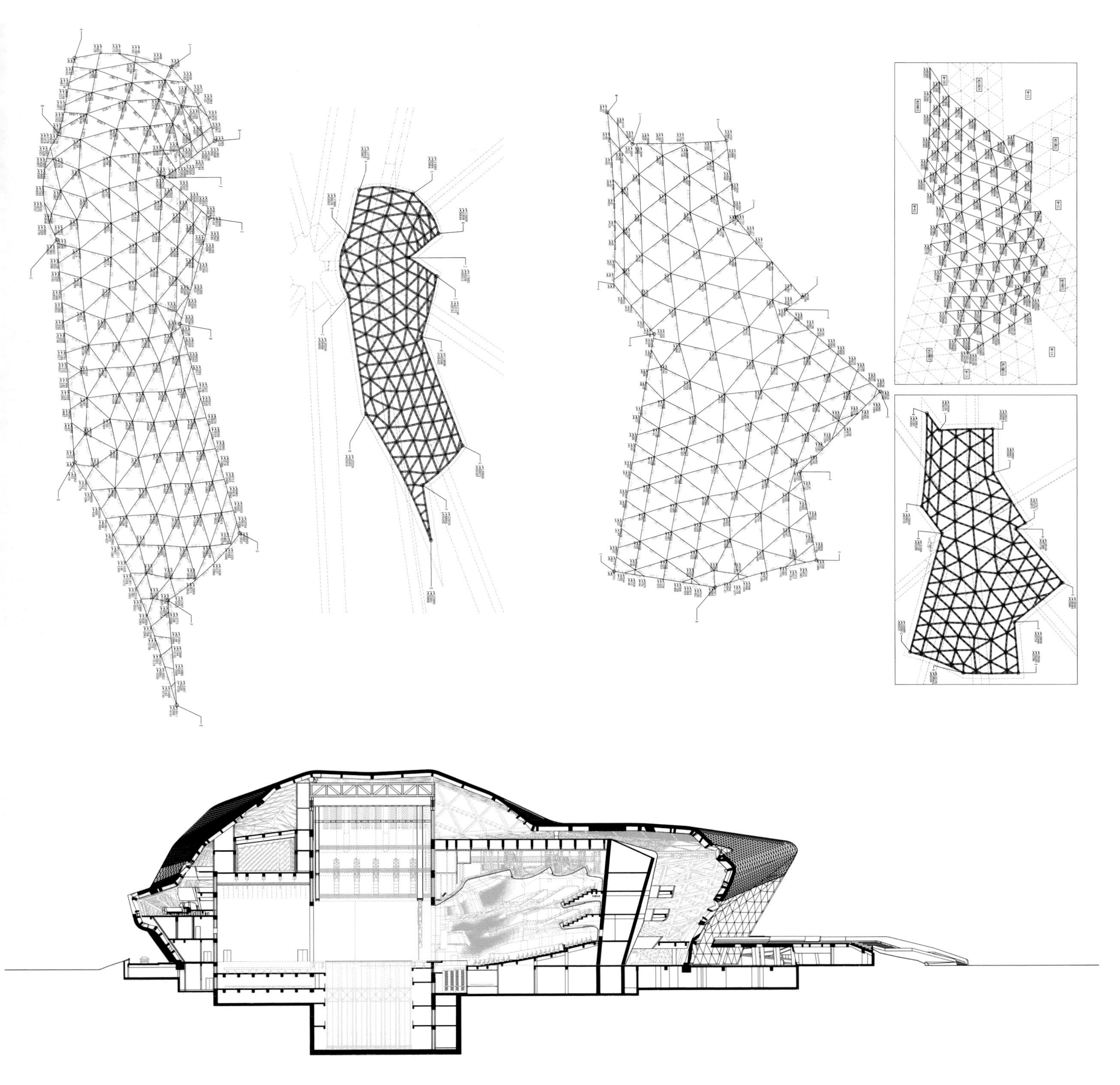

Cross section

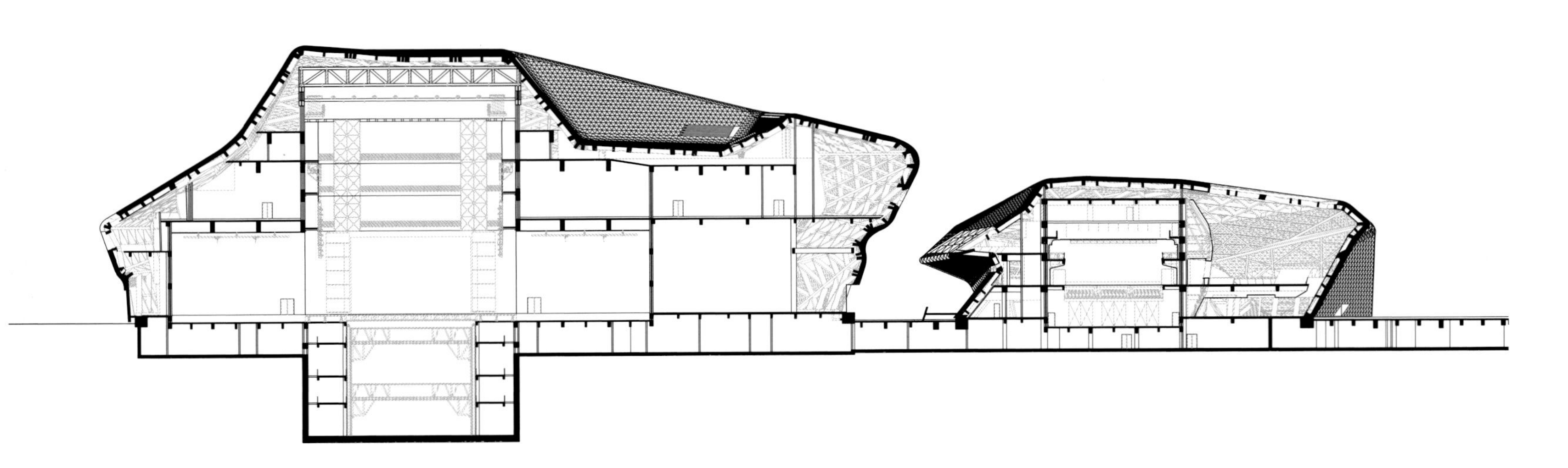

Longitudinal Section

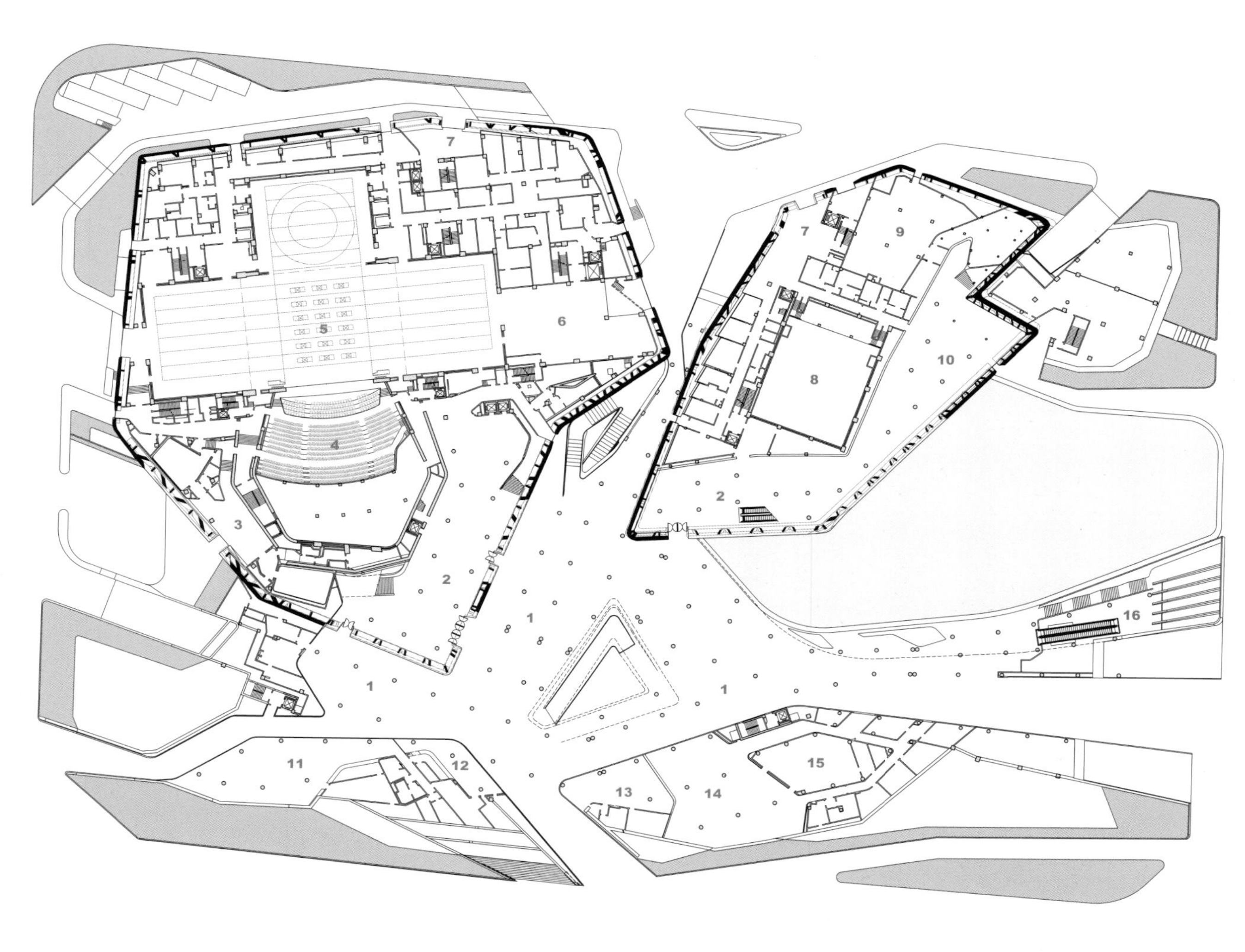

1. UNDER-PLAZA SPACE
2. ENTRANCE LOBBY
3. VIP LOUNGE
4. AUDITORIUM
5. STAGE
6. SCENERY ASSEMBLY
7. STAFF ENTRANCE
8. STAGE STORAGE
9. KITCHEN
10. DINING AREA
11. CAFETERIA
12. TICKET OFFICE
13. GIFT SHOP
14. RESEARCH CENTER
15. PRESS CONFERENCE ROOM
16. UNDERGROUND ACCESS

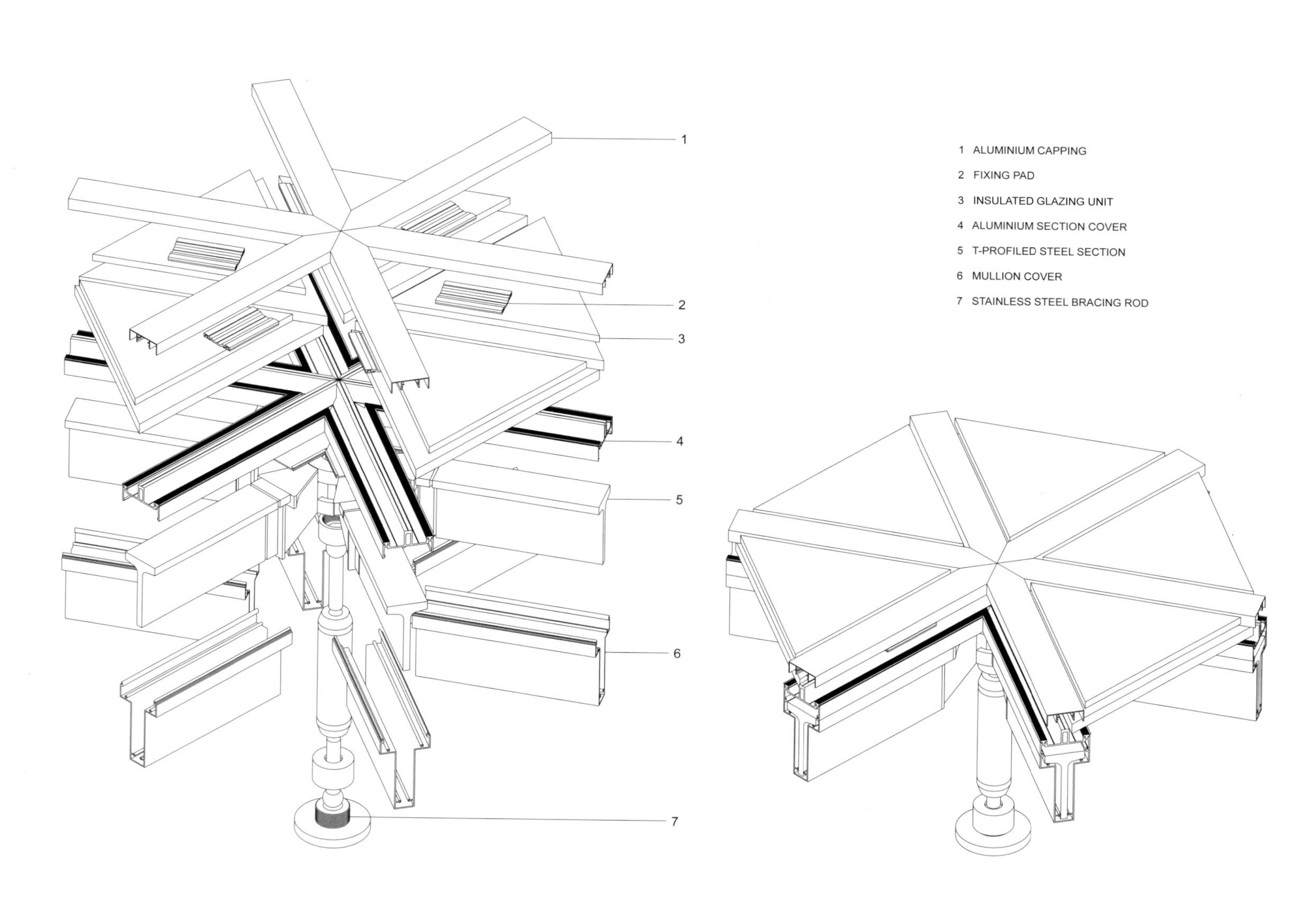

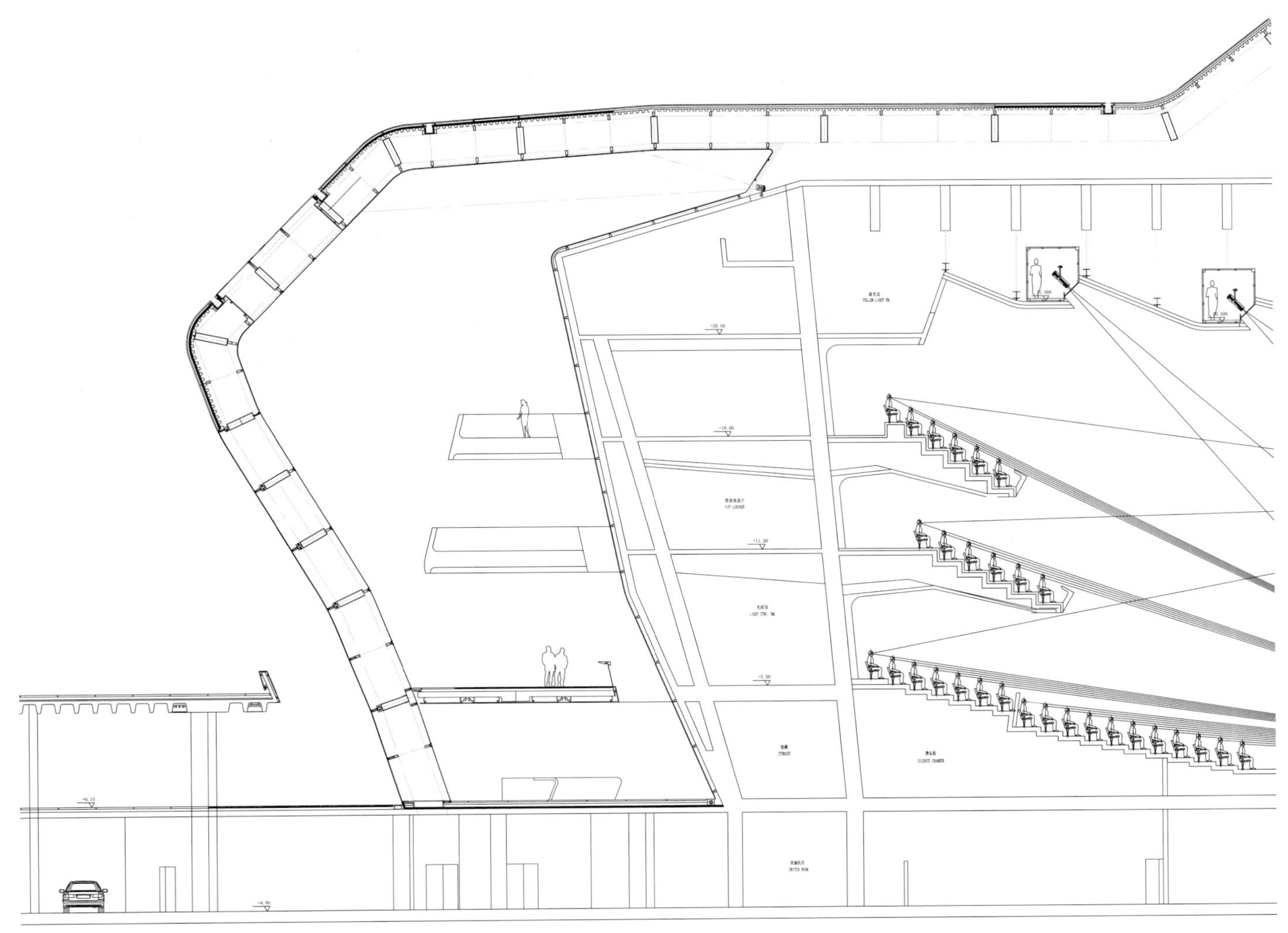

DETAIL Section 1

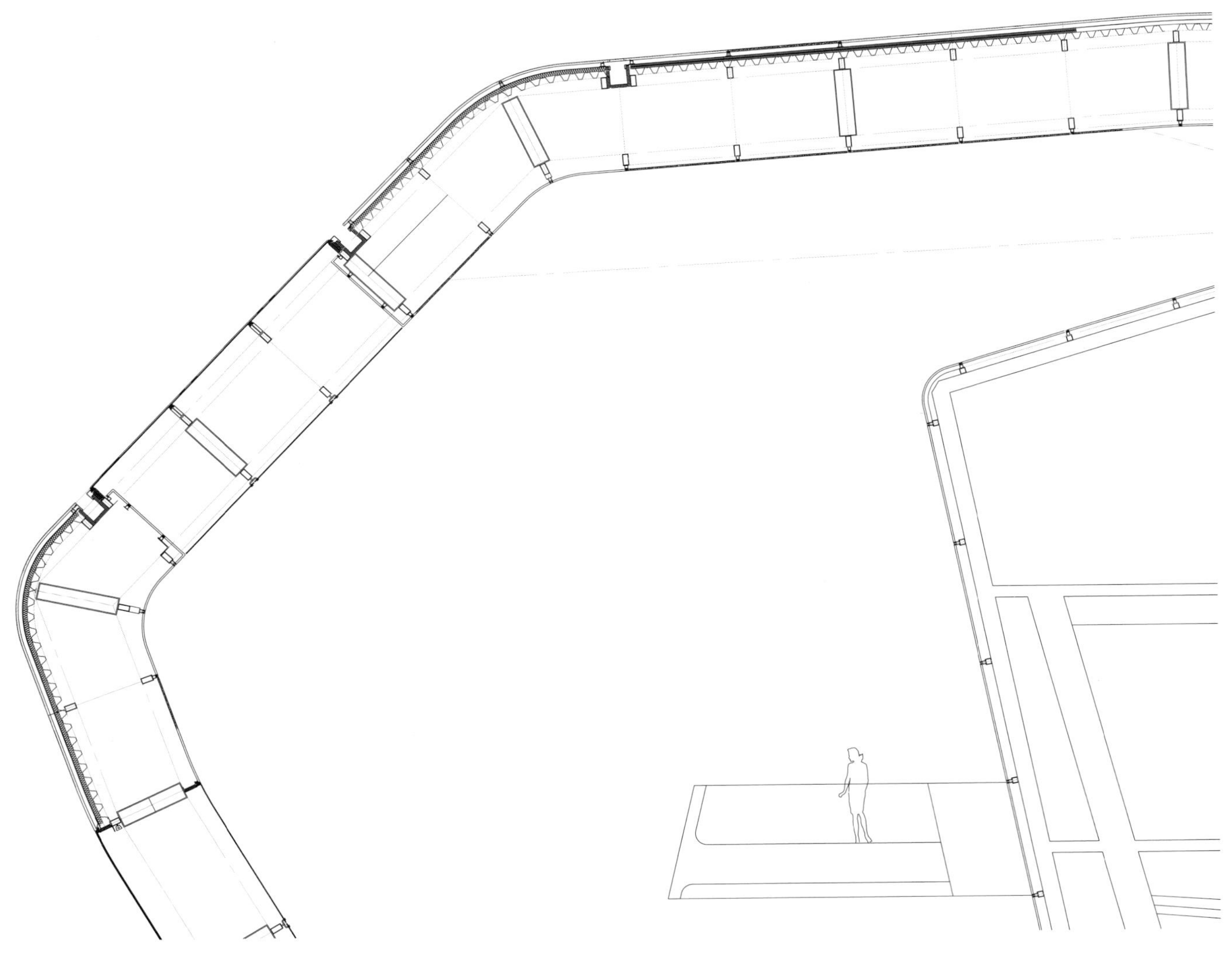

DETAIL Section 2-1

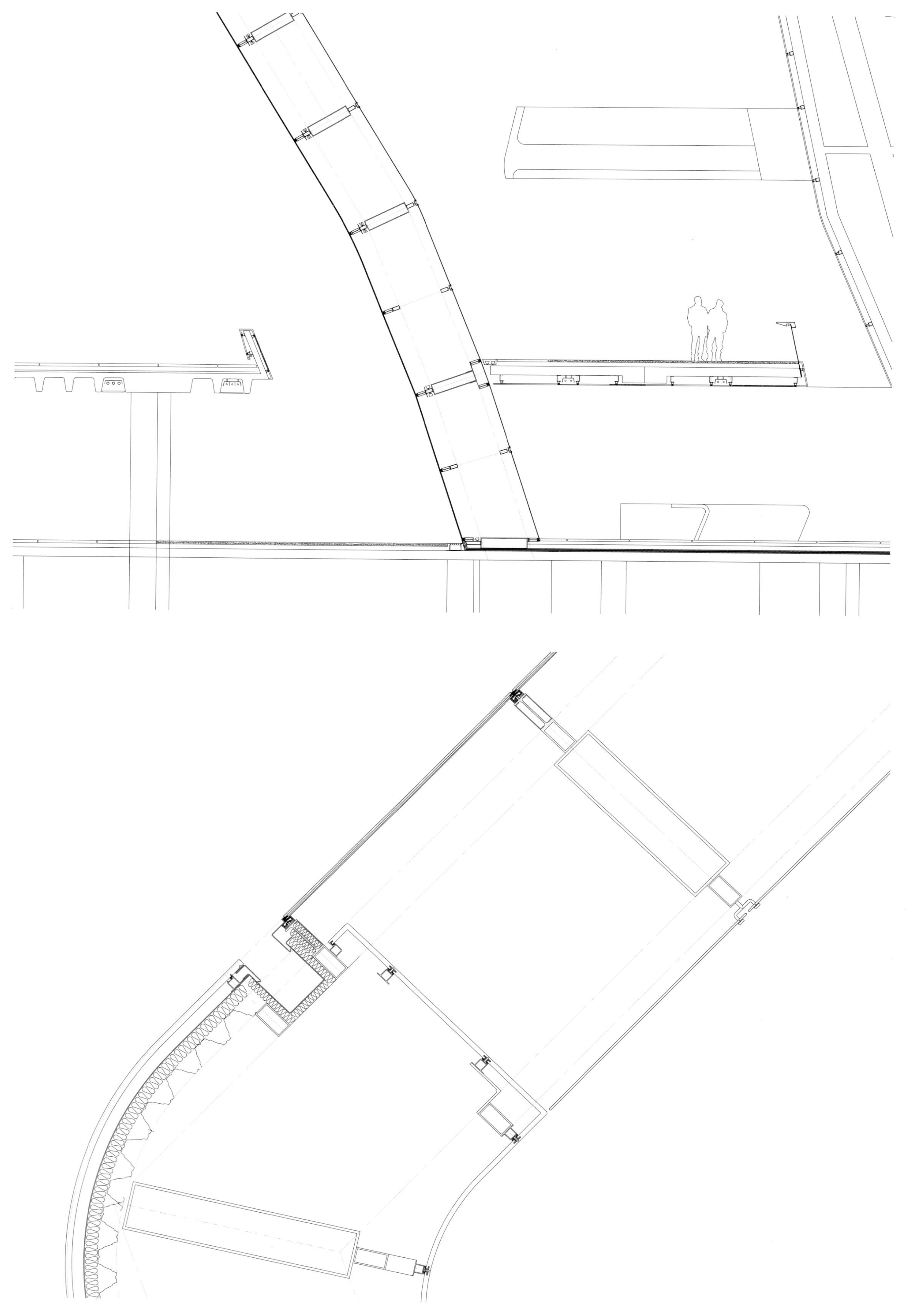

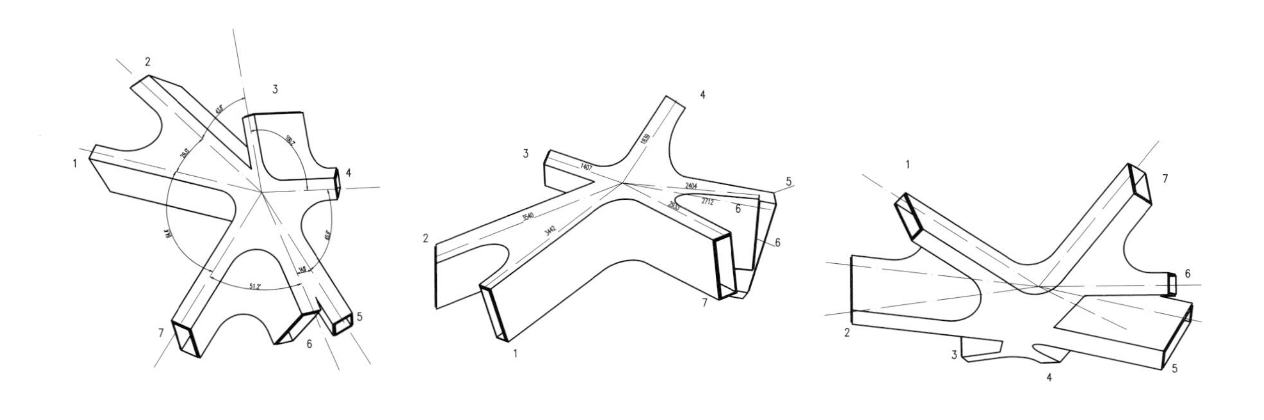

DJD - 9 AXONOMETRIC DRAWING

DJD - 9 TOP VIEW

DJD - 9 FRONT VIEW

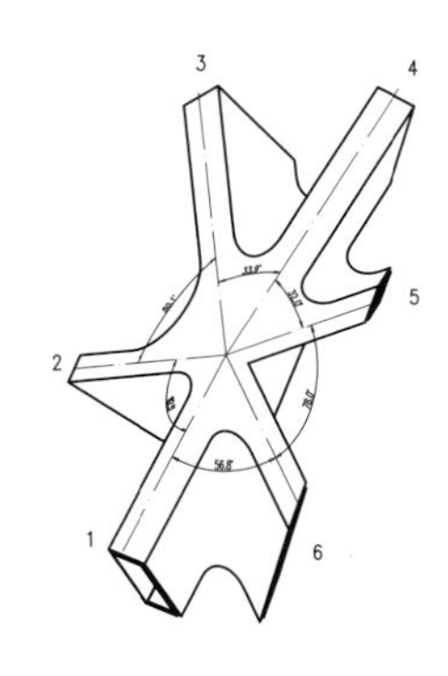

DJD - 10 AXONOMETRIC DRAWING

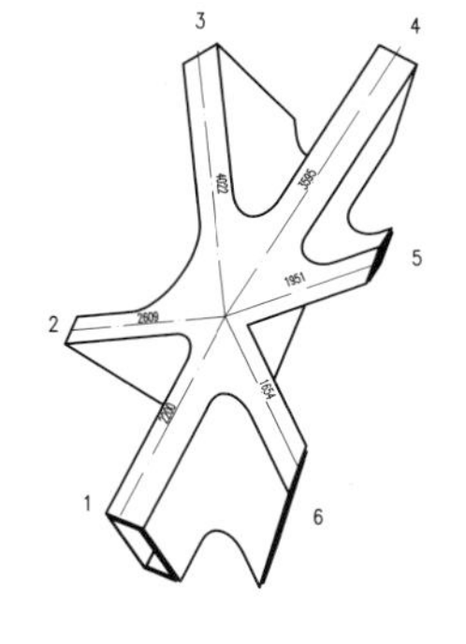

DJD - 10 AXONOMETRIC DRAWING

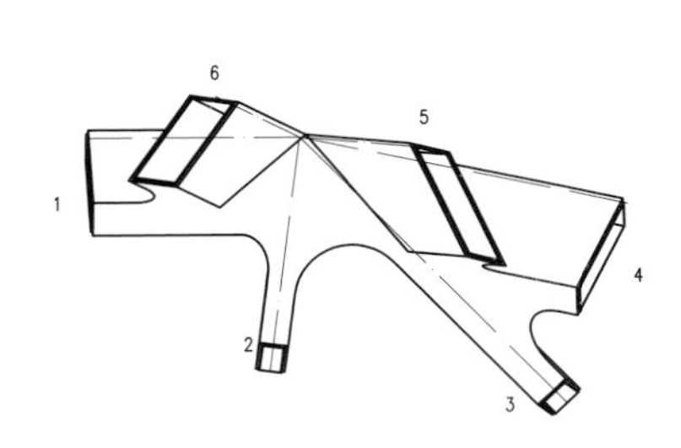

DJD - 10 FRONT VIEW

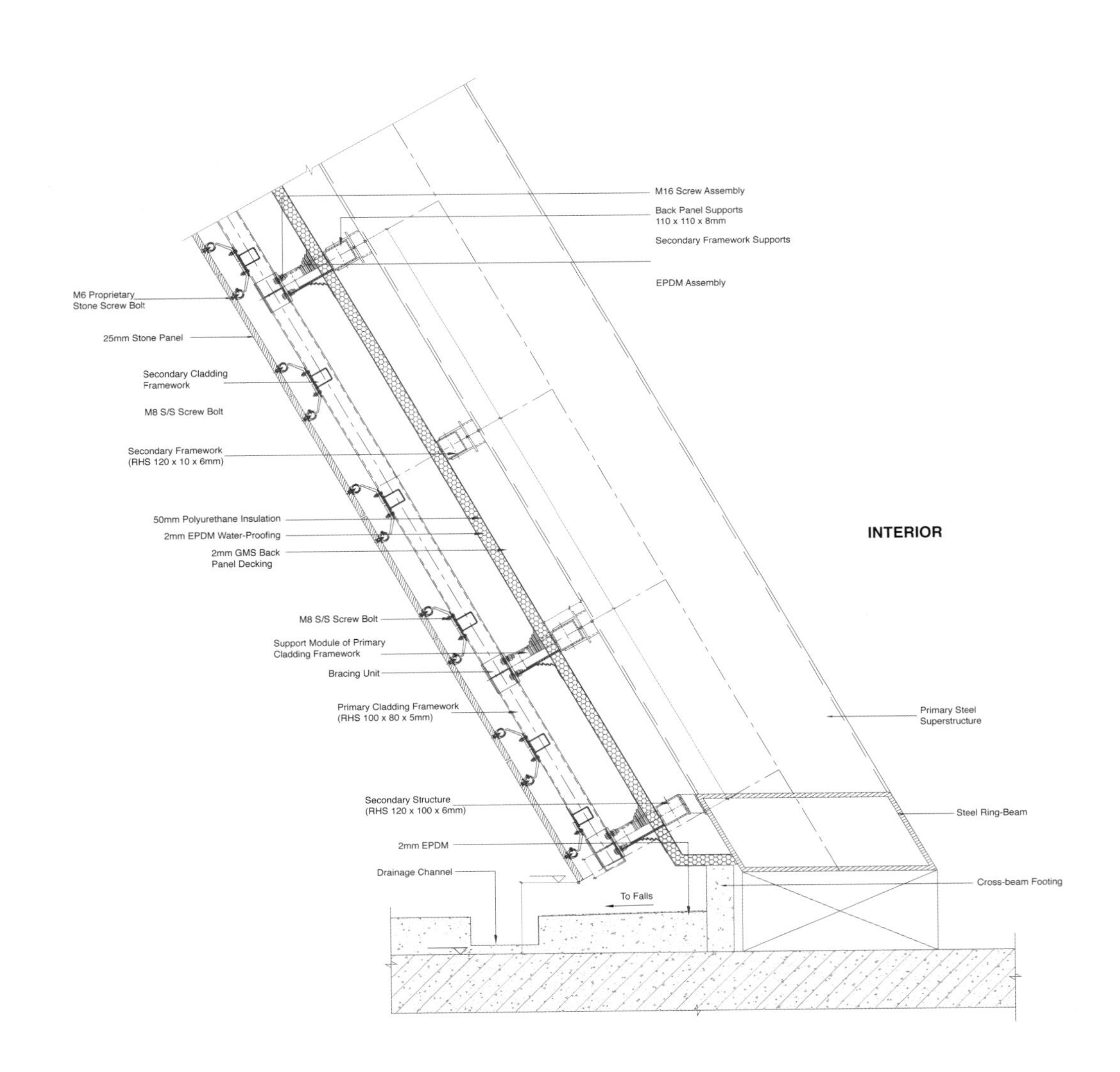

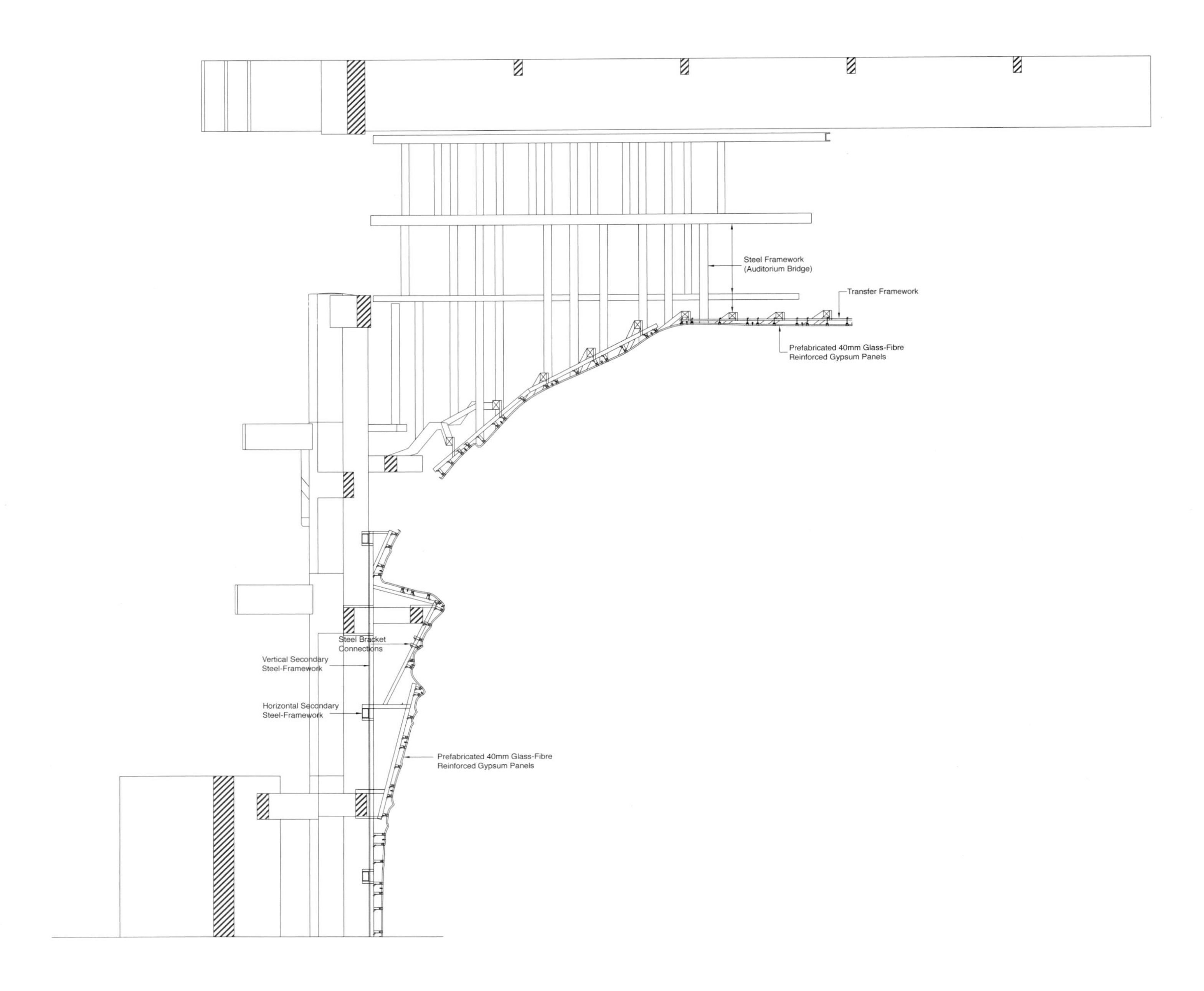

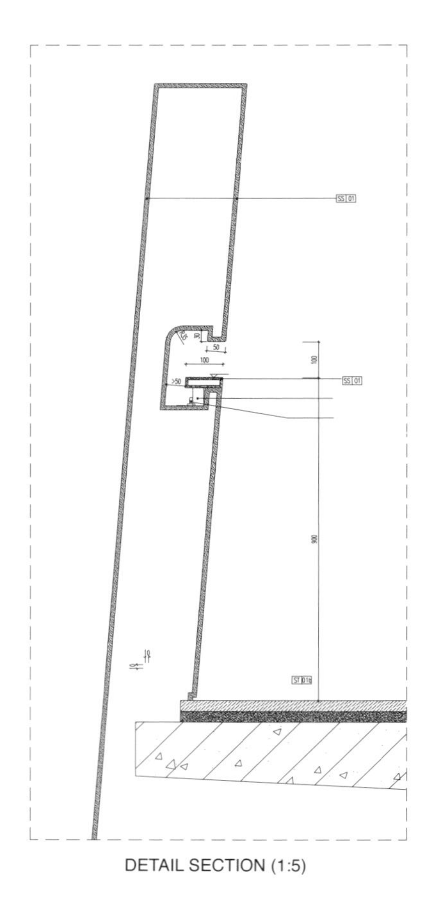

DETAIL SECTION (1:5)

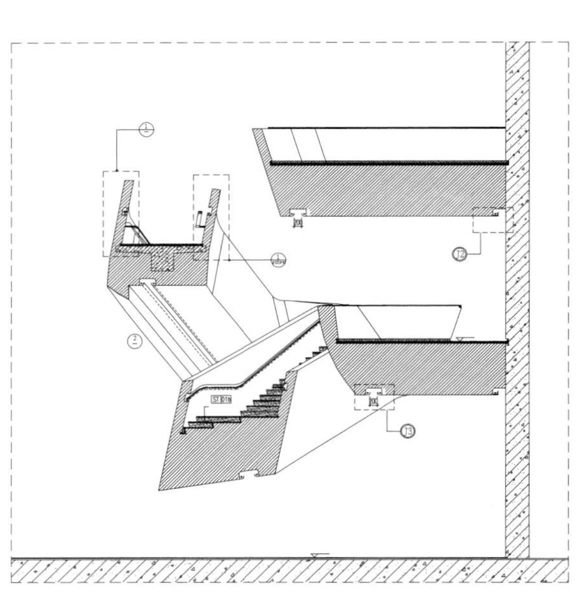

A-LT10 STAIRWAY - SECTION 1-1 (1:50)

DETAIL SECTION (1:10)

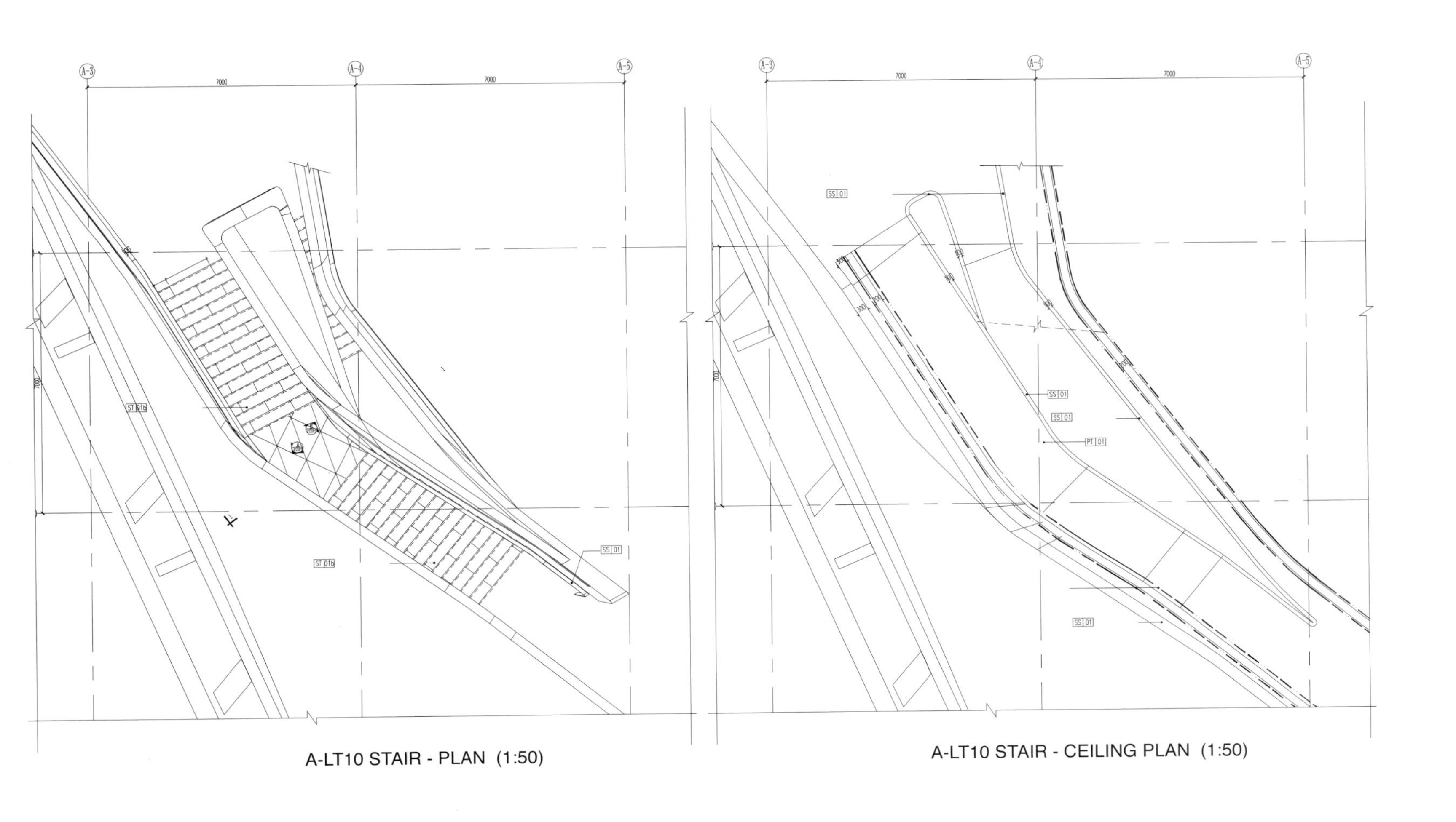

A-LT10 STAIR - PLAN (1:50)

A-LT10 STAIR - CEILING PLAN (1:50)

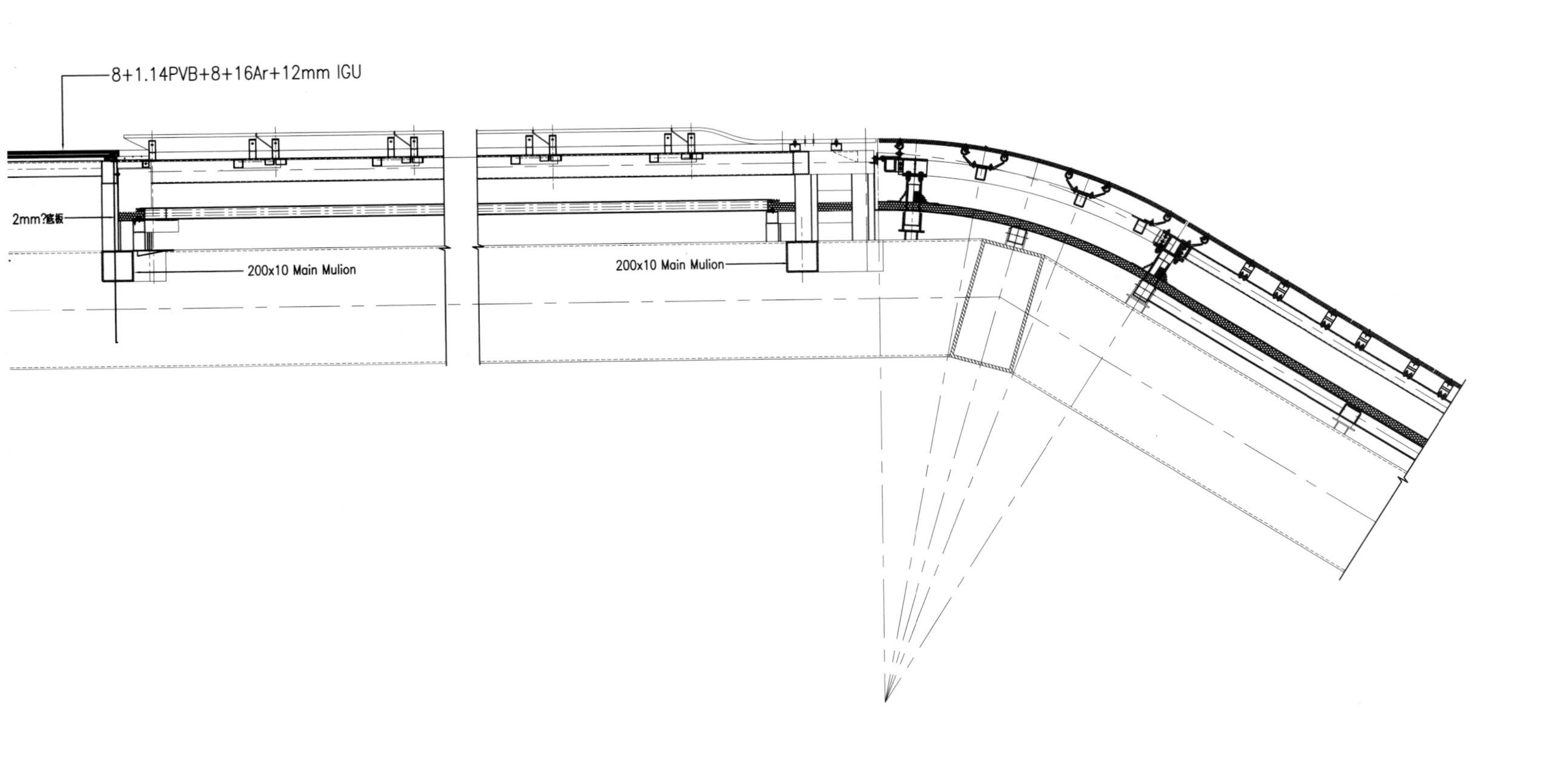

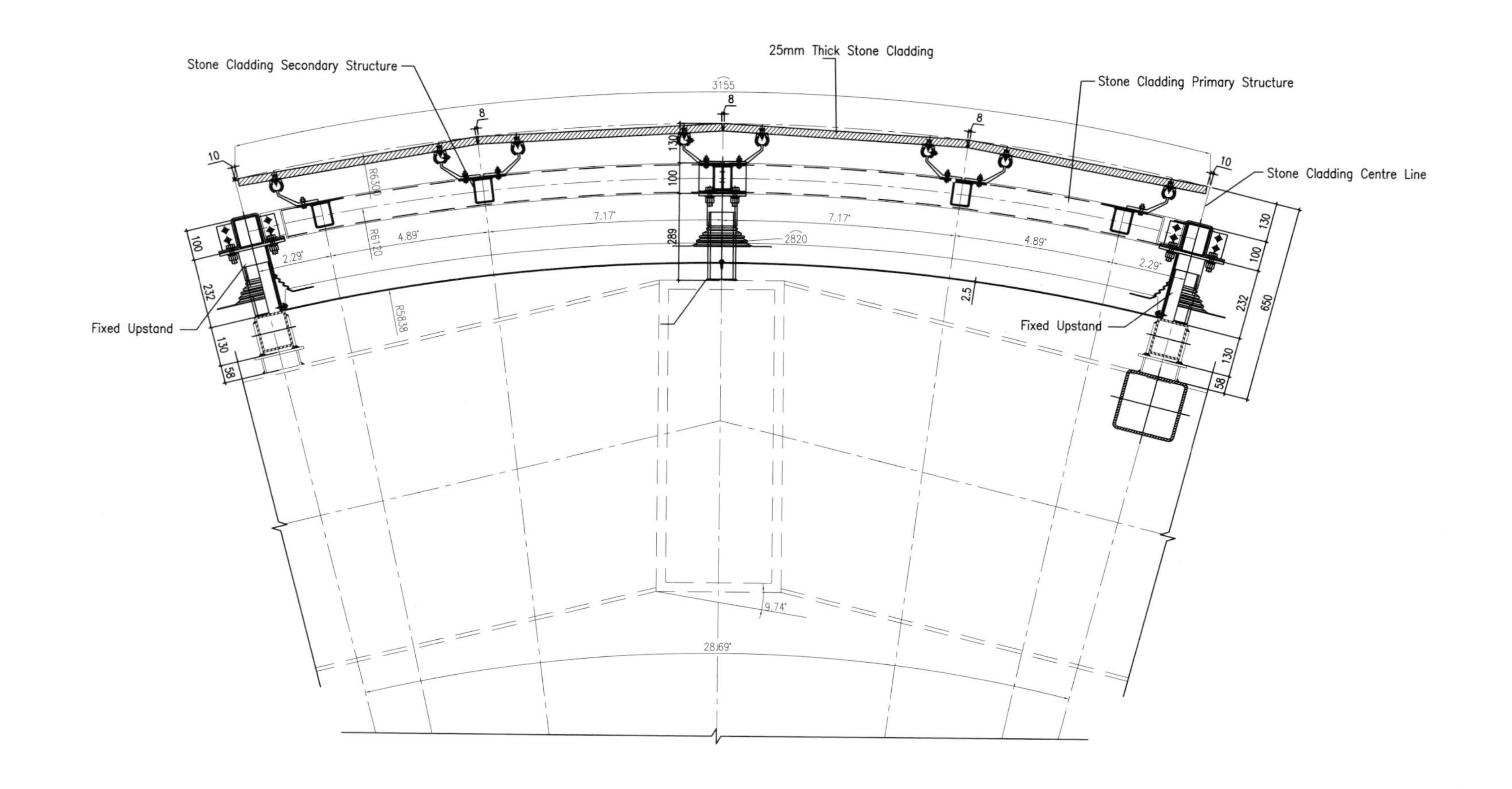
25mm Thick Stone Cladding
Stone Cladding Secondary Structure
Stone Cladding Primary Structure
Stone Cladding Centre Line
Fixed Upstand
Fixed Upstand
3155
2820
R6300
R6120
R5838
7.17°
7.17°
4.89°
4.89°
2.29°
2.29°
9.74°
28.69°
650

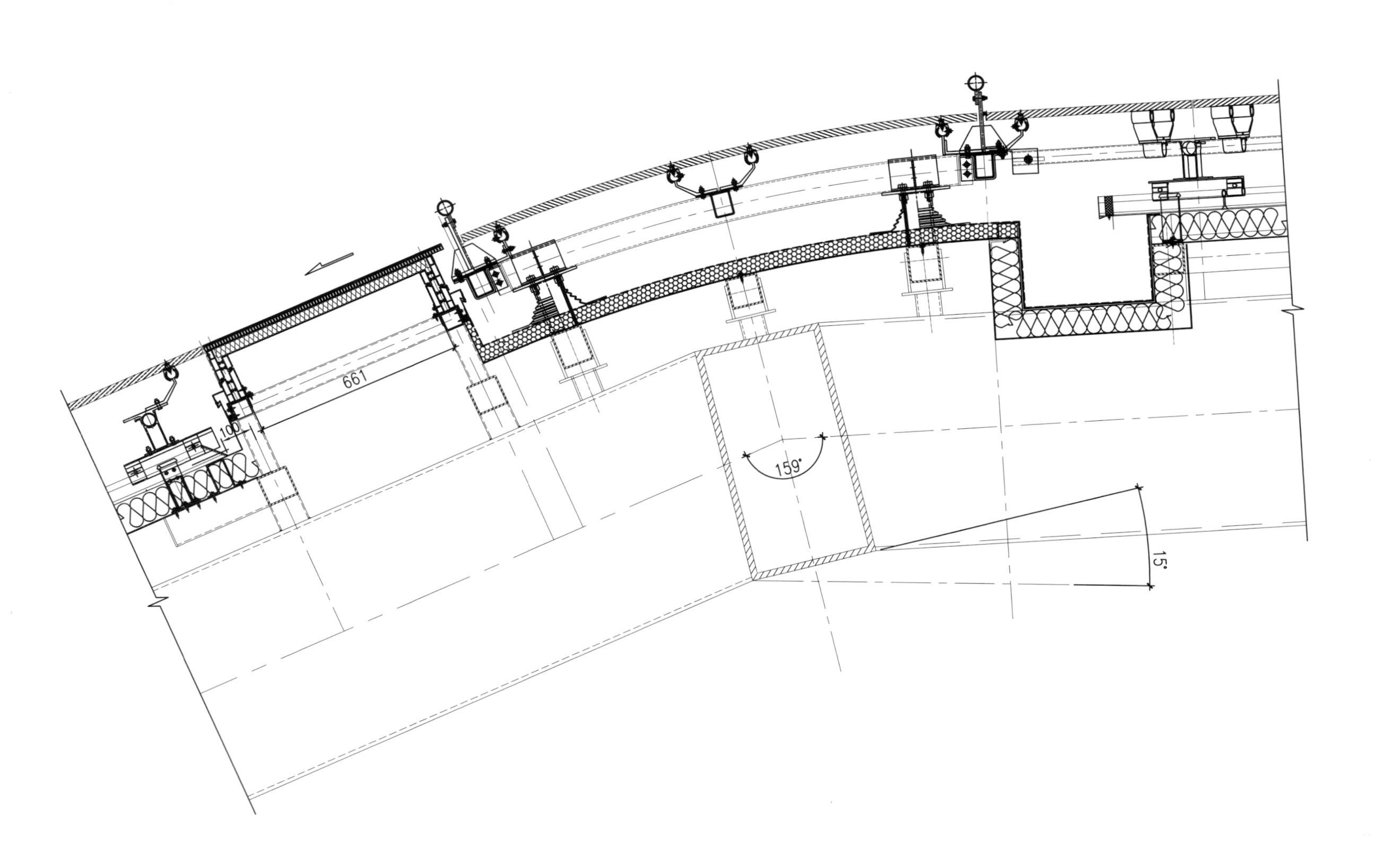
661
159°
15°

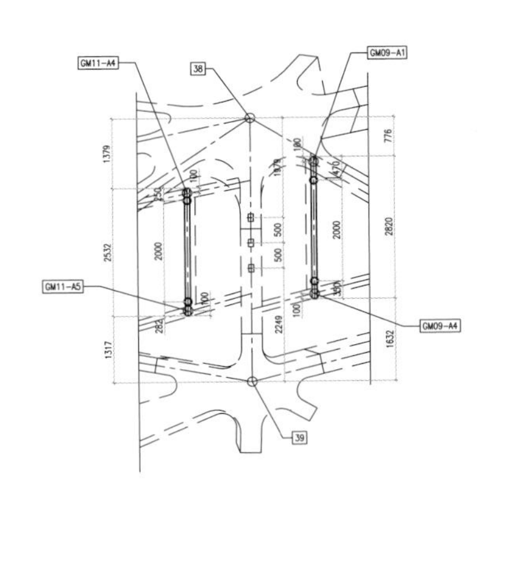
GM11-A4
38
GM09-A1
GM11-A5
GM09-A4
39
1379
2532
1317
2000
500
776
2820
1632
2249

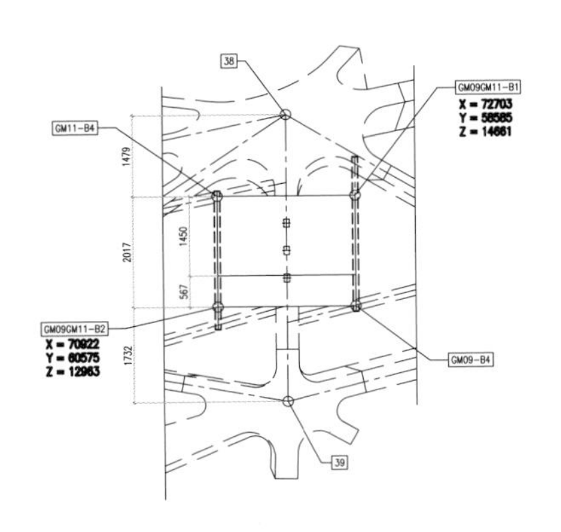
38
GM11-B4
GM09GM11-B1
X = 72703
Y = 58585
Z = 14881
GM09GM11-B2
X = 70922
Y = 60575
Z = 12963
GM09-B4
39
1479
2017
1450
567
1732

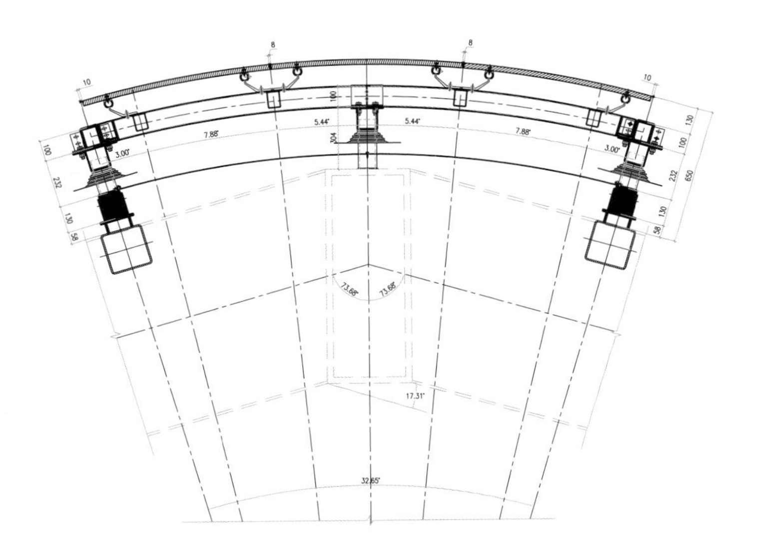
7.88°
5.44°
5.44°
7.88°
3.00°
3.00°
73.68°
73.68°
17.31°
32.65°

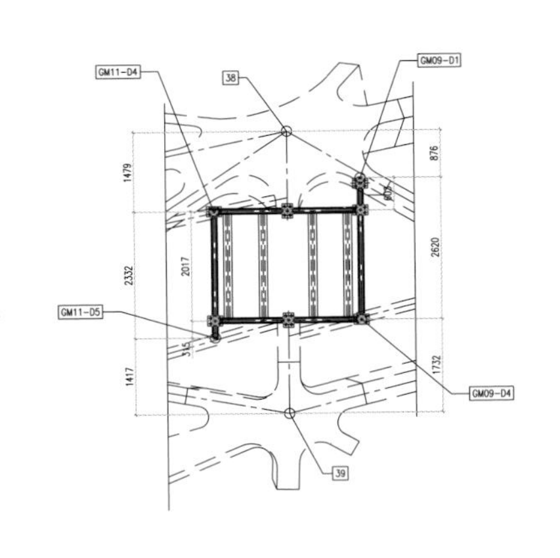
GM11-D4
38
GM09-D1
GM11-D5
GM09-D4
39
1479
2332
1417
2017
876
2620
1732

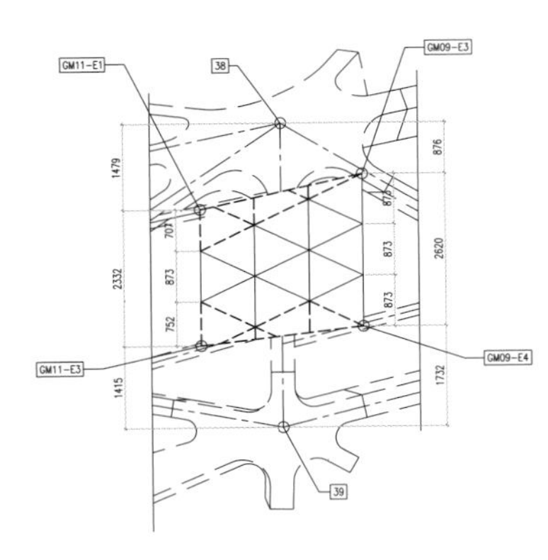
GM11-E1
38
GM09-E3
GM11-E3
GM09-E4
39
1479
2332
1415
873
873
752
876
2620
1732

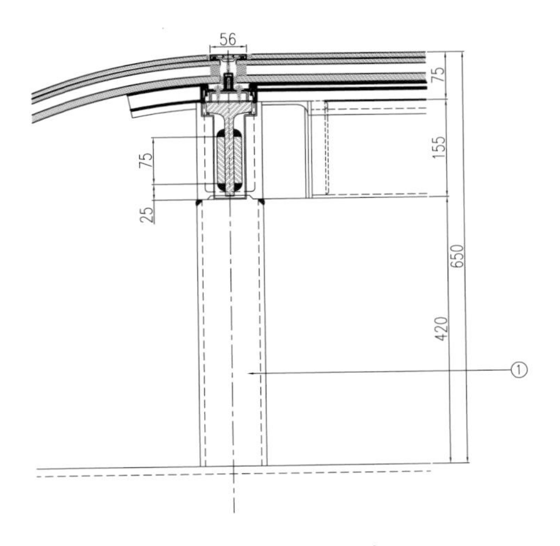
56
75
155
75
25
650
420
1

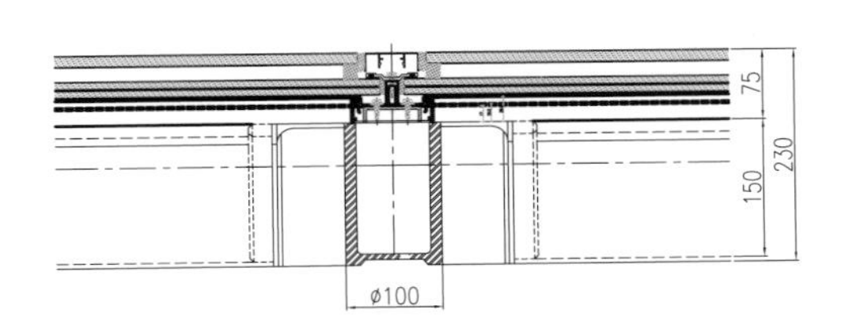
75
150
230
ø100

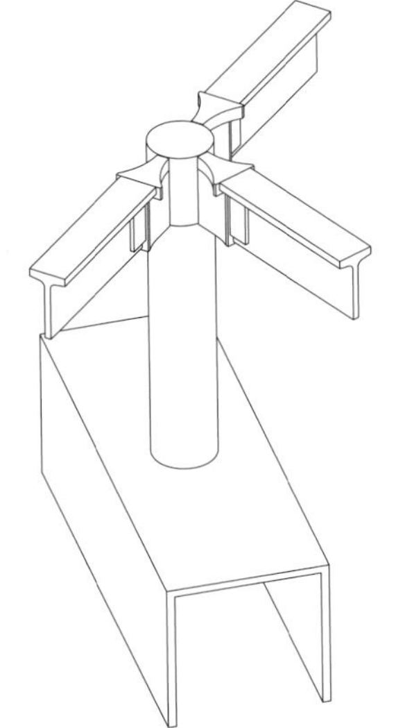

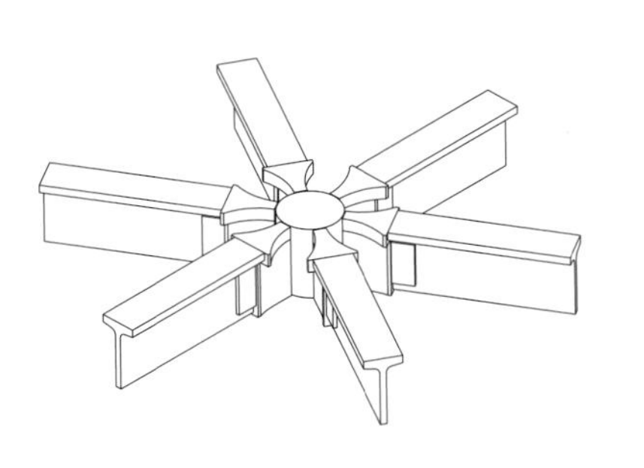

Education

King Fahad National Library

Location: Riyadh, Kingdom of Saudi Arabia
Architect: Gerber Architekten
Client: Kingdom of Saudi Arabia
Gross Floor Area: 86,632 m^2
Completion Date: 2013

The client's requirements are the basis for the development of design concepts. Deductions from the site's specific characteristics and history, the surrounding structures, climate and access paths, as well as energy and ecological needs - aspects of particular importance today - are all parameters of design process. Ultimately, the building or urban landscaped ensemble, should, as a unique element, reflect the genius loci, formulating a distinctive and memorable idea of structure and space in all its parts.

By reducing design concepts developed in this way, all important ideas should - in the form of a small logo - be visually imparted, make an impression and set a sign. When the building is used and experienced, this conceptual sign should be perceptible as a structure and in all its parts recognizable through its clarity and conclusiveness, right down to the use of materials and color. This is only possible, however, when the design's basis is a rational, functionally intelligent, innovative and thus formally expressive concept.

The goal, as a team of architects, interior designers, engineers, urban planners and landscape designers, is to create built environments that affect people and stir their desires; places people like to visit and linger in; spaces that are tangible and logically accessible. These should be structures that improve the urban and landscaped environment with their beauty and in their simplicity, as well as being exciting in their spatial arrangements, which are clear and self-evident with regards to the orientation between inside and out.

The main aspect of designers' efforts is to focus within the existing multiplicity of what is right, to distill that to a well-proportioned concentrate and to connect things with one another aesthetically, thus creating solutions to the tasks at hand for our fellow beings.

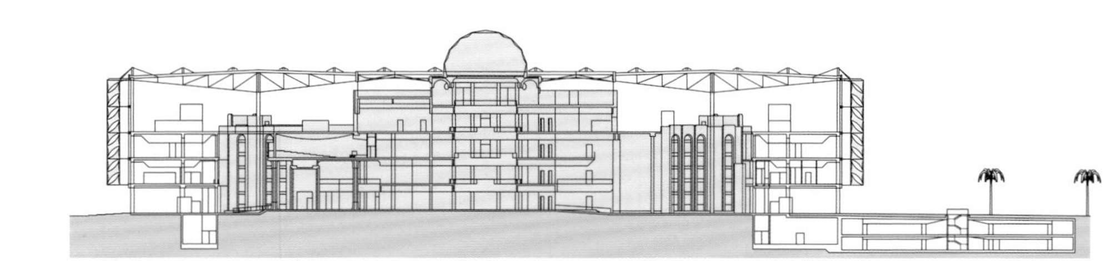

King Fahad Library section

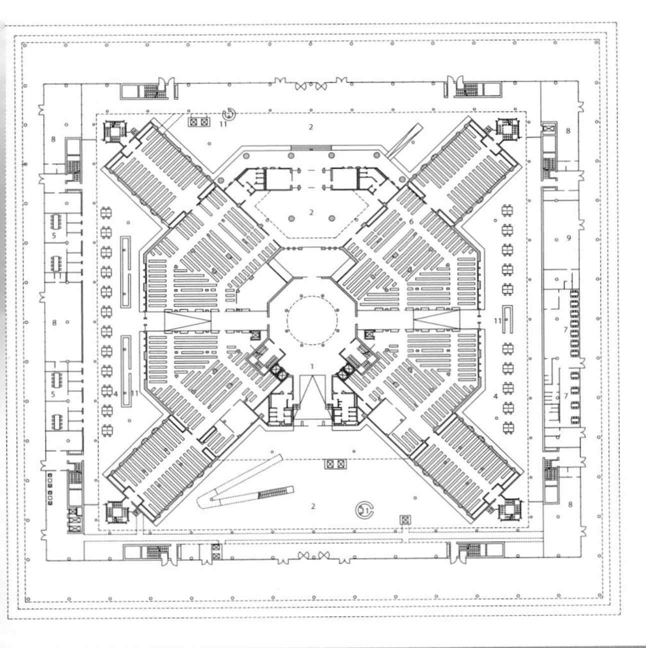

1 Rotunda
2 Entrance Hall
3 Public Halls
4 Reading Area
5 Studyroom
6 Closed Access Stack
7 Restaurant
8 Shop
10 Vault Room
11 Information and Service

Makkah

King Fahad Library
ground floor

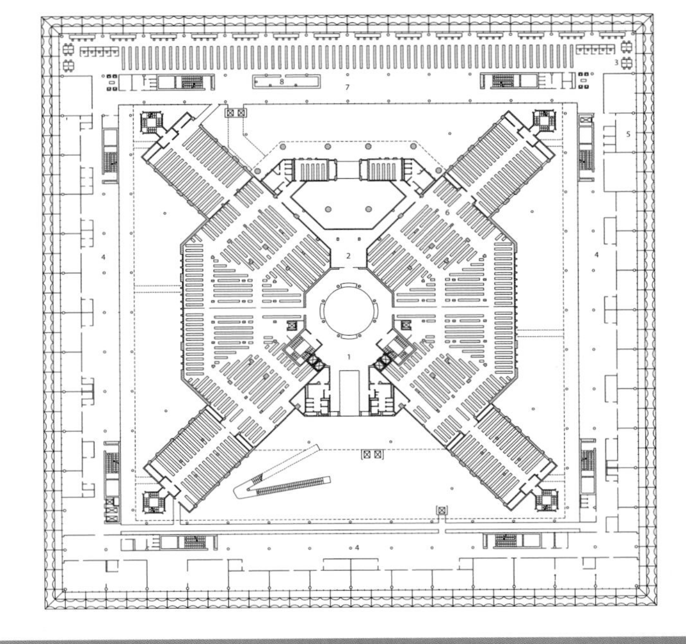

1 Rotunda
2 Entrance Hall
3 Reading Area
4 Administration
5 Seminar Room
6 Closed Access Stack
7 Women's Library
8 Information and Service

King Fahad Library
1th floor

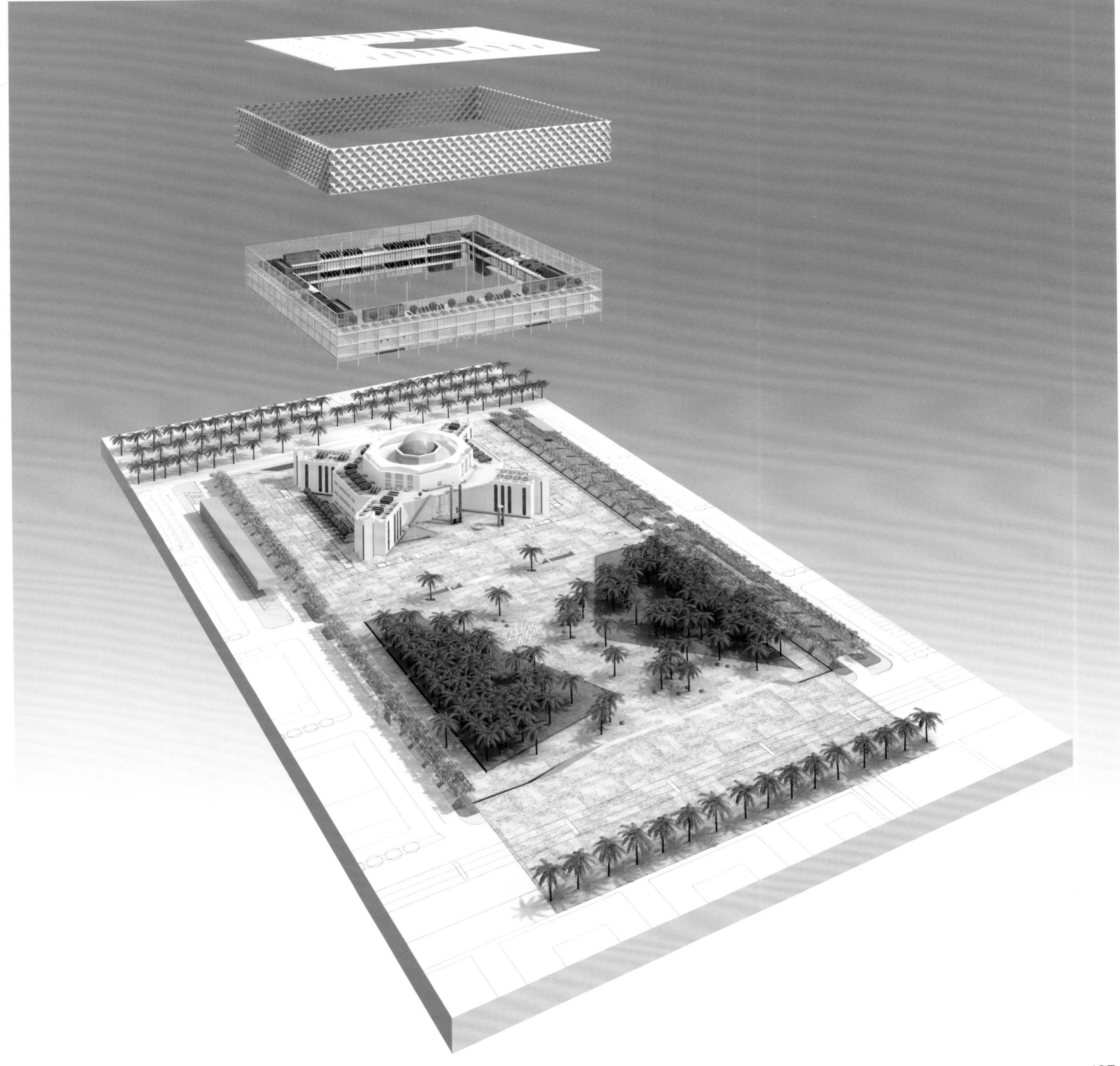

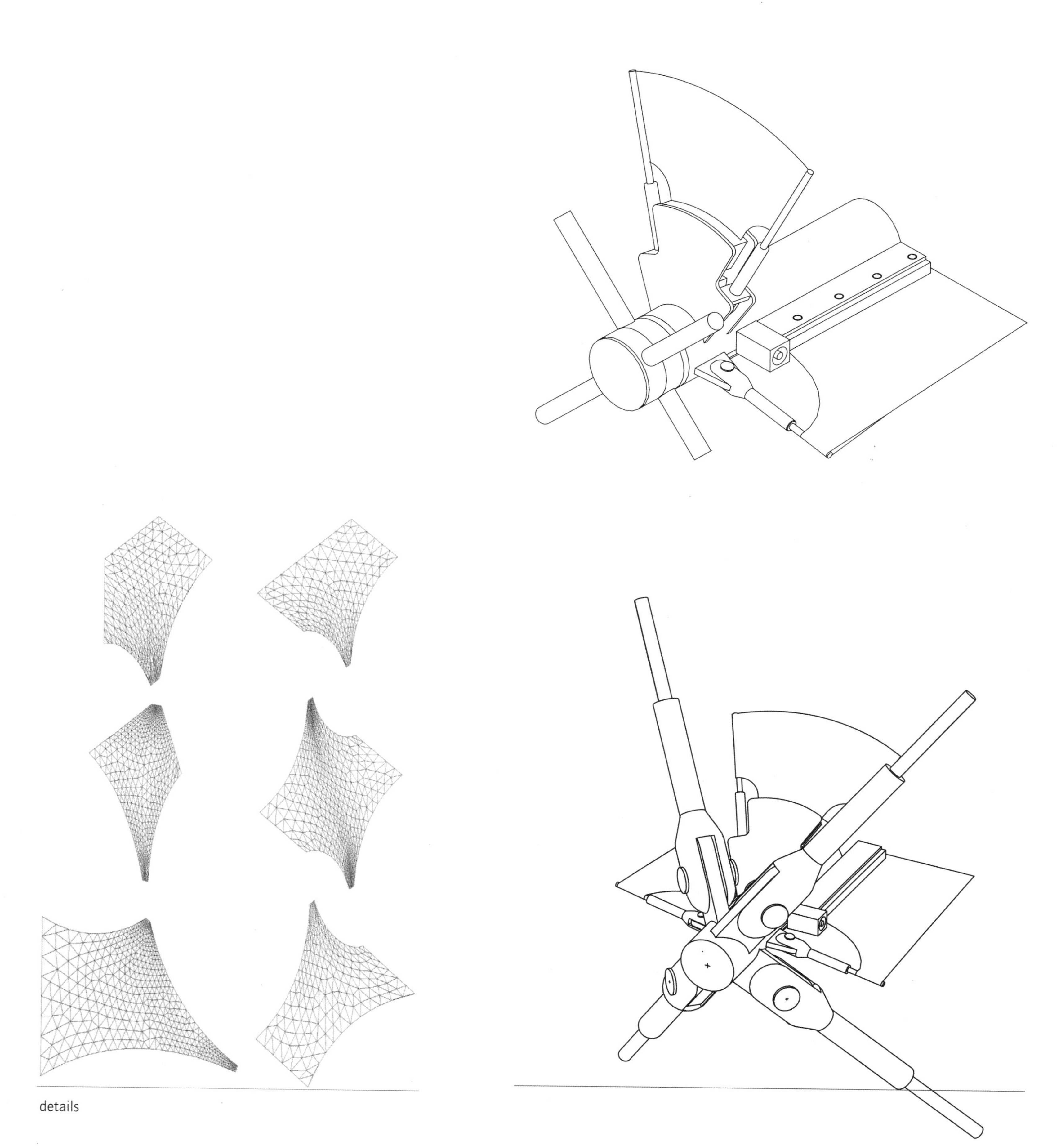

details

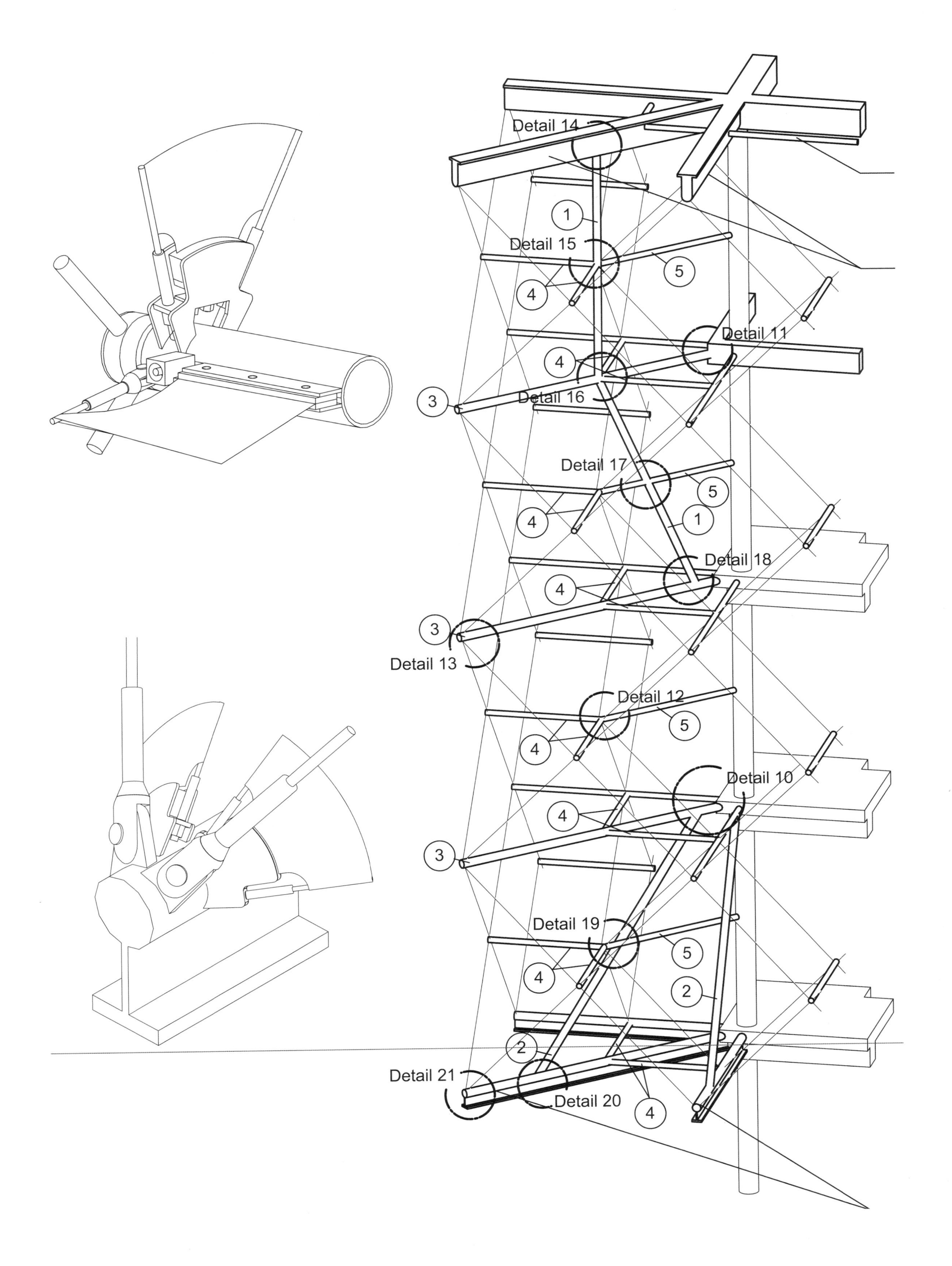
Detail 14
1
Detail 15
5
4
Detail 11
4
3
Detail 16
Detail 17
5
4
1
Detail 18
4
3
Detail 13
Detail 12
5
4
Detail 10
4
3
Detail 19
5
4
2
2
Detail 21
Detail 20
4

Extension of Université de Provence

Location: Provence, France
Architect: Dietmar Feichtinger Architectes
Client: Aix-Marseille Université
Site Area: 50,336 m²
Completion Date: 2013
Photographers: Barbara Feichtinger-Felber, Sergio Grazia

A major gesture is not always needed. In further developing a campus that dates originally from the 1960s Dietmar Feichtinger shows that added value can be better achieved through restraint, carefully positioned cubes, and subtle facade rhythms than by the use of showy sculptural symbols.

In approaching the project for the extension of the humanities faculty of Aix-en-Provence University Dietmar Feichtinger found himself confronted with a contradiction. On the one hand the extension, by nature of its position, had to mark an entrance to the campus, but on the other neither the brief nor the context suggested a specific form of architectural expression. The brief, relatively modest in terms of floor area (8000 m²) and banal in terms of content (lecture halls and offices), contained no elements such as an auditorium or large library that would have called for strong architectural expression. In addition the dreary mediocrity of the context spoke against a too striking kind of architecture. Given the featureless nature of the periphery of this provincial town with its groups of little houses and allotment gardens, banal workshops, pizzerias, and here and there a social housing block, even a minor architectural gesture would have seemed unsuitable, indeed even provocative.

Consequently, Feichtinger opted for restraint, if not to say architectural neutrality. He prefers a space to a building, an empty volume to a full one, and a place to a piece of architecture. By breaking up the brief into a number of small buildings he allows an entrance to the campus to develop. Through the considered way in which he positions the buildings and the subtle expression of the facades he gives the place an identity. These are two important characteristics of this project which Feichtinger achieved by appropriating the characteristics of the campus and reinterpreting them. This proved to be a discrete way of integrating this small extension in one of France's most important university complexes.

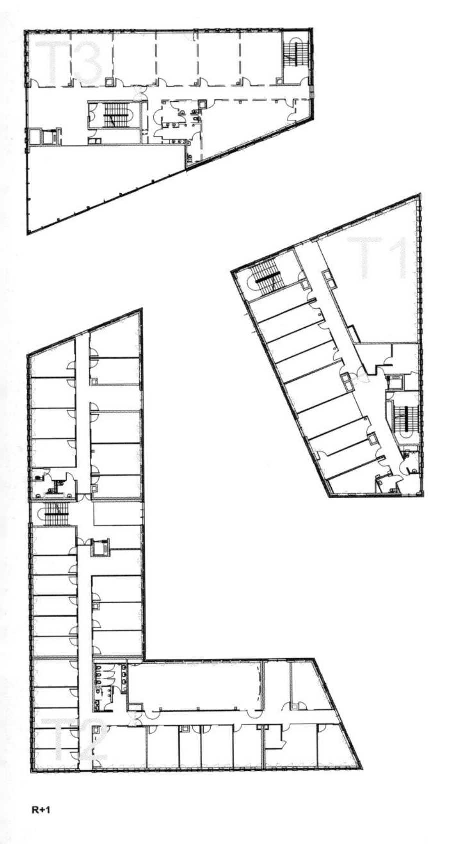

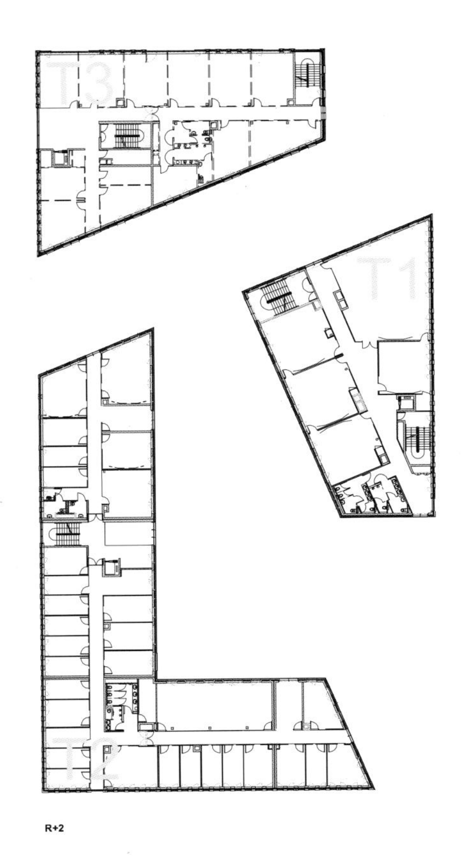

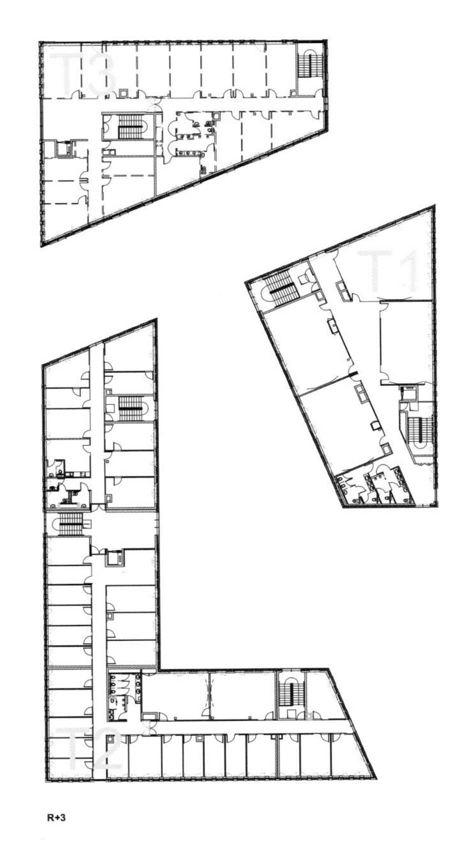

Plan_R+1,R+2,R+3

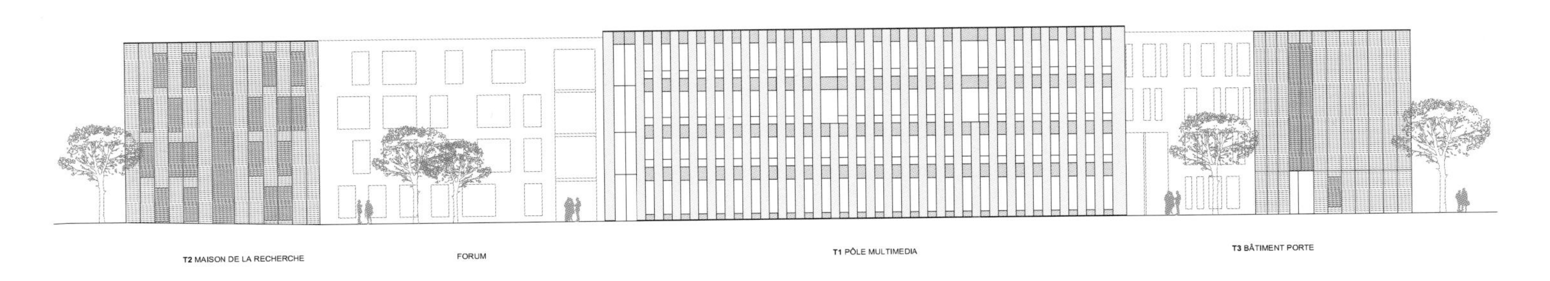

Facade Est Legende

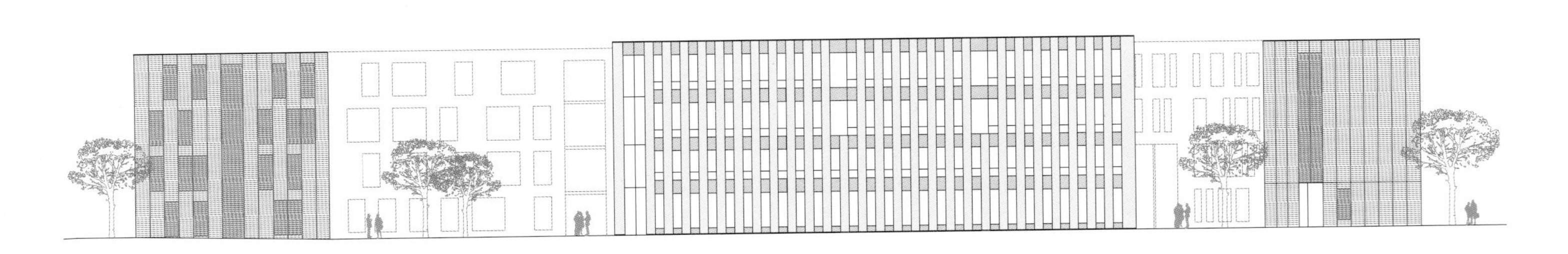

Facade Est Sans Legende

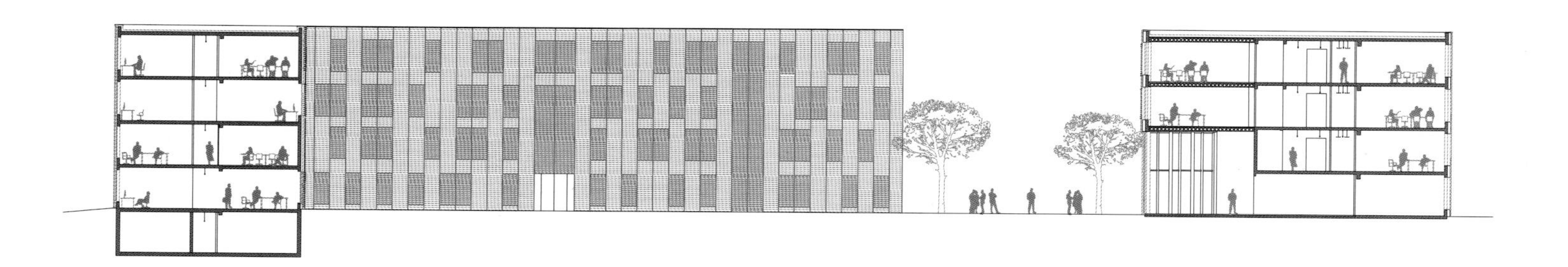

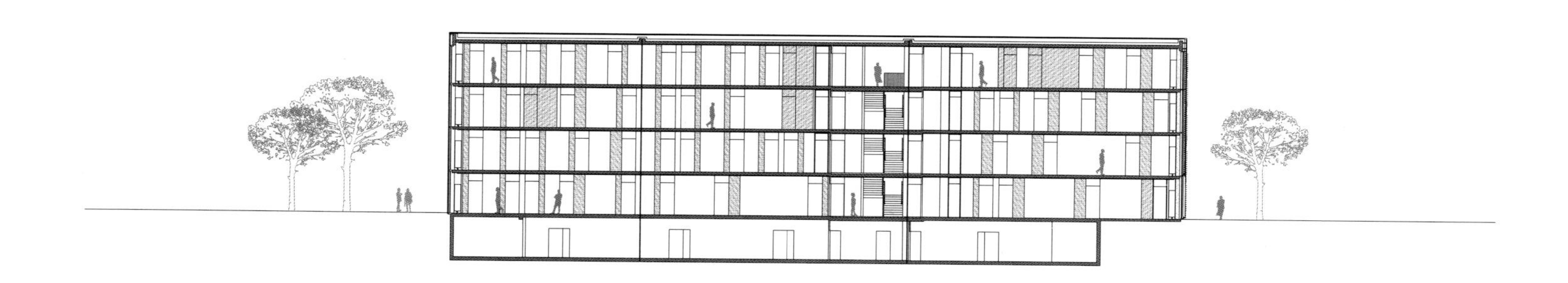

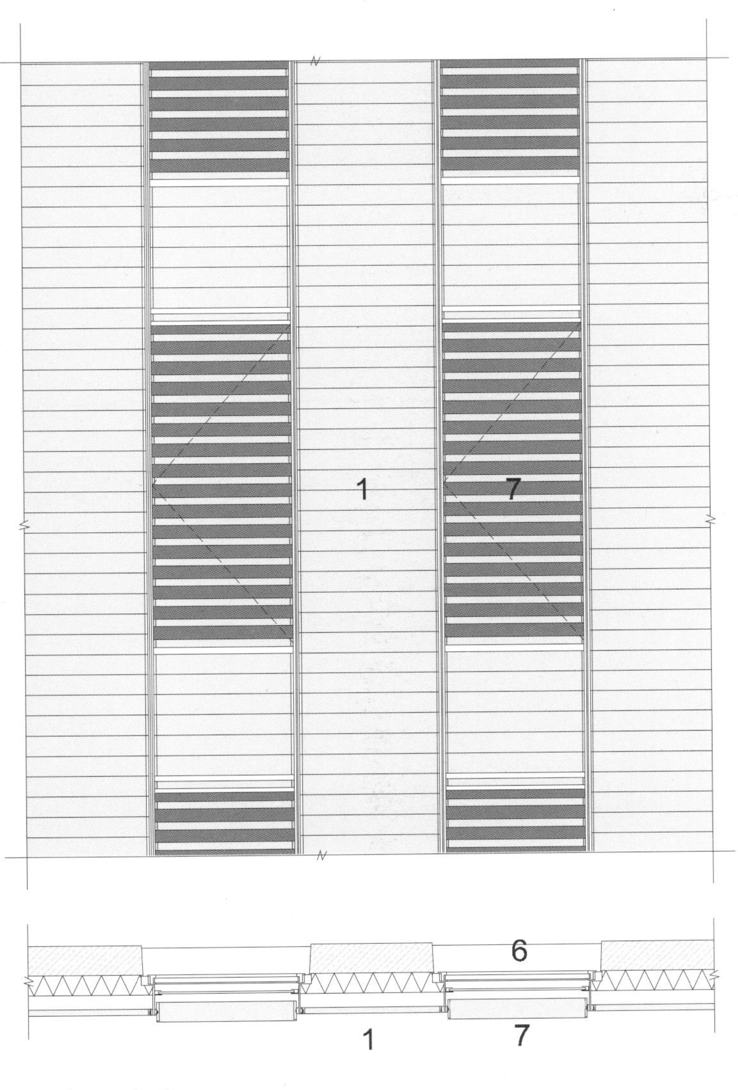

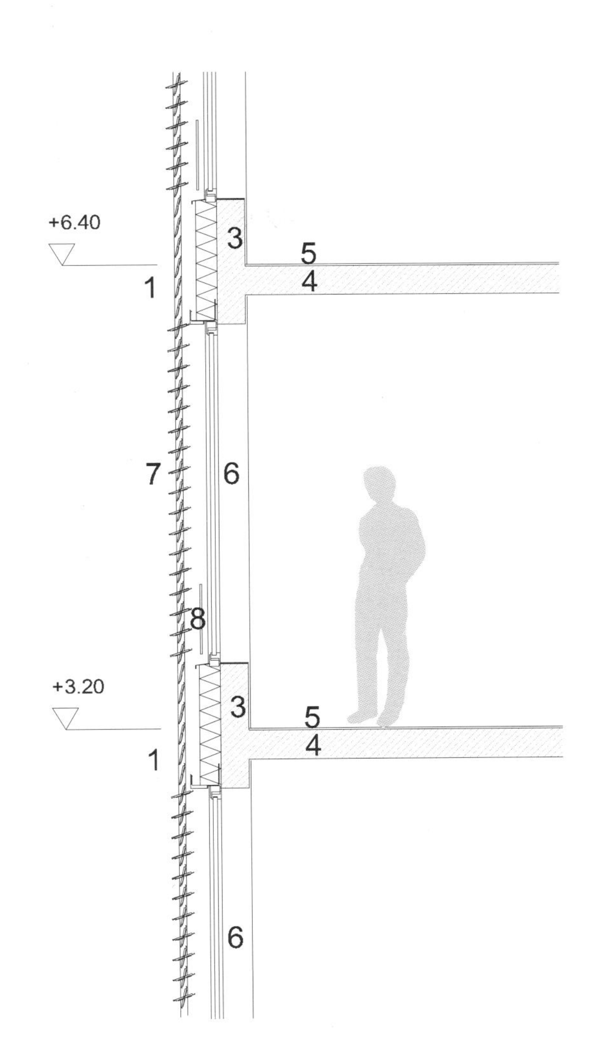

LEGENDE:

1 Lamelles en
système typ
2 Isolation par
3 Allège en be
4 Dalle porteu
5 Revetement
6 Ouvrant dou
7 Lamelles en
système typ
8 Garde corps

Detail1 Facade T3

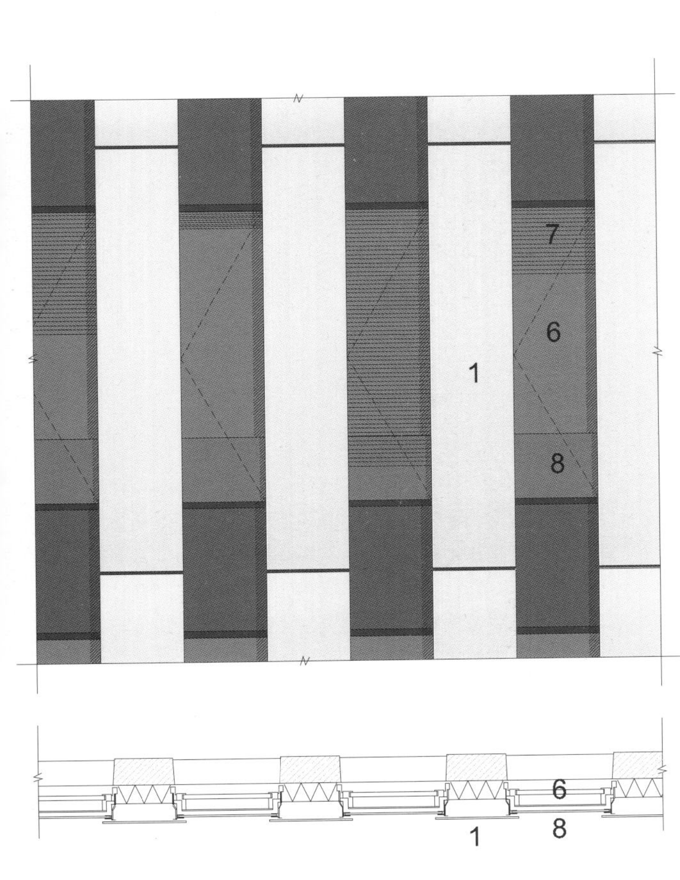

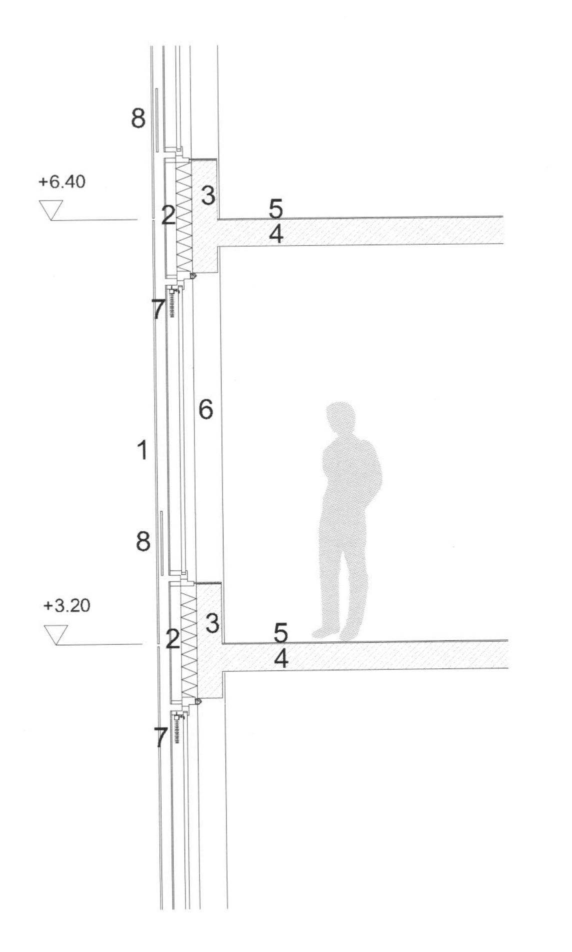

LEGENDE:

1 Beton fibres blanc, isolation
2 Isolation par l'extérieur
3 Allège en beton
4 Dalle porteuse
5 Revetement de sol
6 Ouvrant respirant triple vitrage
7 Store integré
8 Garde corps vitré

Detail2 Facade T3

Monique-Corriveau Library

Location: Québec, Canada
Architect: Dan Hanganu + Côté Leahy Cardas Architects
Client: Ville de Québec, arrondissement Sainte-Foy – Sillery - Cap-Rouge
Size: 4,400 m² (3 levels)
Completion Date: 2013
Photographer: Stéphane Groleau

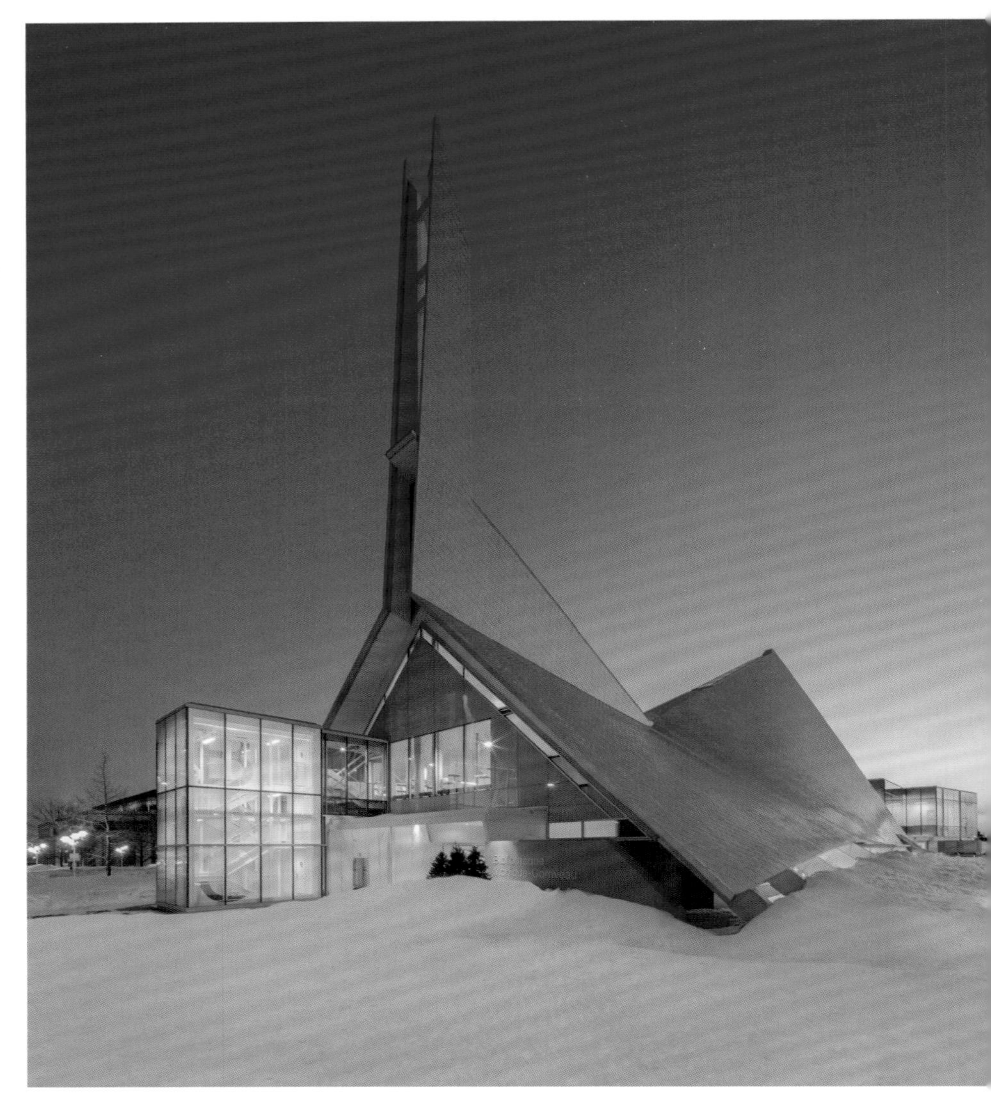

The Monique-Corriveau Library, housed in the Saint-Denys-du-Plateau church, is an exception, and in a rather unusual way. It is a tribute to the career—exceptional for her time—of the Quebec writer whose name it honors. This mother of 10 children, to each of whom she dedicated a book, was the author of numerous children's books and winner of several literary awards.

Converting and expanding such an eloquent example of modern Quebec architectural heritage is a very delicate operation which must be approached with respect and humility. Saint-Denys-du-Plateau Church deserves this special consideration due to its unusual, dynamic volume, which evokes a huge tent inflated by the wind and anchored to the ground with tensioners.

The nave houses the library's public functions, with shelves and work and reading areas, while the addition contains the administration and community hall. This separation of functions means that the community hall can be kept open outside library opening hours, while the spectacular and monumental volume of the nave is preserved, since the architectural concept is to transform the space into a model of spatial appropriation as a reinterpretation of the interior.

To accentuate the fluidity of this volume, the solid soffit above the window has been replaced by glass panel which allows each beam to visually slip seamlessly to its exterior steel base—a revelation of visual continuity.

The volume replacing the presbytery and community hall occupies the same footprint and was executed in clear, silk-screened and colored glass panels. It is separated from the library by a void, marking the transition from old to new. At the front, extending the structure of the choir-screen and the canopy, a code-required emergency staircase is housed in a colored glass enclosure signalling the new place, dominating a new parvis, reconfigured with street furniture, trees and other greenery. Building on transparency and reflection, the architects have made a strong statement with color at the ends of the building, an allusion to the vibrant, bold colors of the 1960s, which contrast the whiteness and brilliance newly captured in the remarkable form of the original church.

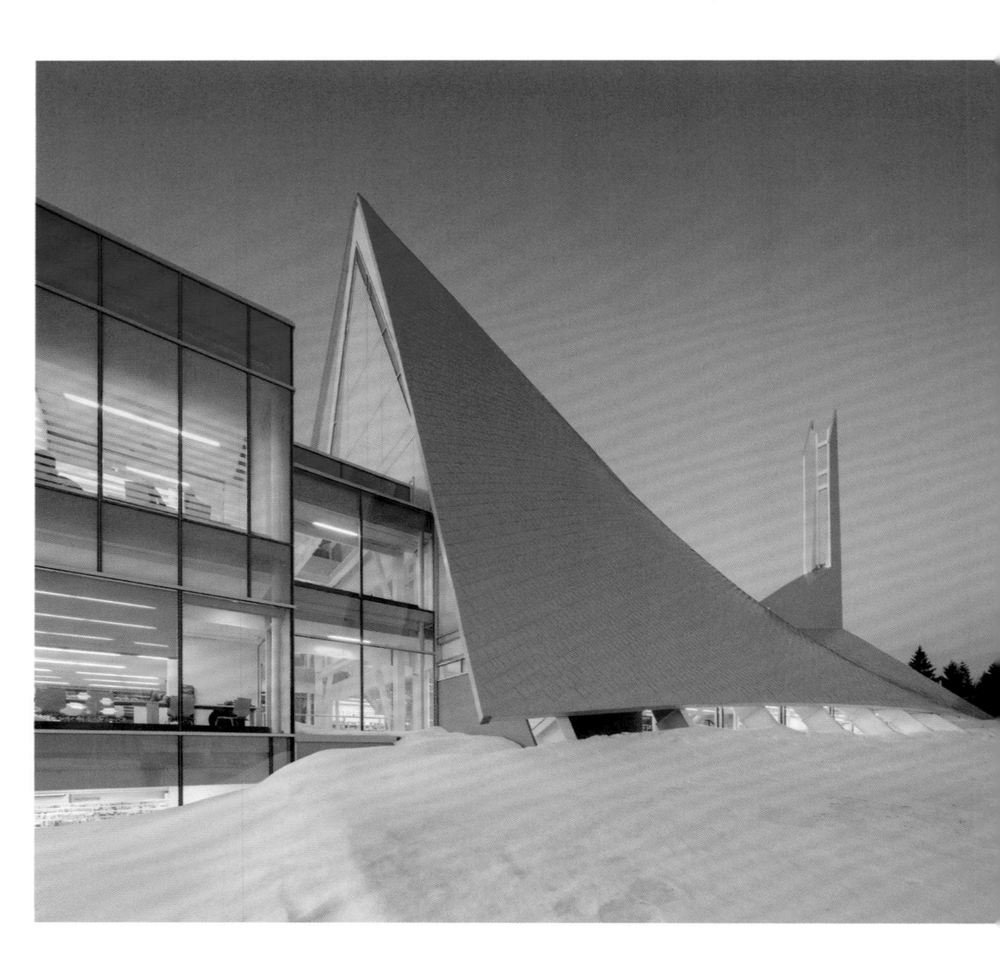

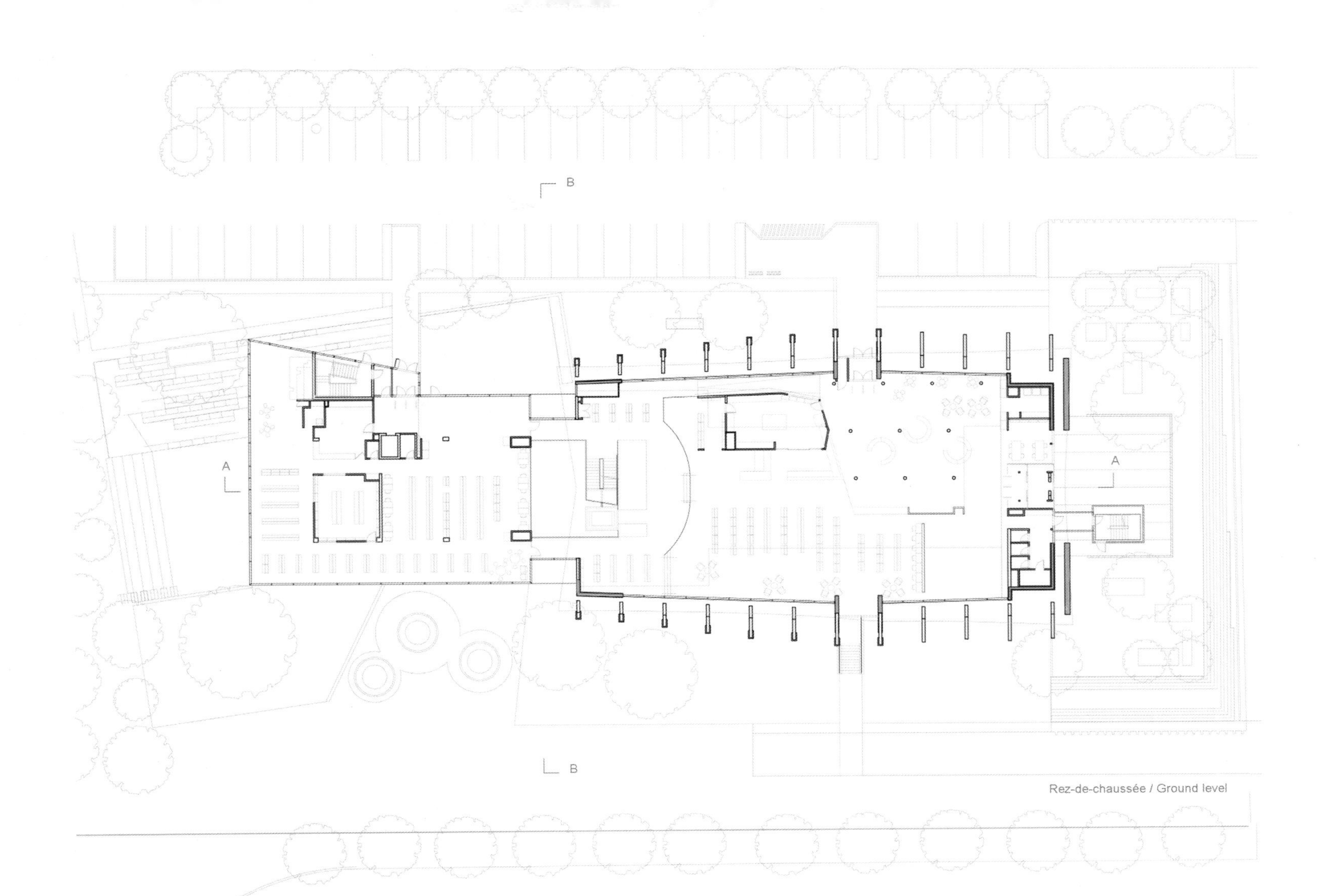

Rez-de-chaussée / Ground level

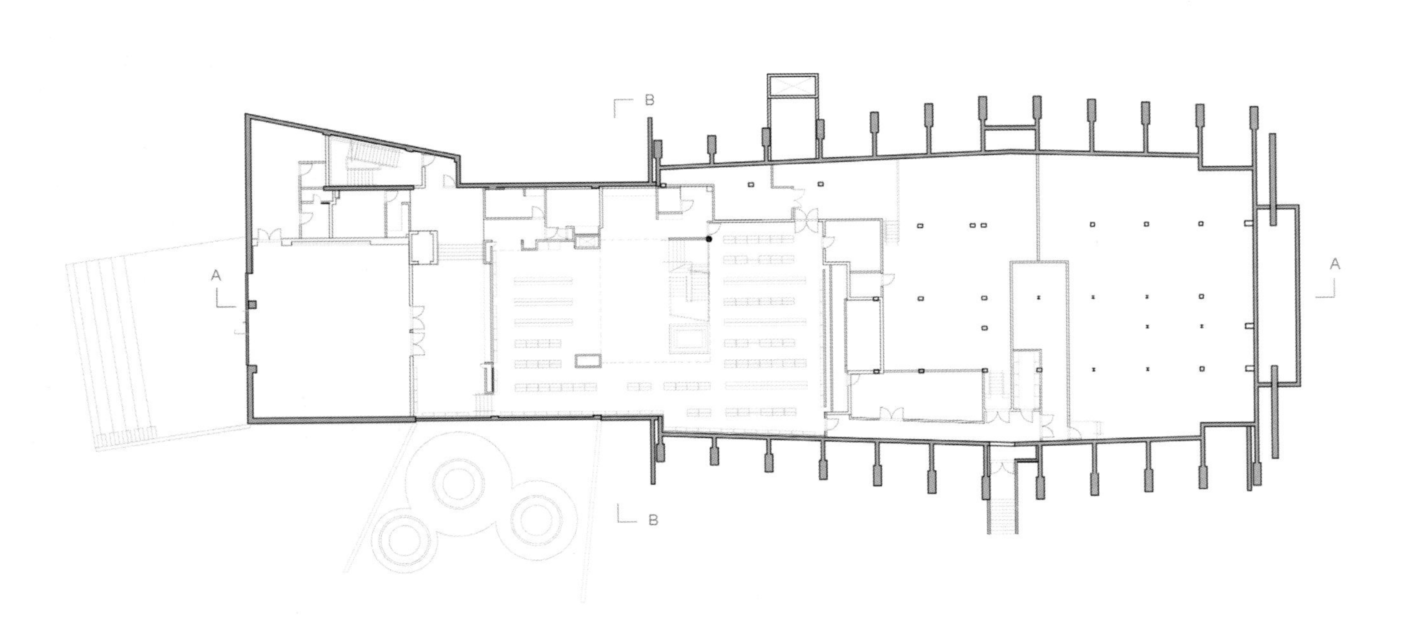

Sous-sol/ Lower level

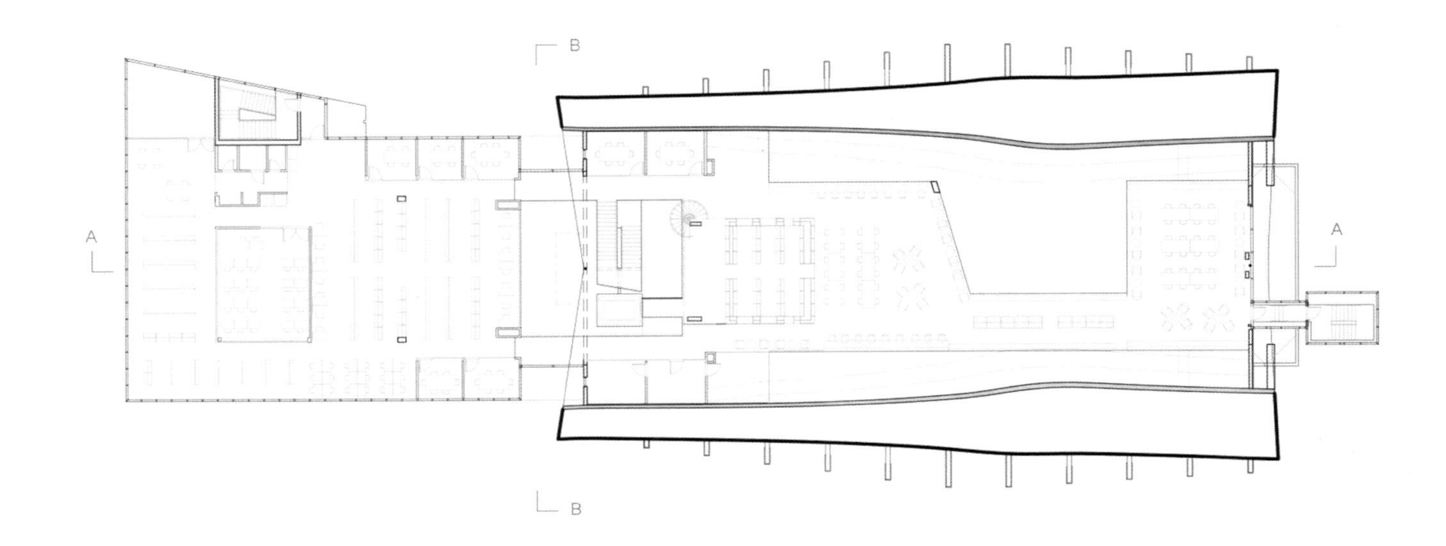

Étage / Upper level

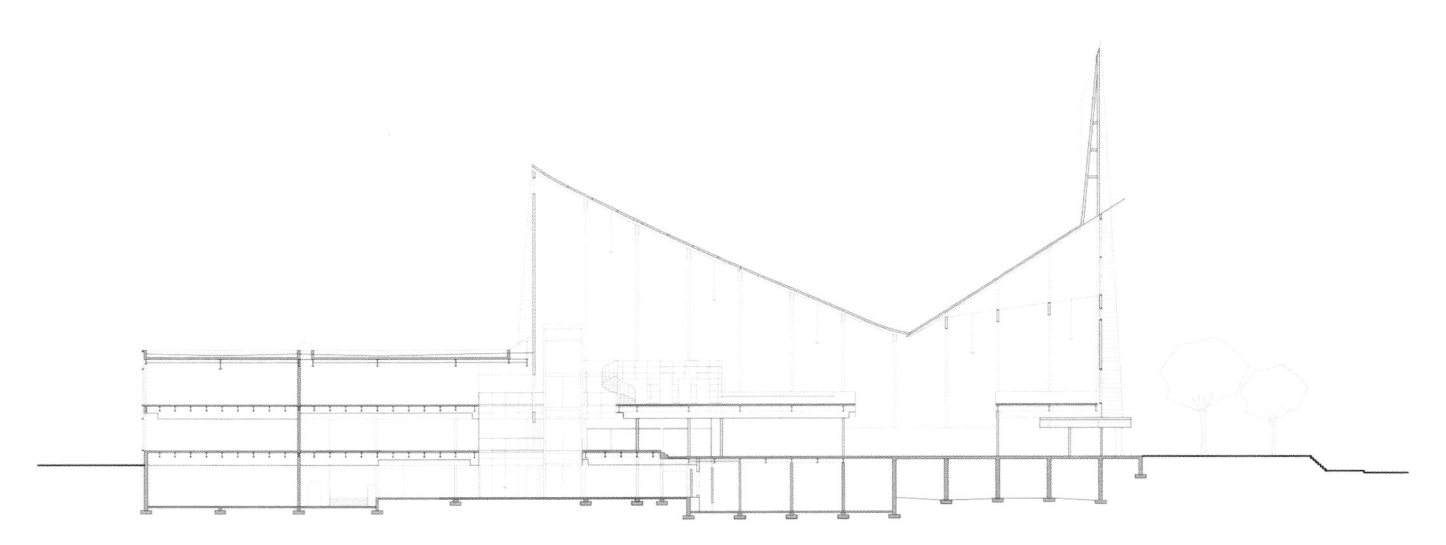

Coupe / Section A

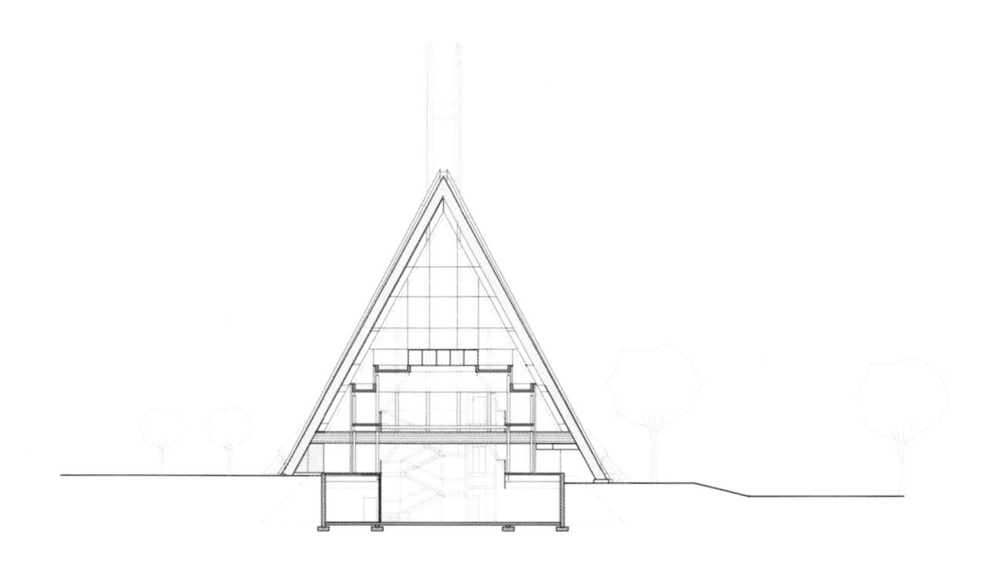

Coupe / Section B

01
Acier Galvalume motif " tôle à la canadienne "
Membrane de toit
Contreplaqué
3 couches de lambourdes
Isolant rigide
Platelage et structure de toit existante
02
Verre clair
03
Nouvelles assises
en acier galvanisé
pour chevrons existants

01
Galvalume pattern " tôle à la canadienne "
Roof membrane
Plywood
3 layers of wood backing strips
Rigid insulation
Existing decking and wood structure
02
Clear glass
03
New galvanized steel
base for
existing wood rafters

Détail / Detail

Ofunato Civic Center and Library

Location: Ofunato city, Iwate, Japan
Architect: Chiaki Arai Urban and Architecture Design
Client: Ofunato City
Site Area: 25,613.70 m^2
Gross Floor Area: 9,290.39 m^2
Completion Date: 2008

Located in Ofunato, Japan, Ofunato Civic Center and Library is a cultural building complex which consists of a main hall with 1100 seats, a library, multi-purpose spaces, an atelier, a tea room, and a studio. In its development, regional workshops and fieldwork have been organized by more than 50 times to make communication among architects, local residents and public officers. Following those workshops, programs of each function were verified, and a library was added as per requests from its locals. How to represent its local form has also been discussed in order to design a new iconic community space. In workshops, architects were required not to adhere to any particular forms or ideas, and to discern architecture, the form of space, from something ambiguous. Through many events, the power in its creating process of architecture motivated the local involvement in the project. Consequently, it has attracted more than 20,000 people every month in a city of 40,000 people since it was inaugurated.

In the exploration of possibilities in this project, the architects interpreted the course of history and consider streams of both consistency and inconsistency in the search for a new order and space. They attempted to construe the "consistent inconsistencies" and "inconsistent inconsistencies" through natural creations. They also interpreted the site-specific locality of the land through workshops and various interactive events with the end users. They cultivated its topophilia and analyzed contemporaneousness to write the design script. Rather than to conform these factors into a consistent space, the architects accepted each inconsistency and subsequently reflected directly into to the architectural form resulting in a dynamic spatial formation of "inconsistent inconsistency".

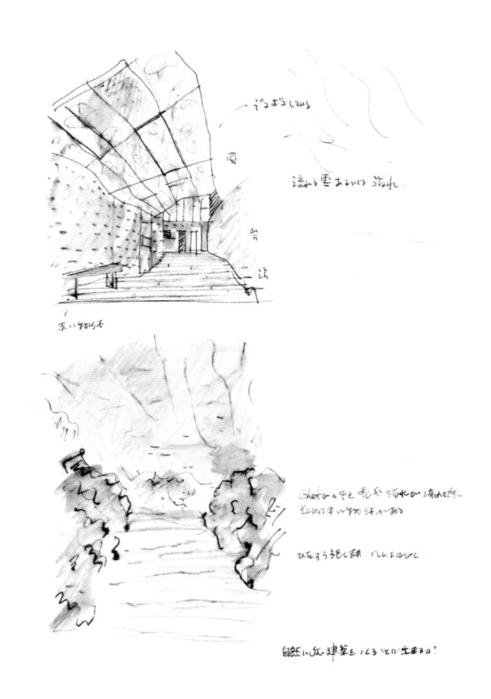

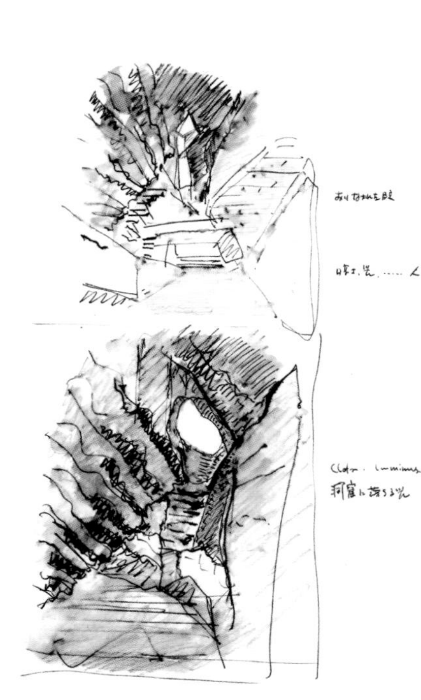

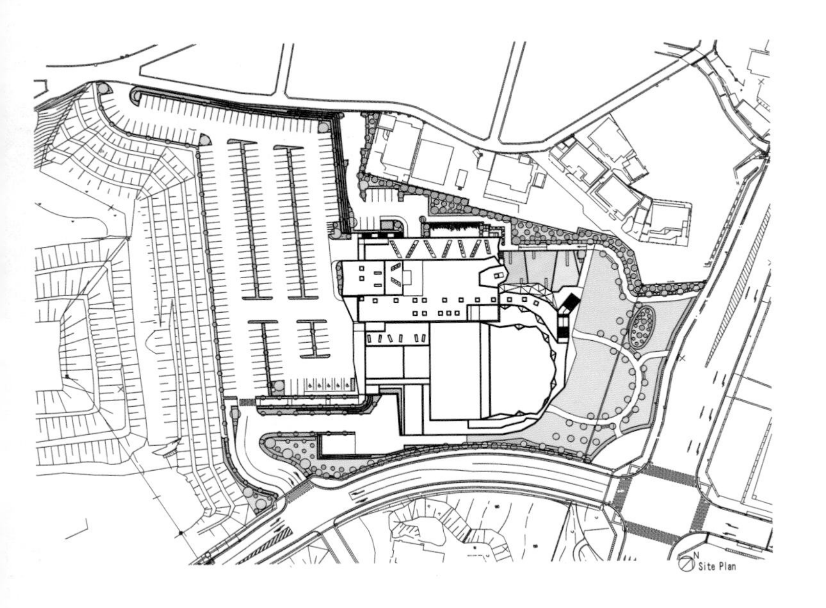
Site Plan

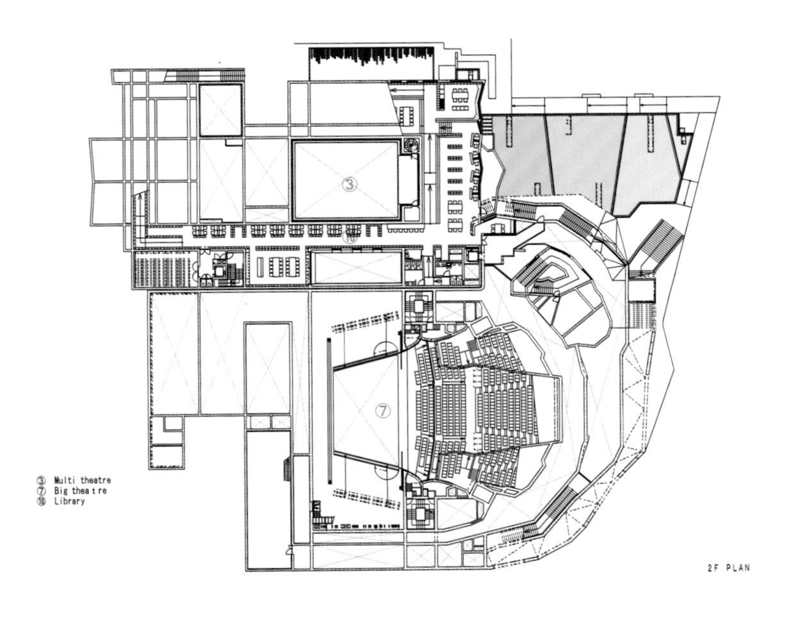
③ Multi theatre
⑦ Big thea t re
⑩ Library
2F PLAN

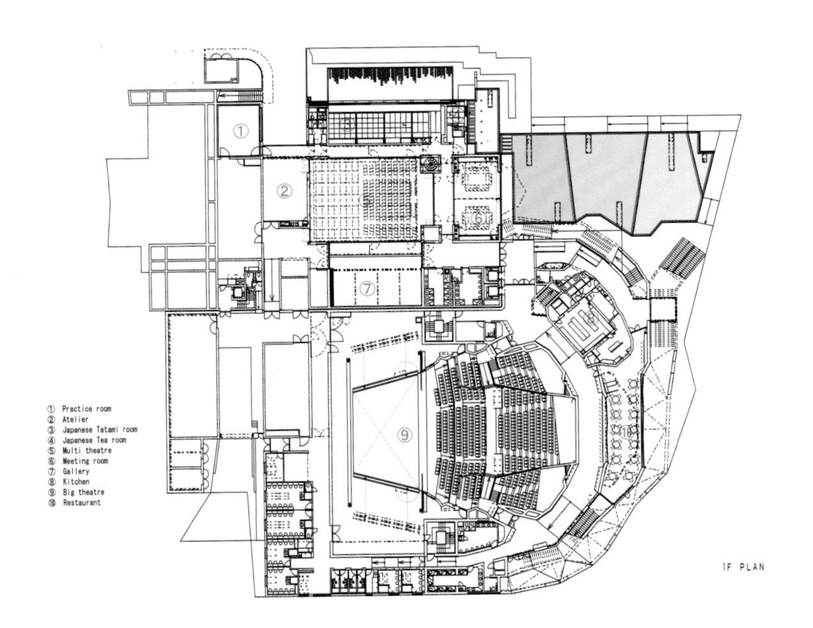
① Practice room
② Atelier
③ Japanese Tatami room
④ Japanese Tea room
⑤ Multi theatre
⑥ Meeting room
⑦ Gallery
⑧ Kitchen
⑨ Big theatre
⑩ Restaurant
1F PLAN

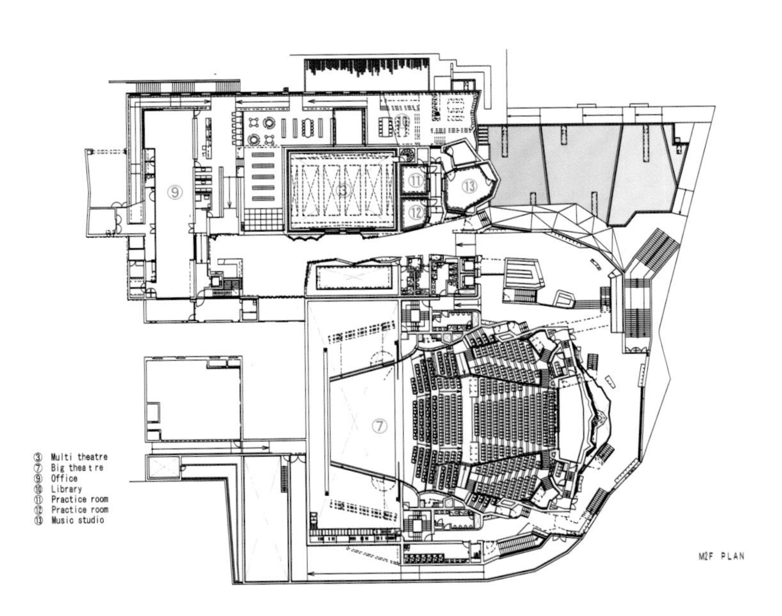
③ Multi theatre
⑦ Big thea t re
⑨ Office
⑩ Library
⑪ Practice room
⑫ Practice room
⑬ Music studio
M2F PLAN

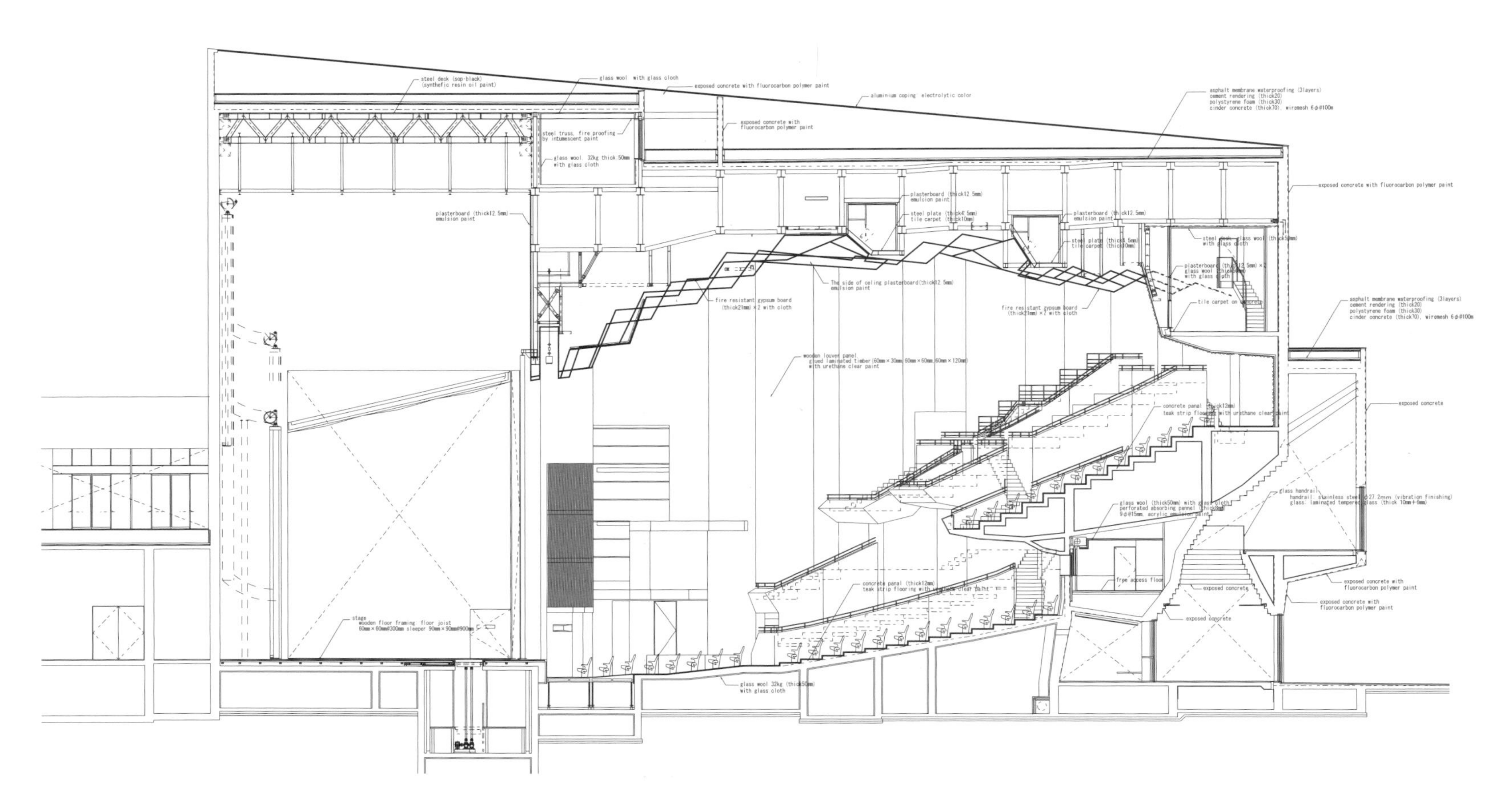

Theatre Section 1

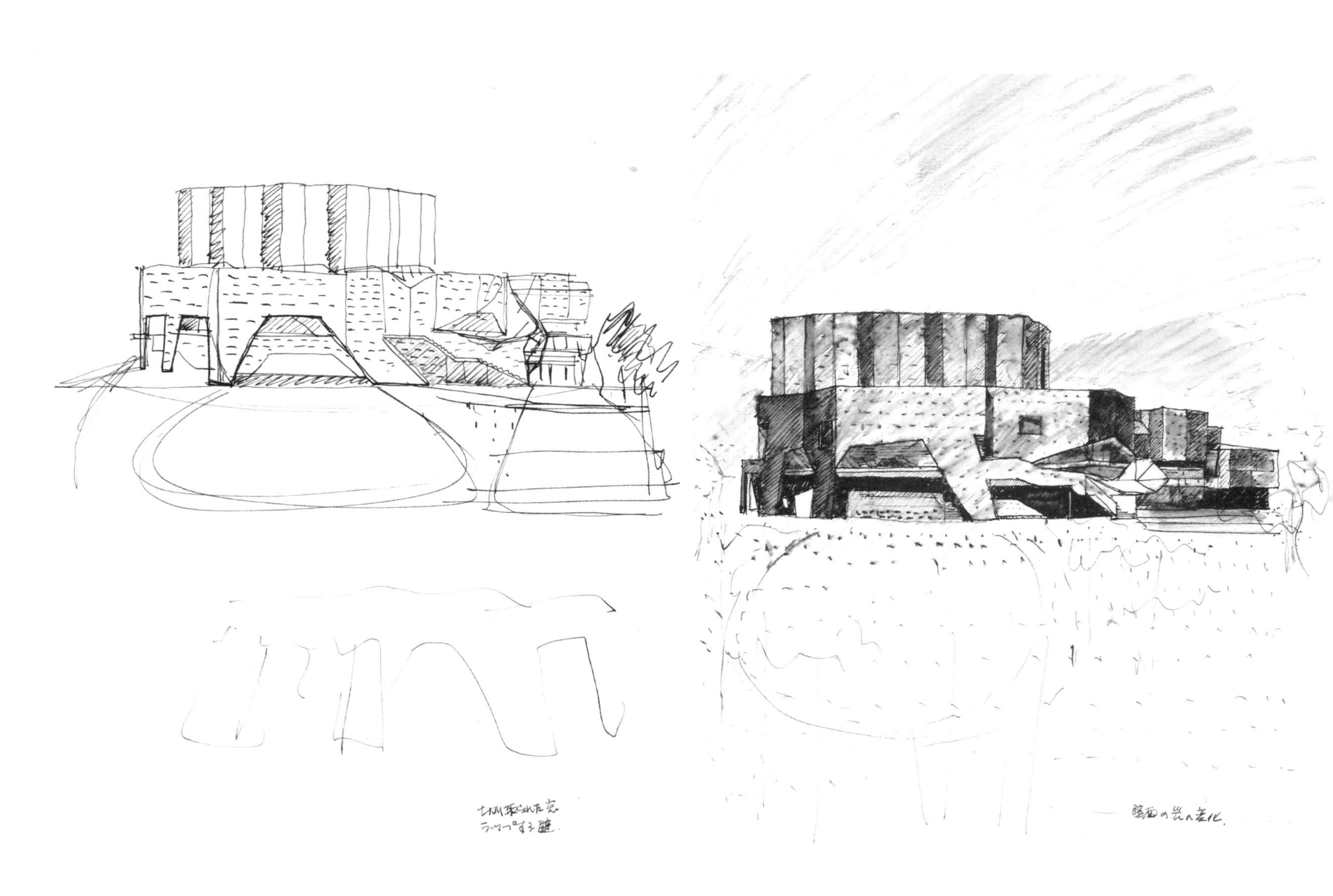

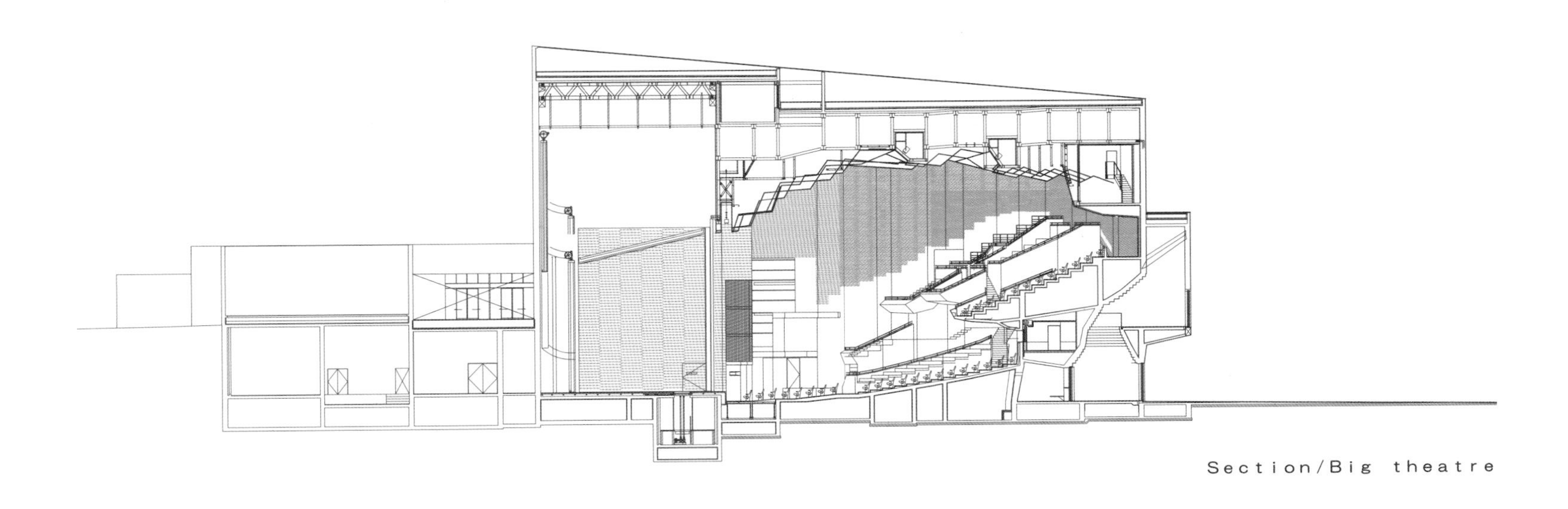

Section/Big theatre

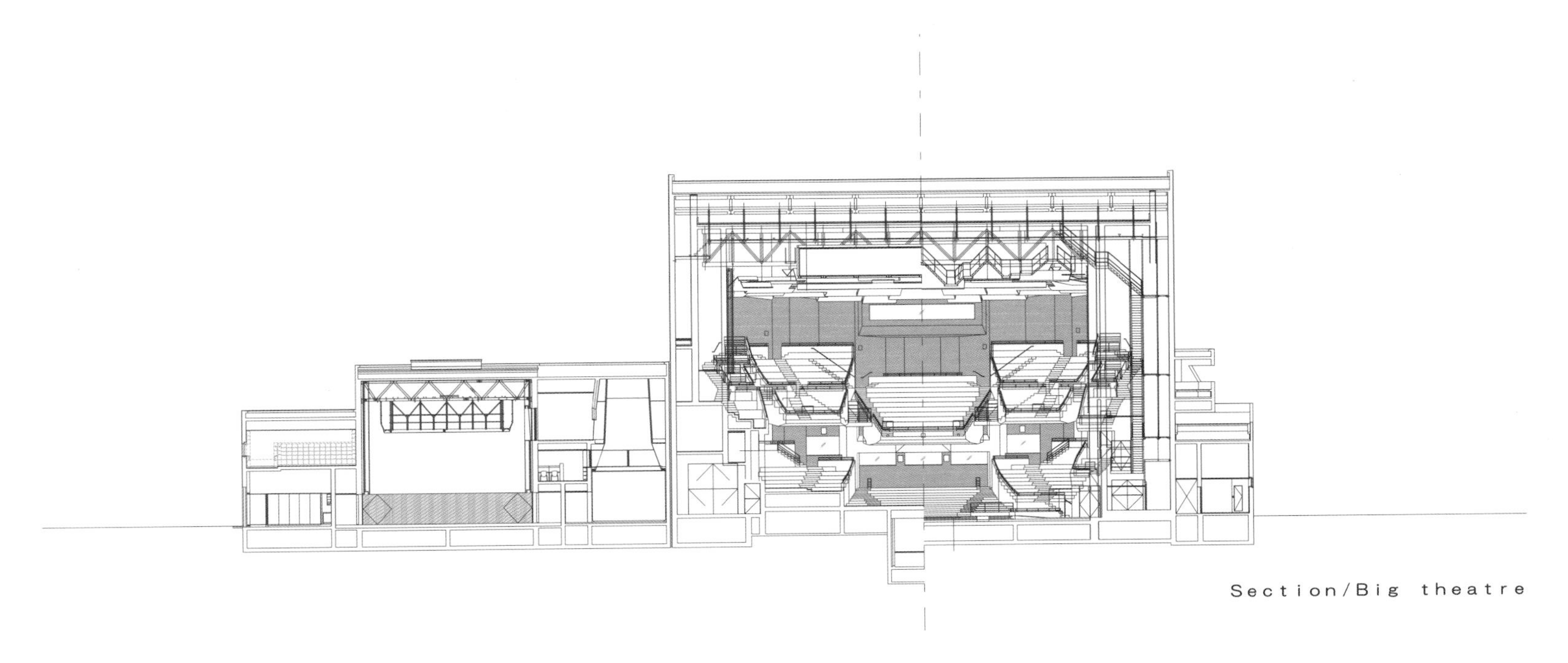

Section/Big theatre

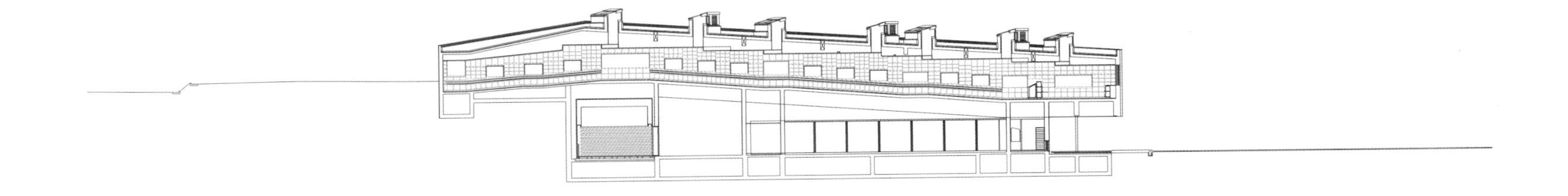

Section/Library

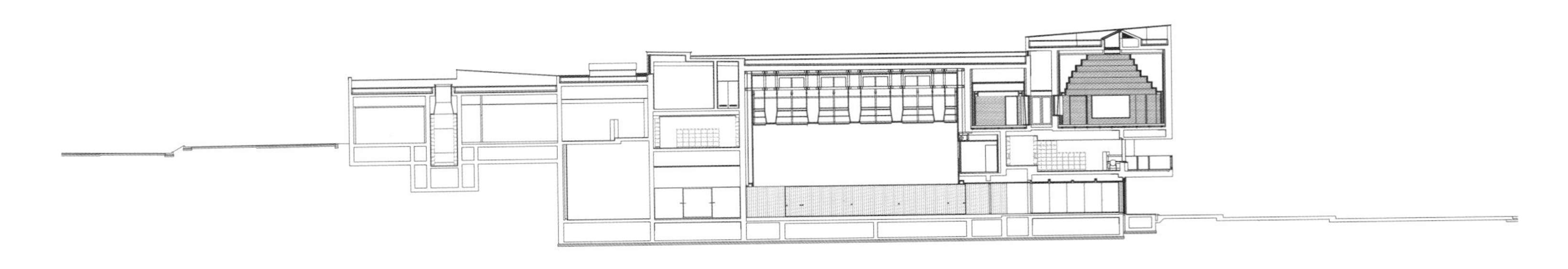

5 10 15m

Section/Muluti theatre

Jockey Club Innovation Tower

Location: Hong Kong, China
Architect: Zaha Hadid Architects
Client: Hong Kong Polytechnic University
Size: 15,000 m^2
Capacity: 1800 students & staff
Completion Date: 2014
Photography: Iwan Baan, Virgile Simon Bertrand, Doublespace

The Jockey Club Innovation Tower (JCIT) is home to the Hong Kong Polytechnic University (PolyU) School of Design and the Jockey Club Design Institute for Social Innovation.

The 15-storey, 15,000 m^2. m. tower accommodates more than 1,800 students and staff, with facilities for design education and innovation that include: design studios, labs and workshops, exhibition areas, multi-functional classrooms, lecture theatre and communal lounge.

The Hong Kong Polytechnic University campus has developed its urban fabric over the last 50 years with the university's many faculties housed in visually coherent, yet very different buildings.

The JCIT creates a new urban space that enriches the diversity of university life and expresses the dynamism of an institution looking to the future.

Located on a narrow, irregular site at the northeastern tip of the university campus (bordered by the university's football ground to the south, and the Chatham Road/ Kowloon Corridor motorway interchange to the north), the JCIT is connected to the heart of the campus; encouraging the university's various faculties and schools to develop multidisciplinary initiatives and engagement with the community, government, industry, NGO's and academia.

The JCIT design dissolves the typical typology of the tower/podium into a more fluid composition. Interior and exterior courtyards create informal spaces to meet and interact, complementing the large exhibition forums, studios, theatre and recreational facilities.

The tower's design promotes a multidisciplinary environment by connecting the variety of programs within the School of Design; establishing a collective research culture where many contributions and innovations can feed off each other.

Students, staff and visitors move through 15 levels of studios, workshops, labs, and exhibition and event areas within the school. Interior glazing and voids bring transparency and connectivity, while circulation routes and communal spaces have been arranged to encourage interaction between the many learning clusters and design disciplines.

With its contribution of HK$249 million towards the construction of JCIT, The Hong Kong Jockey Club Charities Trust also funds the Jockey Club Design Institute for Social Innovation.

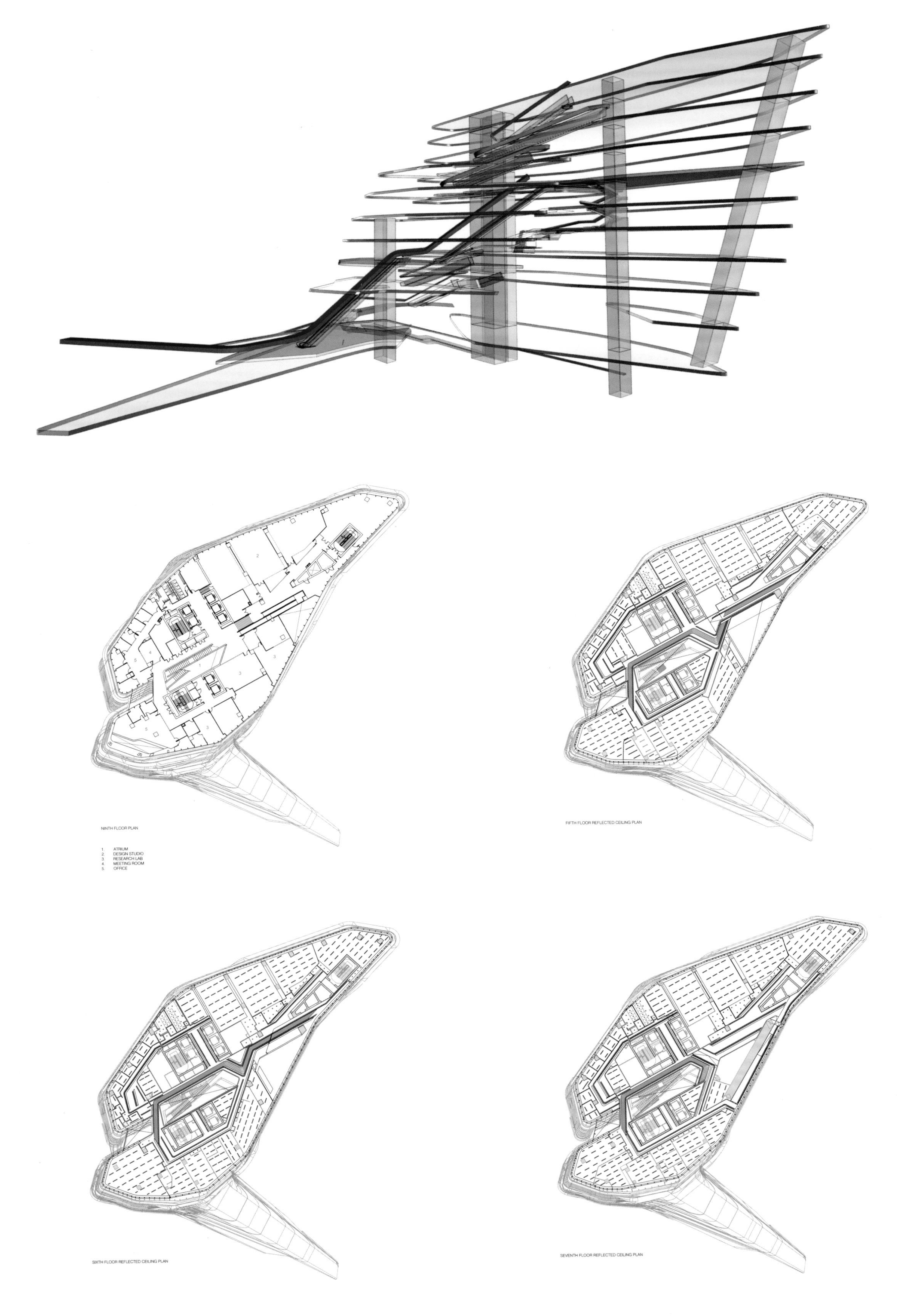
NINTH FLOOR PLAN
1. ATRIUM
2. DESIGN STUDIO
3. RESEARCH LAB
4. MEETING ROOM
5. OFFICE
FIFTH FLOOR REFLECTED CEILING PLAN
SIXTH FLOOR REFLECTED CEILING PLAN
SEVENTH FLOOR REFLECTED CEILING PLAN

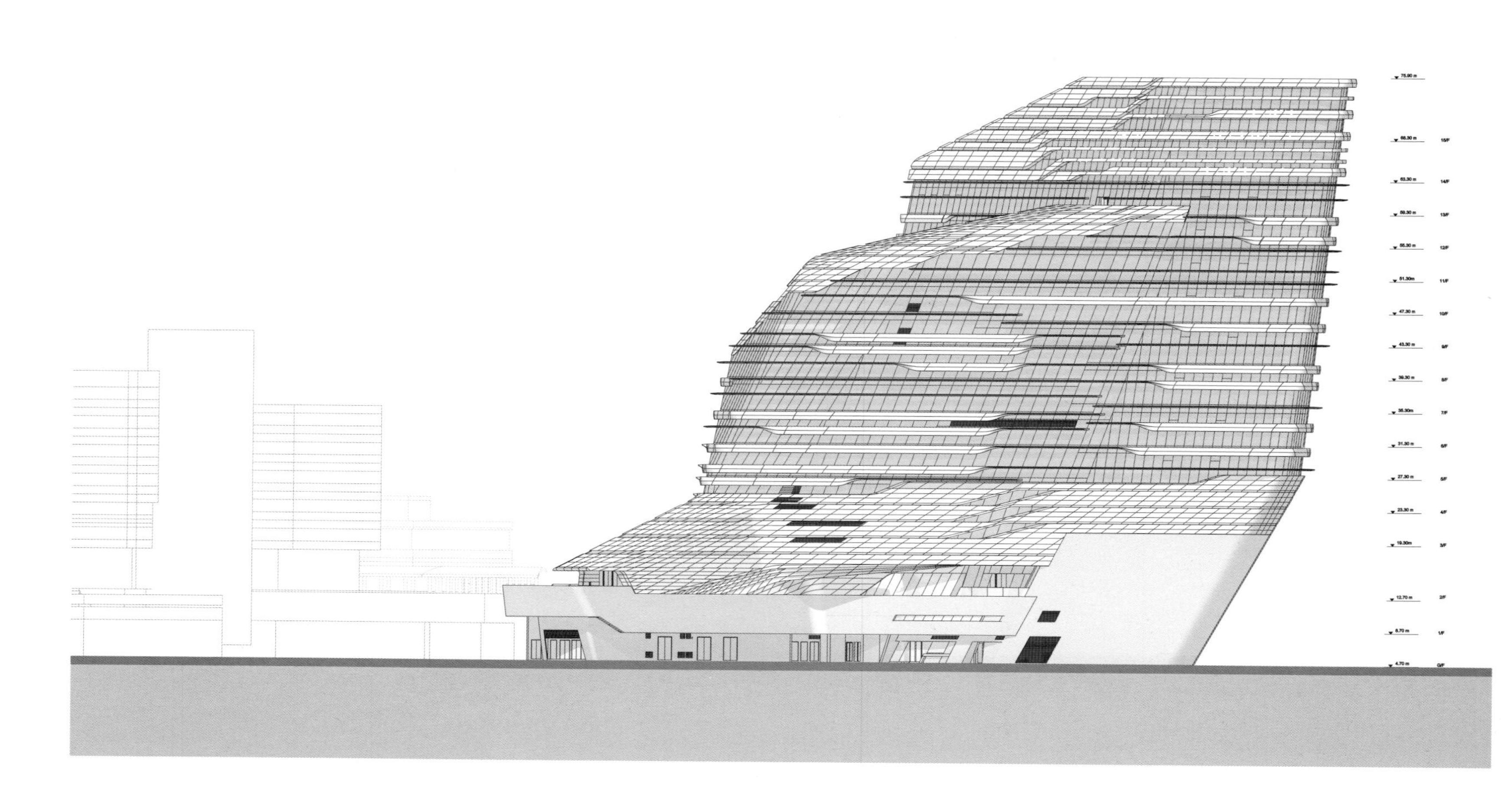
75.90 m
68.30 m 15/F
63.30 m 14/F
59.30 m 13/F
55.30 m 12/F
51.30m 11/F
47.30 m 10/F
43.30 m 9/F
39.30 m 8/F
35.30m 7/F
31.30 m 6/F
27.30 m 5/F
23.30 m 4/F
19.30m 3/F
12.70 m 2/F
8.70 m 1/F
4.70 m G/F

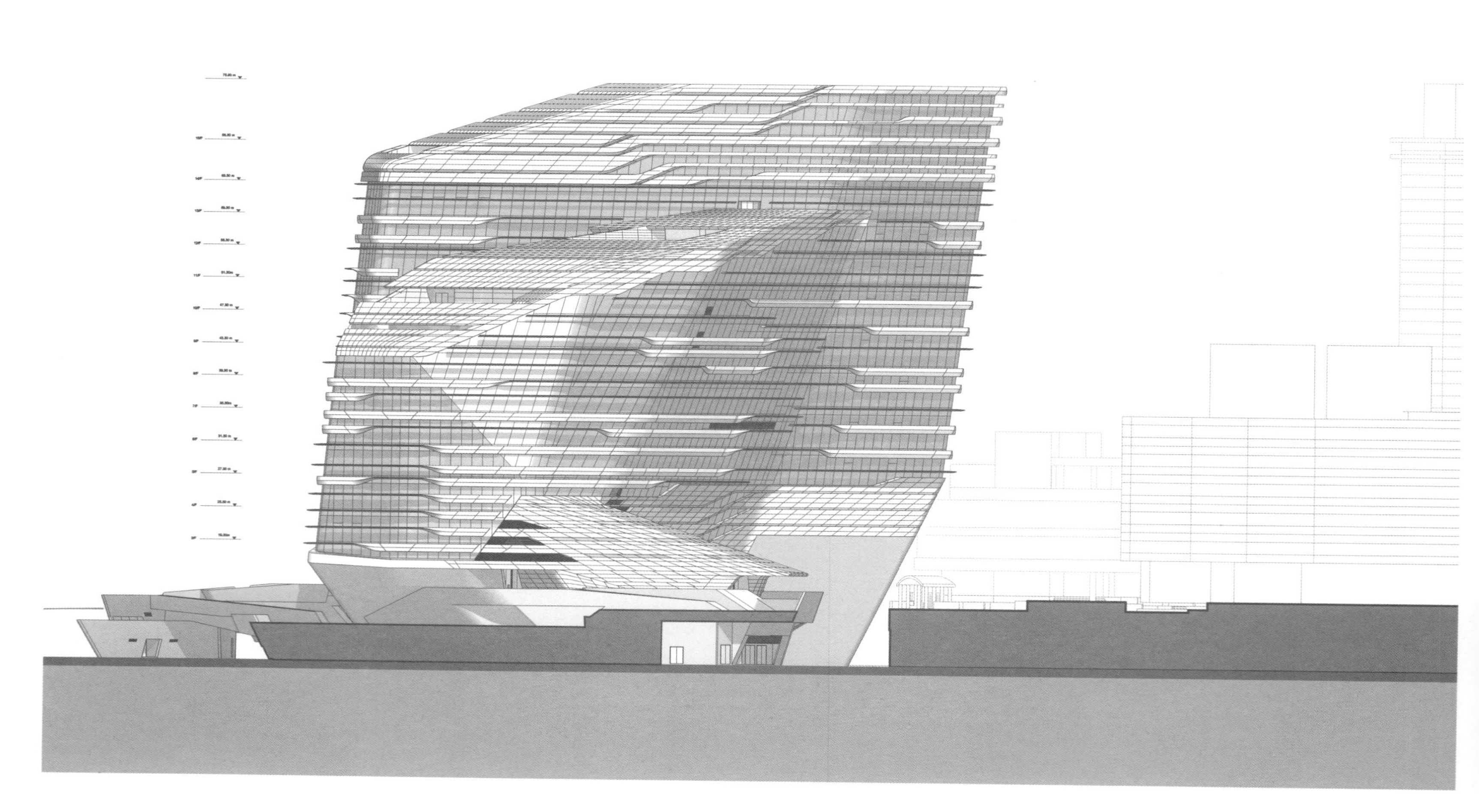

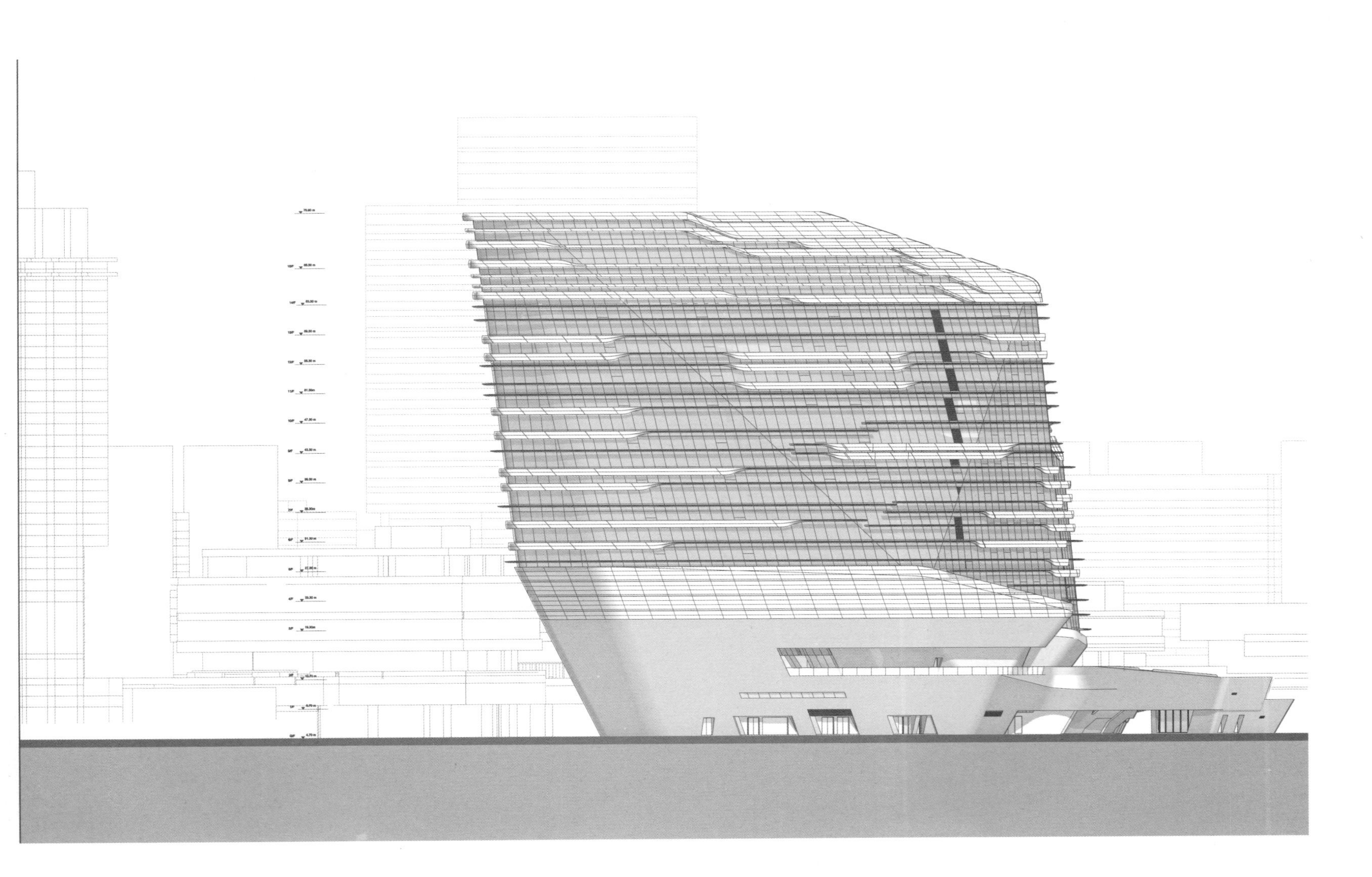

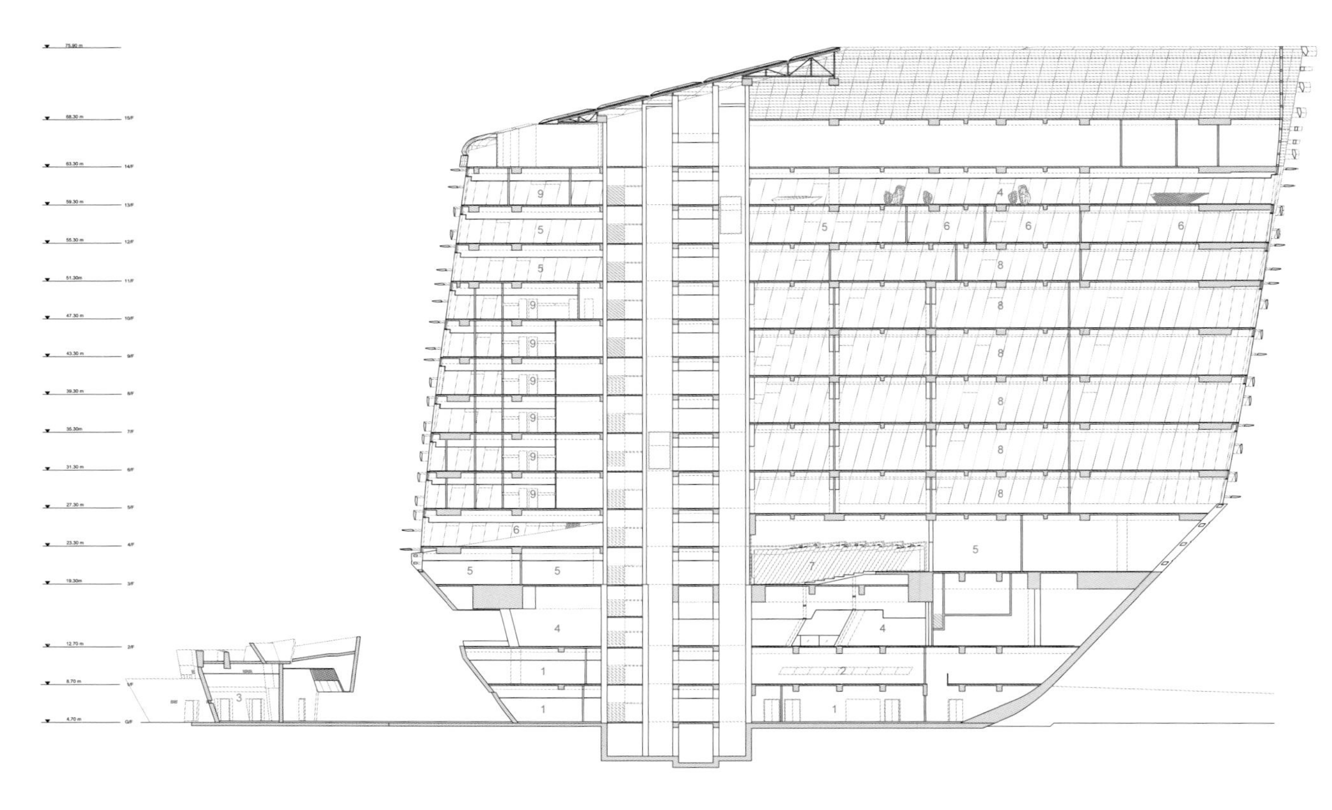

1. Workshop
2. Tri-use studio
3. Car design lab
4. Exhibition
5. Classroom
6. Project space
7. Lecture theatre
8. Design studio
9. Office

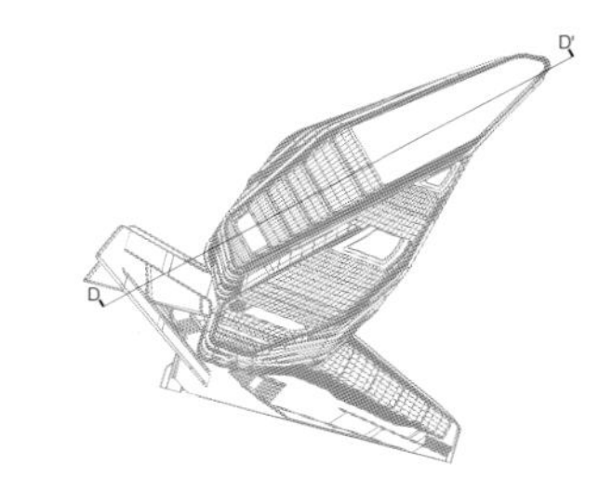

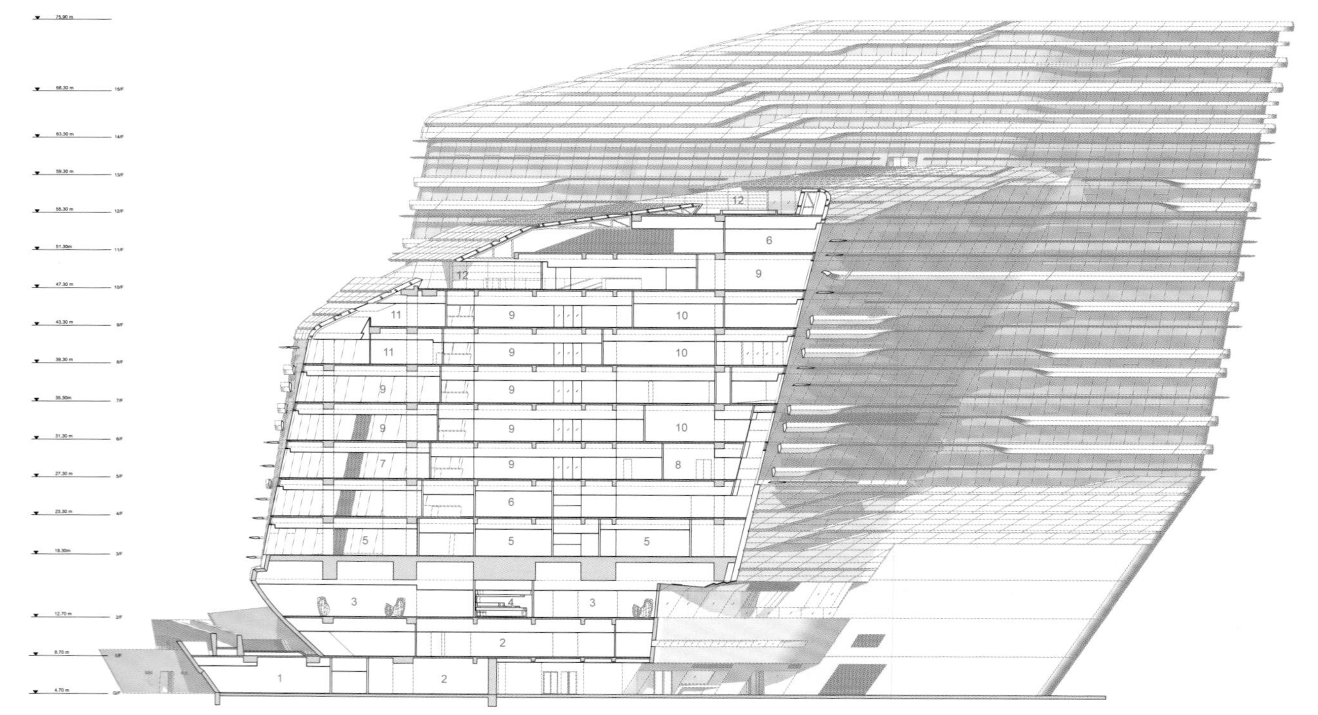

1. Photoshoot studio
2. Workshop
3. Exhibition
4. Shop
5. Classroom
6. Project space
7. Resources library
8. Meeting room
9. Research lab
10. Computer lab
11. Office
12. Terrace

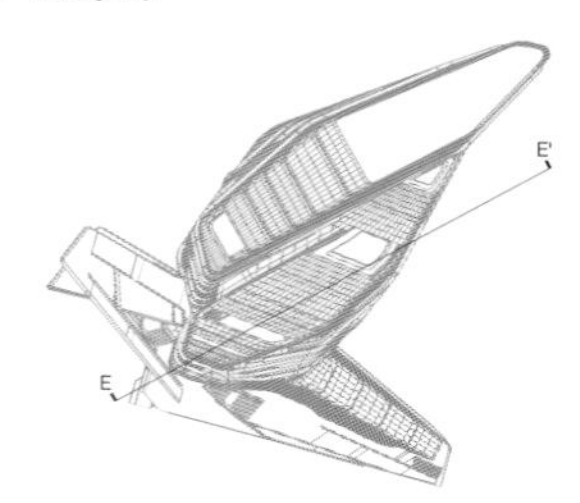

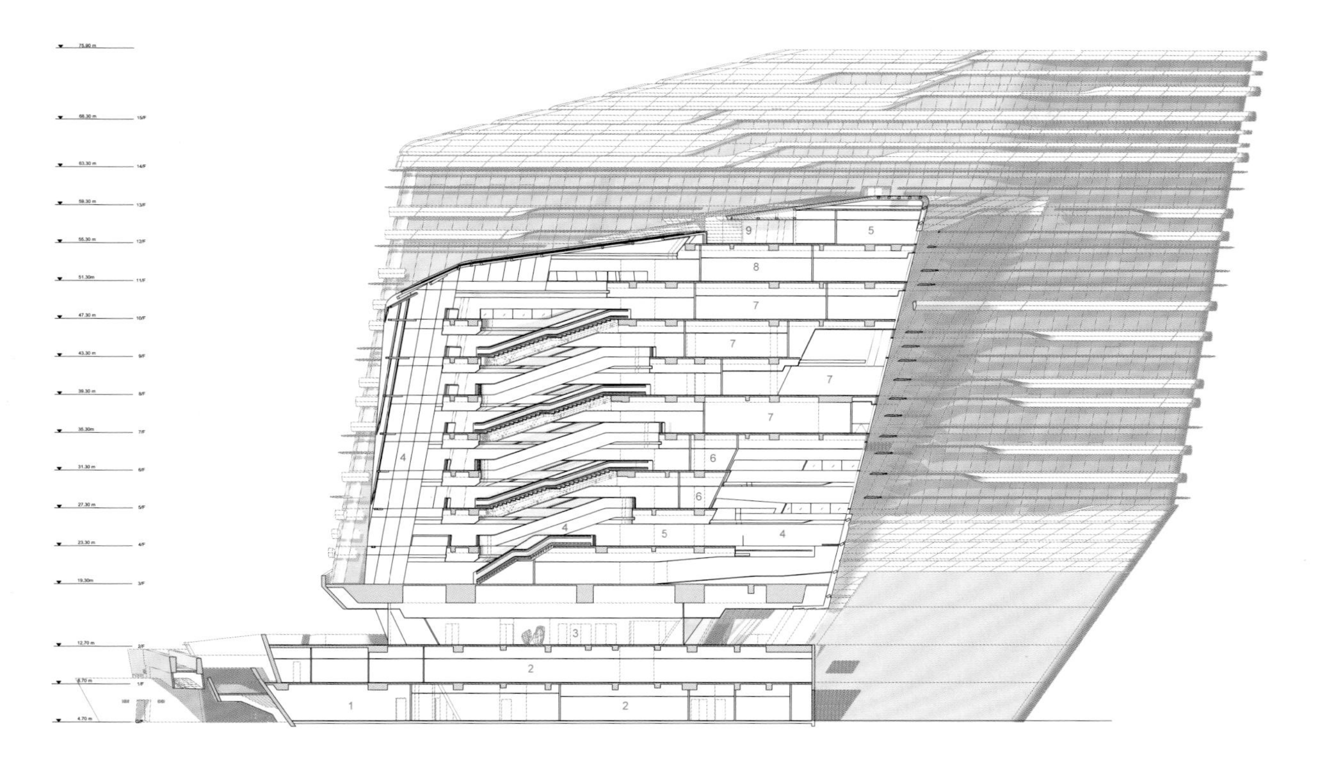

1. Photoshoot studio
2. Workshop
3. Exhibition
4. Atrium
5. Project space
6. Meeting room
7. Design lab
8. Teaching venue
9. Terrace

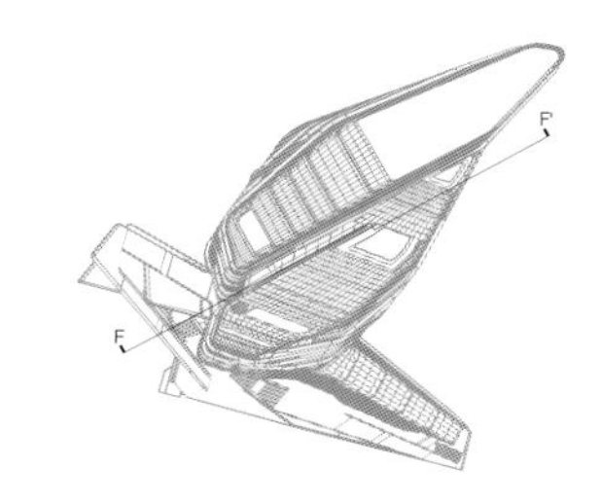

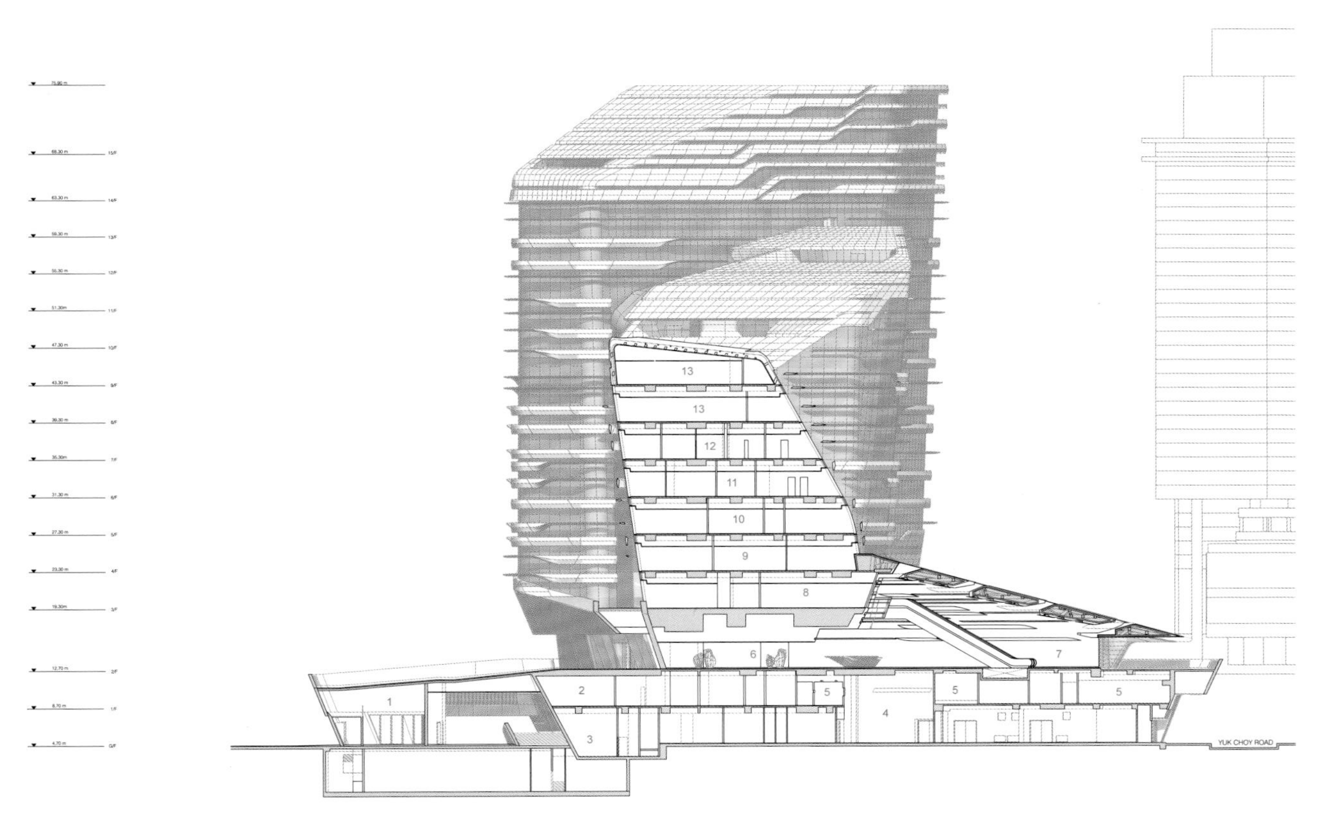

1. Car design lab
2. Clay model shop
3. Photoshoot studio
4. Production workshop/tv studio
5. Audio i video control room
6. Exhibition space
7. Entrance hall
8. Classroom
9. Project space
10. Resources library
11. Design lab
12. Studio
13. Office

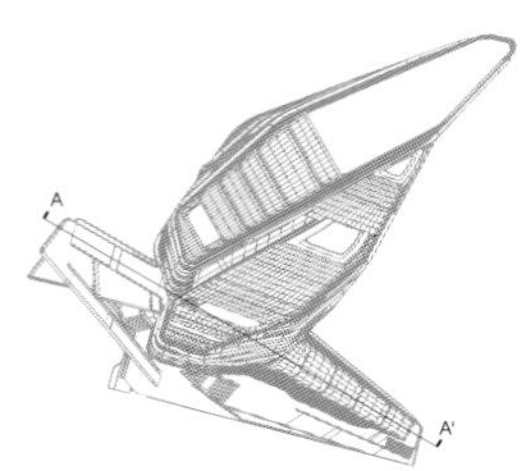

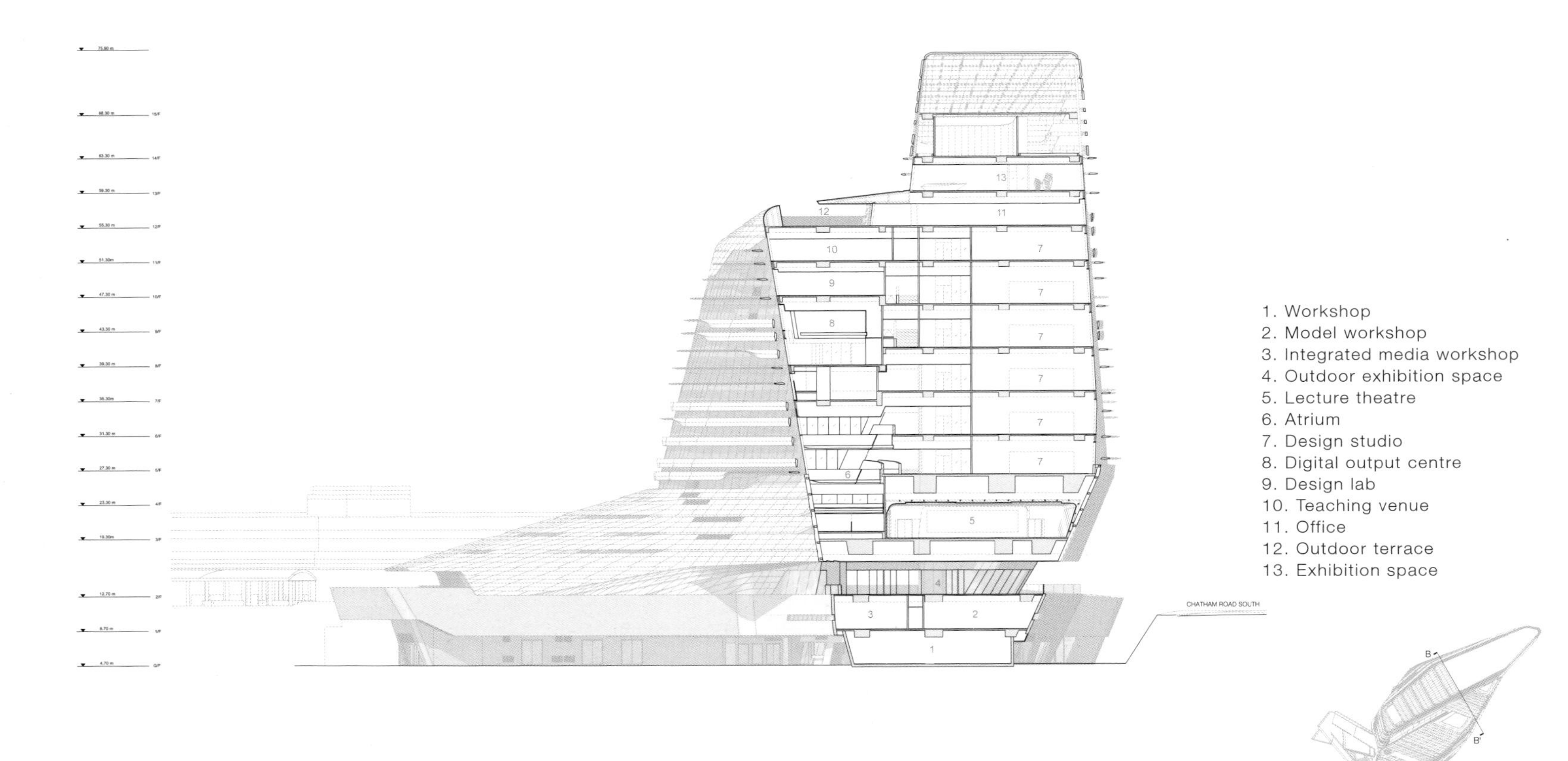

1. Workshop
2. Model workshop
3. Integrated media workshop
4. Outdoor exhibition space
5. Lecture theatre
6. Atrium
7. Design studio
8. Digital output centre
9. Design lab
10. Teaching venue
11. Office
12. Outdoor terrace
13. Exhibition space

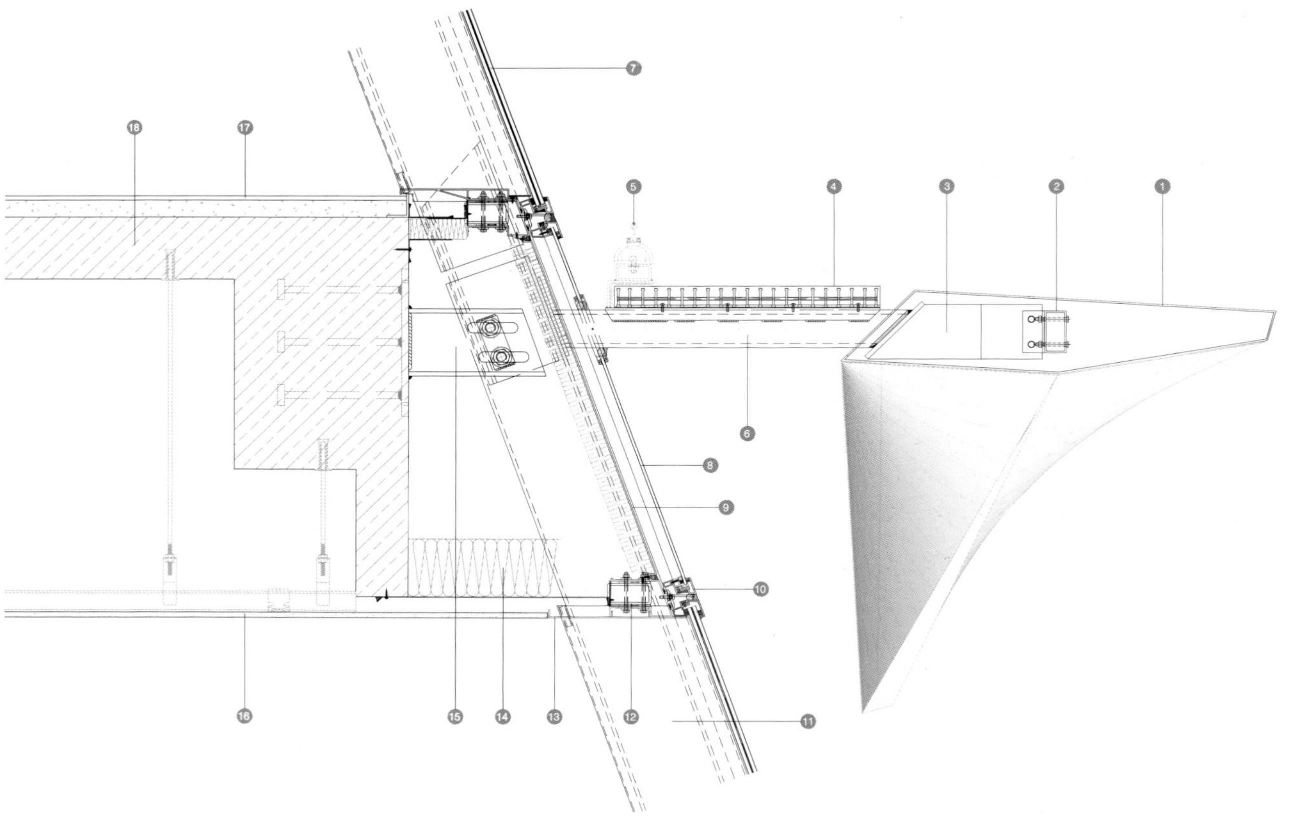

Curtain wall detail

1. 3mm thk. alum, panel; pvf2-3 coats finish
2. 100×50×4mm alum. hollow
3. 8mm thk. alum. bracket
4. 1500×600×45mm alum. gratings
5. stainless steel fall arrest stystem
6.90×10mm galbanized steel c.h.s
7. 12mm grey tinted hst monolithic galss with hard-coat thermal performance coating
8. 8mm grey tinted hst monolithic galss with hard-coat thermal performance coating
9. 3mm thk. alum. backpan with 50mm thk. csr thermal insulation
10. alum. sub-frame with backer fod and weather silicone sealant
11. 200×100×8mm g.m.s. r.h.s
12. 100×50×4mm g.m.s. hollow
13. 3mm thk. alum. caldding
14. 135mm thk. fire stop
15. 160*65*8mm gms channel
16. 12mm thk gypstum board with 5mm skim coat
17. 3mm rubber floor; 7mm self-levering; 40mm 1:3 cement sand screed
18. 150mm thk. reinforced concrete

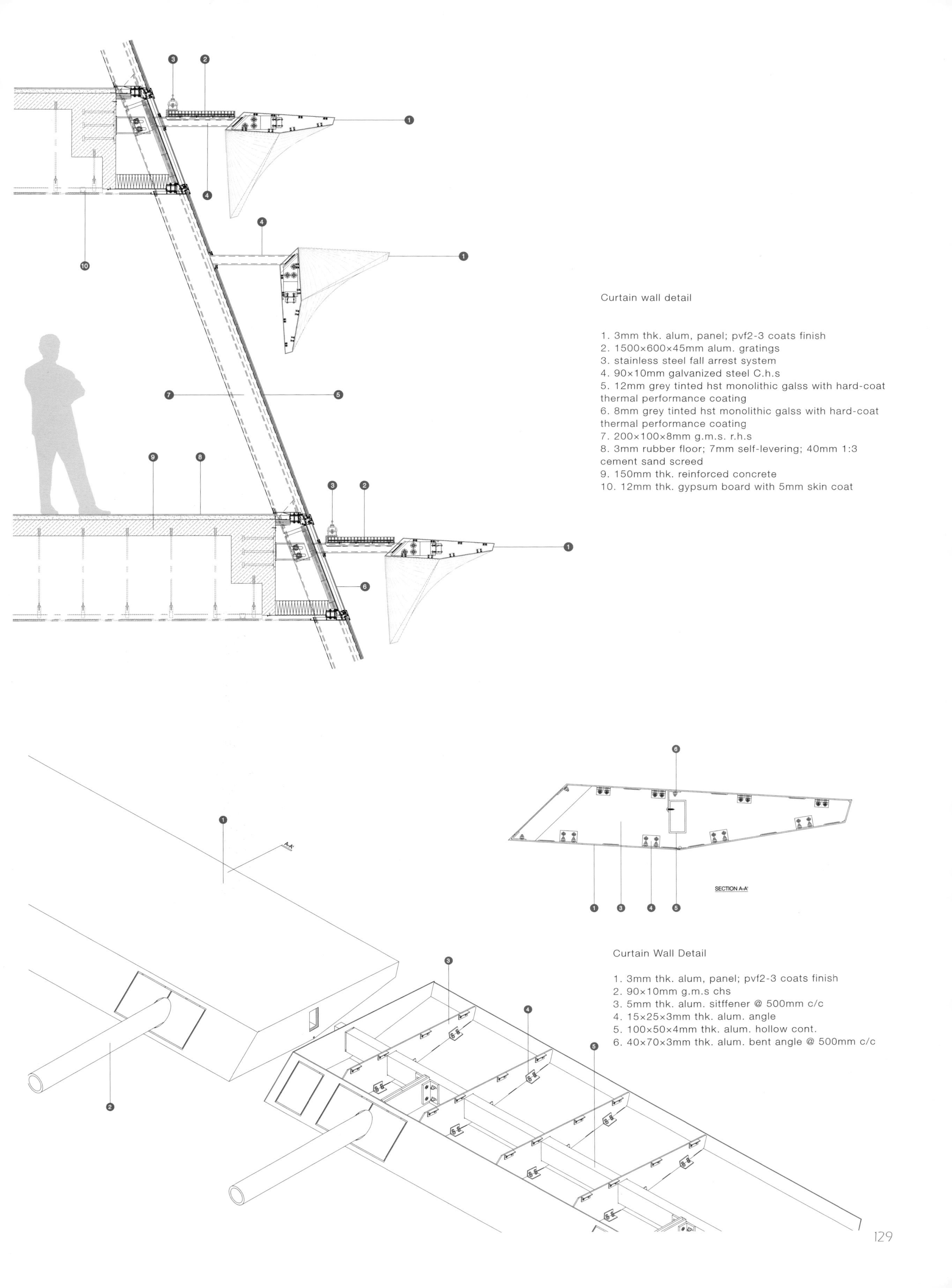
Curtain wall detail
1. 3mm thk. alum, panel; pvf2-3 coats finish
2. 1500×600×45mm alum. gratings
3. stainless steel fall arrest system
4. 90×10mm galvanized steel C.h.s
5. 12mm grey tinted hst monolithic galss with hard-coat thermal performance coating
6. 8mm grey tinted hst monolithic galss with hard-coat thermal performance coating
7. 200×100×8mm g.m.s. r.h.s
8. 3mm rubber floor; 7mm self-levering; 40mm 1:3 cement sand screed
9. 150mm thk. reinforced concrete
10. 12mm thk. gypsum board with 5mm skin coat
SECTION A-A'
Curtain Wall Detail
1. 3mm thk. alum, panel; pvf2-3 coats finish
2. 90×10mm g.m.s chs
3. 5mm thk. alum. sitffener @ 500mm c/c
4. 15×25×3mm thk. alum. angle
5. 100×50×4mm thk. alum. hollow cont.
6. 40×70×3mm thk. alum. bent angle @ 500mm c/c

L'Ourse Library

Location: Dinard, France
Architect: Ricardo Bofill
Completion Date: 2013

At the time of the virtual, the image, the distance between people, L'Ourse Library, a place of culture and trade, is built of stone and glass, thanks to Dinard, its elected officials, to its builders, its architects and engineers. What a good news is that the foundation stone of this project is for a cultural building where the old and the young people can meet. They will perhaps meet New Renan, Oscar Wilde, Isadora Duncan, Jules Verne, illustrious born in Dinard and so on.

Central part of the overall composition, it is architectural counterpoint with its environment. Some simple lines, a large roof wing-shaped, frosted glass frontage and a bright vertical signal are the core of the new district. With a gross floor area of 1,500 m^2, all its functions are grouped on one level ground floor with a floor space of 1,200 m^2. Crossed by an internal street, acting as the hall, it is divided into three areas, a bright and generous reading room, a consultation room and an open living room of conviviality. The main entrance, located at the intersection of Rue des Mimosas and extended by the internal street connects the Library and cultural plaza treated as an amphitheater. This provision allows to extend outdoor the Library's activities and make performances with the building's backdrop stage.

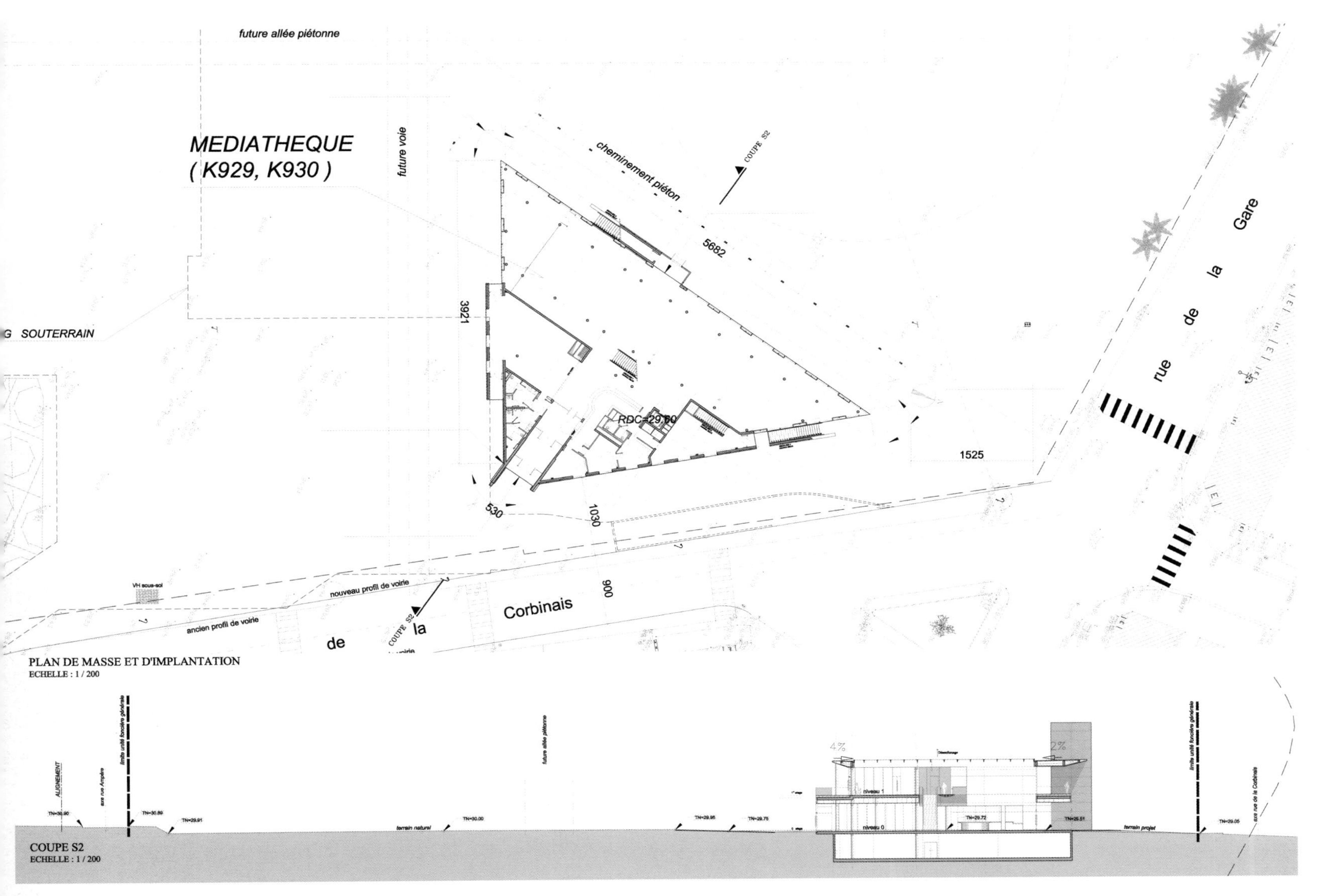

Site Plan-02

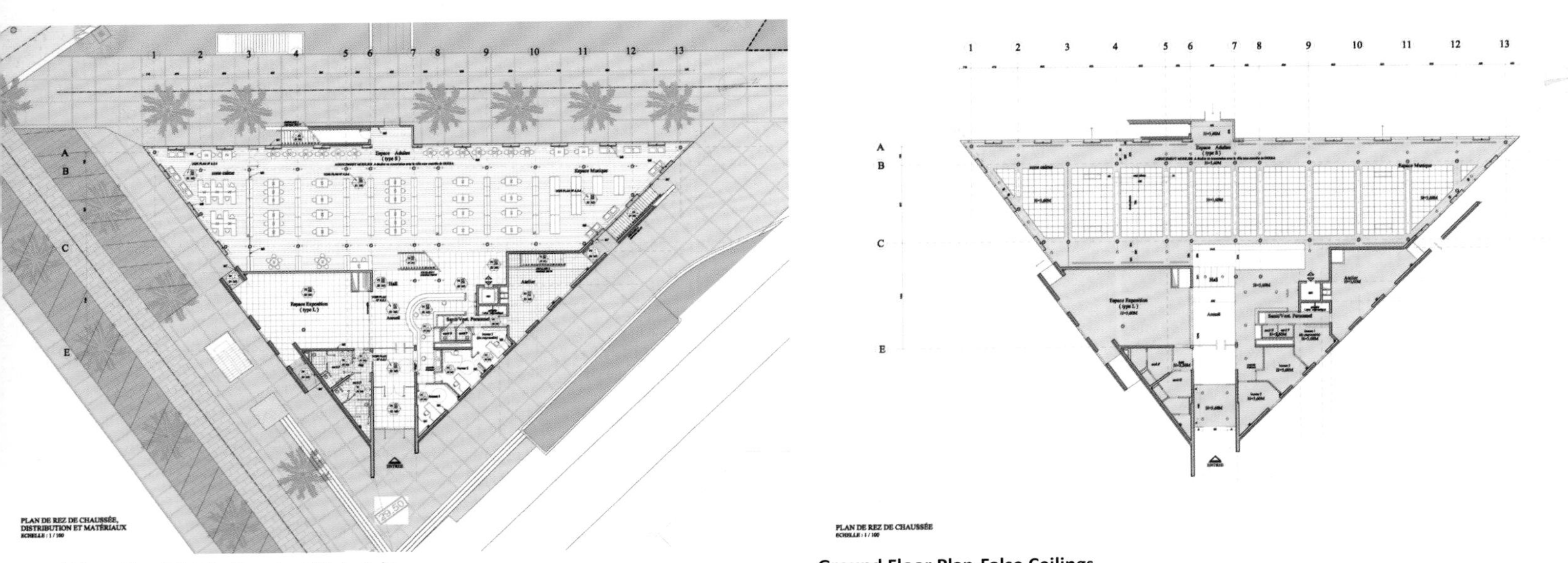

Ground Floor Plan-Distribution And Materials

Ground Floor Plan-False Ceilings

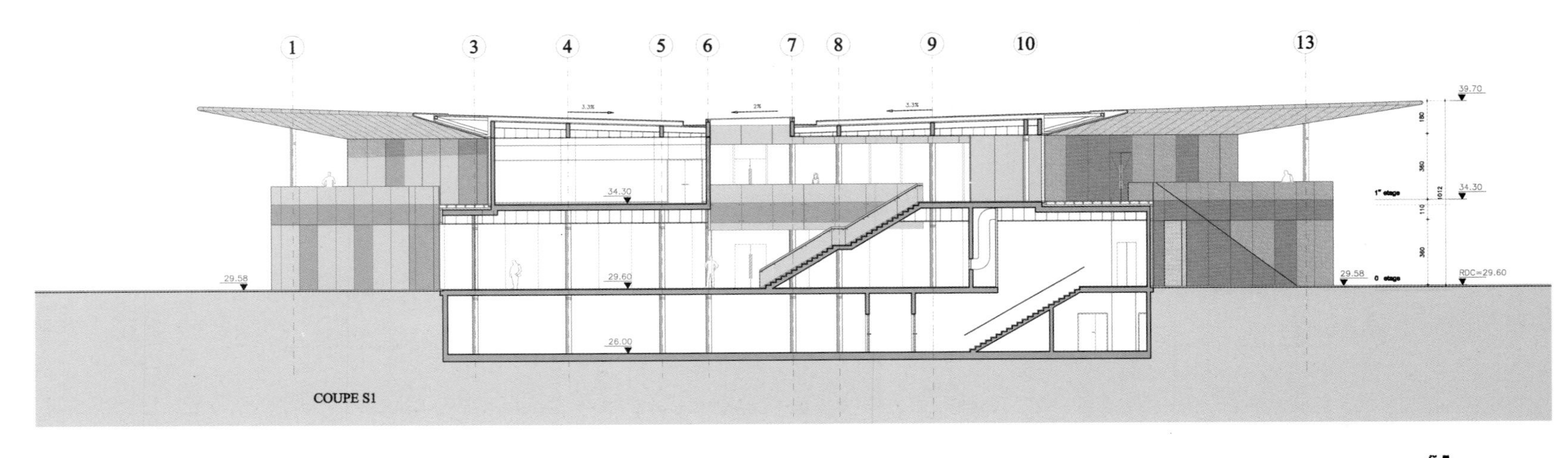

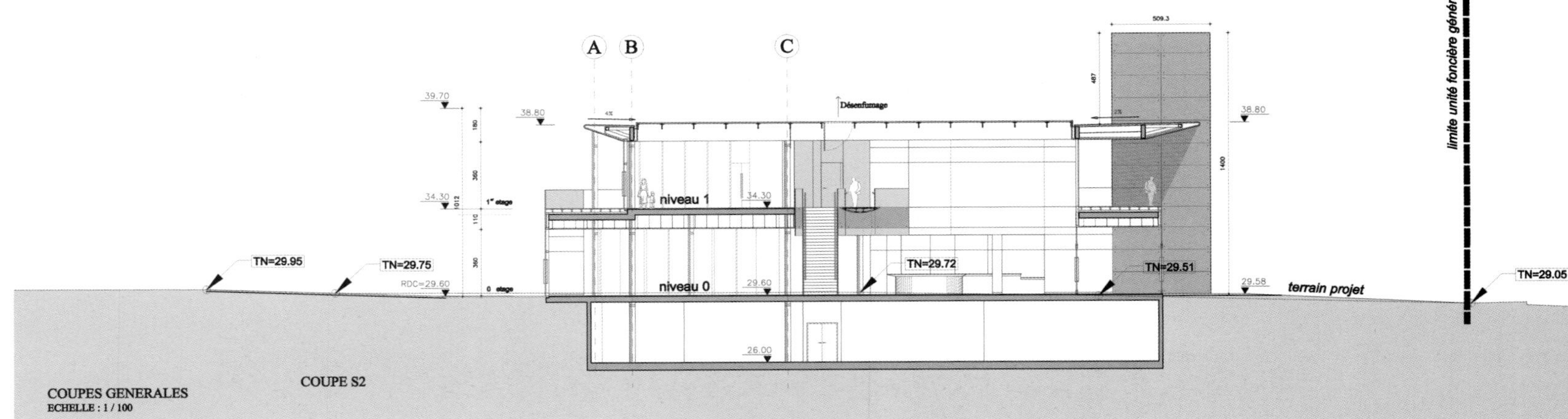

General Sections-S1 And S2

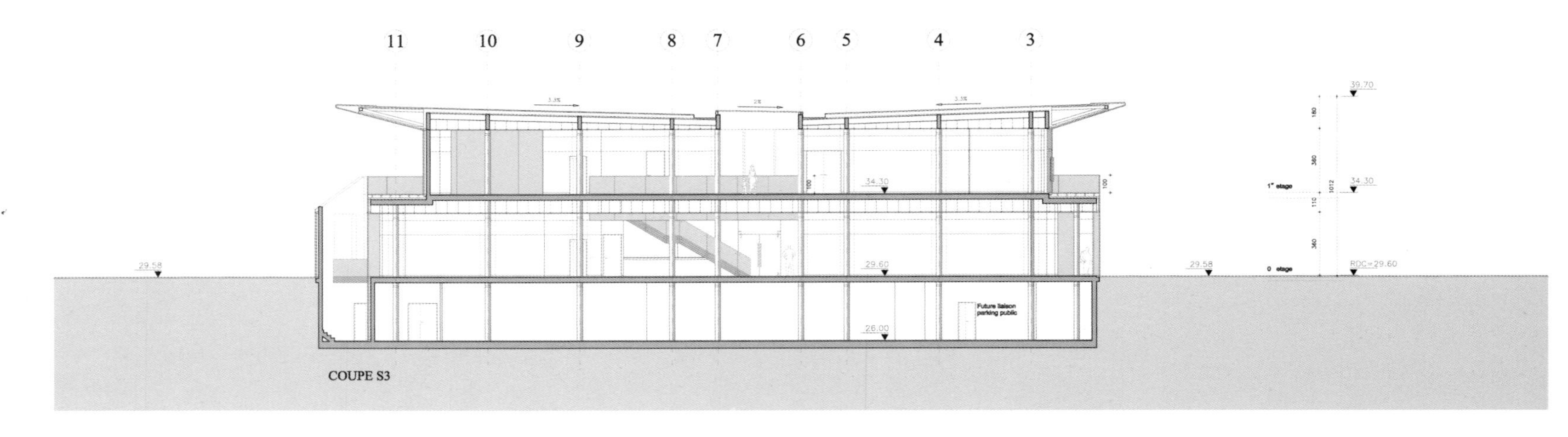

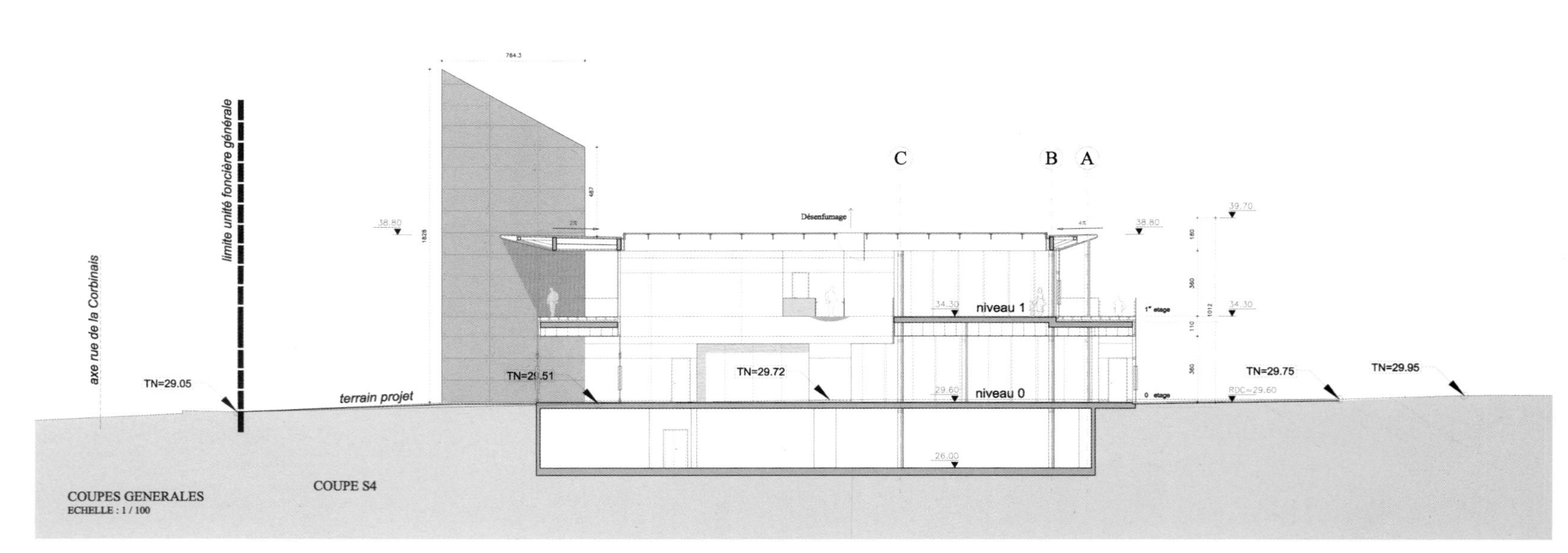

General Sections-S3 And S4

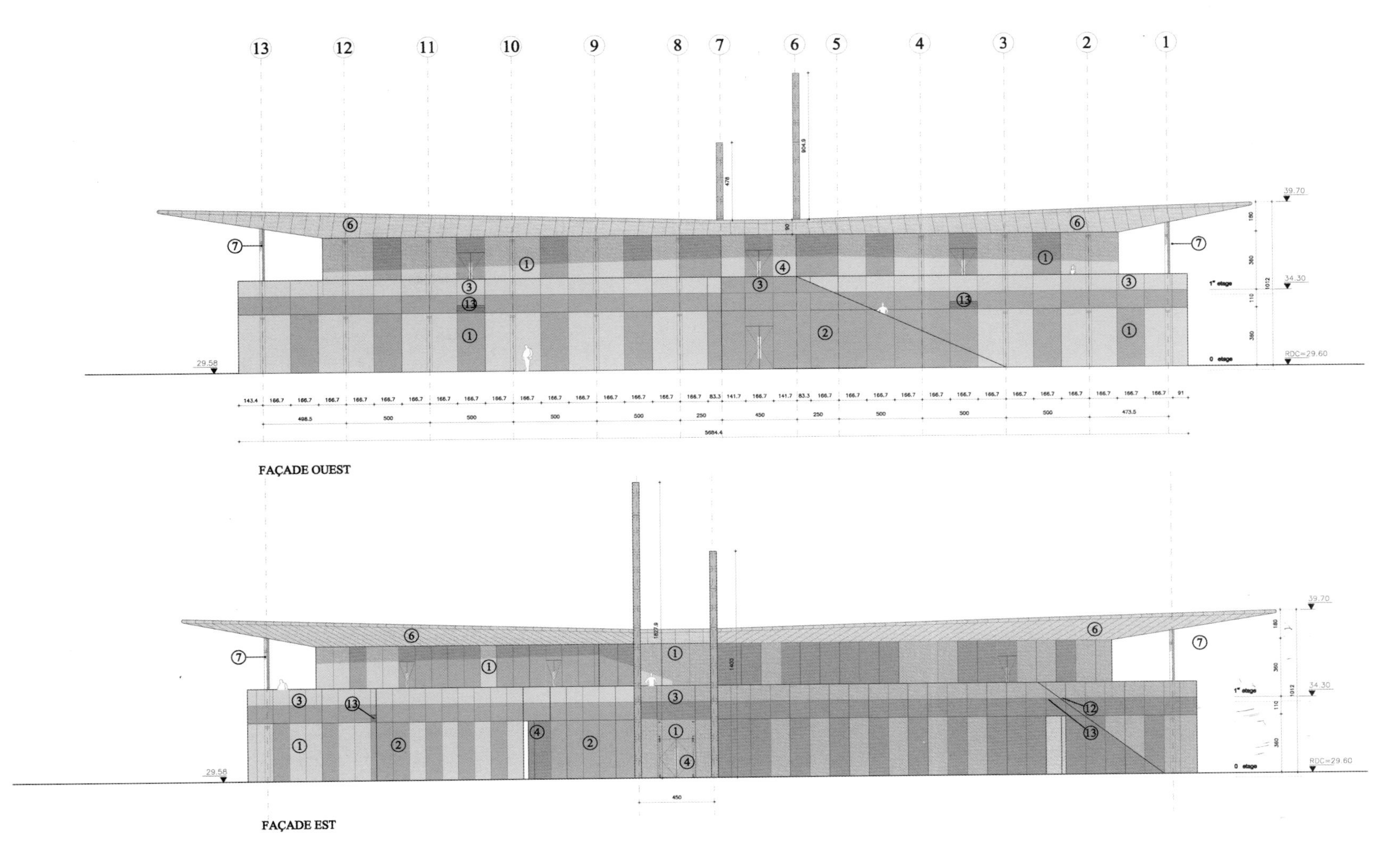

West And East Facades

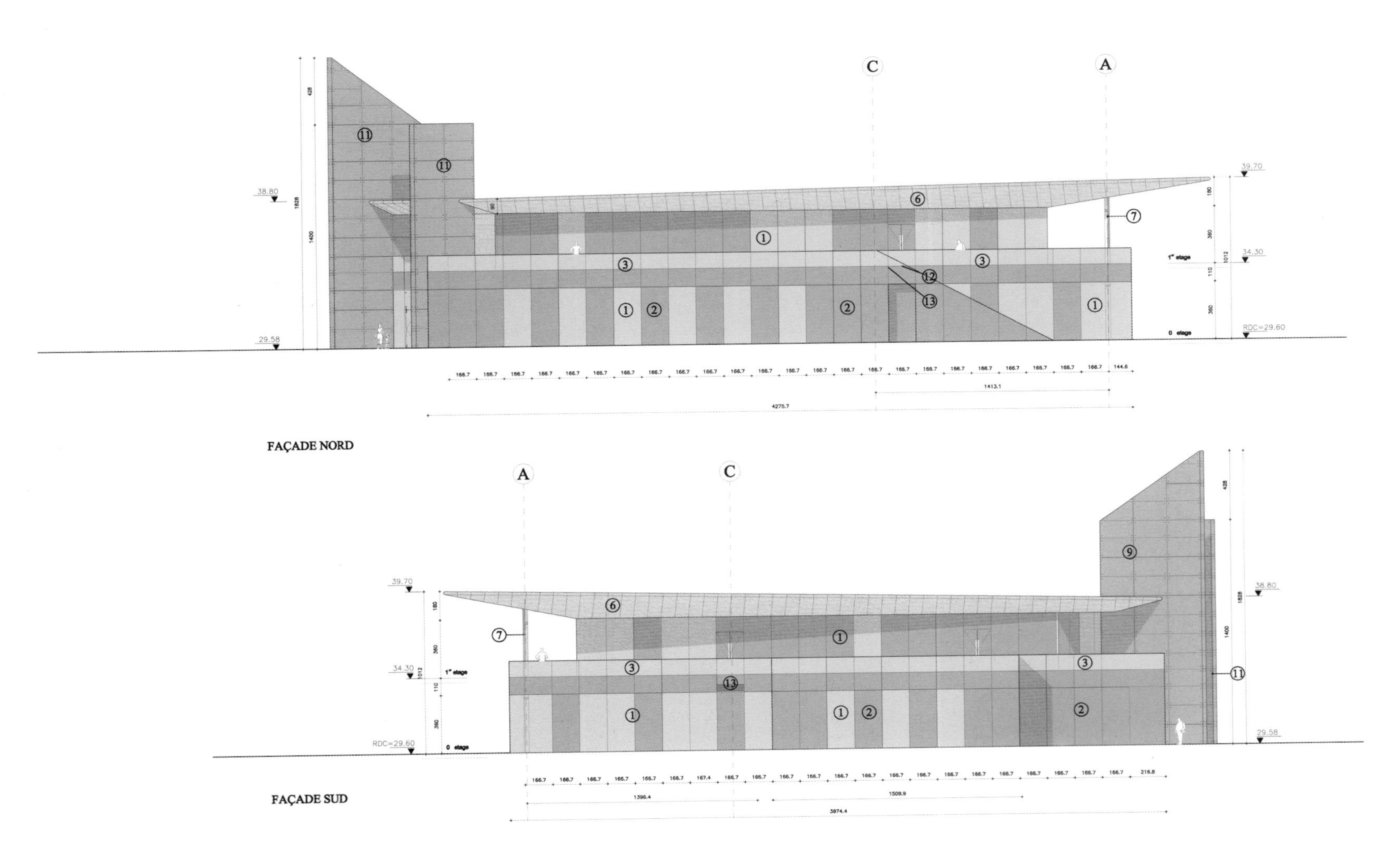

Nord And South Facades

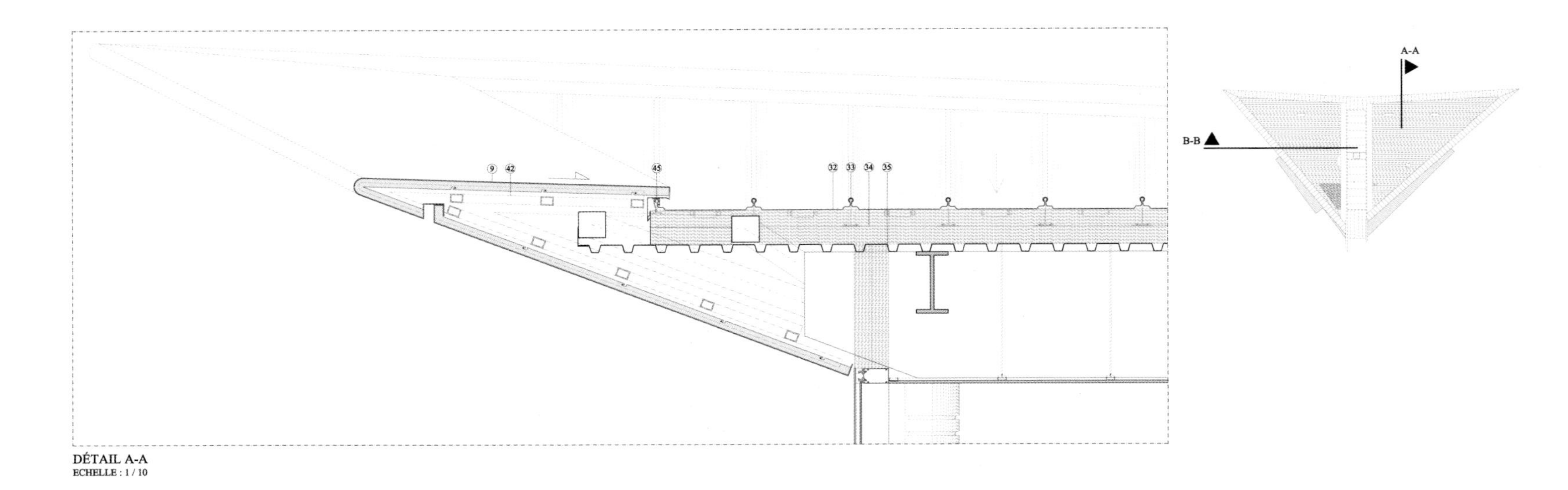

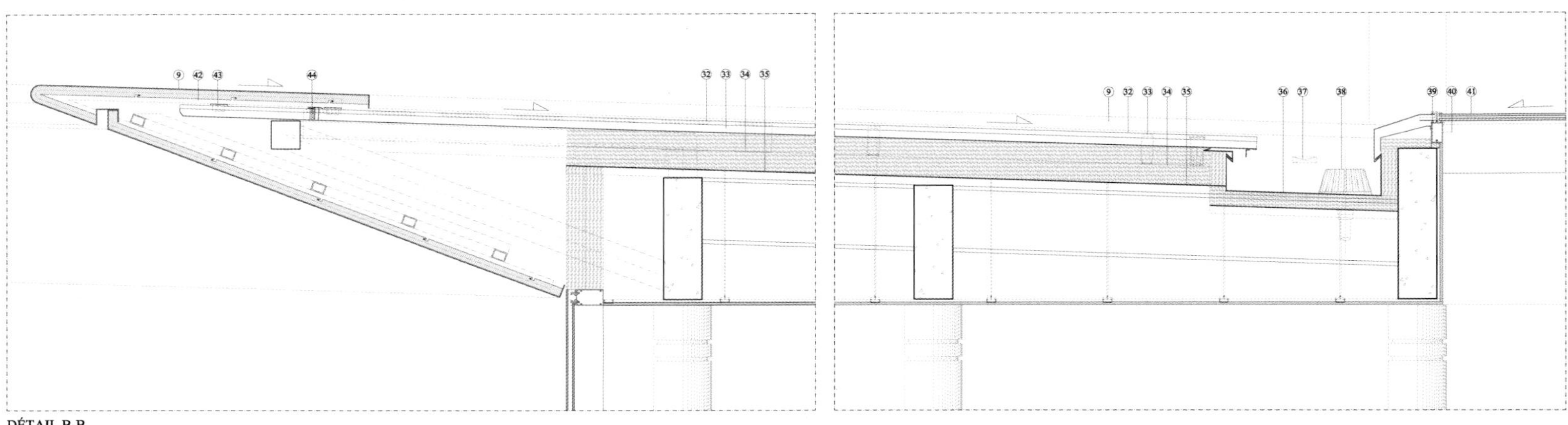

Detail Roof Cornice, Glass Roof And Gutter

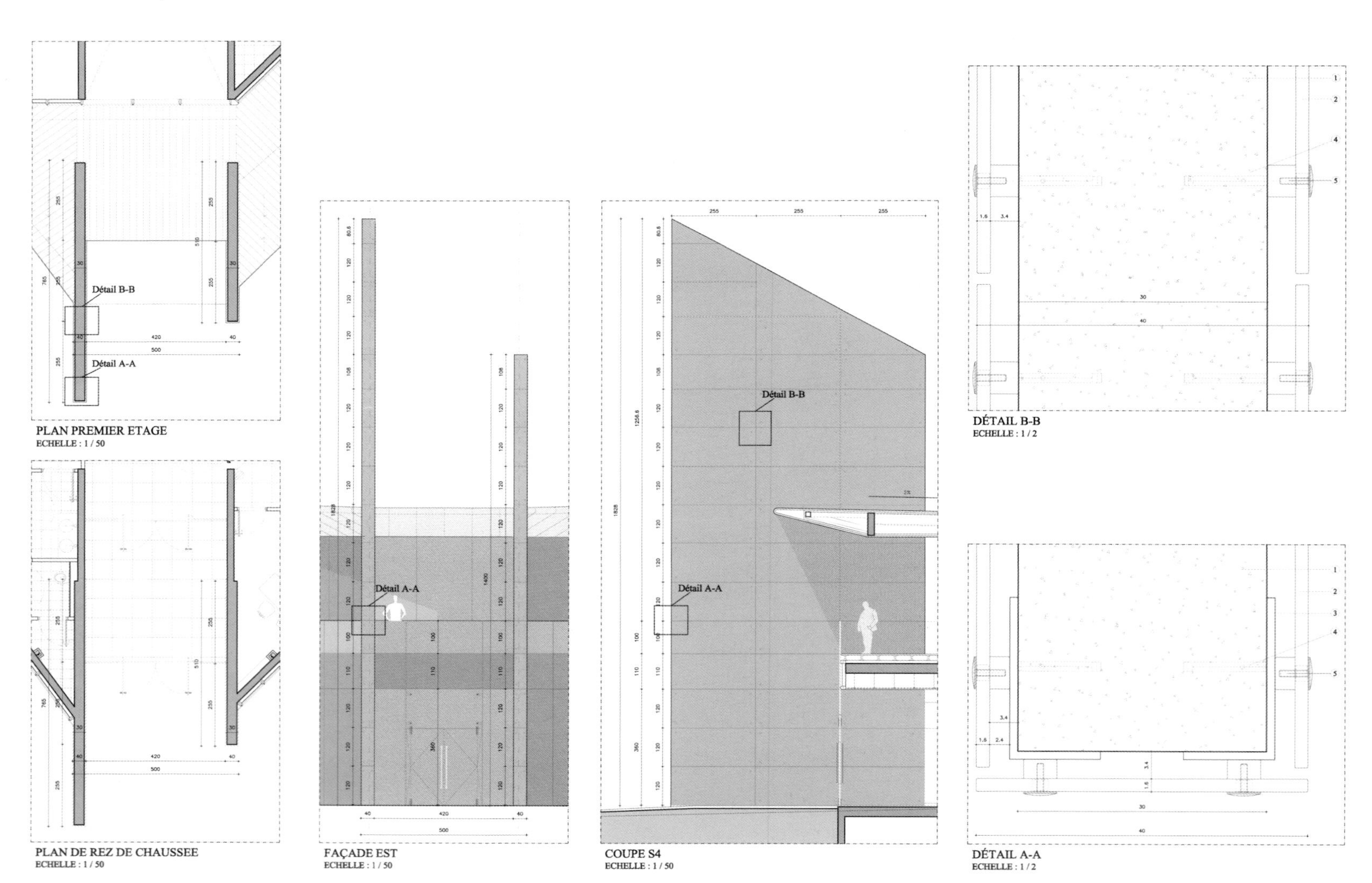

Glass Detail 01

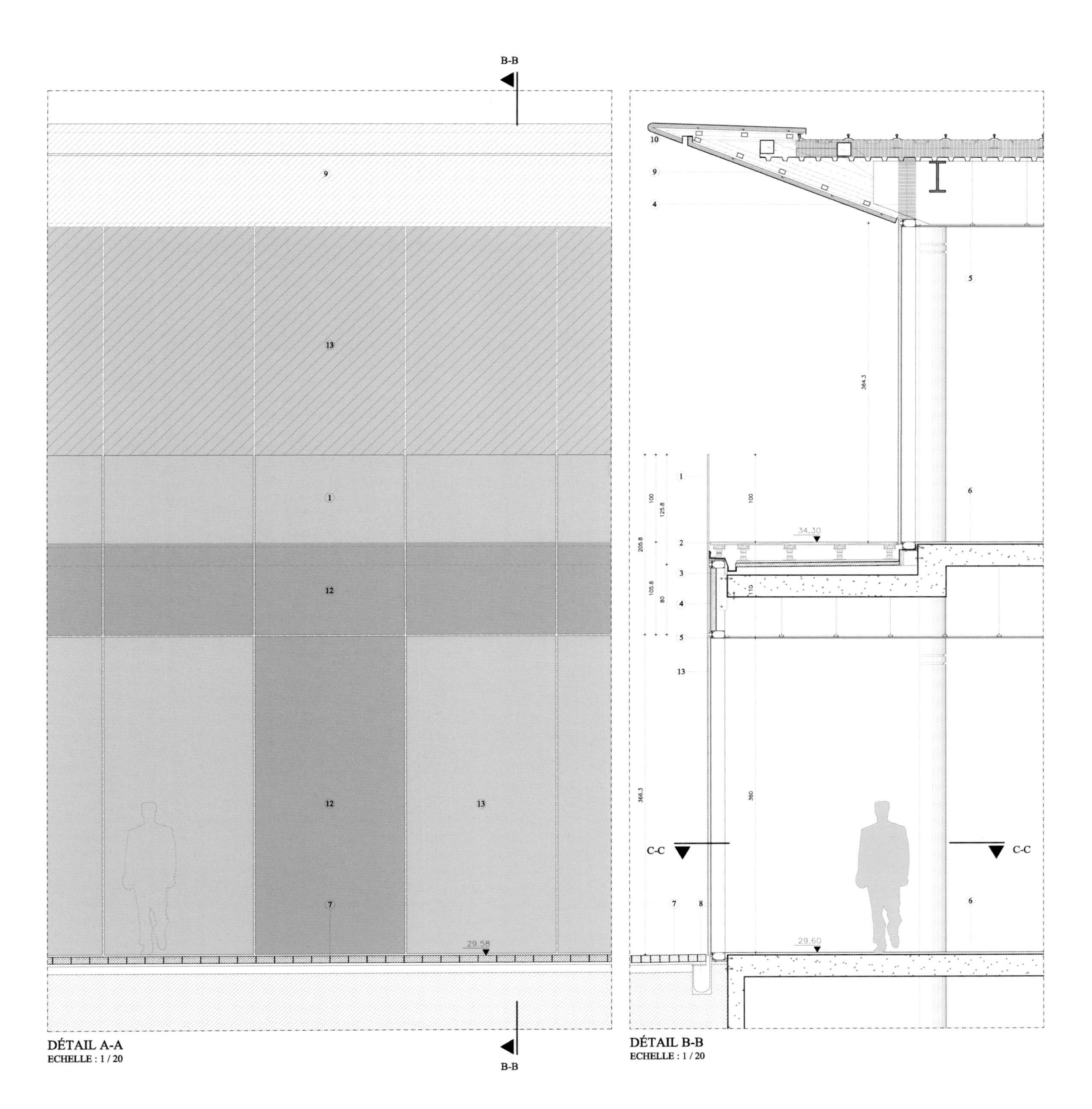

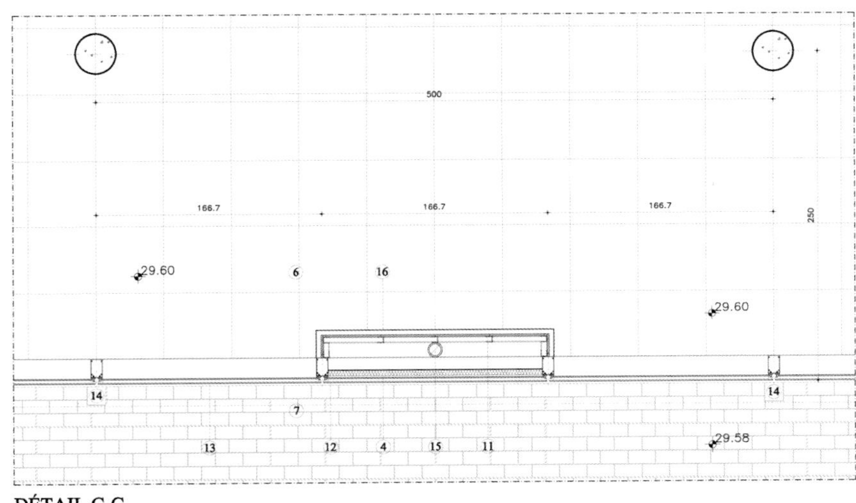

Detail Facade Transparent Glass

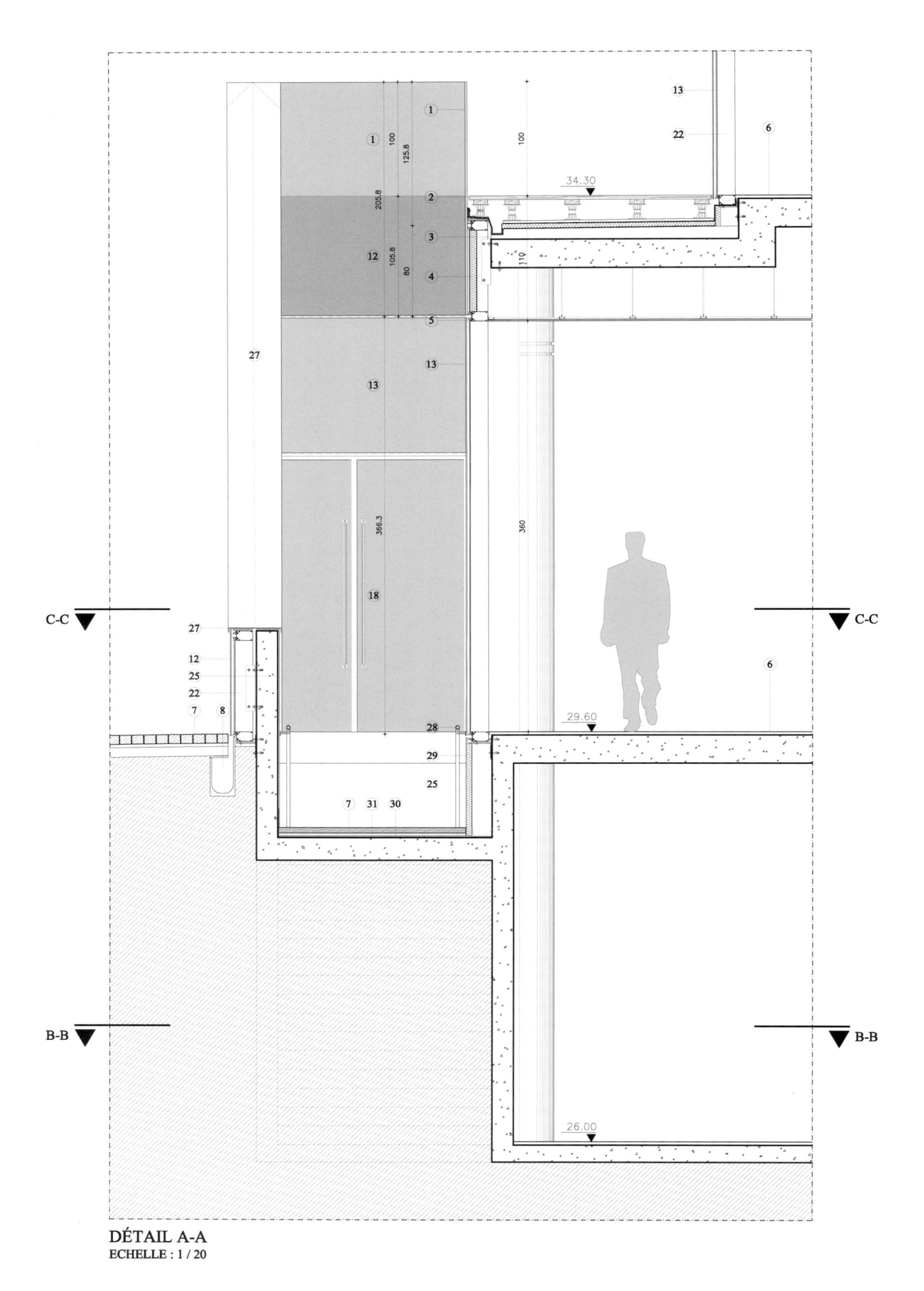

Detail Exterior Stairs

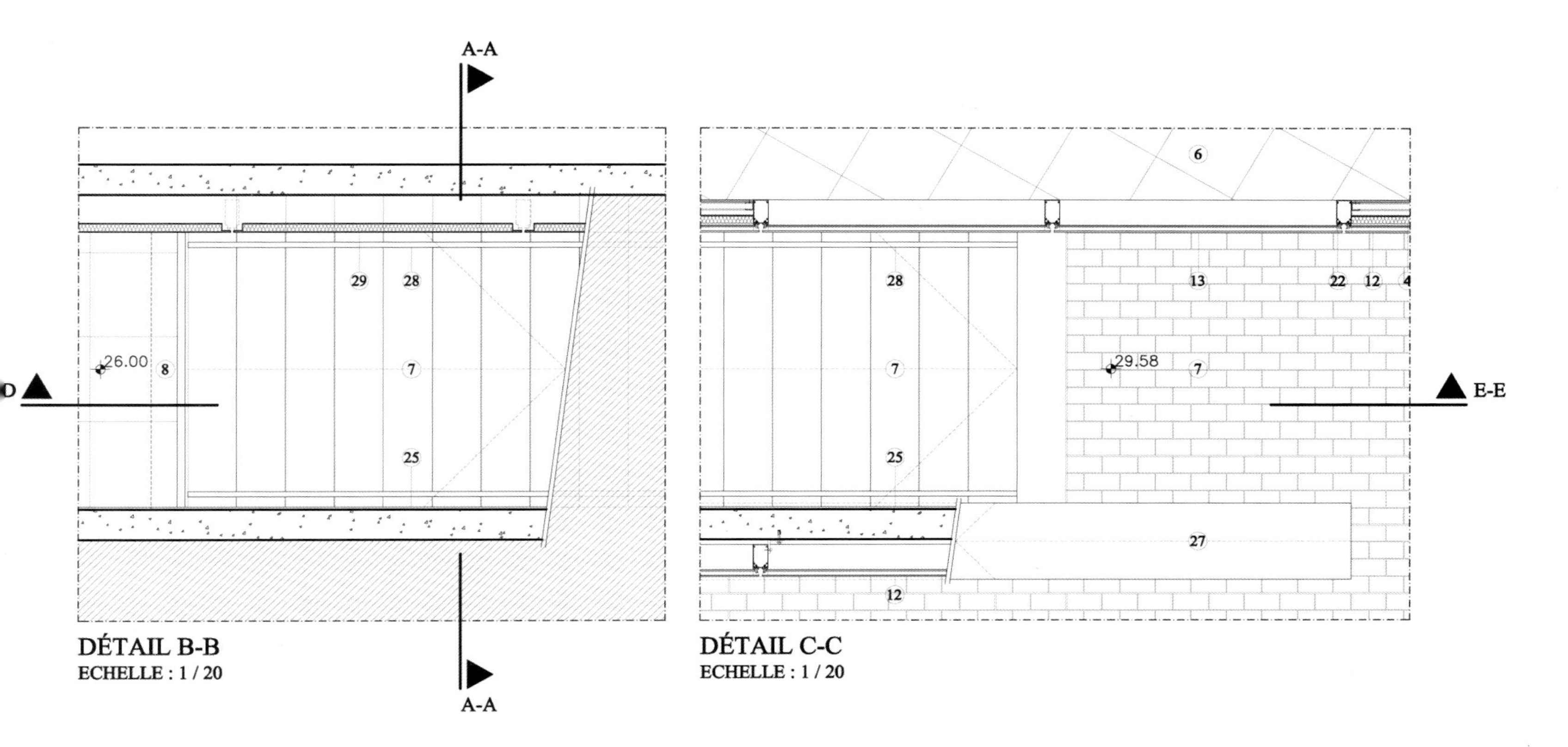

DÉTAIL B-B
ECHELLE : 1 / 20

DÉTAIL C-C
ECHELLE : 1 / 20

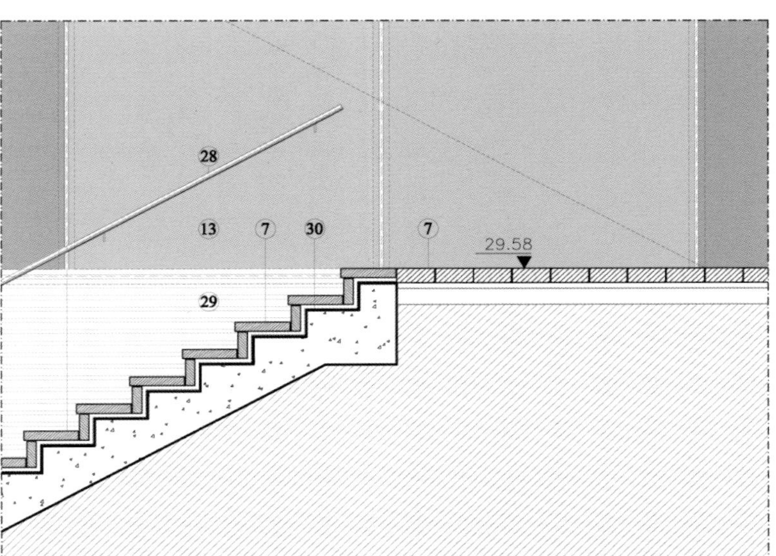

DÉTAIL E-E
ECHELLE : 1 / 20

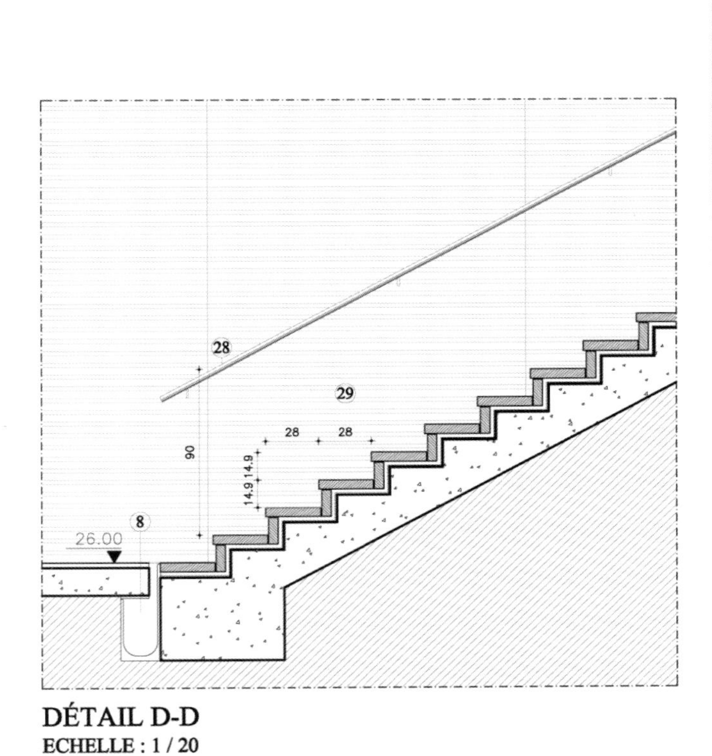

DÉTAIL D-D
ECHELLE : 1 / 20

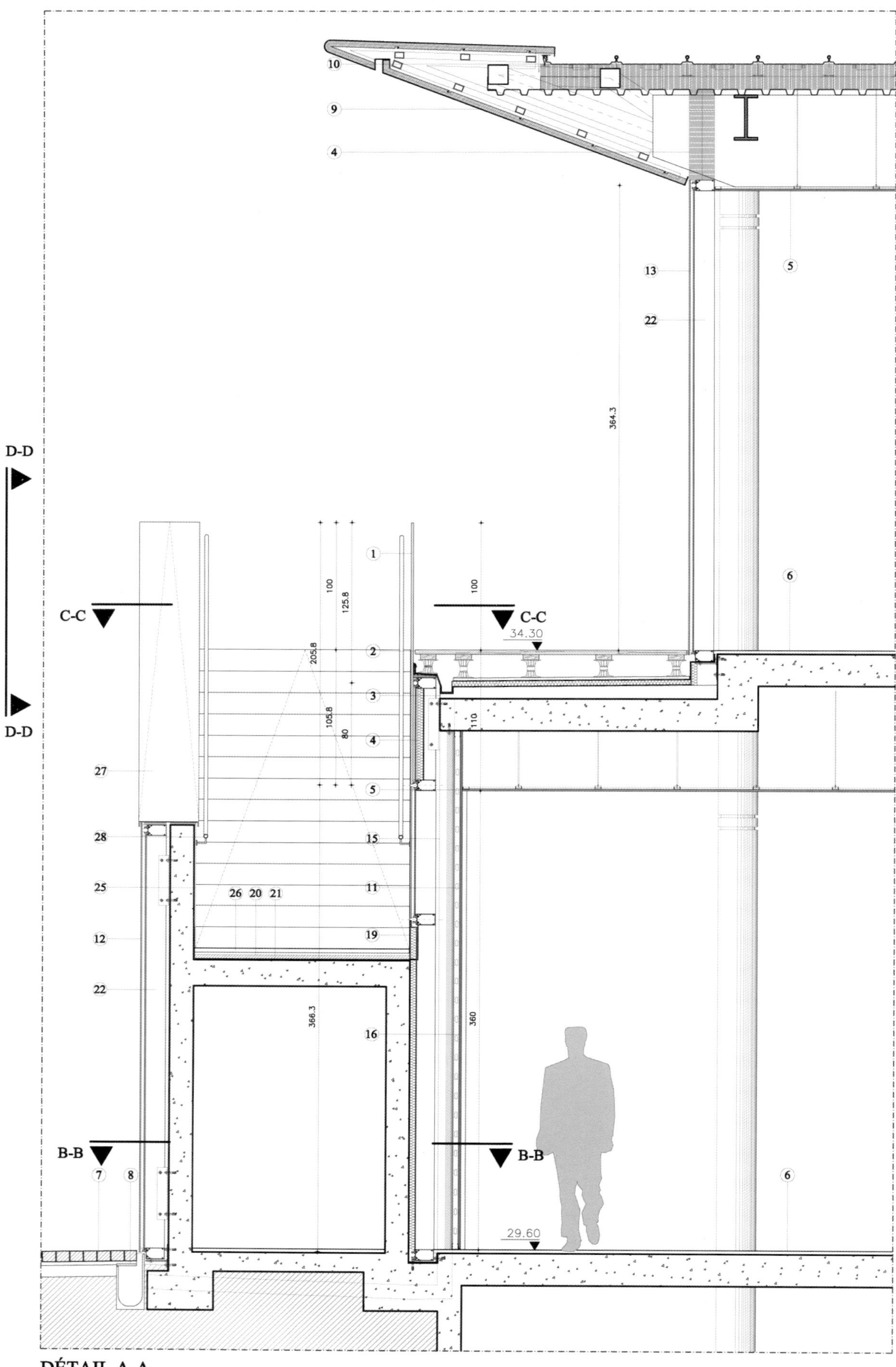

DÉTAIL A-A
ECHELLE : 1 / 20

Detail exterior facade stairs

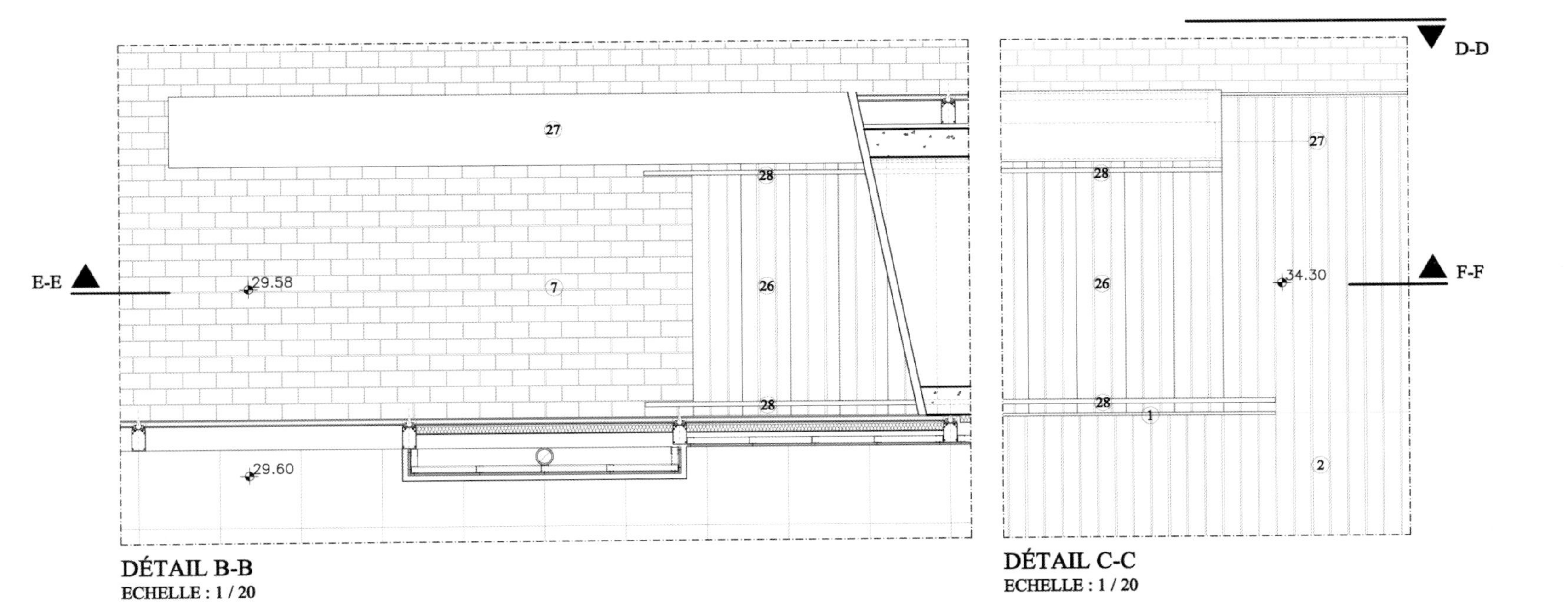

DÉTAIL B-B
ECHELLE : 1 / 20

DÉTAIL C-C
ECHELLE : 1 / 20

DÉTAIL D-D
ECHELLE : 1 / 20

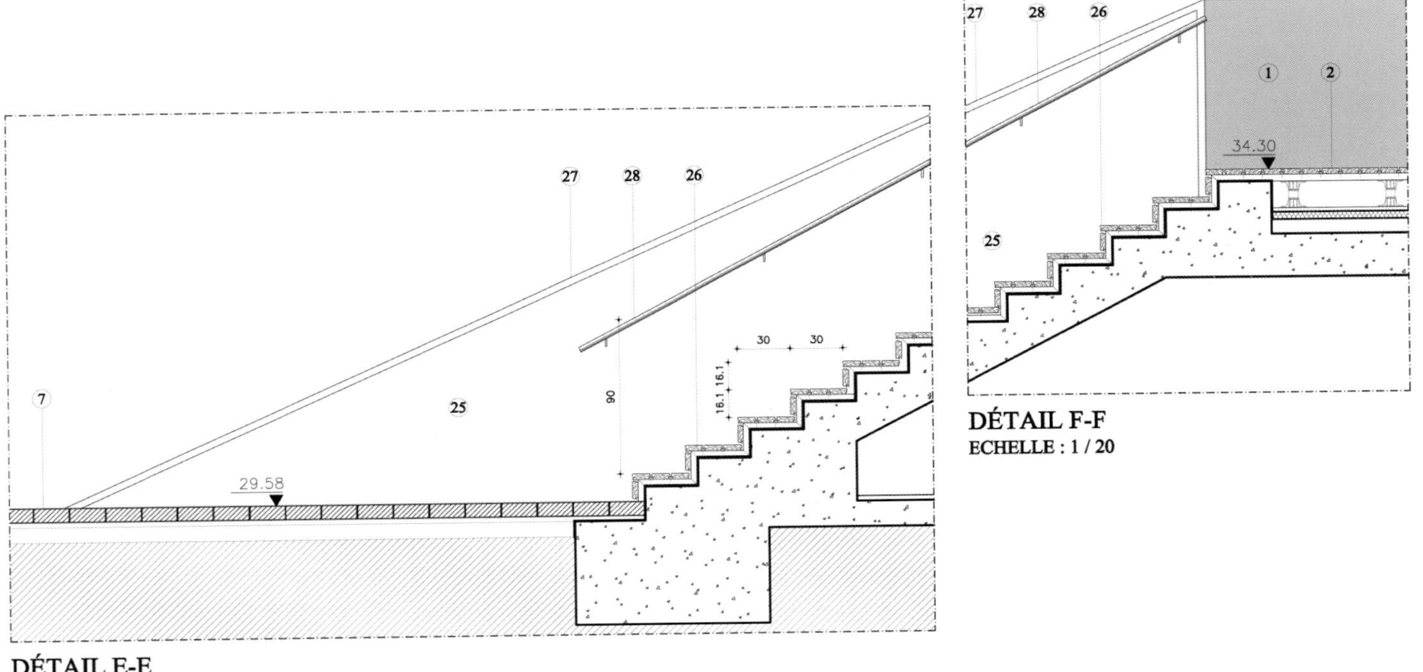

DÉTAIL E-E
ECHELLE : 1 / 20

DÉTAIL F-F
ECHELLE : 1 / 20

Office

Lille Offices

Location: Paris, France
Architect: LAN Architecture
Land Area: 1 387m²
Floor Area: 3 486 m²
Completion Date: 2014

The parcel's strategic position, located at the intersection of different axes, pushed the research towards a sophisticated solution that acts as a hub, as a stitch that brings together the elements gravitating around it. The architects strove for a "multiform" architecture whose geometry could provide a specific response to the various challenges tied to the project's scale, geography, and program.

By extending and crossing the axes within the parcel, the initial extrusion was carved to obtain a kind of small tower. By completing the Avenue Le Corbusier, this vertical element is also a corner building on the Place Valladolid and it signals the city to drivers coming up from the beltway below.

This architecture has created a new urban space that combines private and public, vertical and horizontal. The base of the project provides inhabitants and office workers a public space that fosters social interaction; it functions on a human scale. Due to the prohibition from building out to the edge of the parcel, a kind of portico provides a sense of porosity as well as protection from inclement weather. It is a lively outdoor space where people who live and work in the building can mingle with passers-by and shop customers.

This office project has a very flexible program; form dictates use, and not vice-versa. Each level is organized around a central core that holds all the servant spaces and vertical circulations. The office floor areas were conceived to allow for a flexible, rational layout and to encourage the division of the floor spaces into two equivalent surfaces.

Moreover, the tower's geometric faceting at once frees up the views and opens up the entire intersection to the wooded background of the cemetery to the north.

In order to complete this process of interrelation, the facades were designed to become a series of windows that provide a 360-degree panorama of the city, framing views of the city's newer parts, its green spaces, and the downtown.

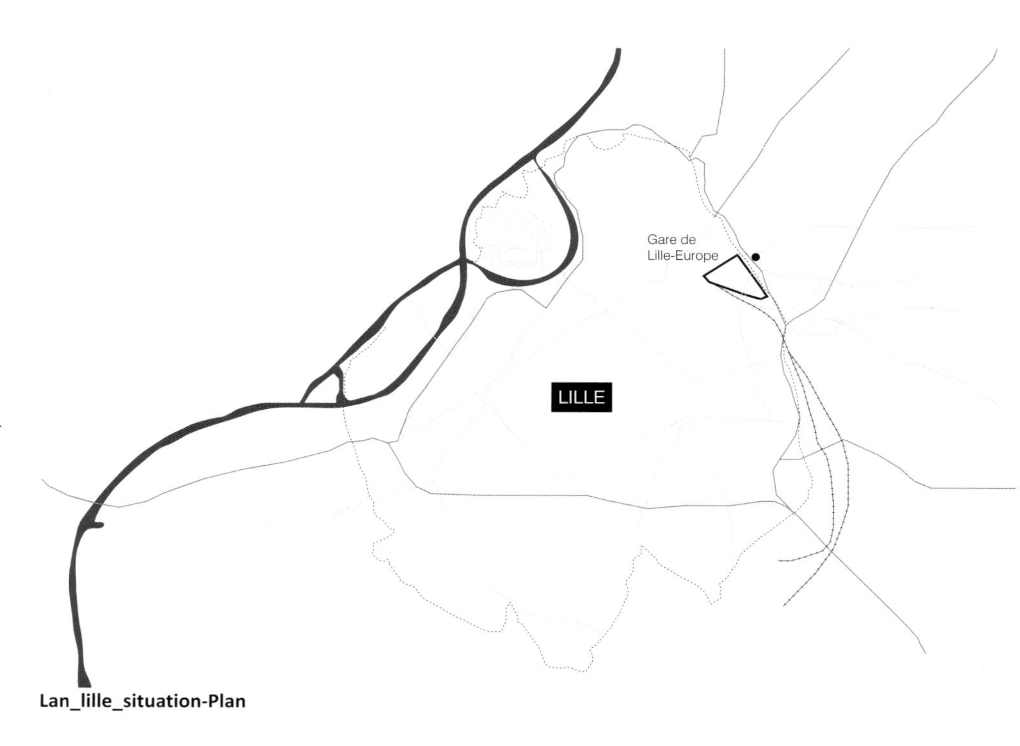

Lan_lille_situation-Plan

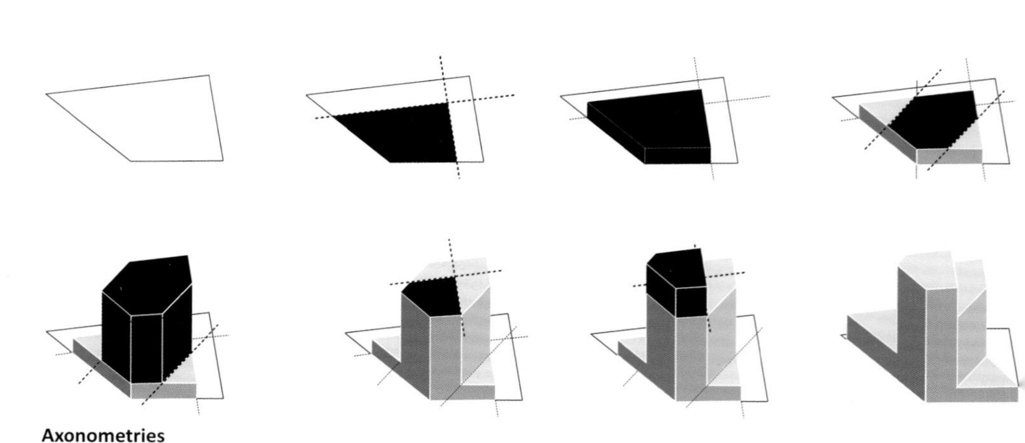

Axonometries

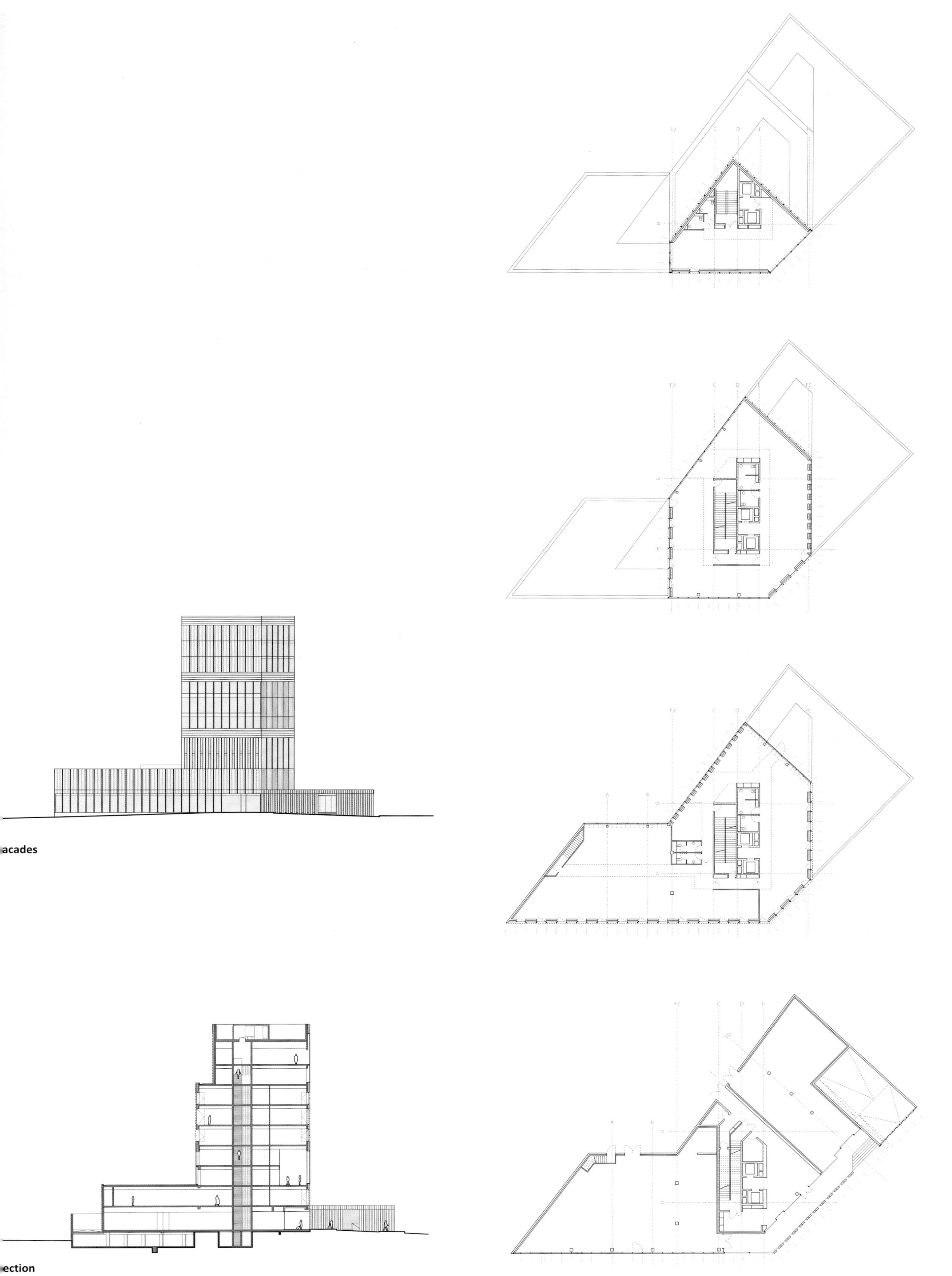

acades

ection

Plans Standard

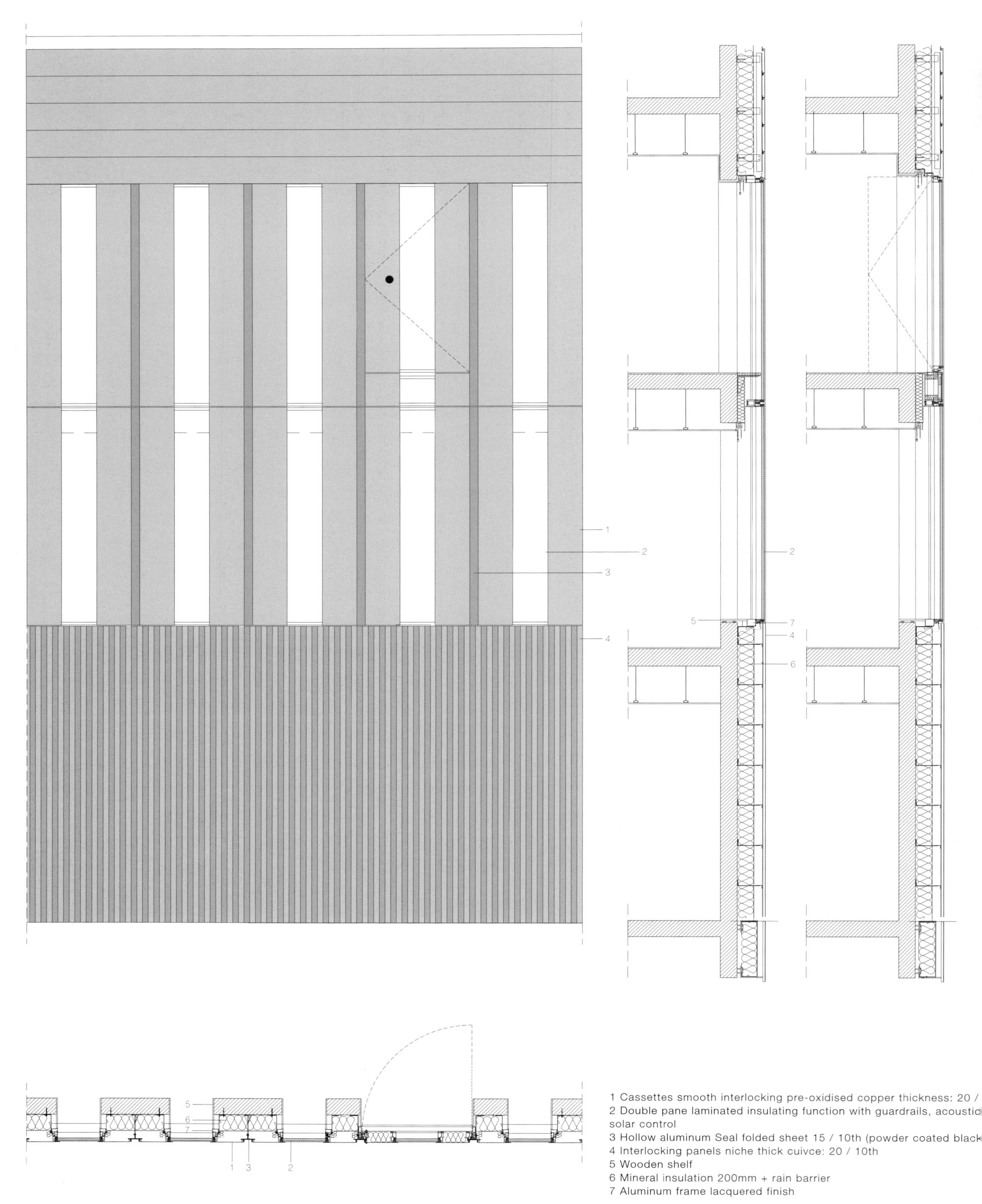

1 Cassettes smooth interlocking pre-oxidised copper thickness: 20 /
2 Double pane laminated insulating function with guardrails, acoustic
solar control
3 Hollow aluminum Seal folded sheet 15 / 10th (powder coated black
4 Interlocking panels niche thick cuivce: 20 / 10th
5 Wooden shelf
6 Mineral insulation 200mm + rain barrier
7 Aluminum frame lacquered finish

Facades 1

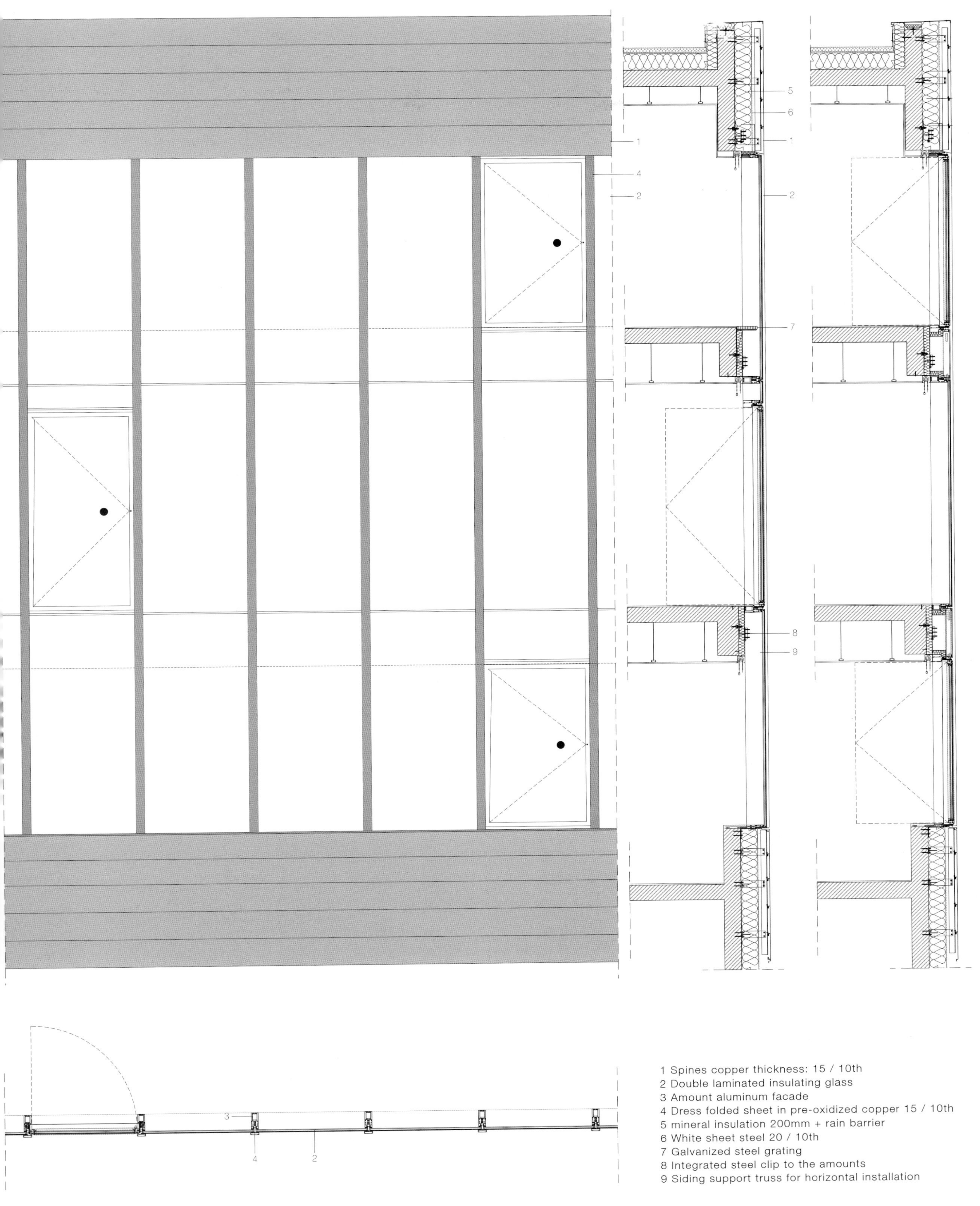

1 Spines copper thickness: 15 / 10th
2 Double laminated insulating glass
3 Amount aluminum facade
4 Dress folded sheet in pre-oxidized copper 15 / 10th
5 mineral insulation 200mm + rain barrier
6 White sheet steel 20 / 10th
7 Galvanized steel grating
8 Integrated steel clip to the amounts
9 Siding support truss for horizontal installation

Facades 2

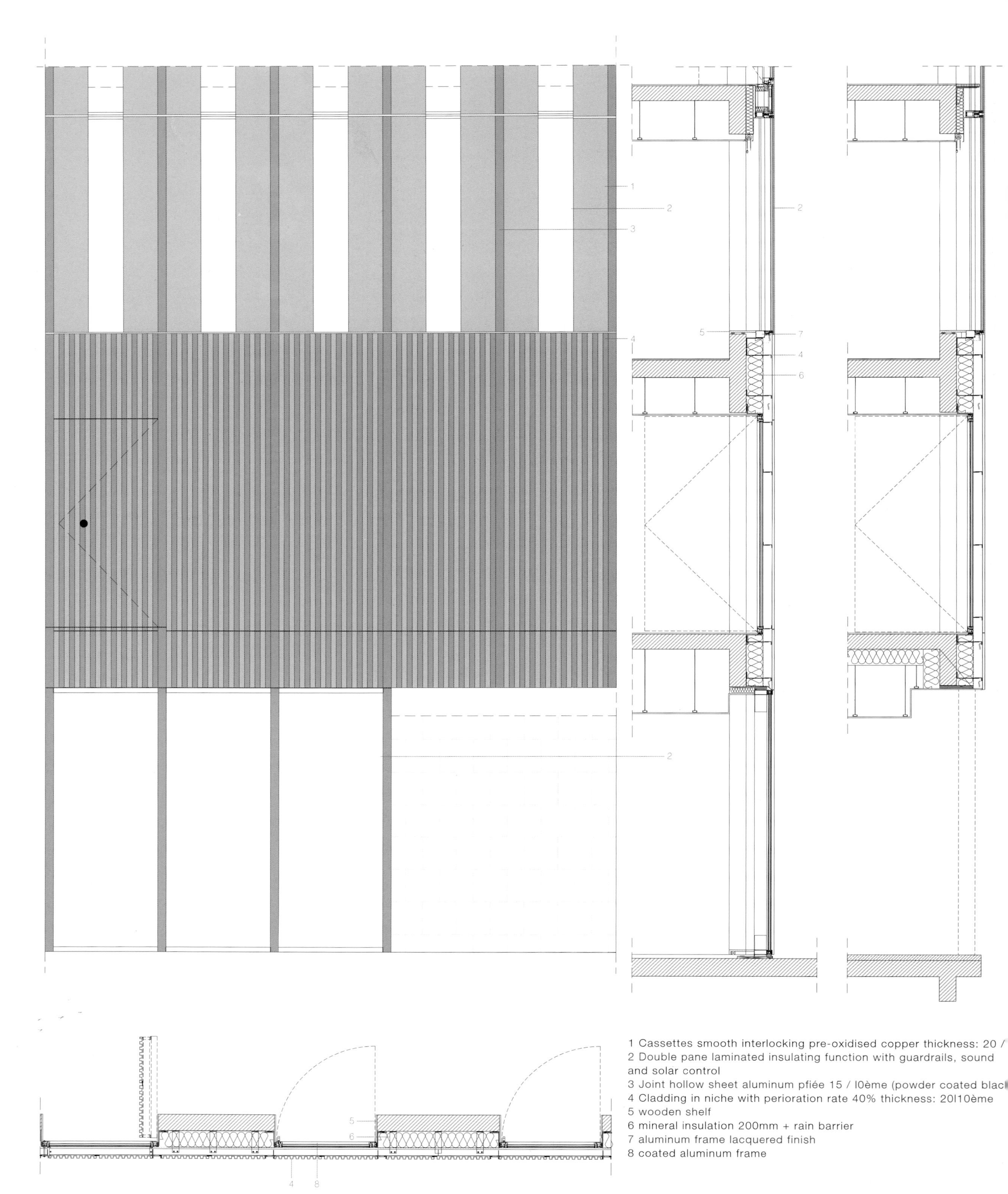

1 Cassettes smooth interlocking pre-oxidised copper thickness: 20 /
2 Double pane laminated insulating function with guardrails, sound and solar control
3 Joint hollow sheet aluminum pfiée 15 / I0ème (powder coated blac
4 Cladding in niche with perioration rate 40% thickness: 20I10ème
5 wooden shelf
6 mineral insulation 200mm + rain barrier
7 aluminum frame lacquered finish
8 coated aluminum frame

Facades 3

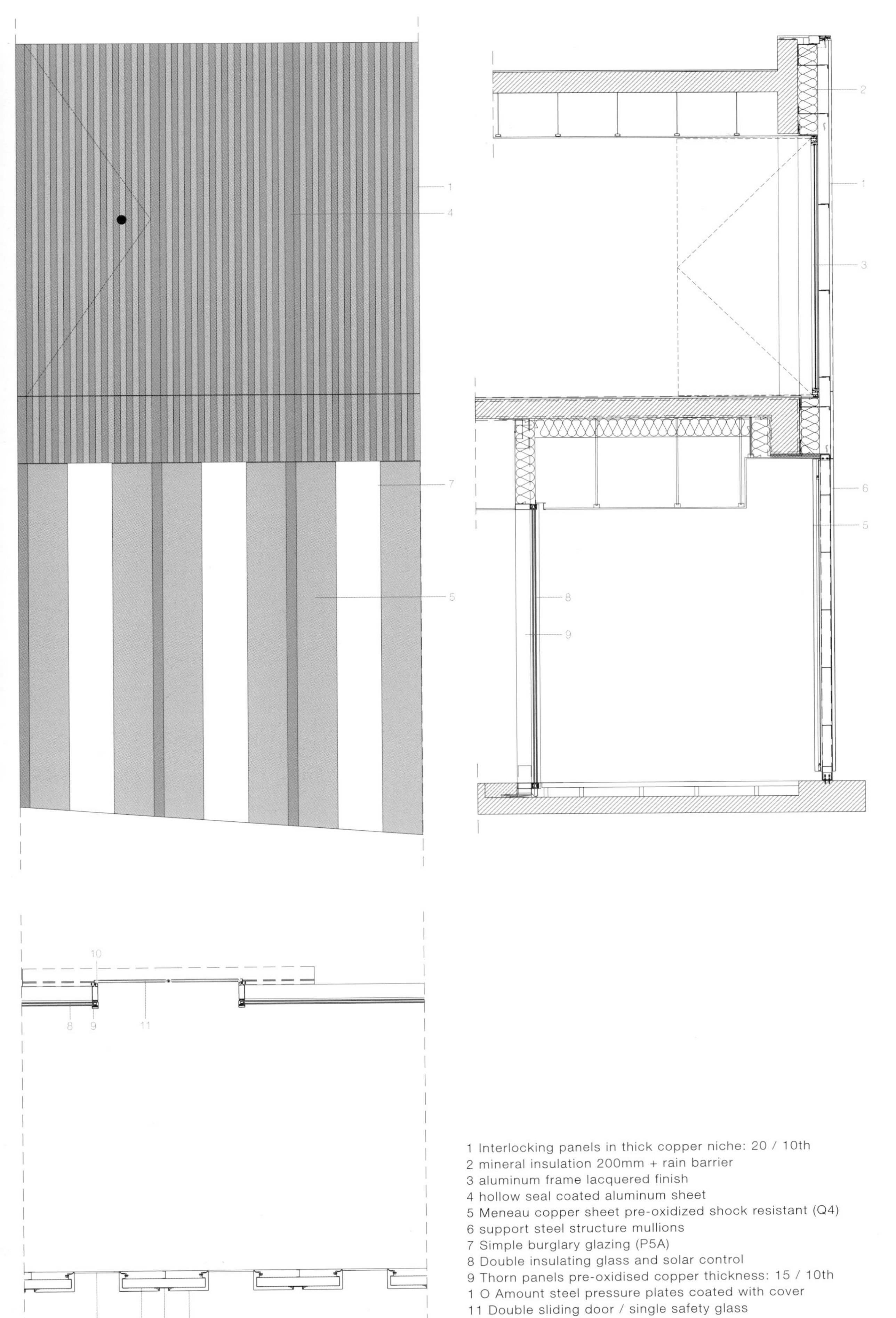

1 Interlocking panels in thick copper niche: 20 / 10th
2 mineral insulation 200mm + rain barrier
3 aluminum frame lacquered finish
4 hollow seal coated aluminum sheet
5 Meneau copper sheet pre-oxidized shock resistant (Q4)
6 support steel structure mullions
7 Simple burglary glazing (P5A)
8 Double insulating glass and solar control
9 Thorn panels pre-oxidised copper thickness: 15 / 10th
1 O Amount steel pressure plates coated with cover
11 Double sliding door / single safety glass

Pannais fitting into each other, in crenels,
Mineral insulating 200mm + rainscreen
Aluminum chassis (finish: lacquered)
Shadowgap in folded sheet of aluminum
Mullion in a pre-Oxidised copper sheet, in
Steel framework holding the mullions up
Simple glazing, burglar-proof (P5A)
Laminated and insulating double glazing.
Mullions in pre-oxidised copper pannels 1
Steel stiles with tighting cover (finish: then
Double sliding doors / simple safety glazing

Facades 4

Dear Ginza

Location: Tokyo, Japan
Architect: Amano Design Office Inc.
Building Area: 155.55 m^2
Total Floor Area: 1,300.02 m^2
Completion Date: March 2013
Photography: Nacasa & Partners Inc.

The designer desired to provide a “slight feeling of strangeness” to the passersby that would attract them to the building. Considering the views from the inside, simply obtaining openness with glass seems futile, since the outside scenery is hopeless. Therefore, a double skin structure is employed, which consists of glass curtain walls and graphically treated aluminum punched metal.

The facade becomes a part of the interior decoration and obviates the need for window treatments such as blinds or curtains. By using a double skin, reduction of the air conditioning load and the glass cleaning burden was also intended. The irregular facade design was determined by computing a design to avoid arbitrary forms and to approximate forms in nature. Designer thought that a well-made incidental form would likely be a less-disagreeable design.

In the neighborhood of mostly modernist architecture with horizontal and vertical or geometric shapes, the building has a proper feeling of strangeness, attracts special attention, and has an appeal as a commercial building. The abstract flower graphic is used to balance the impression of the façade, i.e., to free it up from becoming too edgy.

By computing the design, individual aluminum punched panels are irregular with different angles and shapes, yet all fit into a standard size, resulting in excellent material yield. To avoid being clunky, an extremely lightweight structure is required. Therefore, much caution was taken in its details. The colored LED upper lighting, which is installed inside the double skin, entertains the passersby with different programs depending on the season.

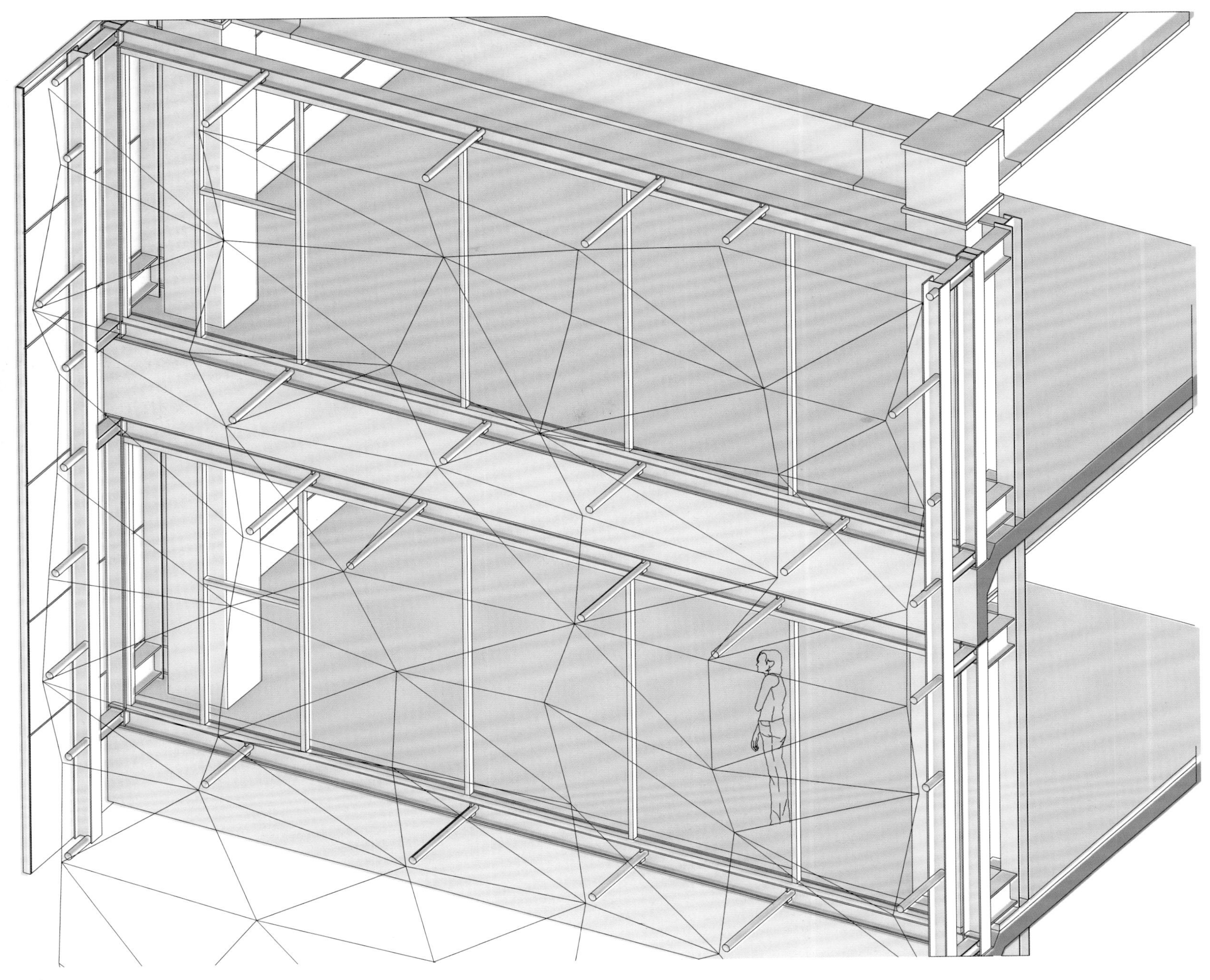

Structural concept

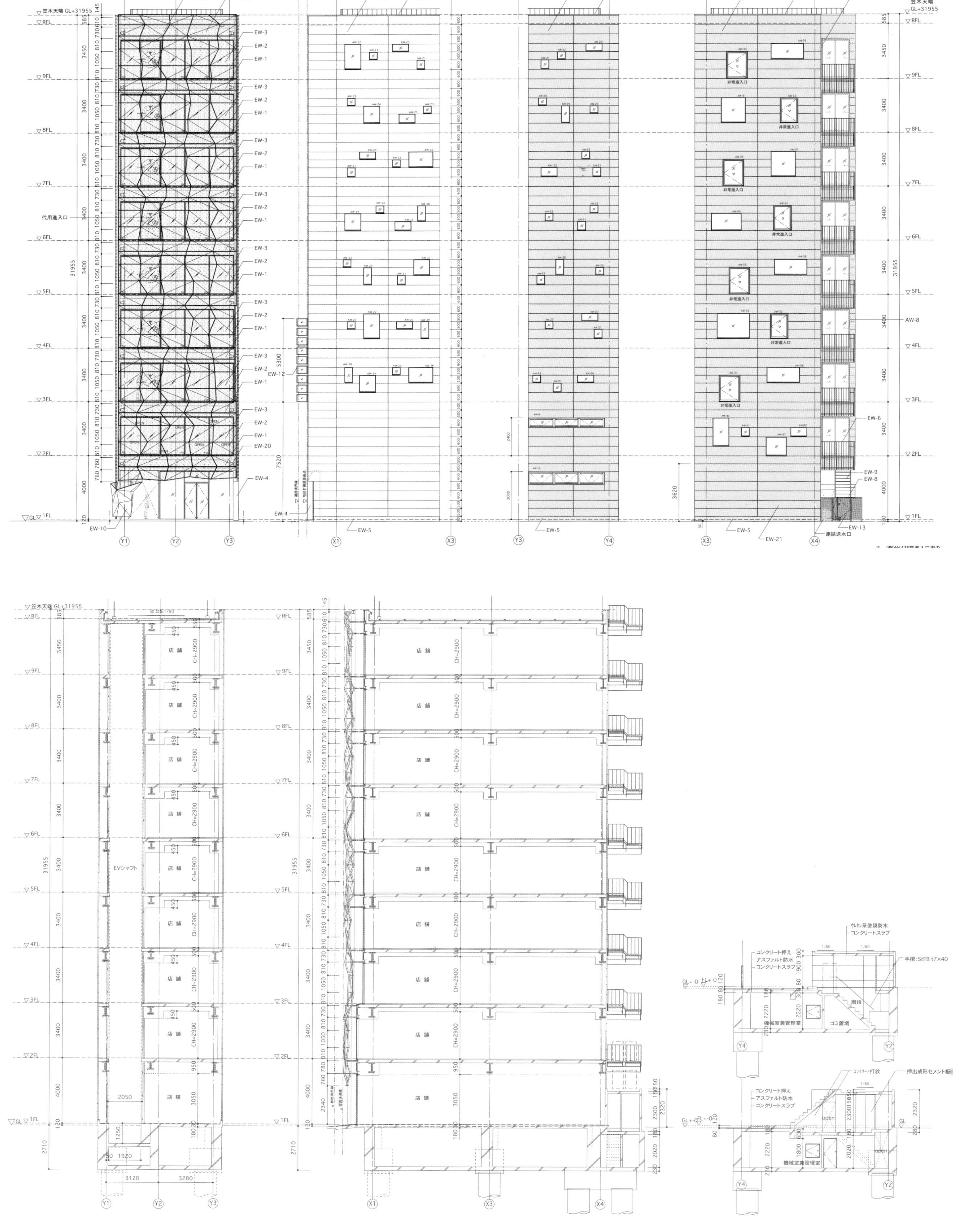

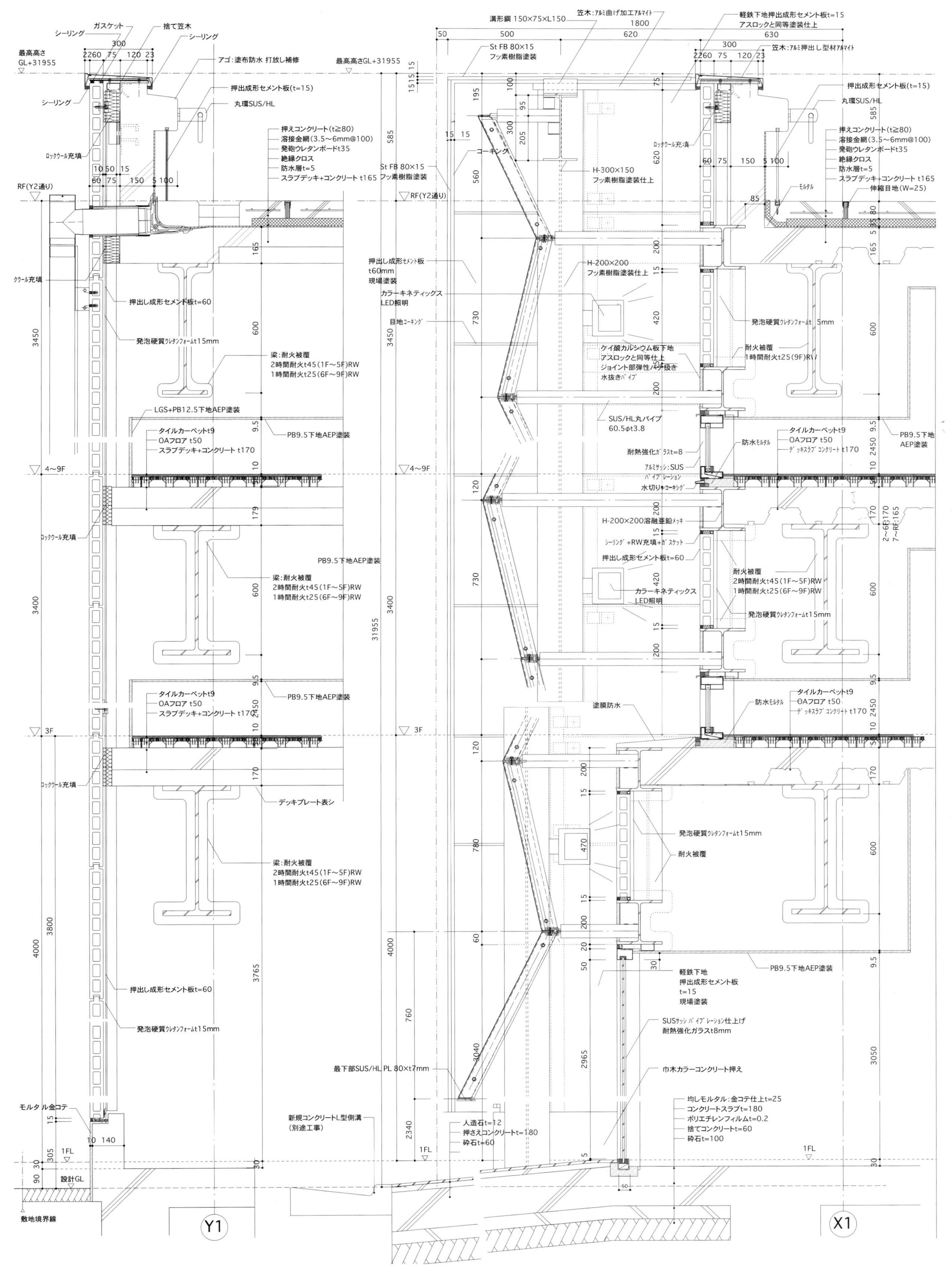

ガスケット
捨て笠木
シーリング
シーリング
最高高さ
GL+31955
アゴ:塗布防水 打放し補修
押出成形セメント板(t=15)
丸環SUS/HL
押えコンクリート(t≧80)
溶接金網(3.5~6mm@100)
発砲ウレタンボードt35
絶縁クロス
防水層t=5
スラブデッキ+コンクリート t165
RF(Y2通り)
ロックウール充填
押出し成形セメント板t=60
発泡硬質ウレタンフォームt15mm
梁:耐火被覆
2時間耐火t45(1F~5F)RW
1時間耐火t25(6F~9F)RW
LGS+PB12.5下地AEP塗装
タイルカーペットt9
OAフロア t50
スラブデッキ+コンクリート t170
PB9.5下地AEP塗装
4~9F
3F
デッキプレート表シ
モルタル金コテ
1FL
設計GL
敷地境界線
最高高さGL+31955
溝形鋼 150×75×L150
笠木:アルミ曲げ加工アルマイト
St FB 80×15
フッ素樹脂塗装
コーキング
H-300×150
フッ素樹脂塗装仕上
軽鉄下地押出成形セメント板t=15
アスロックと同等塗装仕上
笠木:アルミ押出し型材アルマイト
伸縮目地(W=25)
モルタル
押出し成形セメント板
t60mm
現場塗装
カラーキネティックス
LED照明
目地コーキング
H-200×200
フッ素樹脂塗装仕上
ケイ酸カルシウム板下地
アスロックと同等仕上
ジョイント部弾性パテ拭き
水抜きパイプ
SUS/HL丸パイプ
60.5φt3.8
耐熱強化ガラスt=8
アルミサッシ:SUS
バイブレーション
水切り+コーキング
H-200×200溶融亜鉛メッキ
シーリング+RW充填+ガスケット
押出し成形セメント板t=60
耐火被覆
1時間耐火t25(9F)RW
防水モルタル
デッキスラブコンクリート t170
塗膜防水
軽鉄下地
押出成形セメント板
t=15
現場塗装
SUSサッシ バイブレーション仕上げ
耐熱強化ガラスt8mm
巾木カラーコンクリート押え
均しモルタル:金コテ仕上t=25
コンクリートスラブt=180
ポリエチレンフィルムt=0.2
捨てコンクリートt=60
砕石t=100
最下部SUS/HL PL 80×t7mm
新規コンクリートL型側溝
(別途工事)
人造石t=12
押さえコンクリートt=180
砕石t=60
Y1
X1

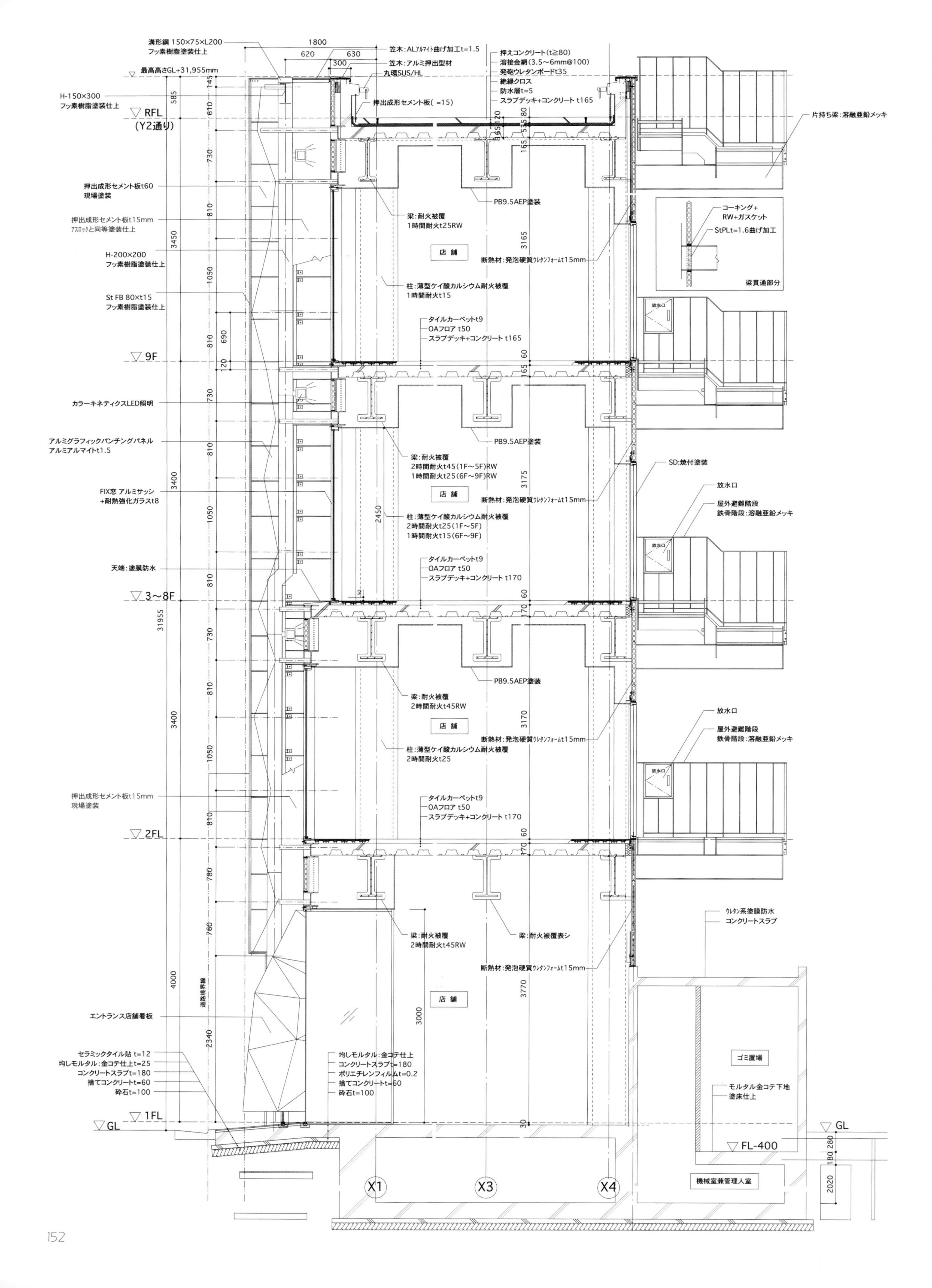
溝形鋼 150×75×L200
フッ素樹脂塗装仕上
1800
620
630
300
笠木：ALアルマイト曲げ加工t=1.5
笠木：アルミ押出型材
丸環SUS/HL
押出成形セメント板(=15)
押えコンクリート(t≧80)
溶接金網(3.5～6mm@100)
発砲ウレタンボードt35
絶縁クロス
防水層t=5
スラブデッキ+コンクリート t165
最高高さGL+31,955mm
H-150×300
フッ素樹脂塗装仕上
RFL
(Y2通り)
片持ち梁：溶融亜鉛メッキ
押出成形セメント板t60
現場塗装
押出成形セメント板t15mm
アスロックと同等塗装仕上
H-200×200
フッ素樹脂塗装仕上
St FB 80×t15
フッ素樹脂塗装仕上
PB9.5AEP塗装
梁：耐火被覆
1時間耐火t25RW
店舗
断熱材：発泡硬質ウレタンフォームt15mm
コーキング+
RW+ガスケット
StPLt=1.6曲げ加工
梁貫通部分
柱：薄型ケイ酸カルシウム耐火被覆
1時間耐火t15
タイルカーペットt9
OAフロア t50
スラブデッキ+コンクリート t165
放水口
9F
カラーキネティクスLED照明
アルミグラフィックパンチングパネル
アルミアルマイトt1.5
PB9.5AEP塗装
梁：耐火被覆
2時間耐火t45(1F～5F)RW
1時間耐火t25(6F～9F)RW
SD：焼付塗装
FIX窓 アルミサッシ
+耐熱強化ガラスt8
店舗
断熱材：発泡硬質ウレタンフォームt15mm
放水口
屋外避難階段
鉄骨階段：溶融亜鉛メッキ
柱：薄型ケイ酸カルシウム耐火被覆
2時間耐火t25(1F～5F)
1時間耐火t15(6F～9F)
天端：塗膜防水
タイルカーペットt9
OAフロア t50
スラブデッキ+コンクリート t170
3～8F
31955
PB9.5AEP塗装
梁：耐火被覆
2時間耐火t45RW
店舗
断熱材：発泡硬質ウレタンフォームt15mm
放水口
屋外避難階段
鉄骨階段：溶融亜鉛メッキ
柱：薄型ケイ酸カルシウム耐火被覆
2時間耐火t25
押出成形セメント板t15mm
現場塗装
タイルカーペットt9
OAフロア t50
スラブデッキ+コンクリート t170
2FL
ウレタン系塗膜防水
コンクリートスラブ
梁：耐火被覆
2時間耐火t45RW
梁：耐火被覆表シ
断熱材：発泡硬質ウレタンフォームt15mm
道路境界線
店舗
エントランス店舗看板
ゴミ置場
セラミックタイル貼 t=12
均しモルタル：金コテ仕上t=25
コンクリートスラブt=180
捨てコンクリートt=60
砕石t=100
均しモルタル：金コテ仕上
コンクリートスラブt=180
ポリエチレンフィルムt=0.2
捨てコンクリートt=60
砕石t=100
モルタル金コテ下地
塗床仕上
1FL
GL
GL
FL-400
X1
X3
X4
機械室兼管理人室

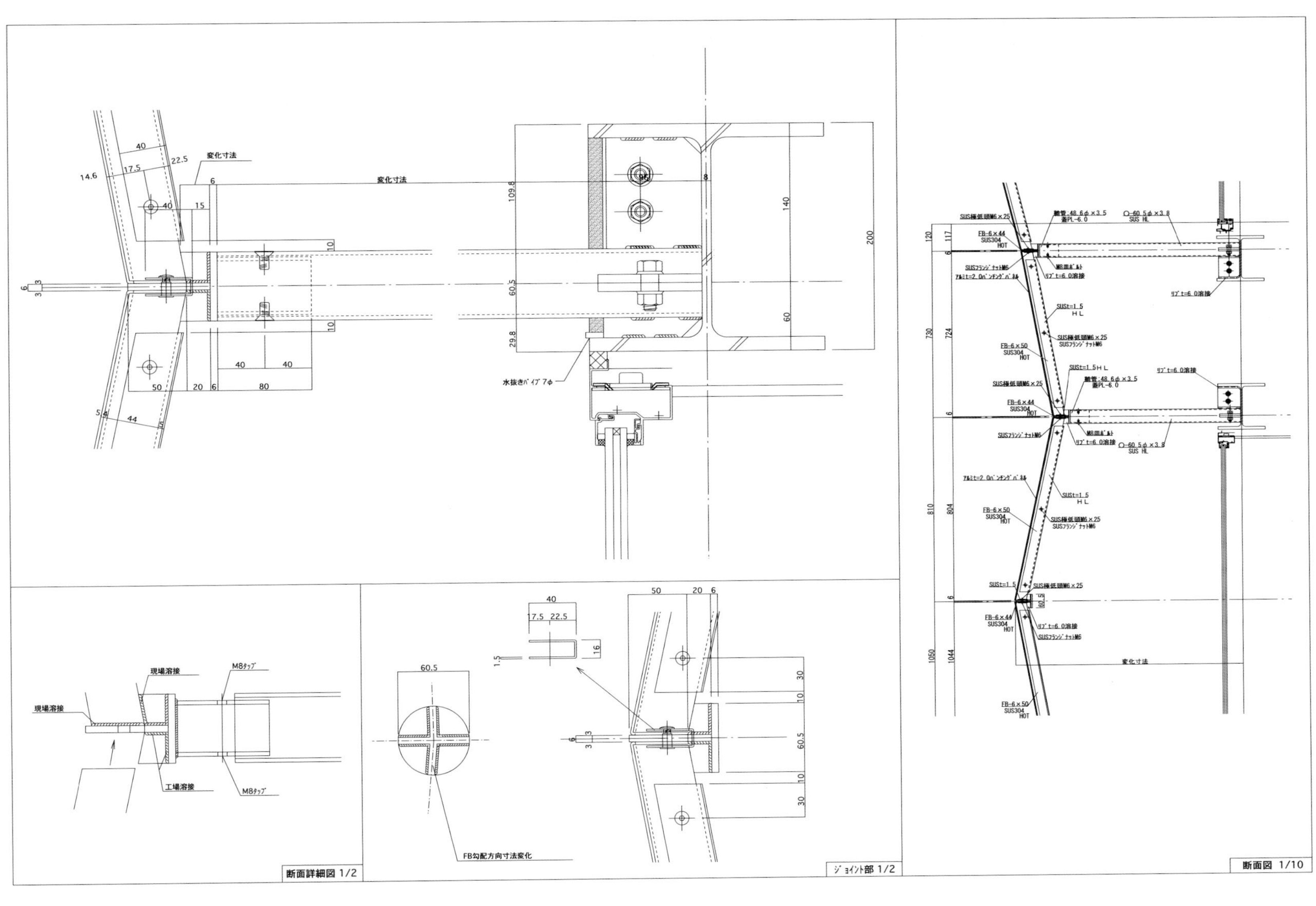

変化寸法
水抜きパイプ7φ
現場溶接
M8タップ
工場溶接
FB勾配方向寸法変化
SUS極低頭M6×25
SUSフランジナットM6
FB-6×44
SUS304
HOT
FB-6×50
鞘管:48.6φ×3.5
蓋PL-6.0
M8皿ボルト
リブt=6.0溶接
アルミt=2.0パンチングパネル
SUSt=1.5
HL
○-60.5φ×3.8
SUS HL
断面詳細図 1/2
ジョイント部 1/2
断面図 1/10

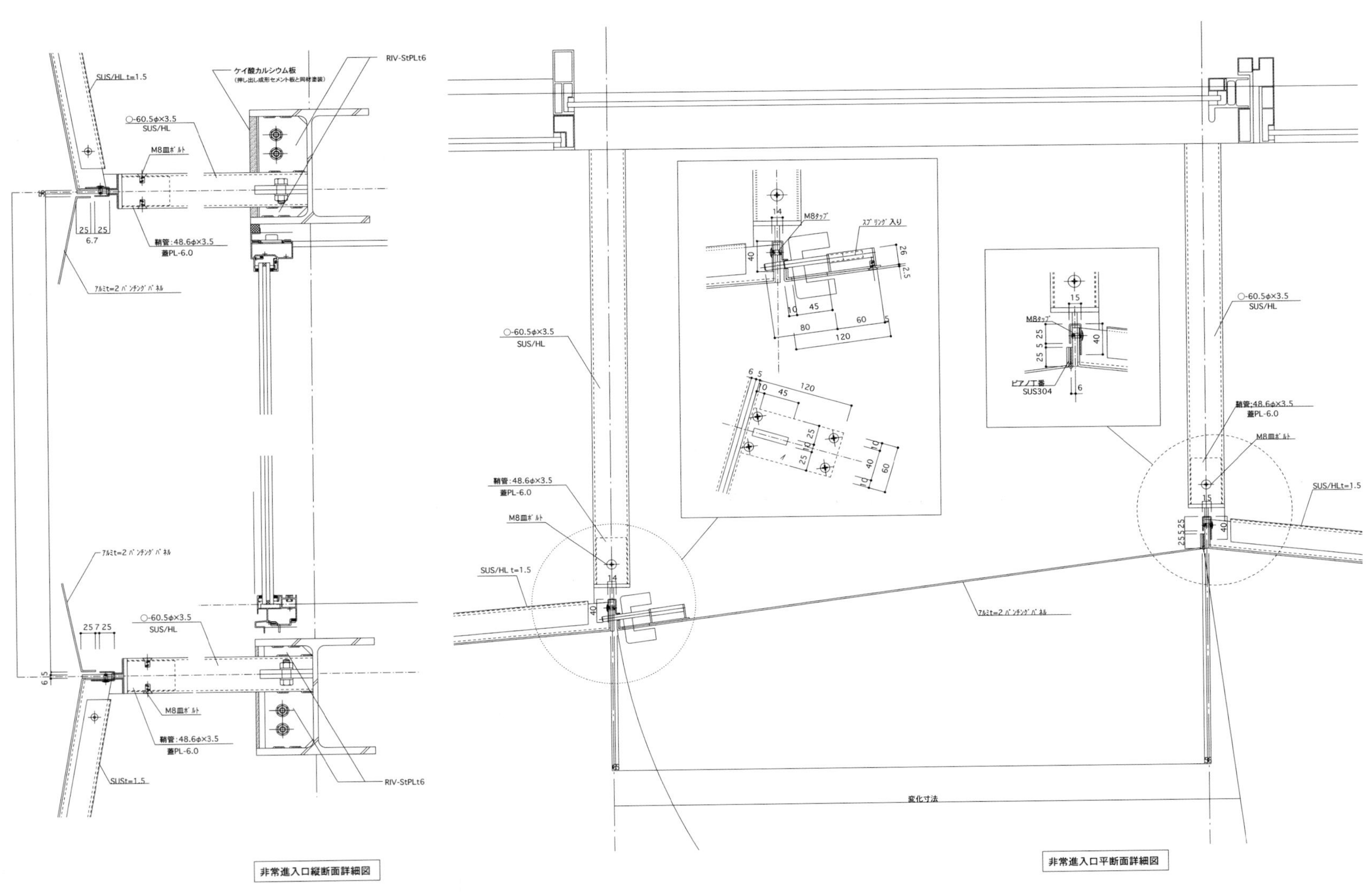

SUS/HL t=1.5
ケイ酸カルシウム板
RIV-StPLt6
○-60.5φ×3.5
SUS/HL
M8皿ボルト
鞘管:48.6φ×3.5
蓋PL-6.0
アルミt=2 パンチングパネル
SUSt=1.5
M8タップ
スプリング入り
ピアノ丁番
SUS304
SUS/HLt=1.5
変化寸法
非常進入口縦断面詳細図
非常進入口平断面詳細図

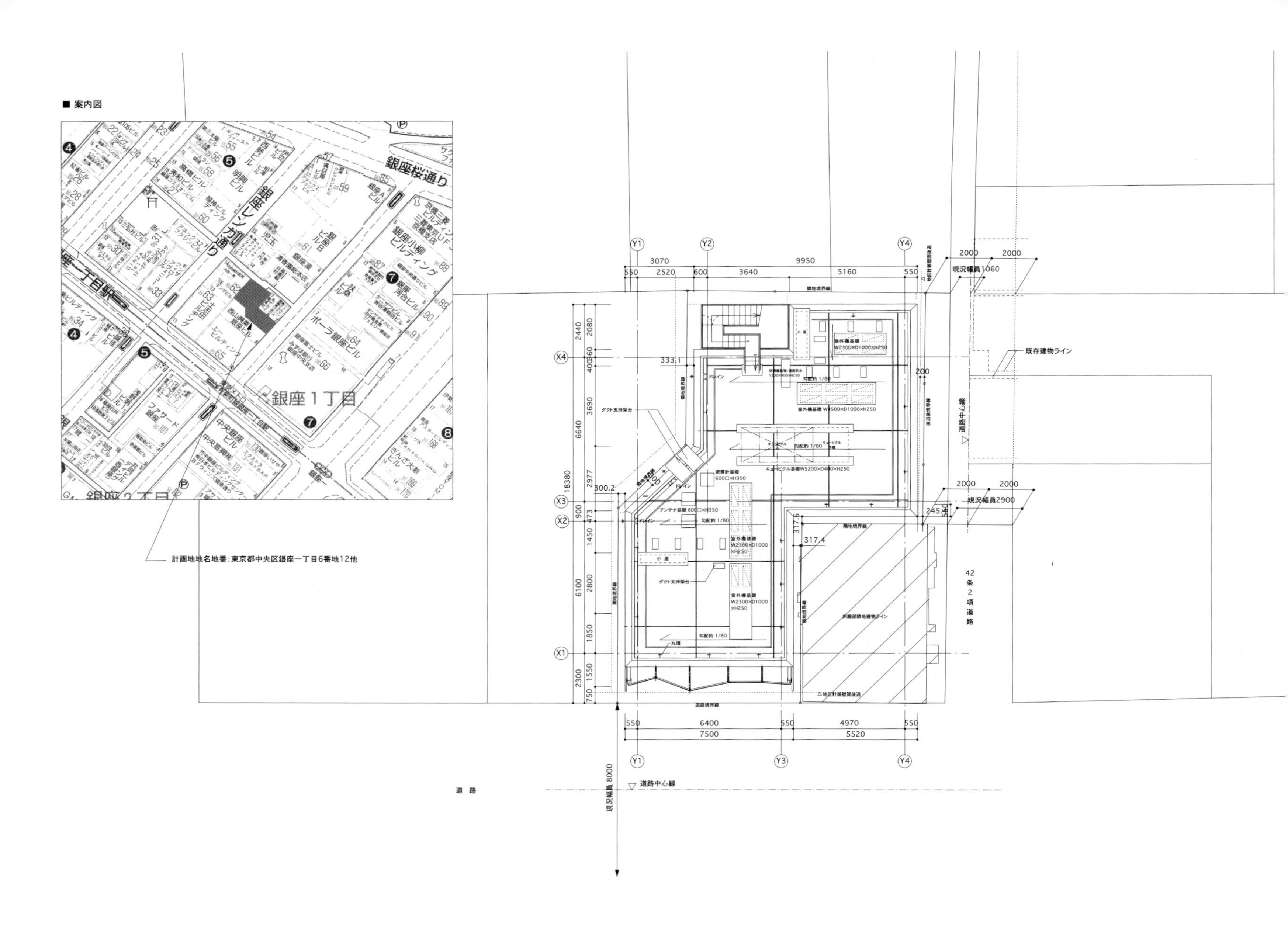
■ 案内図
銀座桜通り
銀座レンガ通り
銀座1丁目
計画地地名地番：東京都中央区銀座一丁目6番地12他
既存建物ライン
道路中心線
現況幅員2900
現況幅員1060
42条2項道路
道路
現況幅員 8000
道路中心線

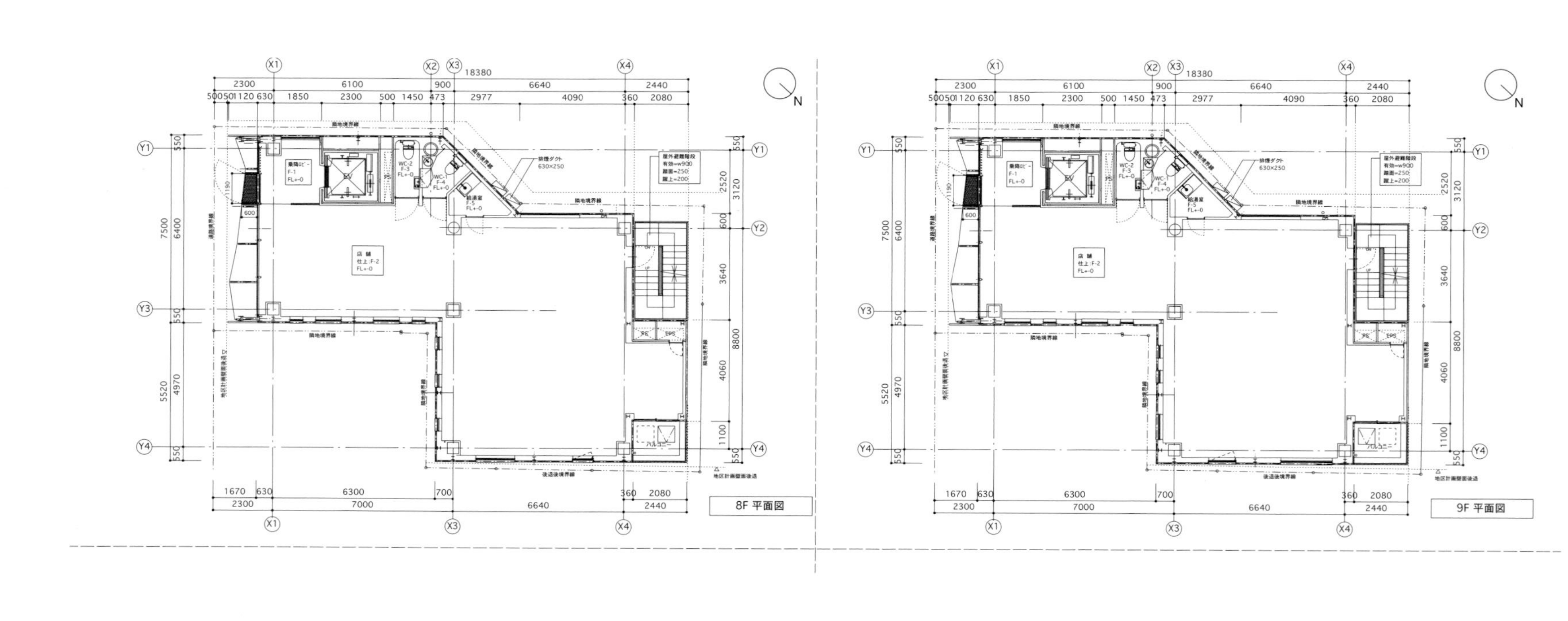
8F 平面図
9F 平面図

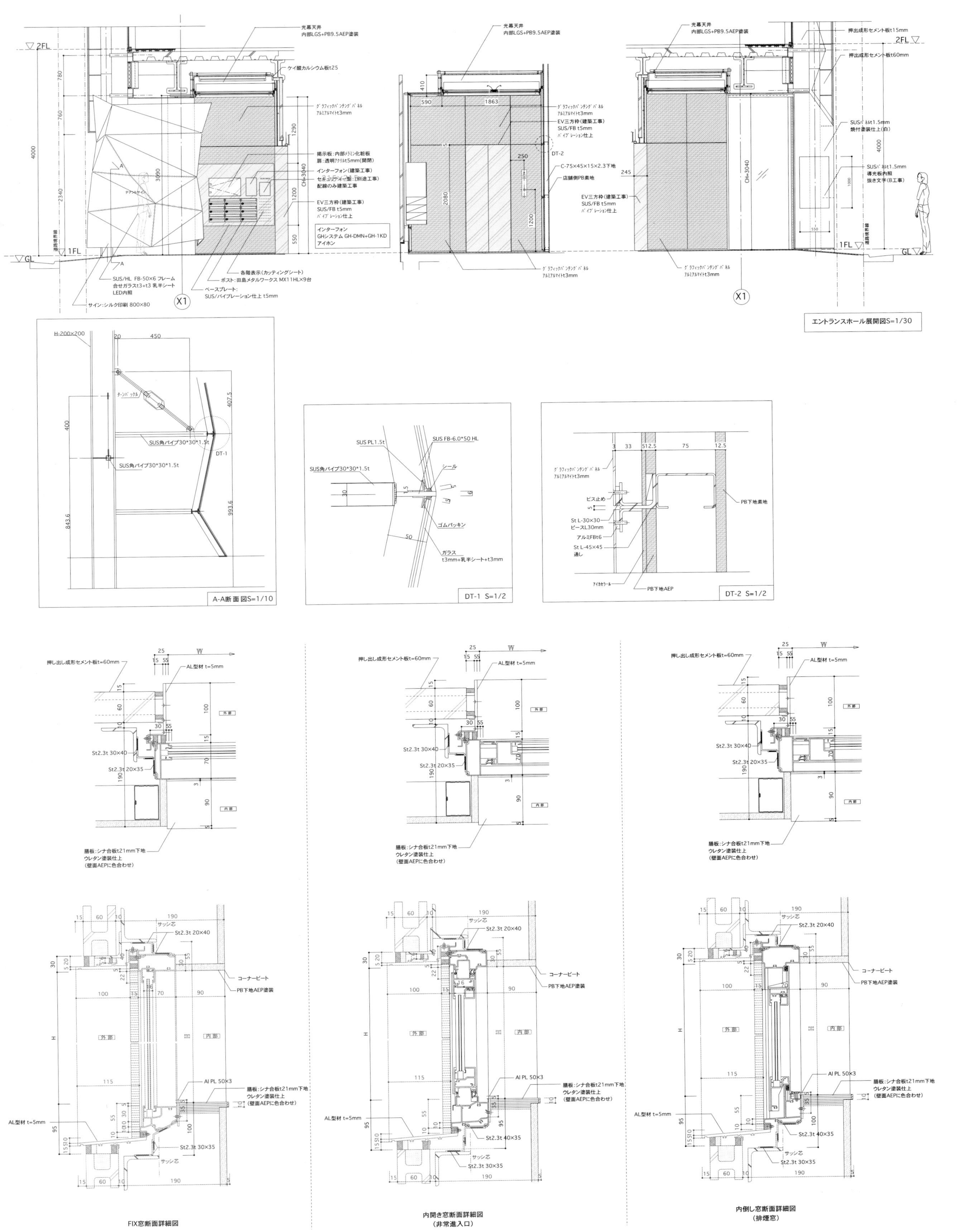

エントランスホール展開図S=1/30

A-A断面図S=1/10

DT-1 S=1/2

DT-2 S=1/2

FIX窓断面詳細図

内開き窓断面詳細図
(非常進入口)

内倒し窓断面詳細図
(排煙窓)

Dear Jingumae

Location: Tokyo,Japan
Architect: Amano Design Office Inc.
Building Area: 121.09 m²
Total Floor Area: 573.73 m²
Completion Date: 2014
Photography: Nacasa & Partners Inc.

This is a conversion project that totally renovated a 25-year old office building, located on a back street of Omotesando in Tokyo, in order to facilitate commercial functions.

The client requested a design that would have a facade expression differentiated from the surrounding buildings, and that would be a part of future tenants' branding.

The architect aimed at a design with a soft expression that would be favorably accepted by passersby, while standing out from the surrounding buildings that tended to have physically hard expressions.

By removing the out-of-fashion decorative frame structure on the existing building frame as much as possible, the primary shape was exposed. Then, metallic louvers were placed by using computer design to add a modern expression that conjures images of soft clothing.

The louvers give a soft expression by connecting three-dimensionally misaligned radiuses. They appear complicated but consist of only two types of radii (700R and 1700R) and straight lines. The louver components are welded to crossed SUS plates, and protrude from the building frame by SUS pipes, resulting in a streamlined workability.

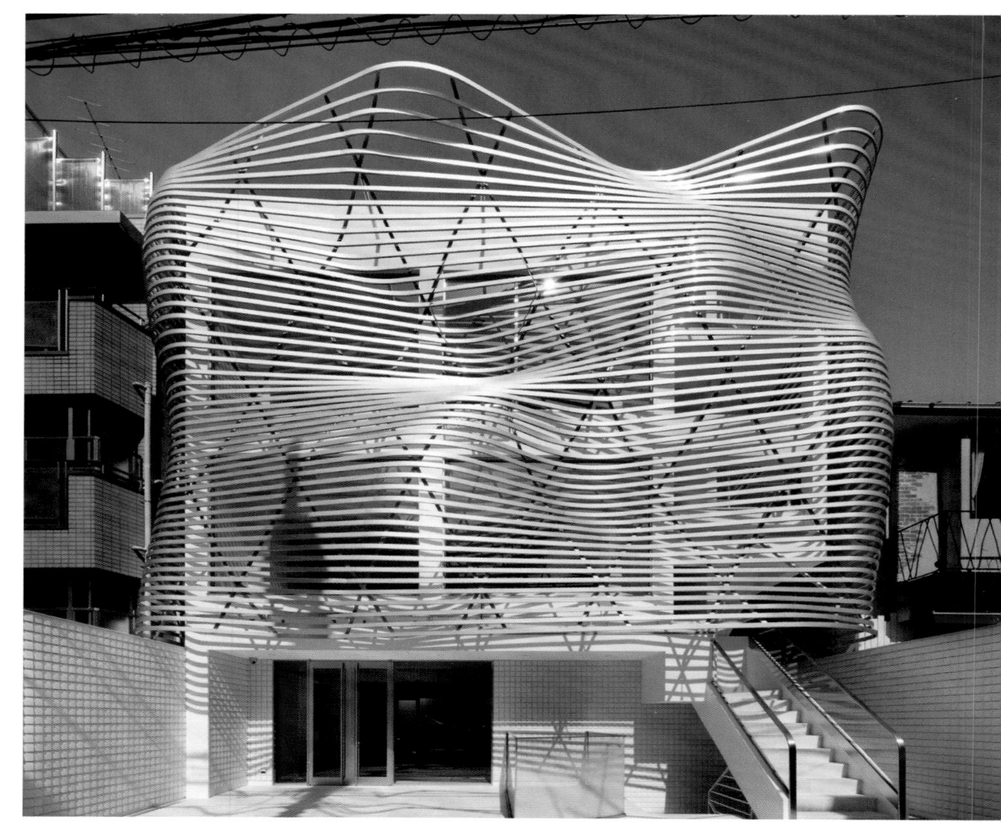

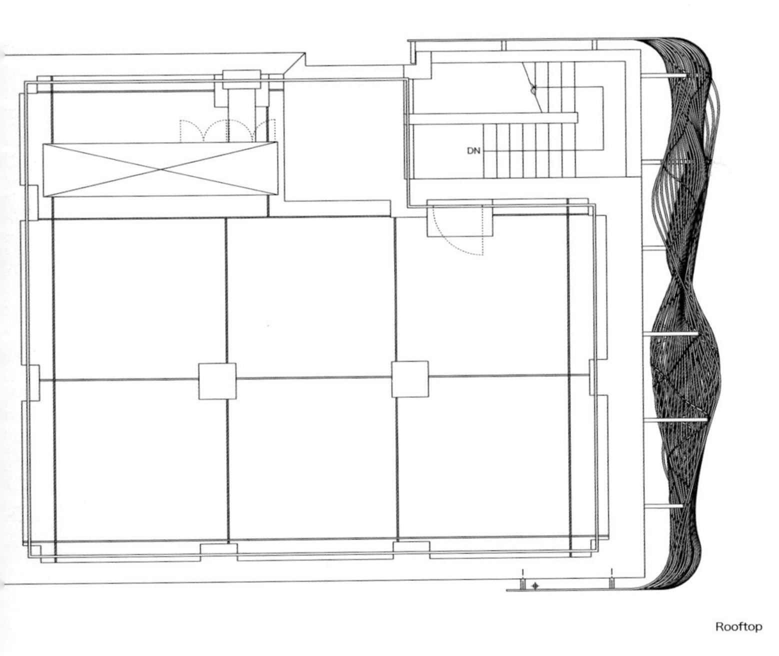

Rooftop

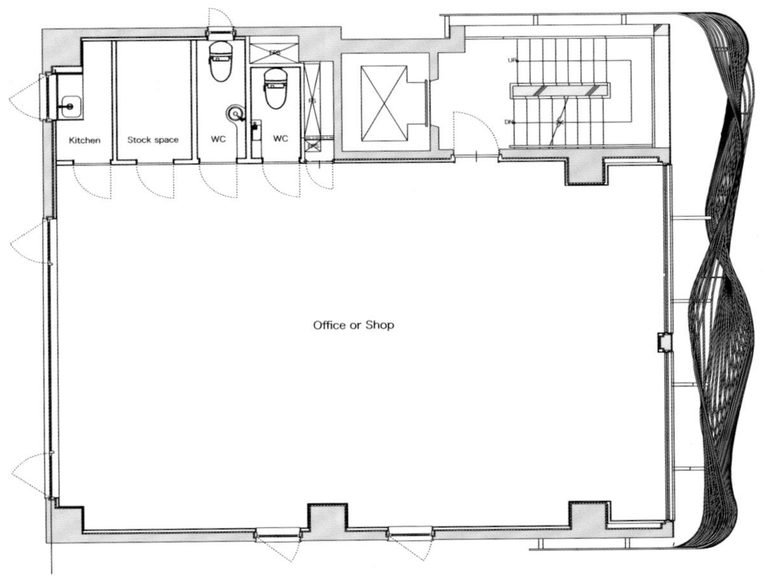

Third floor

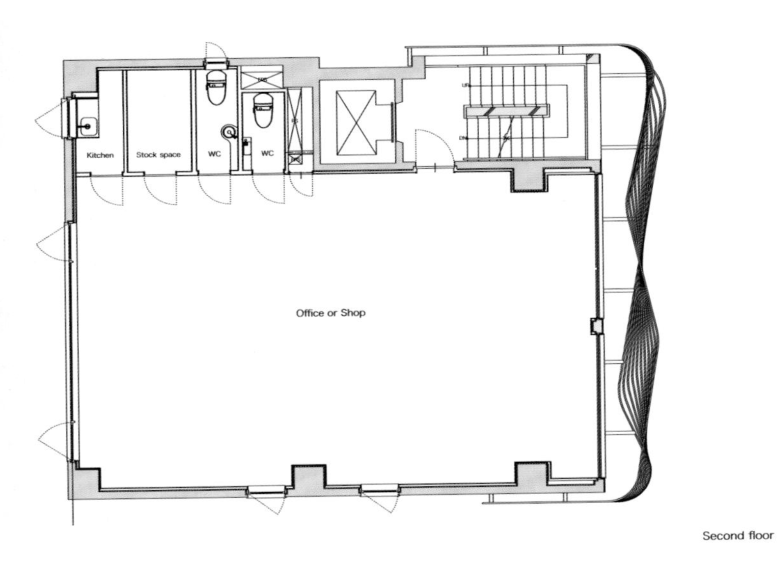

Second floor

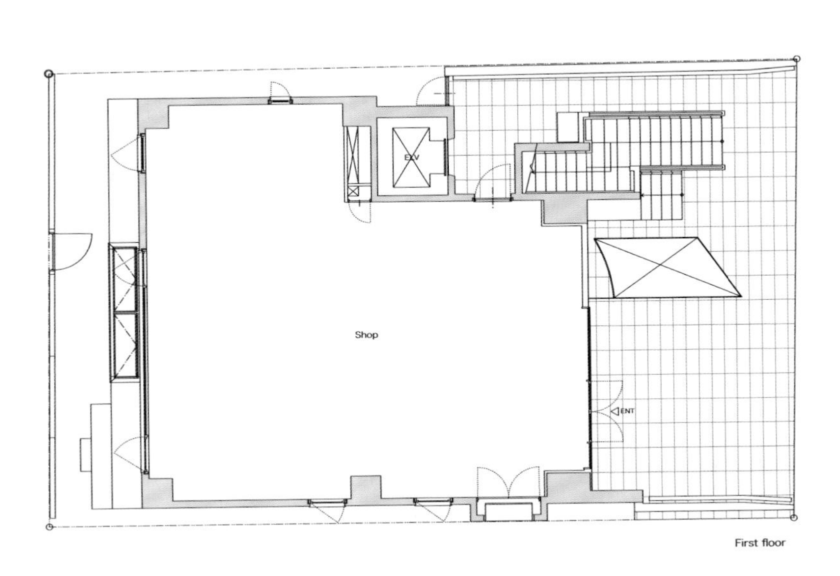

First floor

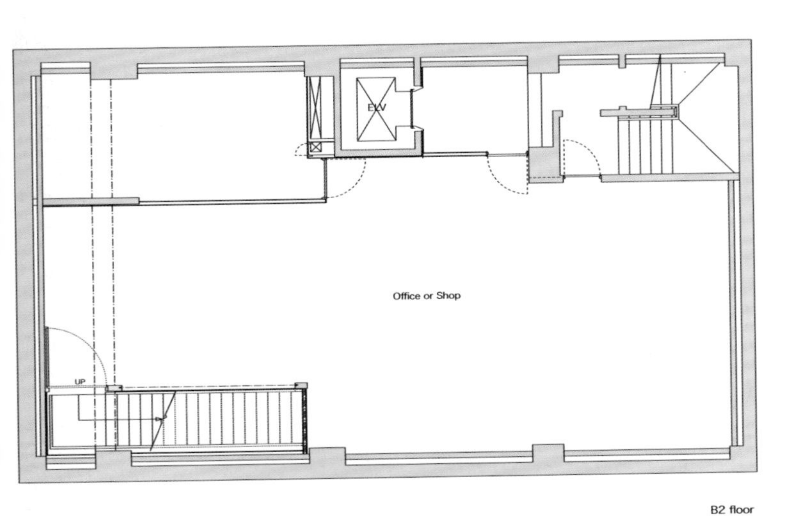

B2 floor

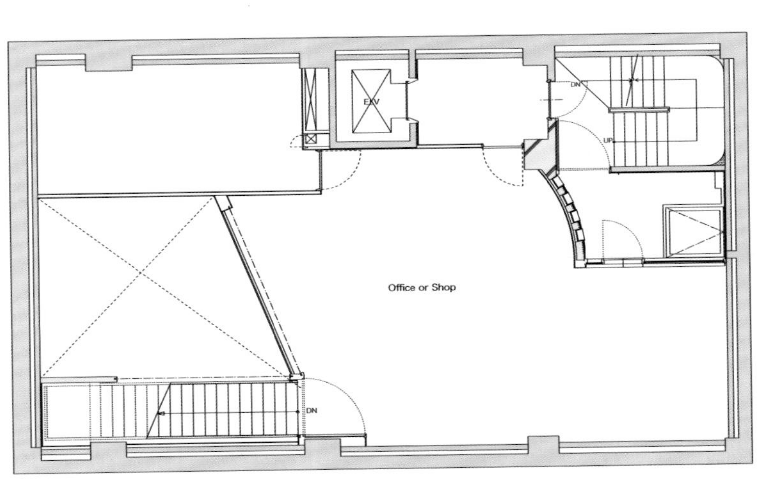

B1 floor

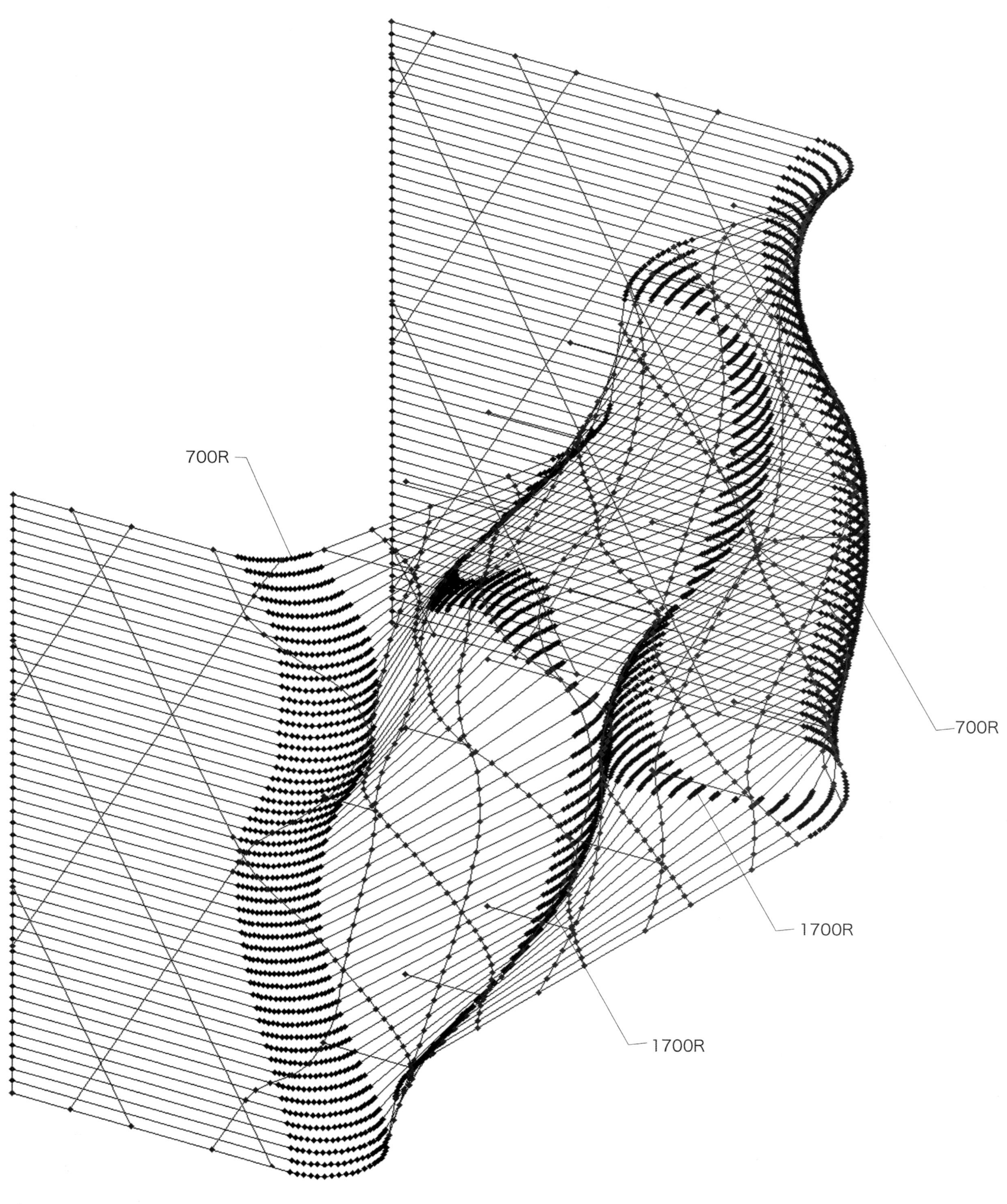

Geometry

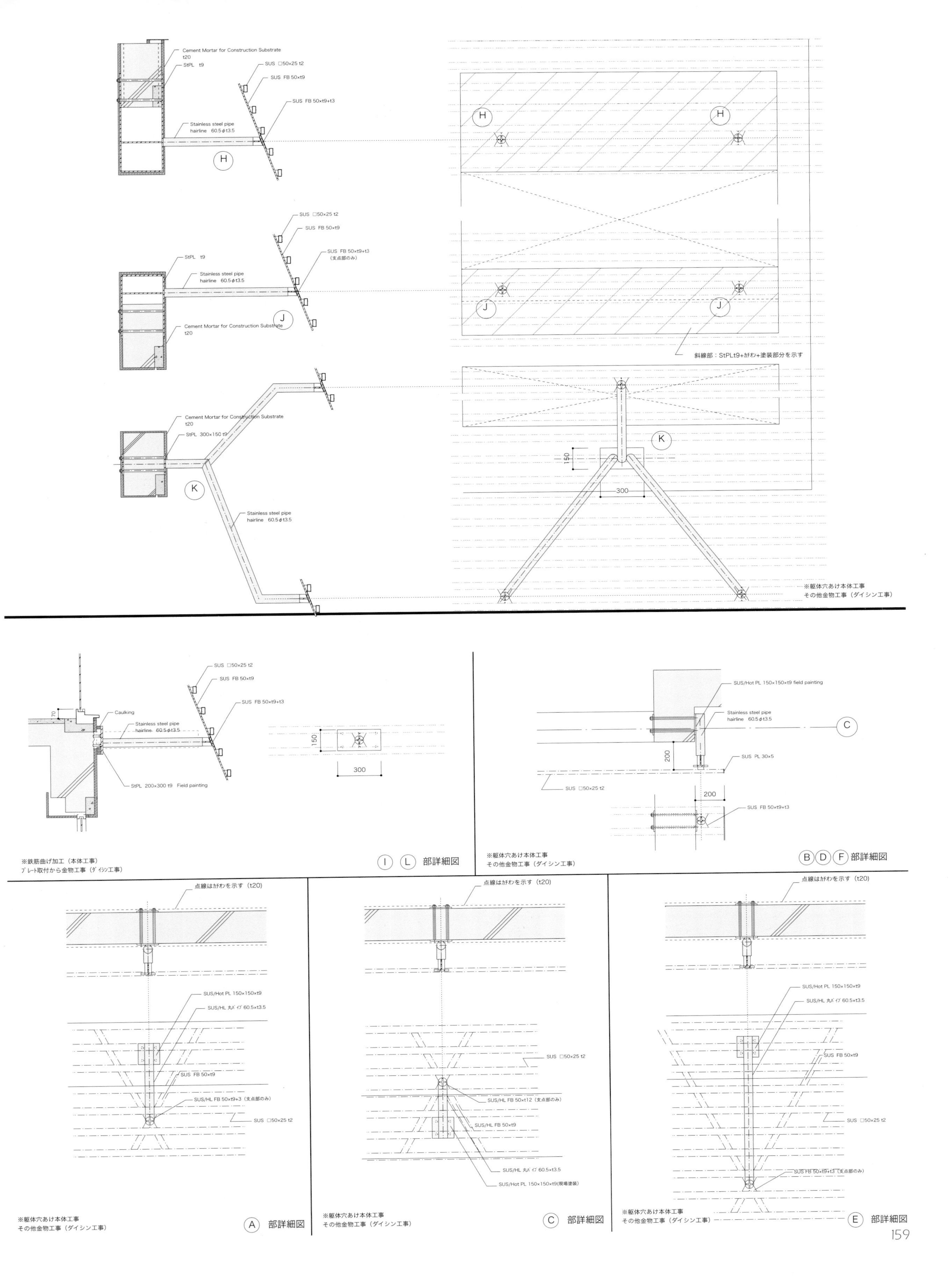

Cement Mortar for Construction Substrate t20
StPL t9
SUS □50×25 t2
SUS FB 50×t9
SUS FB 50×t9+t3
Stainless steel pipe hairline 60.5φt3.5
H
SUS FB 50×t9+t3 (支点部のみ)
J
斜線部：StPLt9+ｶﾁｵﾝ+塗装部分を示す
StPL 300×150 t9
K
150
300
※躯体穴あけ本体工事
その他金物工事（ダイシン工事）
Caulking
70
StPL 200×300 t9 Field painting
※鉄筋曲げ加工（本体工事）
ﾌﾟﾚｰﾄ取付から金物工事（ﾀﾞｲｼﾝ工事）
I L 部詳細図
SUS/Hot PL 150×150×t9 field painting
C
200
SUS PL 30×5
B D F 部詳細図
点線はｶﾁｵﾝを示す（t20）
SUS/Hot PL 150×150×t9
SUS/HL 丸ﾊﾟｲﾌﾟ 60.5×t3.5
SUS/HL FB 50×t9+3（支点部のみ）
A 部詳細図
SUS/HL FB 50×t12（支点部のみ）
SUS/HL FB 50×t9
SUS/Hot PL 150×150×t9(現場塗装)
C 部詳細図
SUS FB 50×t9+t3（支点部のみ）
E 部詳細図

Blaas Company

Location: Bolzano, Italy
Architect: Monovolume Architecture + Design
Volume: 11,000 m³
Completion Date: 2007
Photographer: Oskar Da Riz

The company Blaas in Bolzano is specialized in electro-mechanics. In the new head office the company presents its new product range and offers repair service.

On the ground floor of the building there is the sales division, on the first floor the exposition area and the repair shop. All administration offices are located on the second floor. The overall impression of the structure is a homogenous and closed building. Nevertheless, there exists a separation between the public and the private sector. The client can perceive this clear and formal internal division already from the outside.

The glass facade on the Northern side provides a maximum of visibility and transparency to the exhibition and sales area. The private spaces such as repair offices, stockrooms and offices have their facades exposed to the South, East and West which are protected with a sun screening system.

In order to establish an optimal relation between natural light, development and planning of spaces there has been created a luminous entrance hall in the centre of the building with an inner courtyard. This green open spot permits the administrative sector of the second floor to receive ample natural light and at the same time it generates a protected, quiet recreation area for the staff.

The Blass architecture design entrance area features an inviting free standing stairway with an organic stairway opening and a skylight above. This skylight allows excellent two-sided illumination of the creative office work places. The massive office decor construction style present enough thermal mass to allow the construction to absorb heat during the day and to release it during the night.

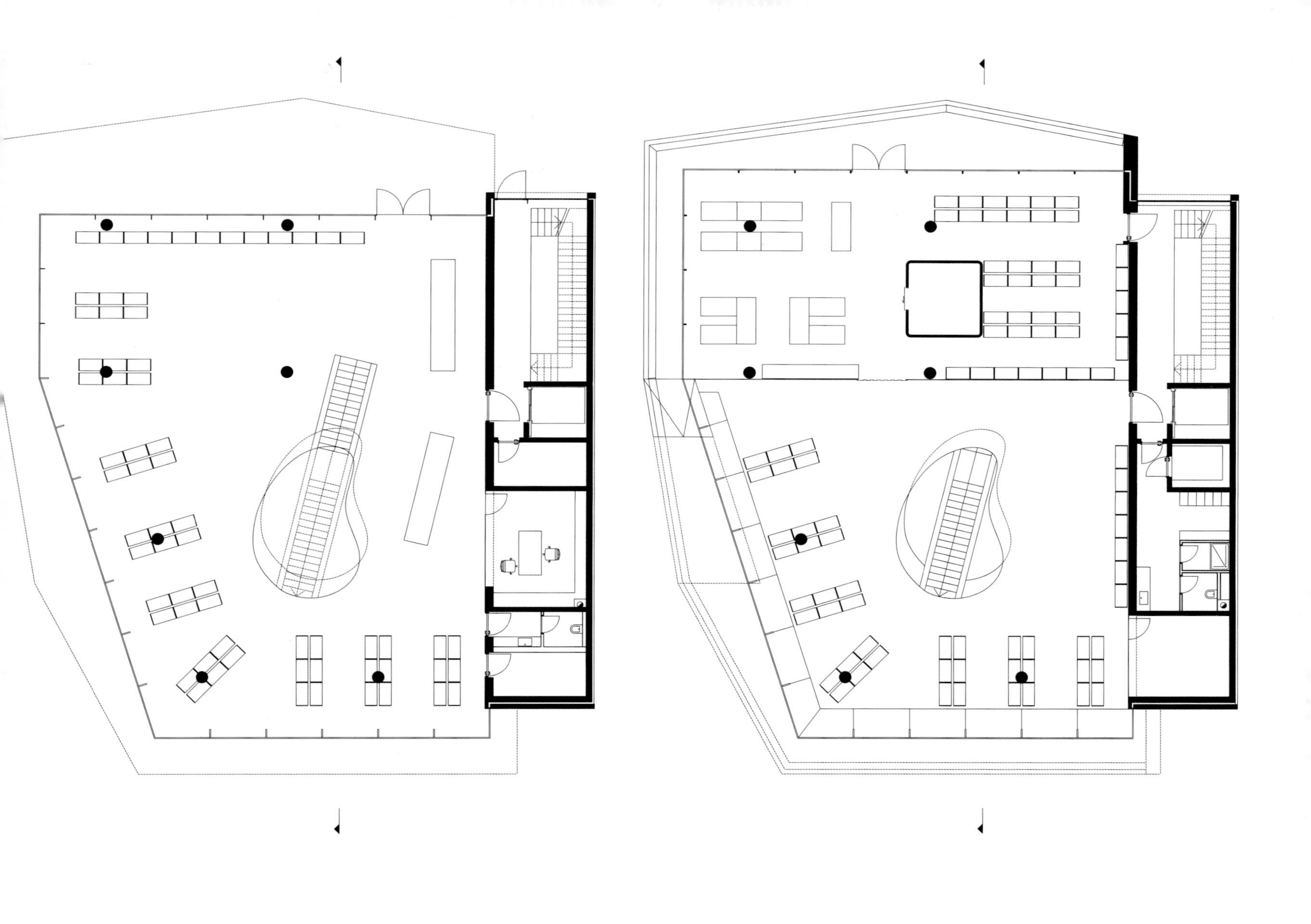

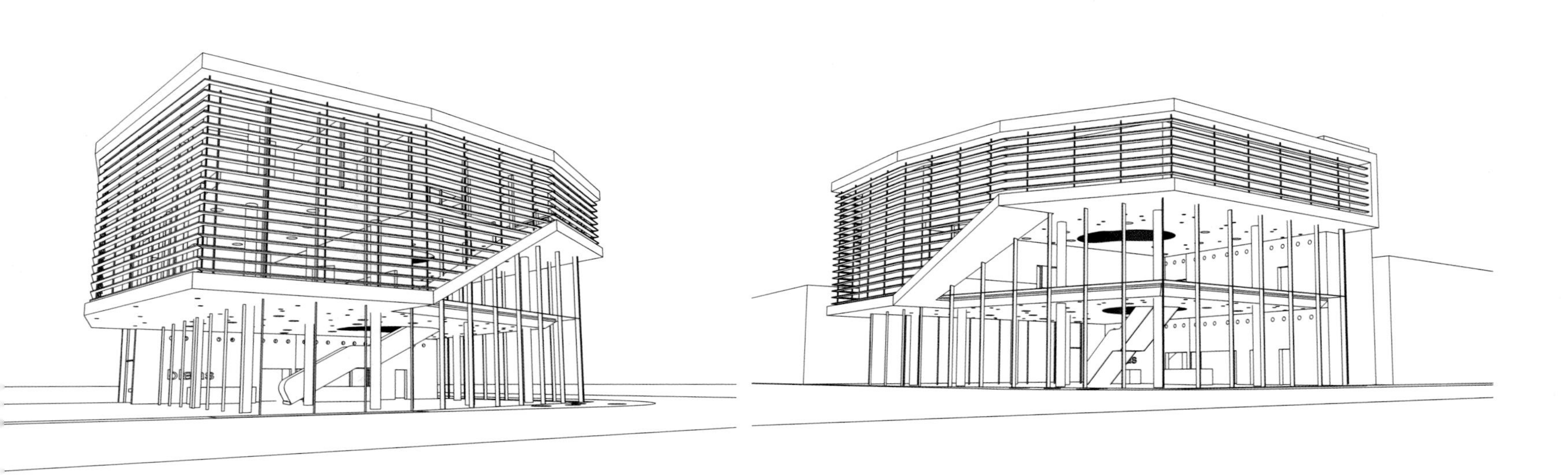

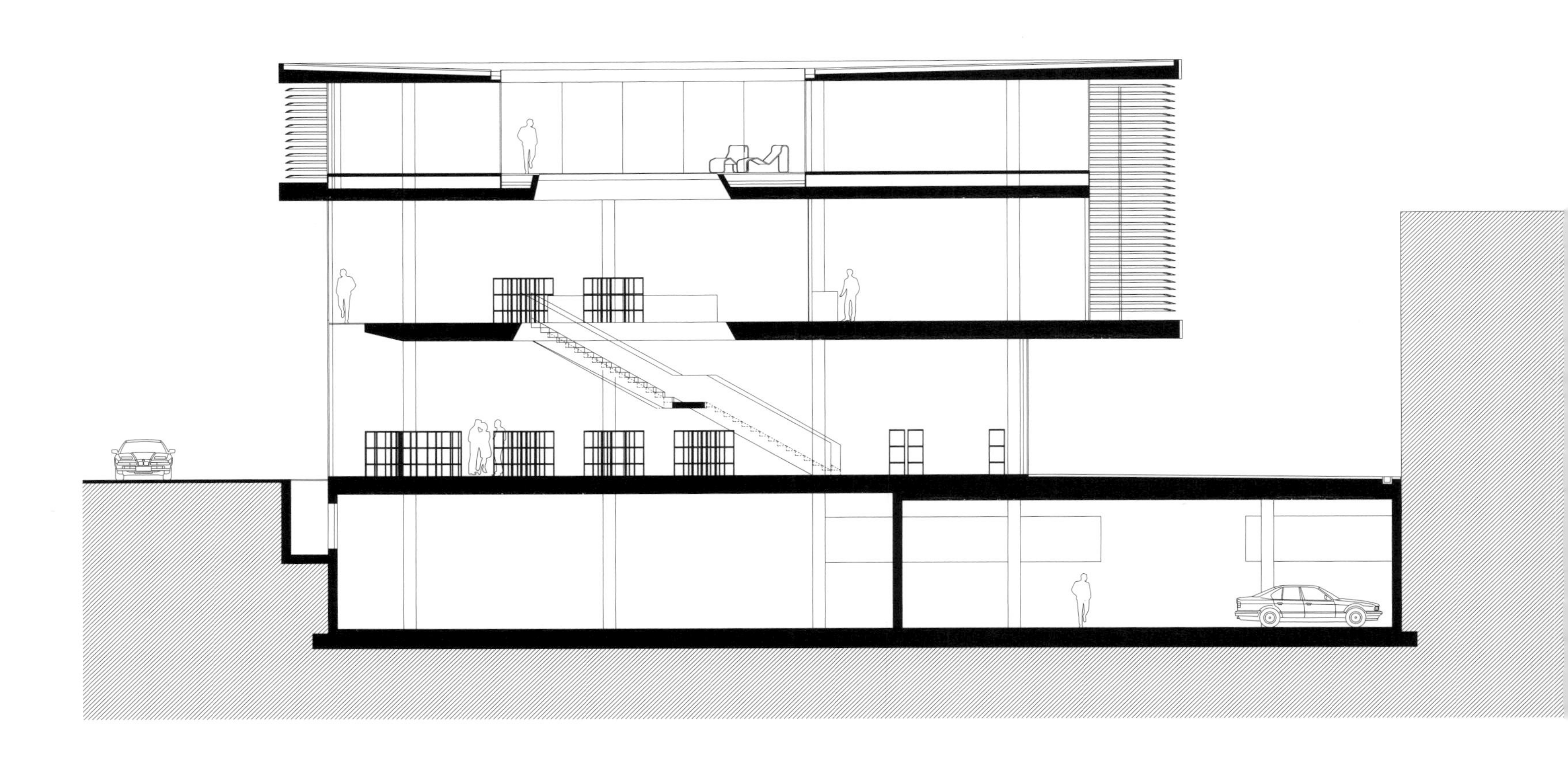

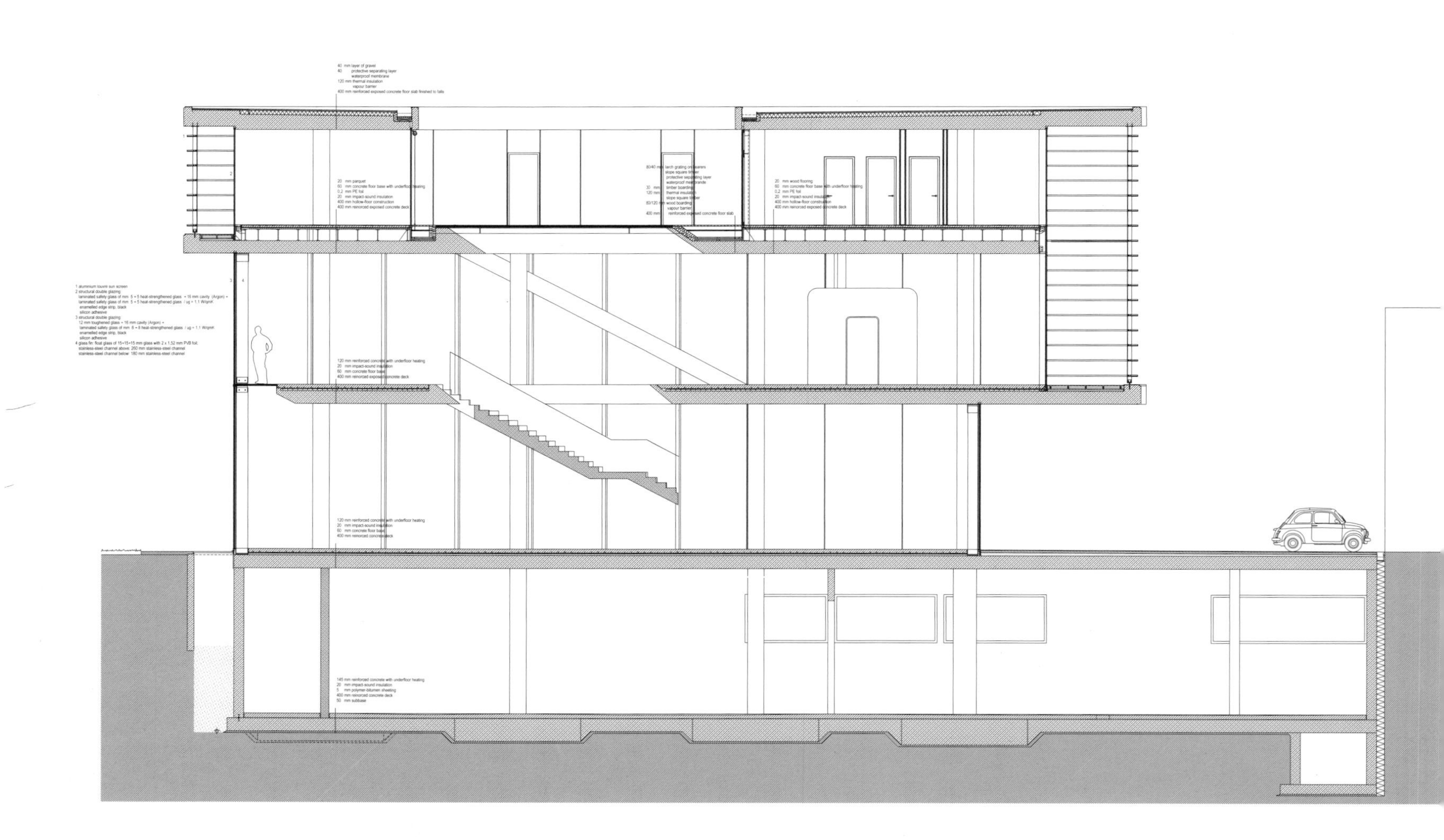

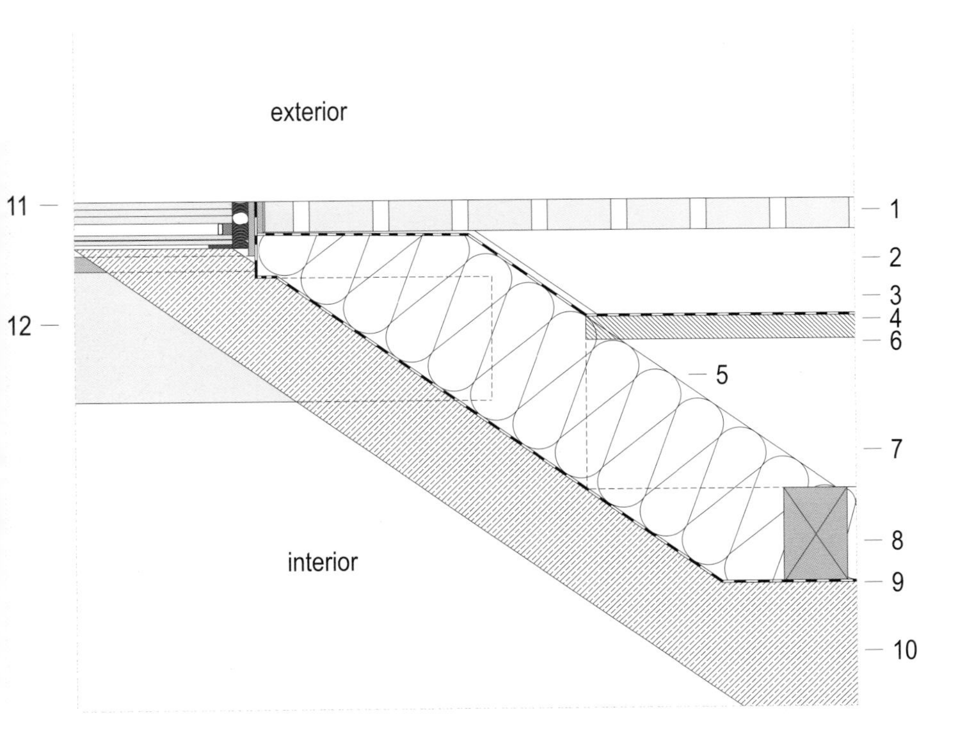

Skylight with horizontal glazing
Detai section
1 80/40 mm larch grating on bearers
2 slope square timber
3 protective separating layer
4 waterproof membrande Samafil G410- 18
5 30 mm timber boarding
6 120 mm thermal insulation
7 slope square timber
8 80/120 mm wood boarding
9 vapour barrier
10 400 mm reinforced exposed concrete floor slab
11 structural glazing: laminated safety glass of heat-strengthened glass 6 + 6 -15 - 15 -5 + 5 mm/ ug = 1.1 W/qmk
enamelled edge strip, black, jointed with black silicone;
12 Steel beam T 60/160/10 mm

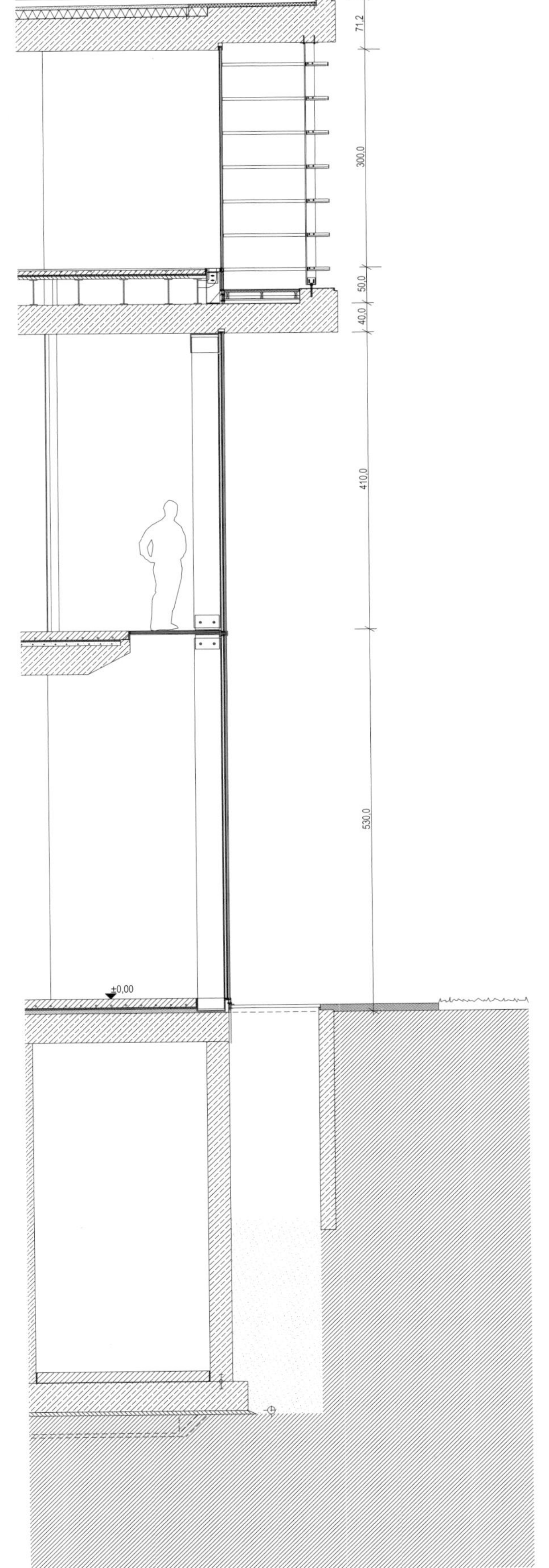

LBG Office

Location: Sicily, Italy
Architect: Architrend Architecture
Size: 10,145 m^2
Completion Date: 2013
Photographer: Moreno Maggi

The modernization of the company, which produces flour for the pharmaceutical and food factory from the seeds of the carob tree (Ceratonia siliqua), a typical cultivation of the local land, involves the reorganization of the entire production cycle with the introduction of new technologies and machinery.

The start-up program is an opportunity for a total redesign of the volumes and the interior spaces, and it is able to reestablish a business image of efficiency and productivity. The production complex is covered with a skin of insulated white aluminum panels which emphasizes the linearity of the volumes, interrupted only by the 25 m tower silos' height.

The office block is incorporated into a new glass volume, 46 meters long, over which is placed a cantilevered volume of 30 meters in a staggered position respect to the main entrance. This gives lightness to the ground floor volume, and the glass curtain wall is defined by the presence of two aluminum string-courses.

The main entrance to the office is marked by a canopy made of a thin cantilever plate. The interior is dominated by light and glossy surfaces. The reception, made by a long reception desk in opal plexiglass and a behind wall made by frosted glass, leads to the offices on the ground floor.

A staircase, made by glass and steel, leads to the first level where there are placed a large boardroom, a meeting room and a testing and quality control laboratory. The laboratory is placed in the northwest corner of the overhanging volume and present a large window through which is possible to see the entire production area.

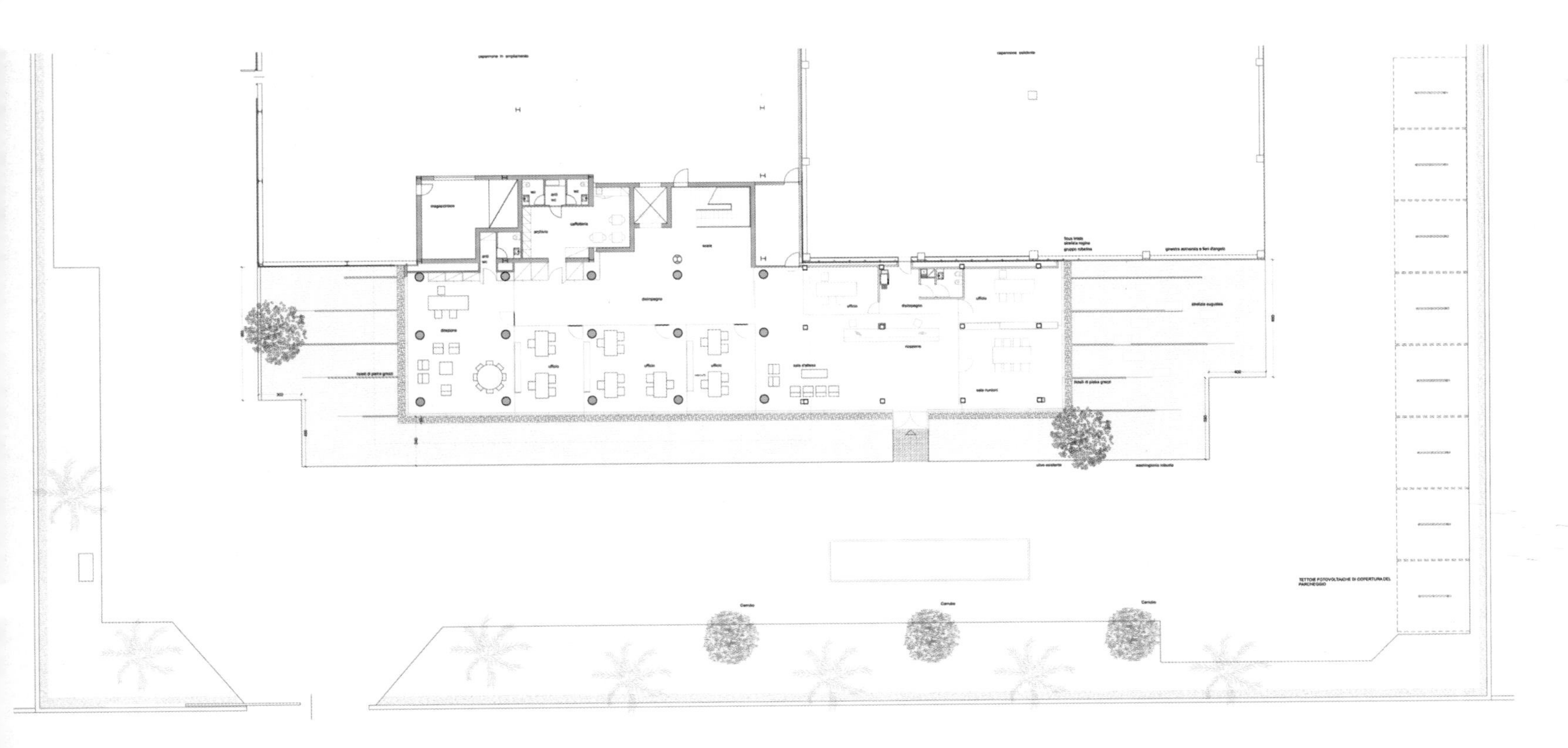

Planimetria Del Lotto

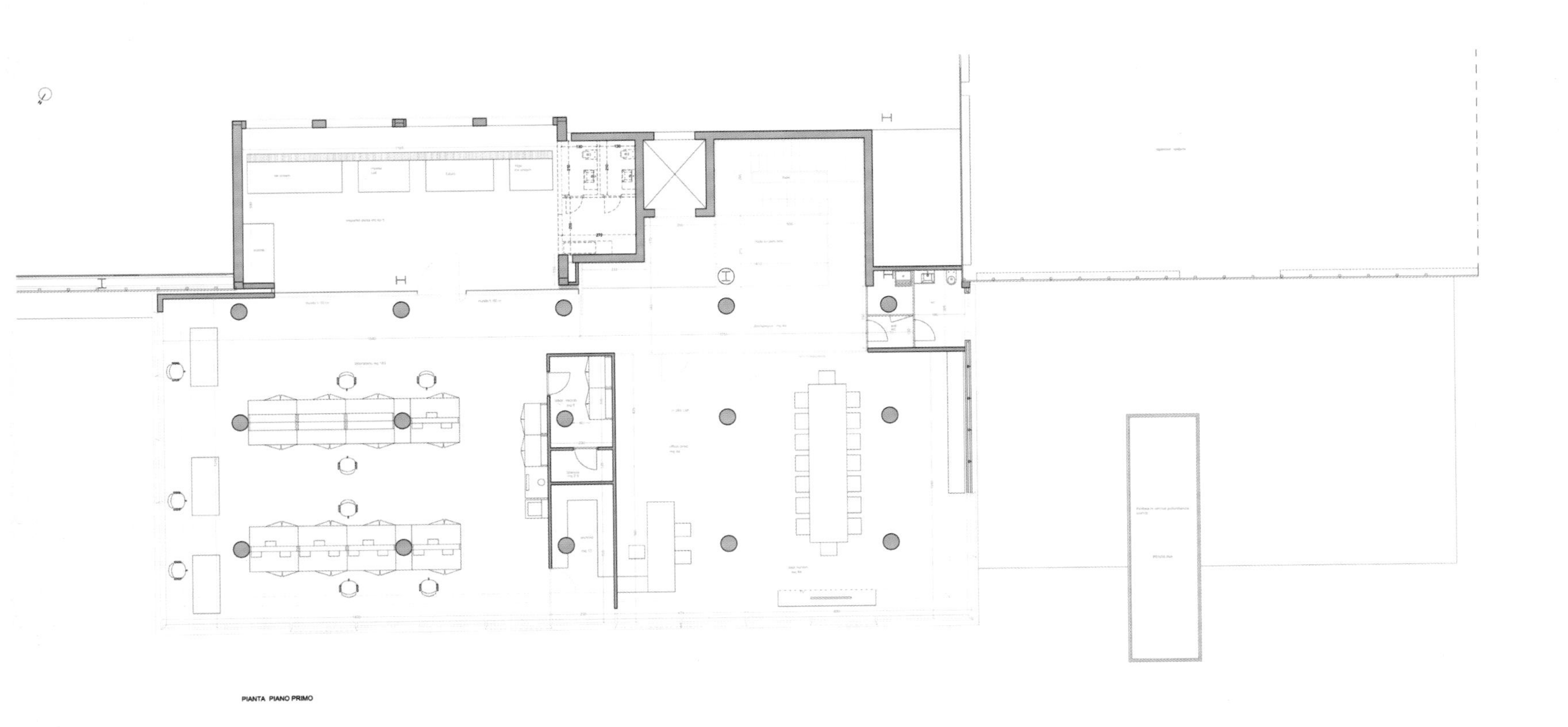

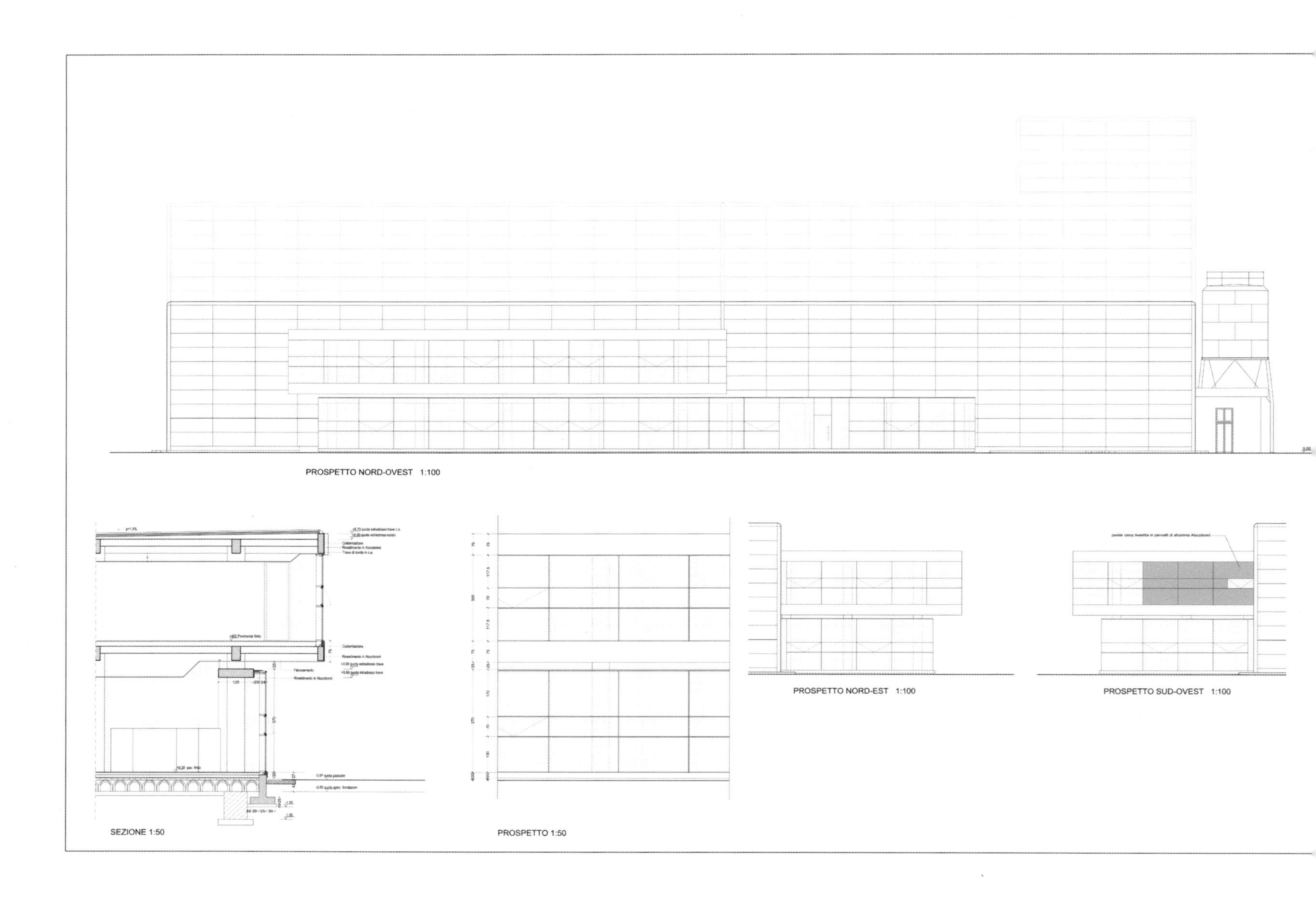
PROSPETTO NORD-OVEST 1:100
SEZIONE 1:50
PROSPETTO 1:50
PROSPETTO NORD-EST 1:100
PROSPETTO SUD-OVEST 1:100

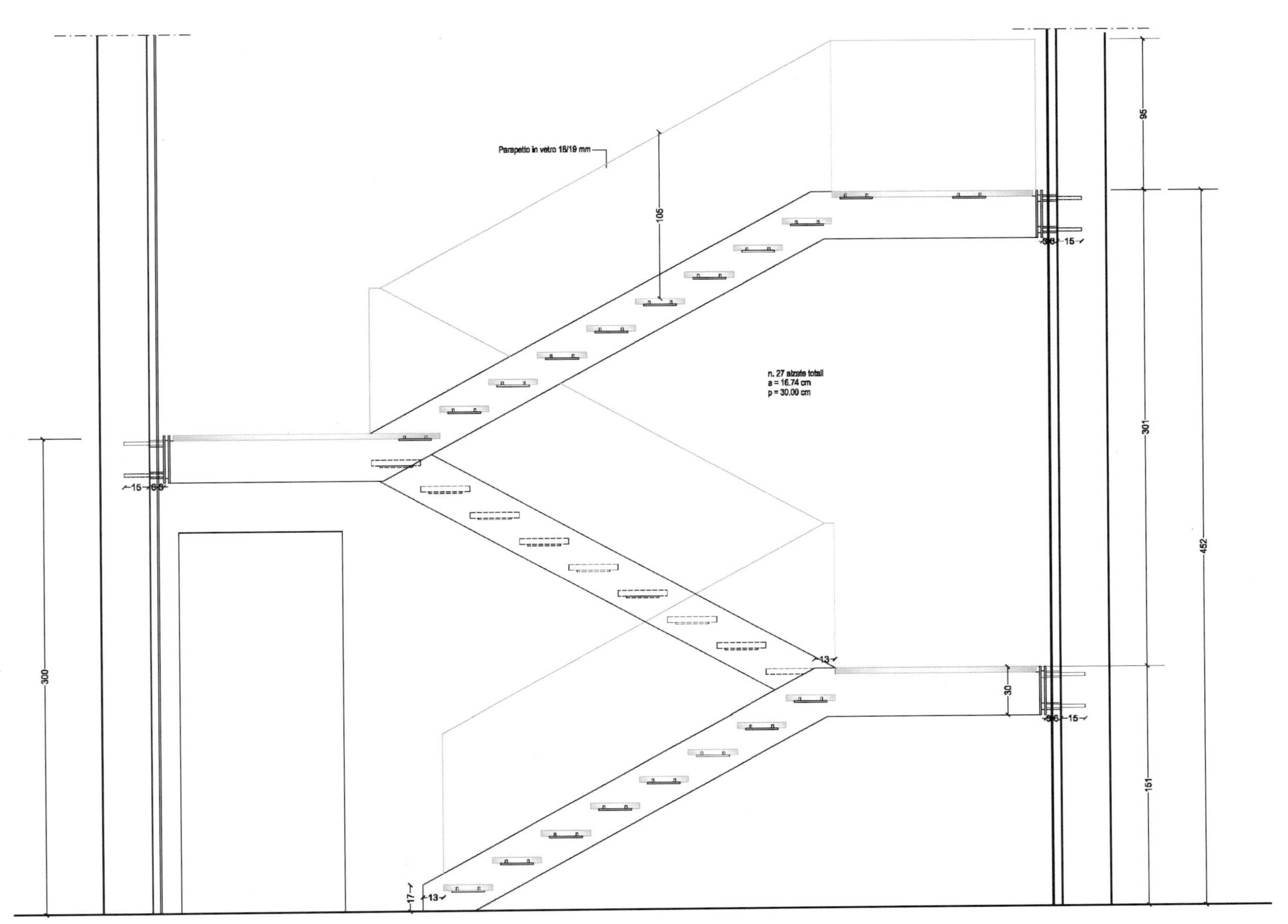

SEZIONE A-A 1:20 PRIMA E TERZA RAMPA

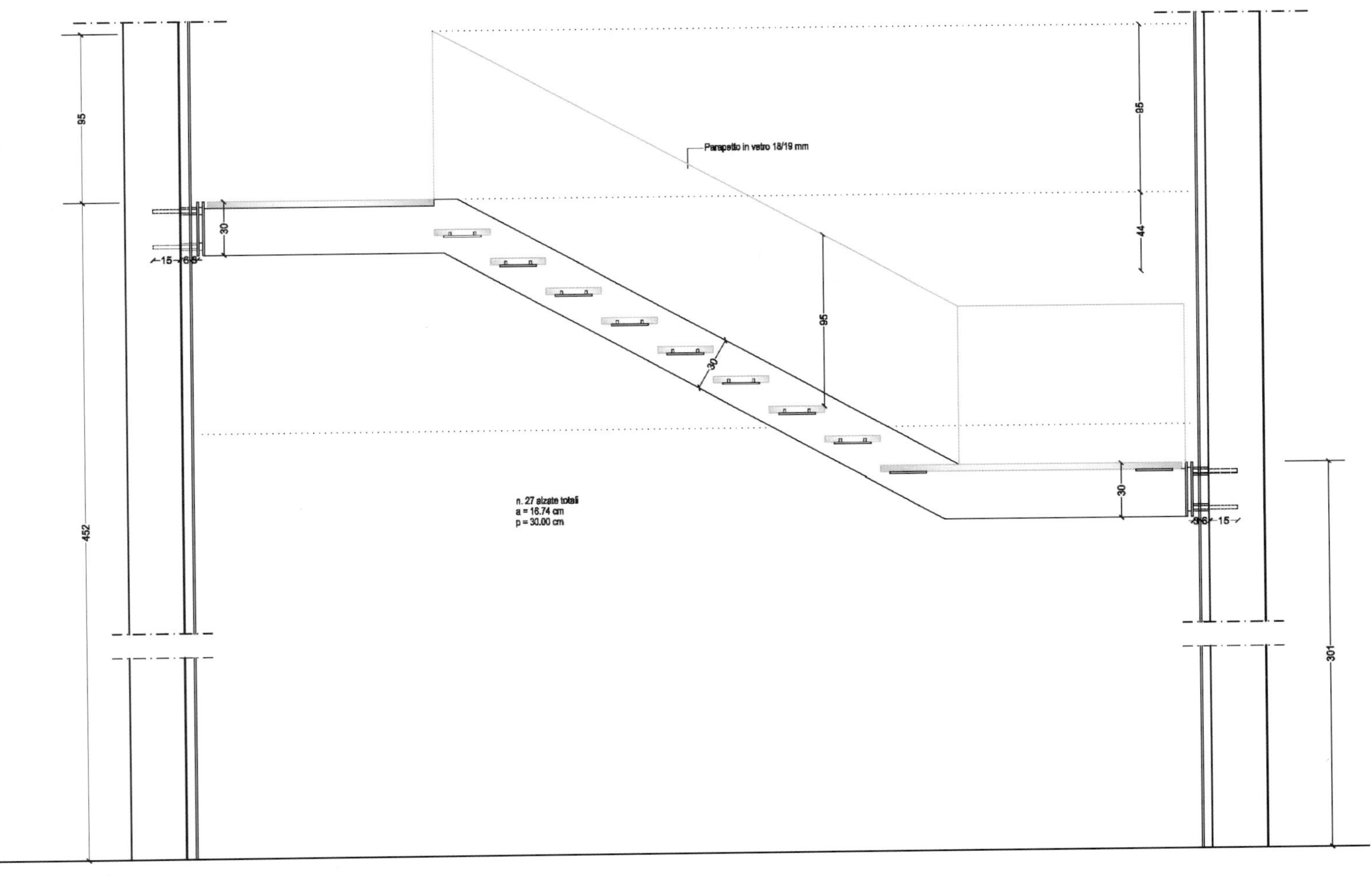

SEZIONE C-C 1:20 ULTIMA RAMPA

Lazika Municipality

Location: Lazika, Georgia
Architect: Architects of Invention
Client: Ministry Of Justice of Georgia
Site Area: 1,225 m²
Building Area: 400 m²
Completion Date: 2012
Photography: Nakanimamasakhlisi Photo Lab

Architects of Invention (AI), alongside engineers Engenuiti, designed a steel-structured building as a part of the development of a new city on the Black Sea coast. The brief was to create a Public Service Hall, a Wedding Hall and Municipal Offices.

The ambition of this project was to make a building as a sculpture made out of one material. The suspended volumes create public spaces, separated from each other, forming a monument, and saliently, a benchmark for the new architecture of new futuristic city. The priority here was the void, not the mass – the architects began with the empty space, since this is a project of the future and there is no memory to which solid mass might refer. Instead of carving void-space from a cube, the architects have done the reverse. It is a void with volumes inserted, and each volume has a fragile connection to the others, via the void. For example, there is a red glass tube (the elevators) connecting the lower floors with the top cubic volume - so the experience is of transition, as though through red lens. When you navigate the building, the voids are felt more than the solid forms, it can be daunting at times.

Paul Grimes, director of Engenuiti, describes the structure: 'The structural frame is constructed from steel, utilizing a series of raking steel columns to produce the dramatic floated upper platform. Challenges include the seismic conditions and the sympathetic employment of local skills to achieve what will be an iconic statement of architectural ambition'.

The foundation solution was for 80 piles 800mm diameter (has been installed through rotary percussion methods) and up to 25m deep. Control of settlements due to organic layers and control of potential liquefaction of the ground during a seismic event were the main concerns.

A key consideration in the design was the need to control costs where possible but also to use local materials and local labour so as to give the maximum benefit to the local economy. Normally this would push the design towards a structural concrete solution in Georgia. However the complexity of the project dictated that the project is best developed using a Structural Steel Framing Solution.

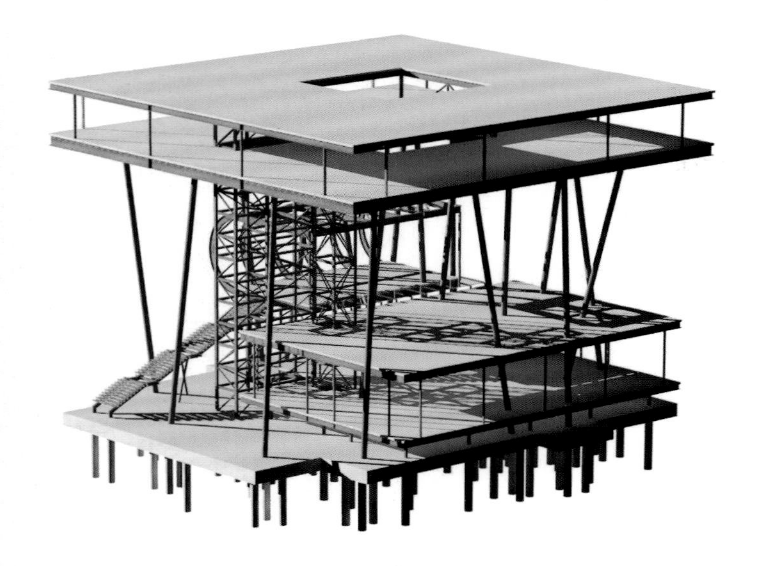

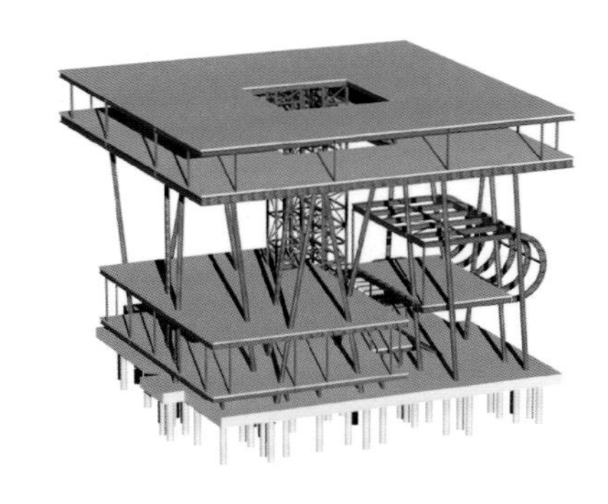

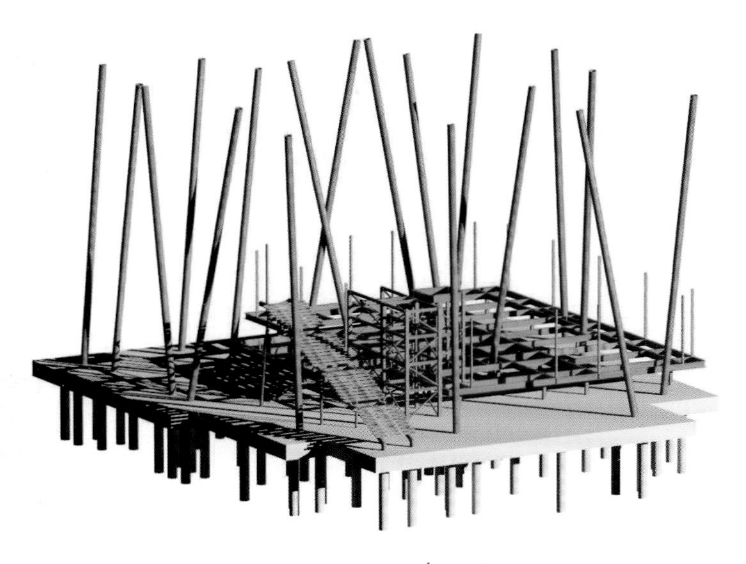

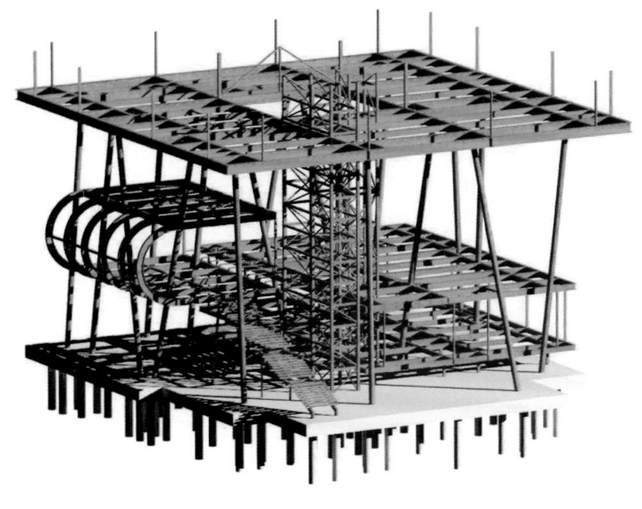

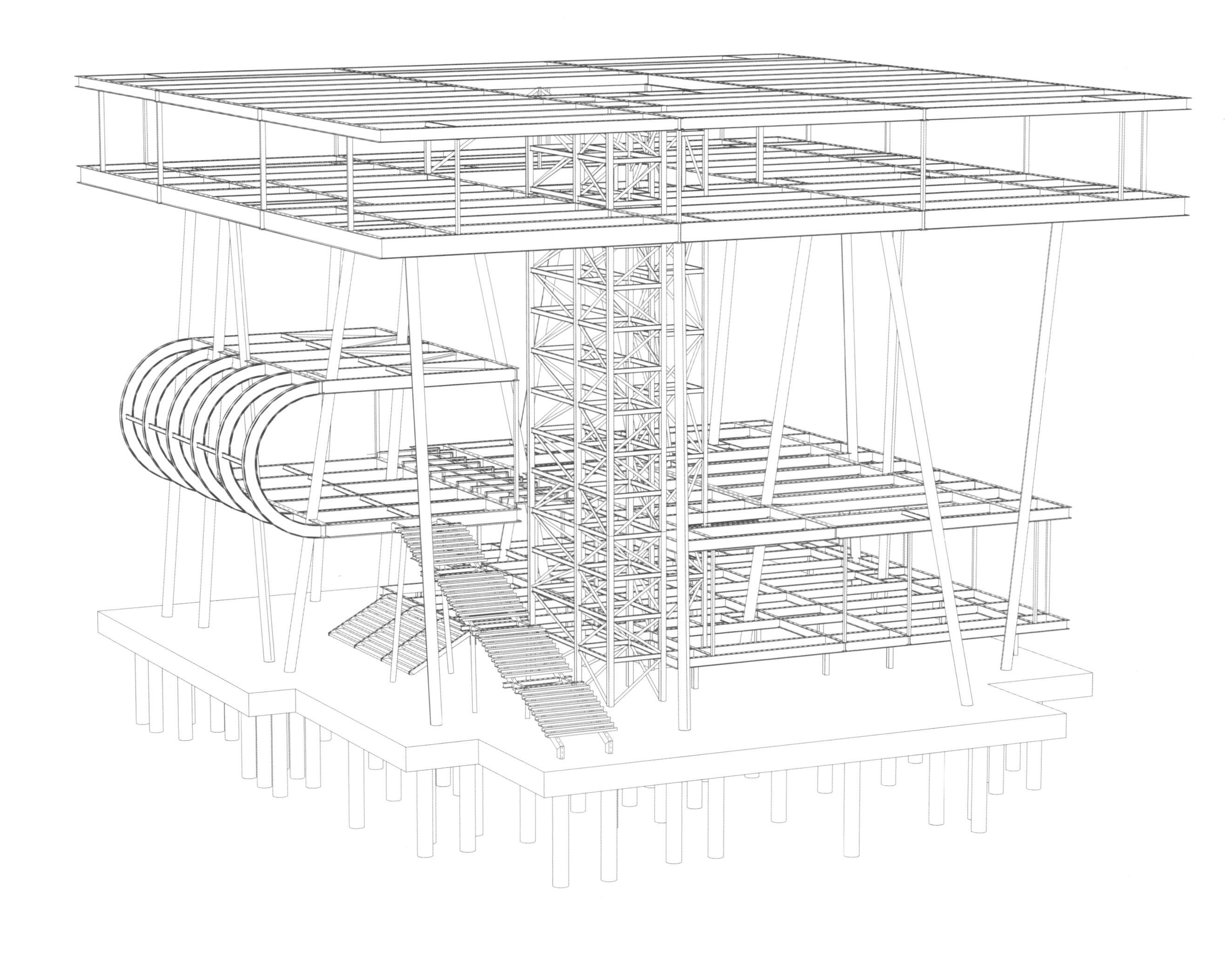

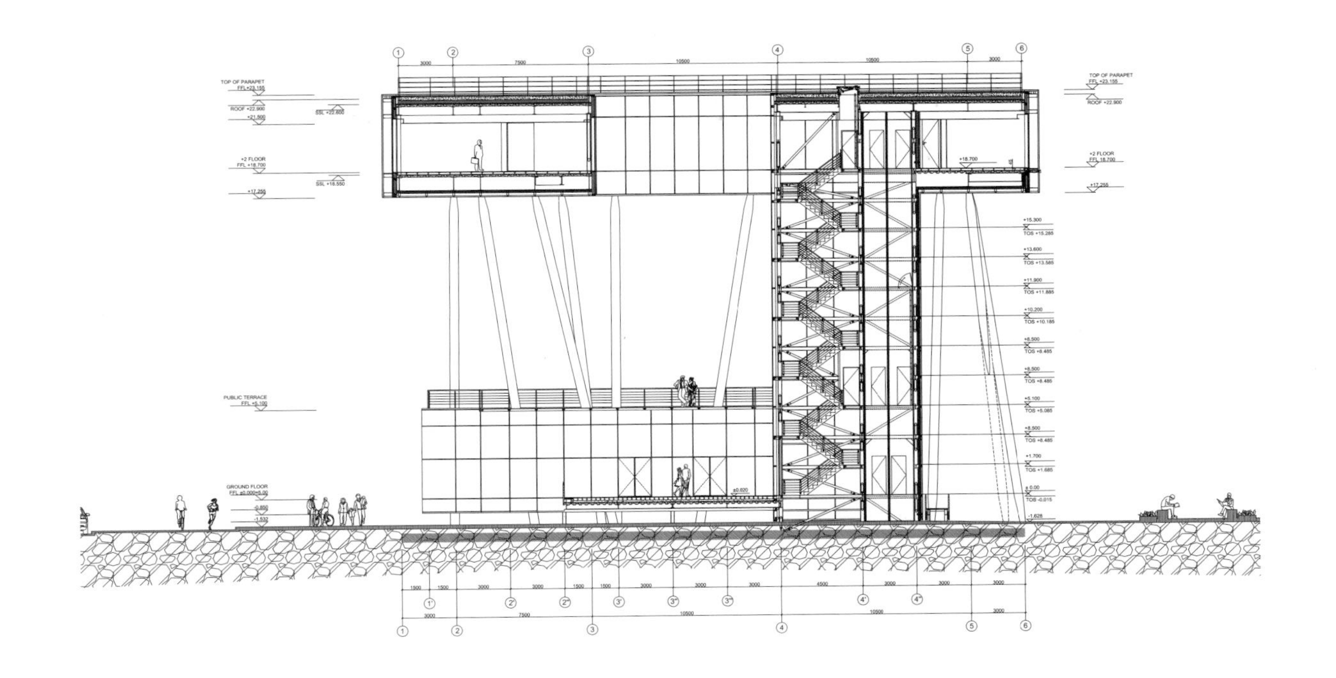

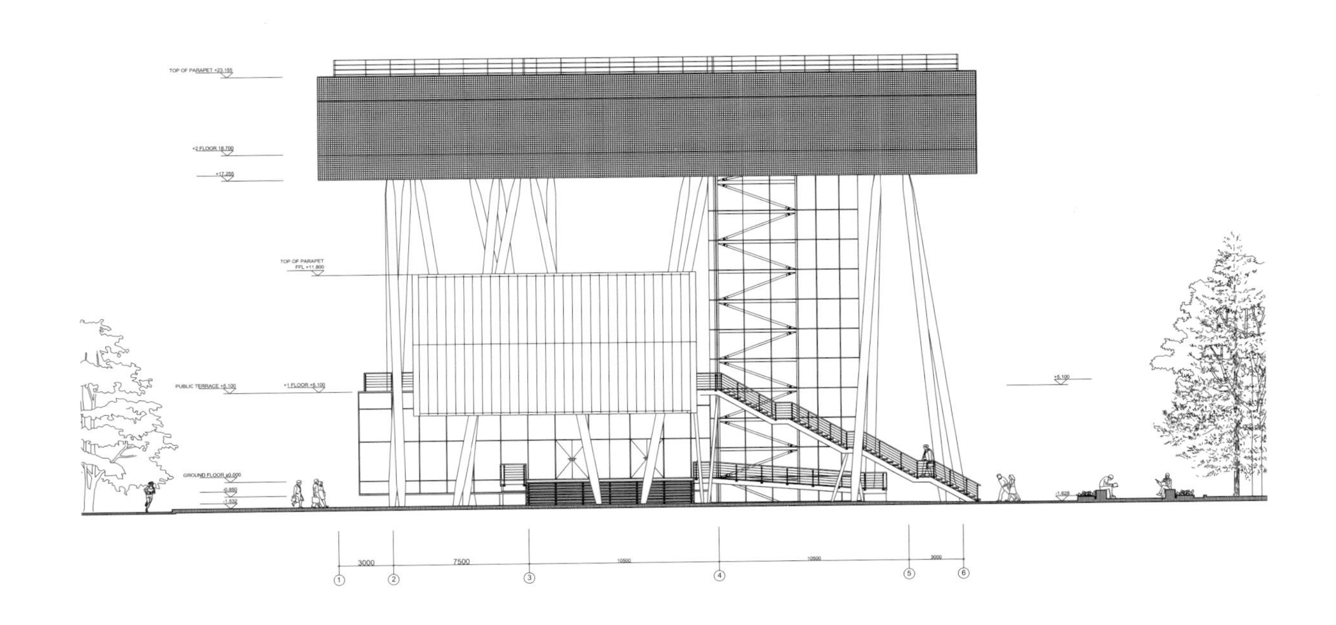

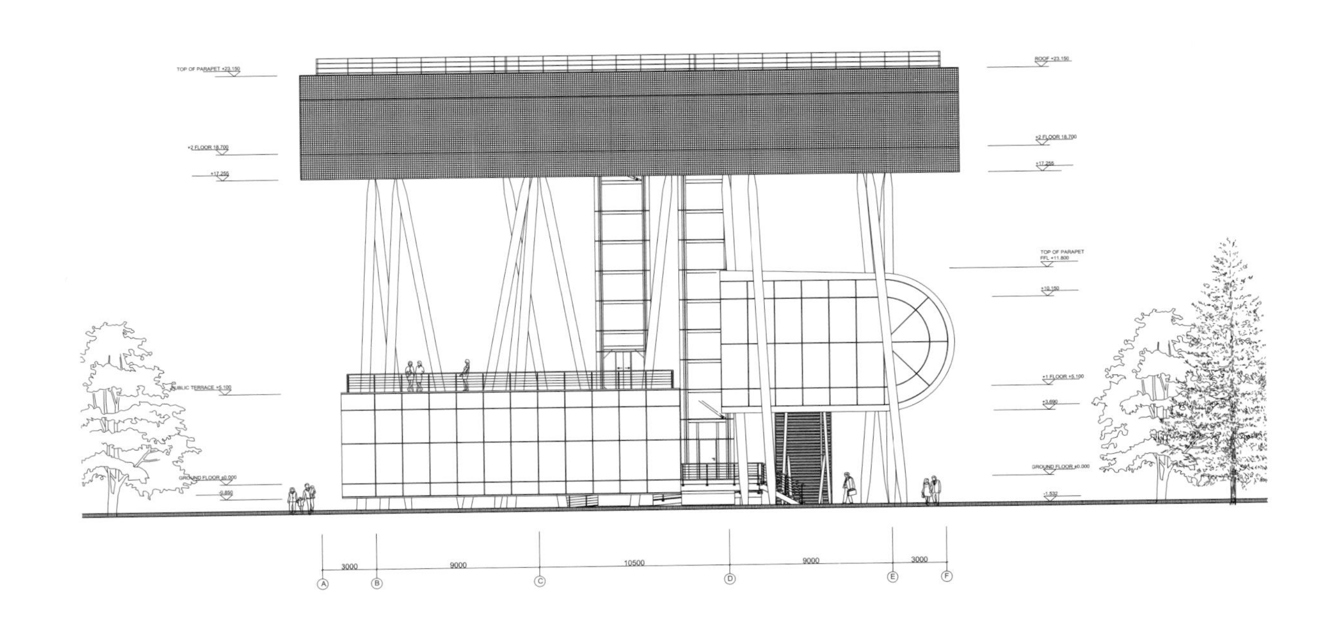

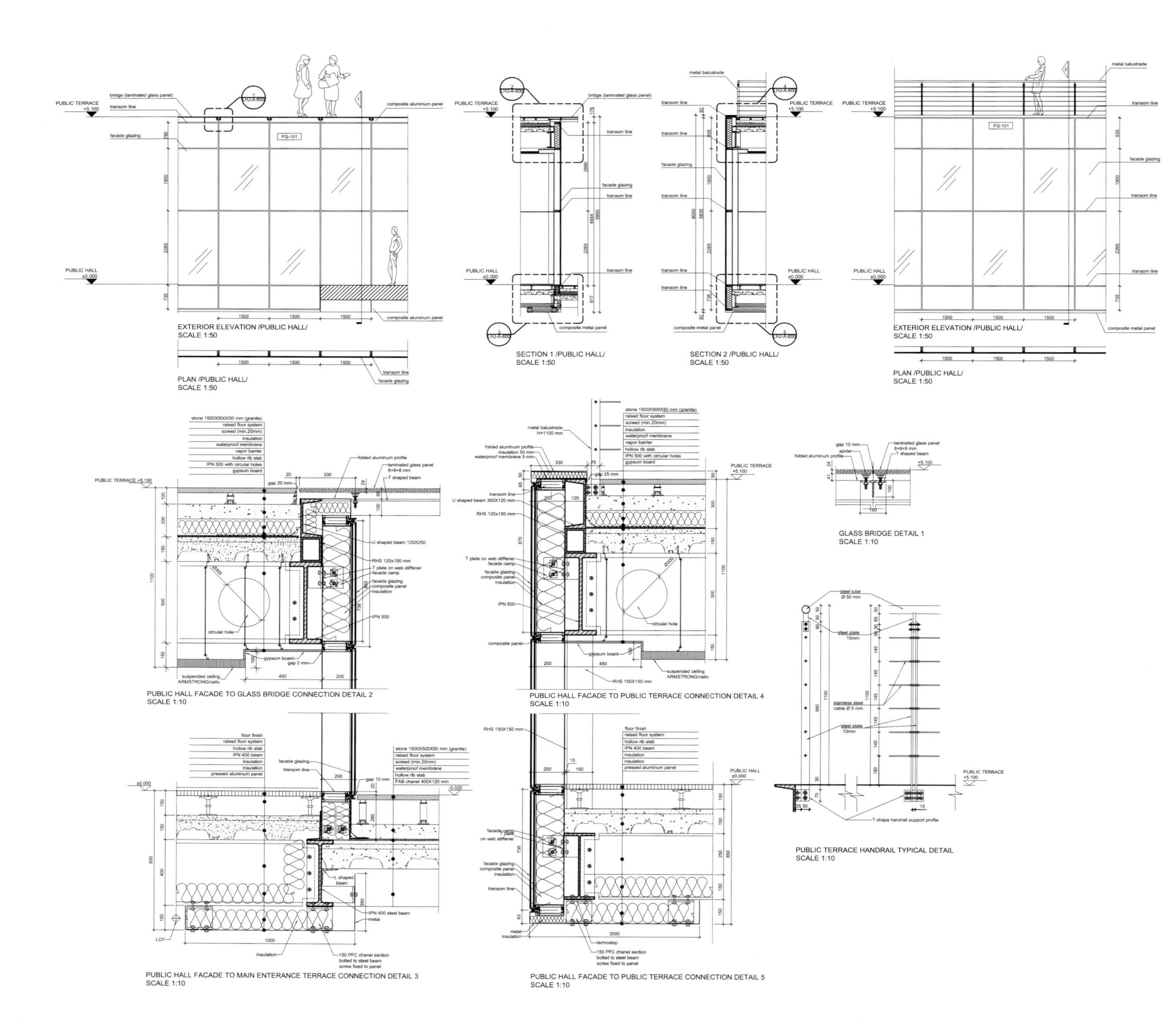
EXTERIOR ELEVATION /PUBLIC HALL/
SCALE 1:50
PLAN /PUBLIC HALL/
SCALE 1:50
SECTION 1 /PUBLIC HALL/
SCALE 1:50
SECTION 2 /PUBLIC HALL/
SCALE 1:50
EXTERIOR ELEVATION /PUBLIC HALL/
SCALE 1:50
PLAN /PUBLIC HALL/
SCALE 1:50
PUBLIC HALL FACADE TO GLASS BRIDGE CONNECTION DETAIL 2
SCALE 1:10
PUBLIC HALL FACADE TO PUBLIC TERRACE CONNECTION DETAIL 4
SCALE 1:10
GLASS BRIDGE DETAIL 1
SCALE 1:10
PUBLIC TERRACE HANDRAIL TYPICAL DETAIL
SCALE 1:10
PUBLIC HALL FACADE TO MAIN ENTERANCE TERRACE CONNECTION DETAIL 3
SCALE 1:10
PUBLIC HALL FACADE TO PUBLIC TERRACE CONNECTION DETAIL 5
SCALE 1:10

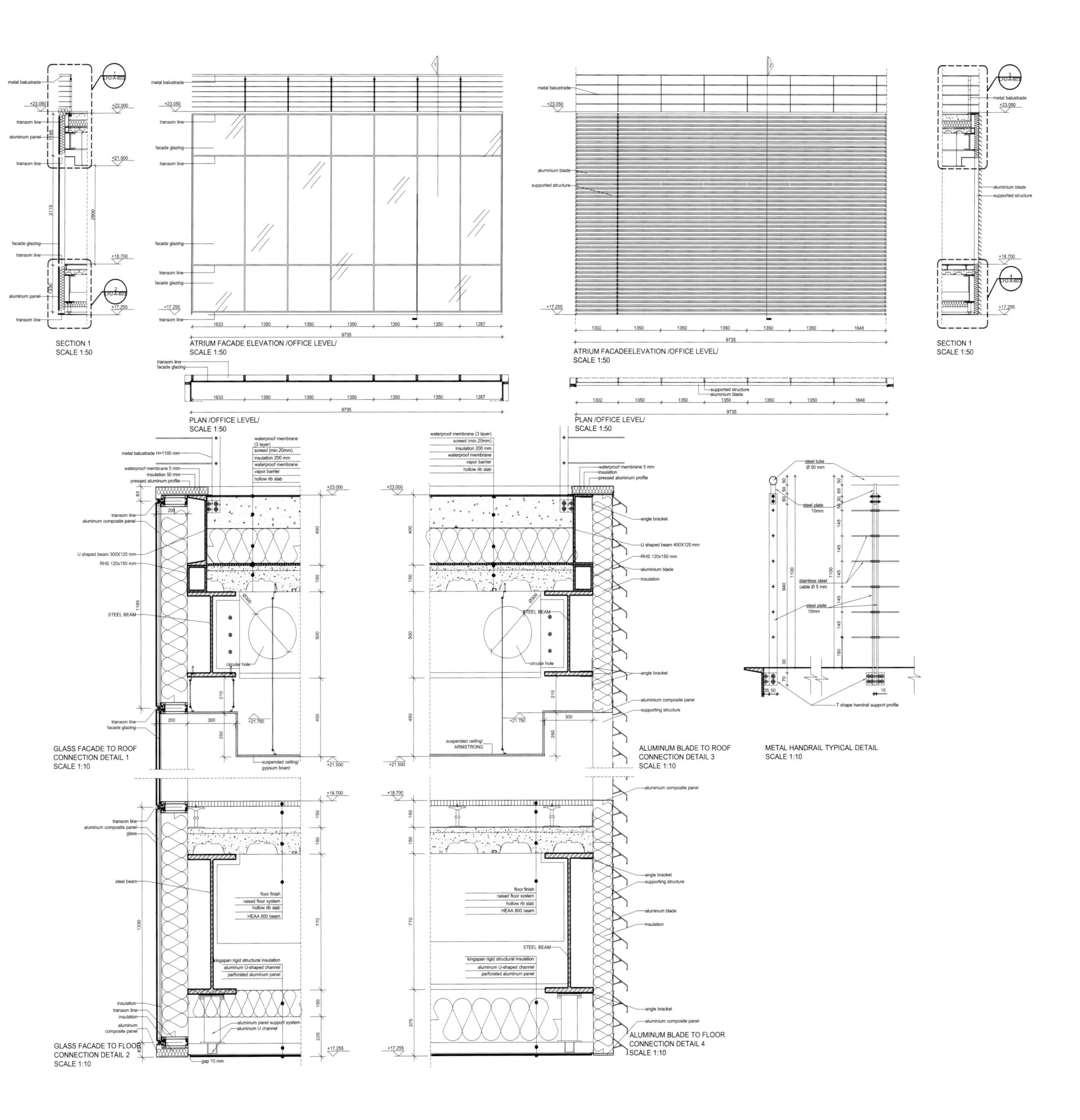
SECTION 1
SCALE 1:50
ATRIUM FACADE ELEVATION /OFFICE LEVEL/
SCALE 1:50
ATRIUM FACADEELEVATION /OFFICE LEVEL/
SCALE 1:50
SECTION 1
SCALE 1:50
PLAN /OFFICE LEVEL/
SCALE 1:50
PLAN /OFFICE LEVEL/
SCALE 1:50
GLASS FACADE TO ROOF
CONNECTION DETAIL 1
SCALE 1:10
ALUMINUM BLADE TO ROOF
CONNECTION DETAIL 3
SCALE 1:10
METAL HANDRAIL TYPICAL DETAIL
SCALE 1:10
GLASS FACADE TO FLOOR
CONNECTION DETAIL 2
SCALE 1:10
ALUMINUM BLADE TO FLOOR
CONNECTION DETAIL 4
SCALE 1:10

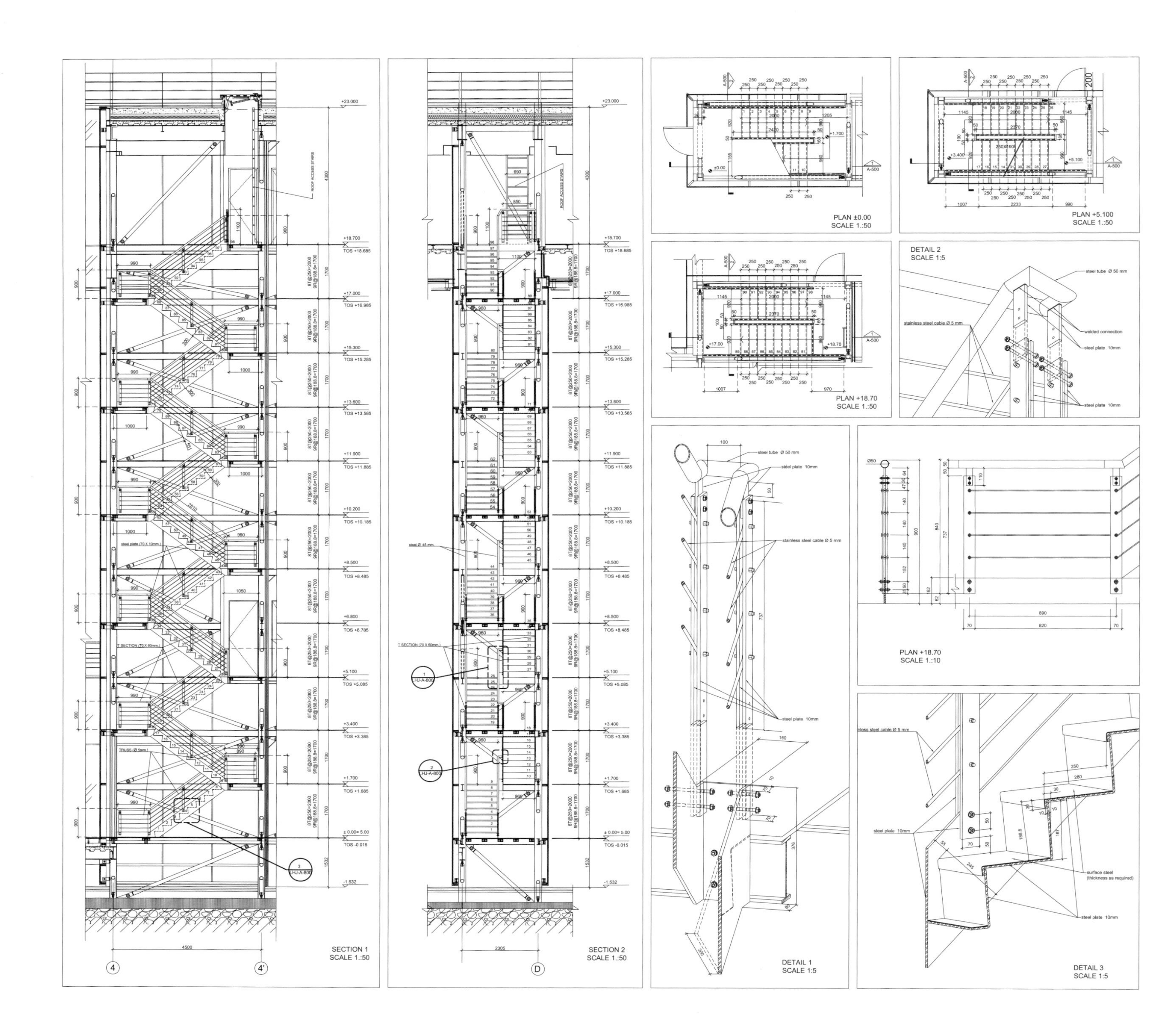
SECTION 1
SCALE 1.:50
SECTION 2
SCALE 1.:50
PLAN ±0.00
SCALE 1.:50
PLAN +5.100
SCALE 1.:50
PLAN +18.70
SCALE 1.:50
DETAIL 2
SCALE 1:5
DETAIL 1
SCALE 1:5
PLAN +18.70
SCALE 1.:10
DETAIL 3
SCALE 1:5
steel tube Ø 50 mm
steel plate 10mm
stainless steel cable Ø 5 mm
welded connection
surface steel
(thickness as required)
ROOF ACCESS STAIRS
T SECTION (70 X 80mm)
TRUSS (Ø 5mm)
+23.000
+18.700
TOS +18.685
+17.000
TOS +16.985
+15.300
TOS +15.285
+13.600
TOS +13.585
+11.900
TOS +11.885
+10.200
TOS +10.185
+8.500
TOS +8.485
+6.800
TOS +6.785
+5.100
TOS +5.085
+3.400
TOS +3.385
+1.700
TOS +1.685
± 0.00= 5.00
TOS -0.015
-1.532
4500
2305

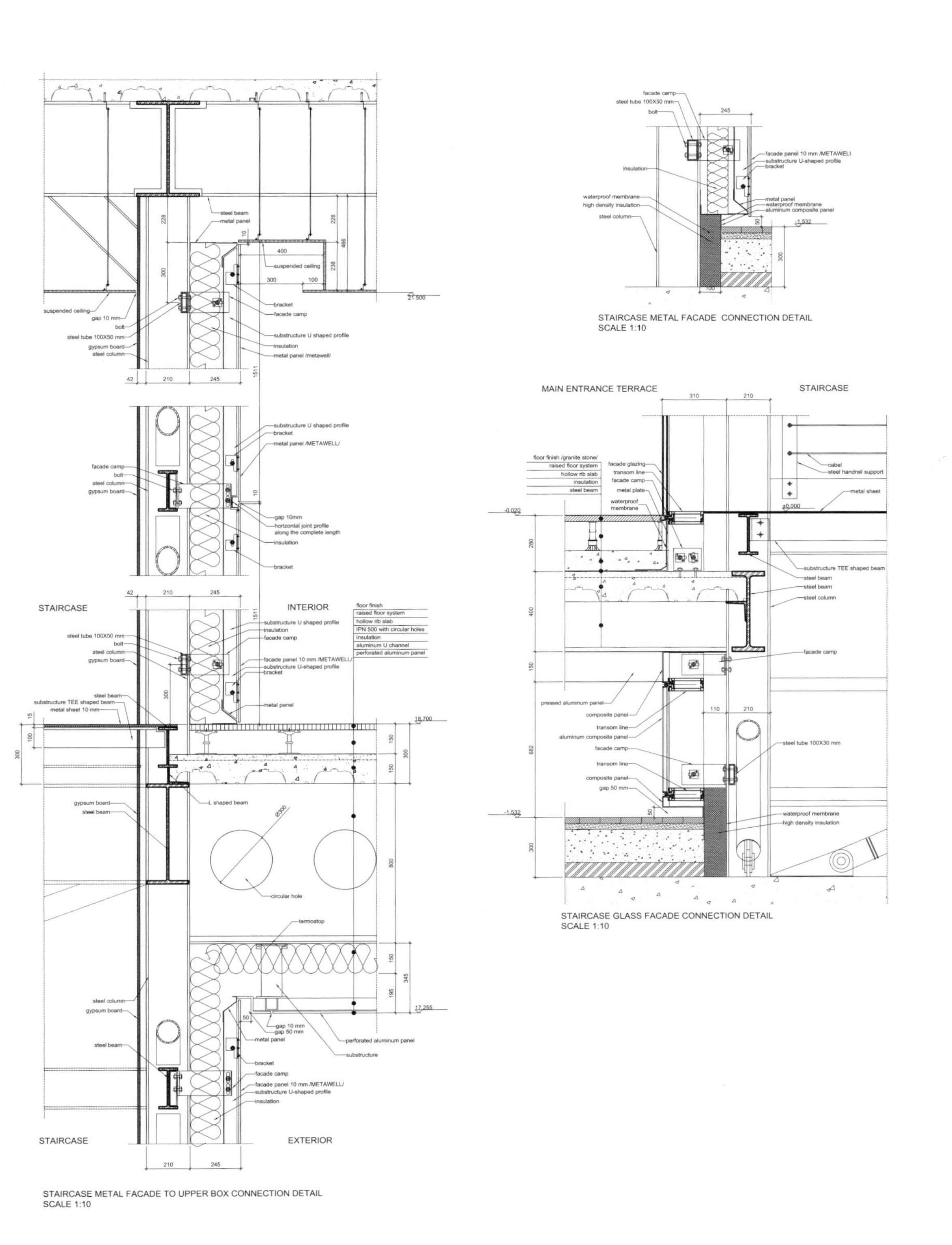

STAIRCASE METAL FACADE CONNECTION DETAIL
SCALE 1:10

STAIRCASE GLASS FACADE CONNECTION DETAIL
SCALE 1:10

STAIRCASE METAL FACADE TO UPPER BOX CONNECTION DETAIL
SCALE 1:10

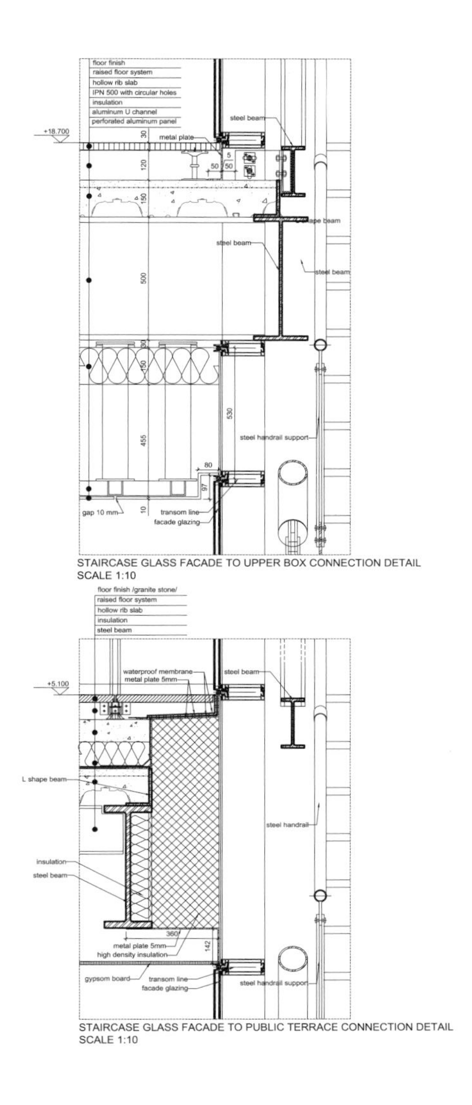

STAIRCASE GLASS FACADE TO UPPER BOX CONNECTION DETAIL
SCALE 1:10

STAIRCASE GLASS FACADE TO PUBLIC TERRACE CONNECTION DETAIL
SCALE 1:10

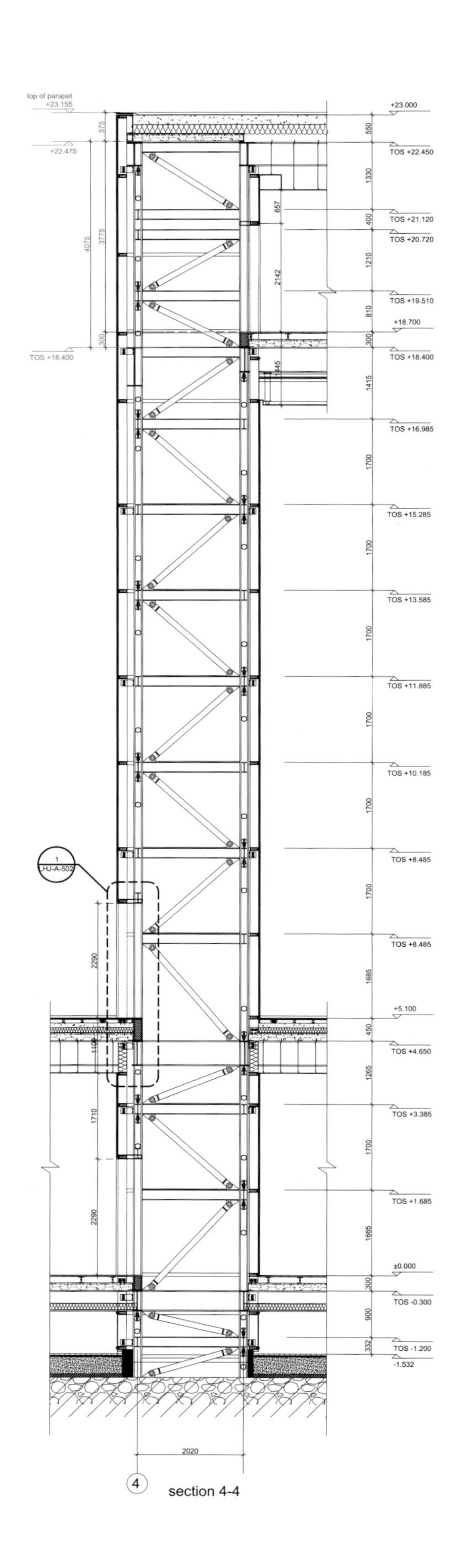
section 4-4

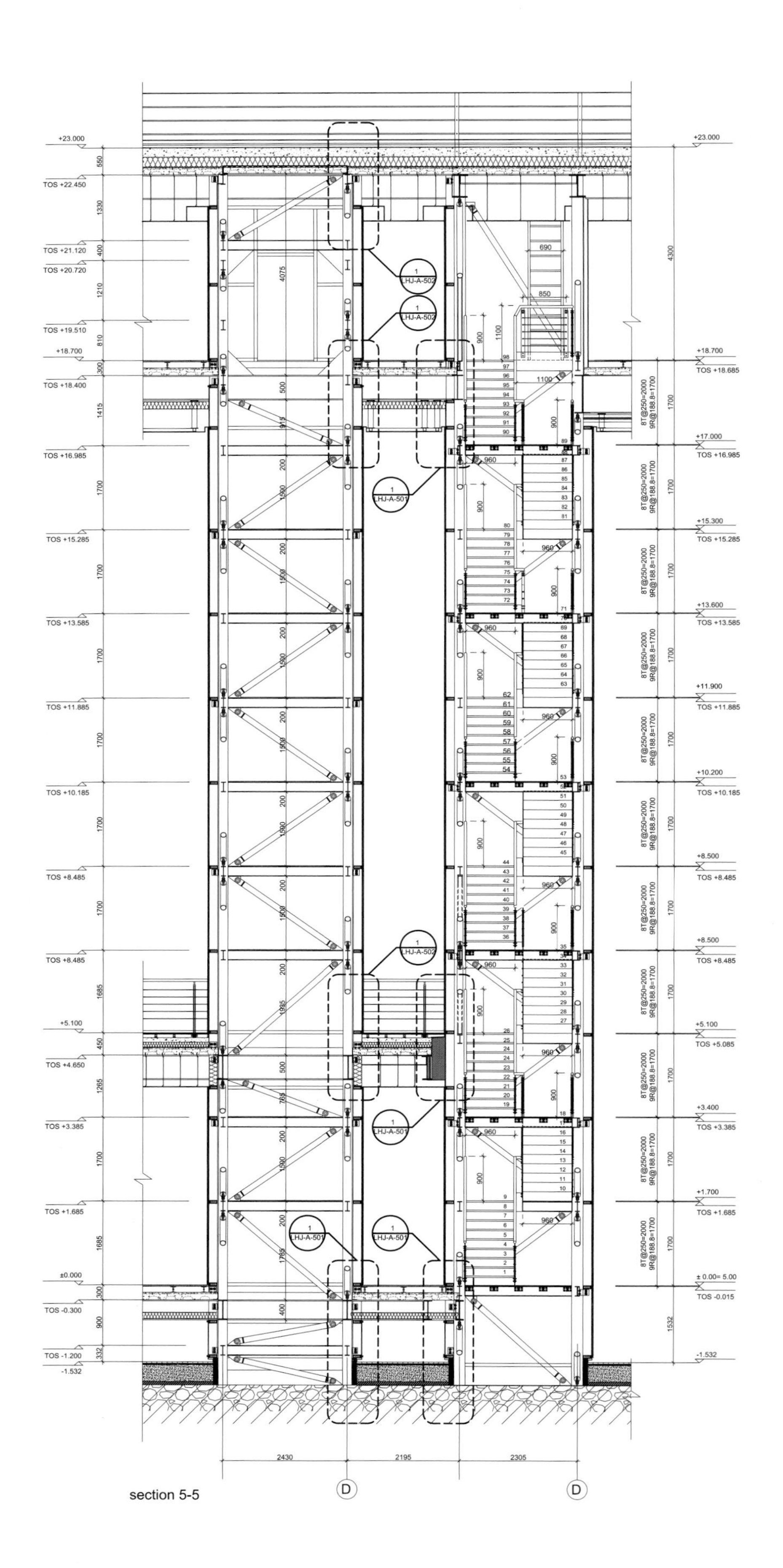
section 5-5

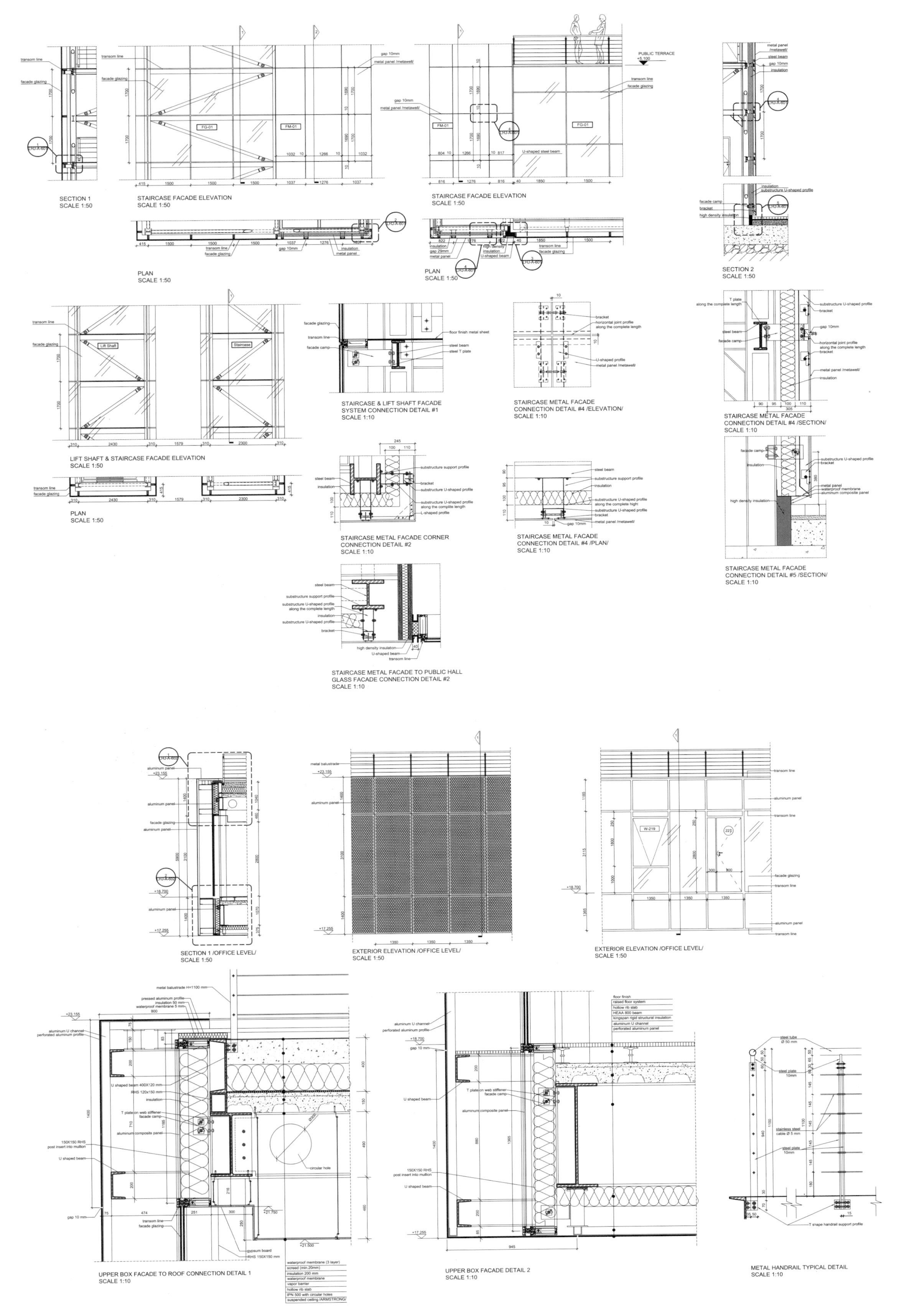

SECTION 1
SCALE 1:50

STAIRCASE FACADE ELEVATION
SCALE 1:50

STAIRCASE FACADE ELEVATION
SCALE 1:50

PLAN
SCALE 1:50

PLAN
SCALE 1:50

SECTION 2
SCALE 1:50

LIFT SHAFT & STAIRCASE FACADE ELEVATION
SCALE 1:50

PLAN
SCALE 1:50

STAIRCASE & LIFT SHAFT FACADE SYSTEM CONNECTION DETAIL #1
SCALE 1:10

STAIRCASE METAL FACADE CONNECTION DETAIL #4 /ELEVATION/
SCALE 1:10

STAIRCASE METAL FACADE CONNECTION DETAIL #4 /SECTION/
SCALE 1:10

STAIRCASE METAL FACADE CORNER CONNECTION DETAIL #2
SCALE 1:10

STAIRCASE METAL FACADE CONNECTION DETAIL #4 /PLAN/
SCALE 1:10

STAIRCASE METAL FACADE CONNECTION DETAIL #5 /SECTION/
SCALE 1:10

STAIRCASE METAL FACADE TO PUBLIC HALL GLASS FACADE CONNECTION DETAIL #2
SCALE 1:10

SECTION 1 /OFFICE LEVEL/
SCALE 1:50

EXTERIOR ELEVATION /OFFICE LEVEL/
SCALE 1:50

EXTERIOR ELEVATION /OFFICE LEVEL/
SCALE 1:50

UPPER BOX FACADE TO ROOF CONNECTION DETAIL 1
SCALE 1:10

UPPER BOX FACADE DETAIL 2
SCALE 1:10

METAL HANDRAIL TYPICAL DETAIL
SCALE 1:10

Hero Group Headquarters

Location: Bintaro, Jakarta, Indonesia
Architect: Atelier TT
Client: PT Hero Supermarket
Site Area: 4,734 m^2
Gross Floor Area: 13,840 m^2
Completion Date: 2014
Photographer: Martin Westlake

The Hero headquarter office is located in Bintaro – an emerging satellite town in the south-western outskirt of Jakarta, Indonesia. The site of the Hero Headquarter office is located in between a junction where two major roads meet, and an existing building, which houses the Giant supermarket – a subsidiary of the Hero Group. The immediate response was for the proposed building to continue the same horizontal language as the existing adjacent building. The mass is extruded to achieve maximum gross floor area (GFA) and then split in the middle in order to allow more natural light into the depths of the offices while maximizing potential views to the surrounding area. The canyon-like split in the middle is then infused with greeneries to act as both a green lung for the building as well as a transitory place of arrival and departure for staff and guests before they access their office quarters. This gesture also allows for greater transparency and visual connectivity between the staff of different units on the different floor levels, thus improving staff interaction on the whole throughout the building. The rooftop, which are traditionally used mainly for building utilities, are re-imagined as a garden space where staff can seek refuge to or as a place for company events to take place. Continuous bands of louvers are employed as a sunshading device that protects the occupants from the harsh tropical sun while visually uniting and elongating the façade of the building as it gently turns the corner following the shape of the site. The use of colour and a single architectural language further accentuates the building's iconic presence within this newly developed area of Bintaro.

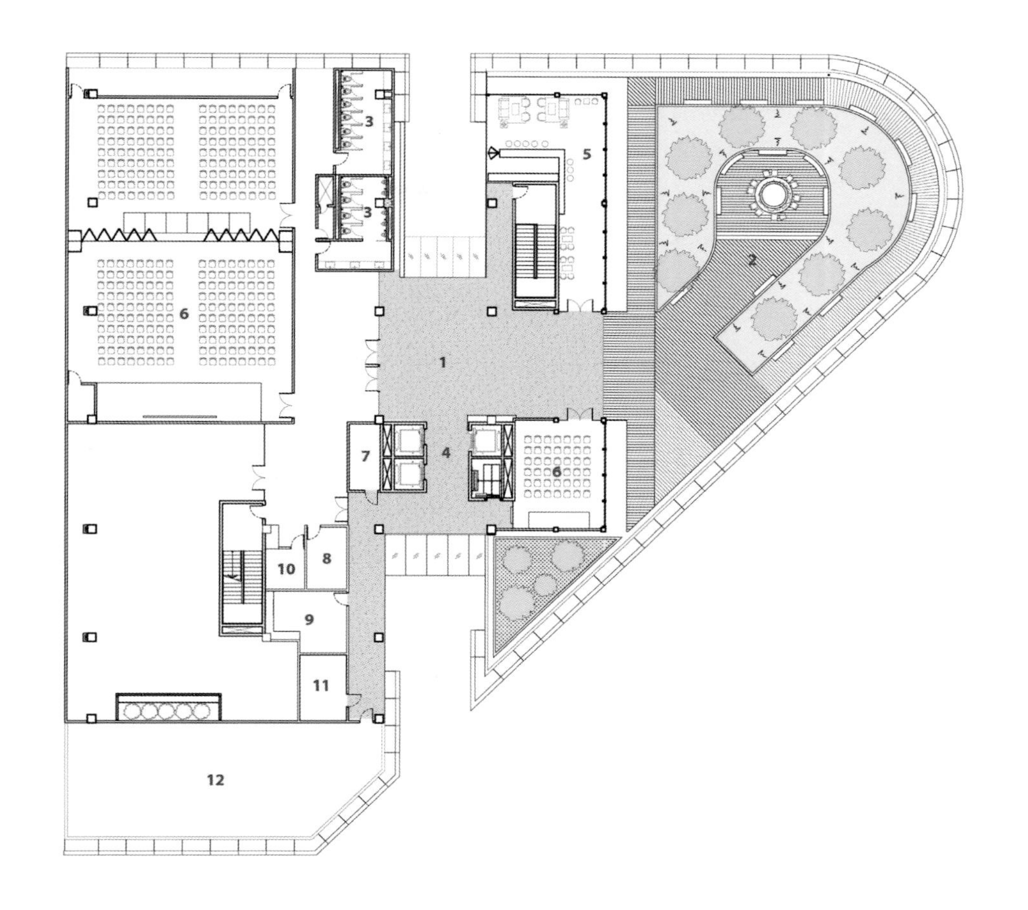

FIFTH FLOOR
1. Foyer
2. Sky Garden
3. Toilet
4. Lift
5. VIP Room
6. Training Room
7. Panel Room
8. Parenting Room
9. Pantry
10. IT Panel
11. Panel Cooling Tower
12. Cooling Tower Location

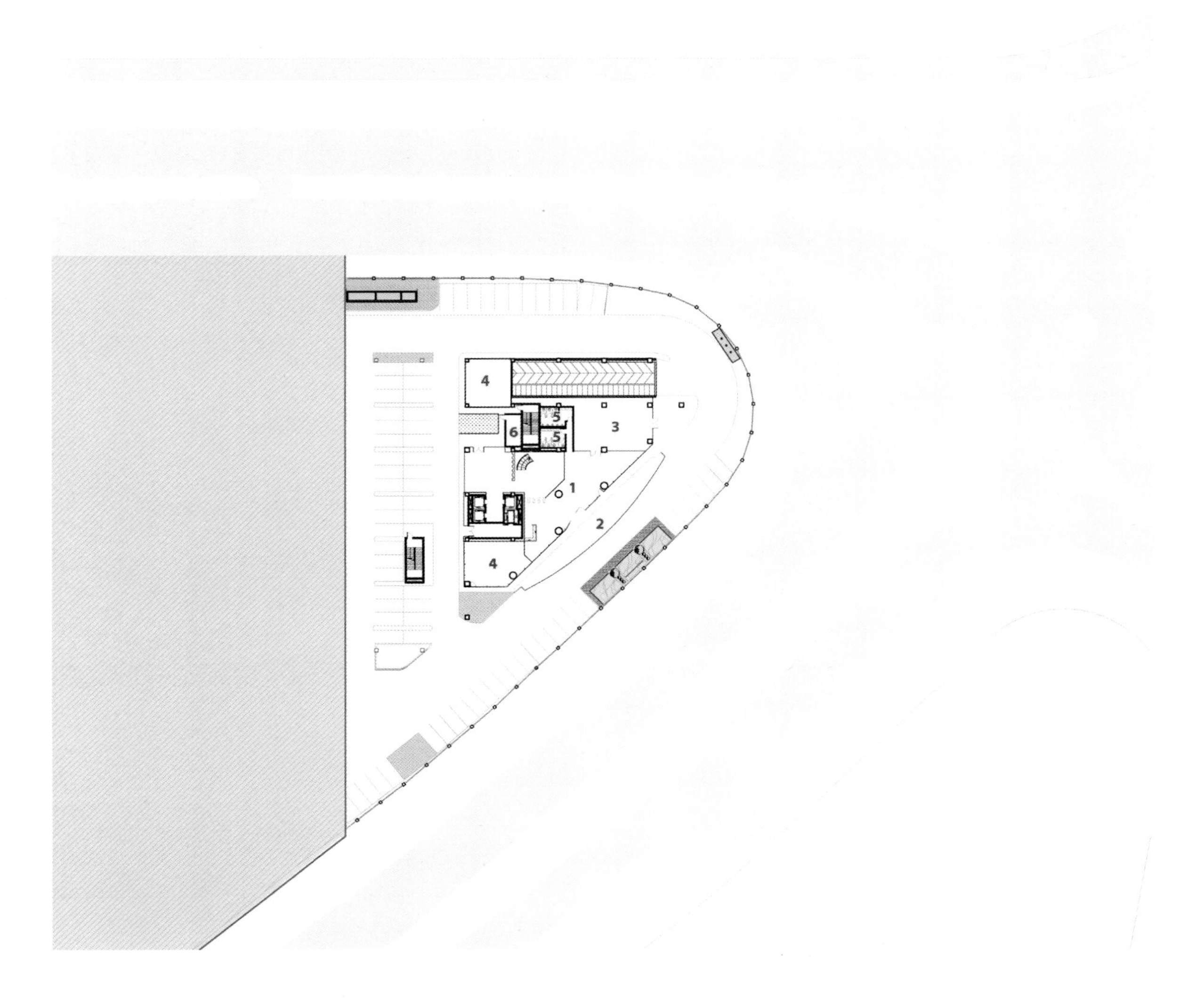

GROUND FLOOR
1. Lobby
2. Drop Off
3. Retail
4. Tenancy
5. Toilet
6. Panel Room

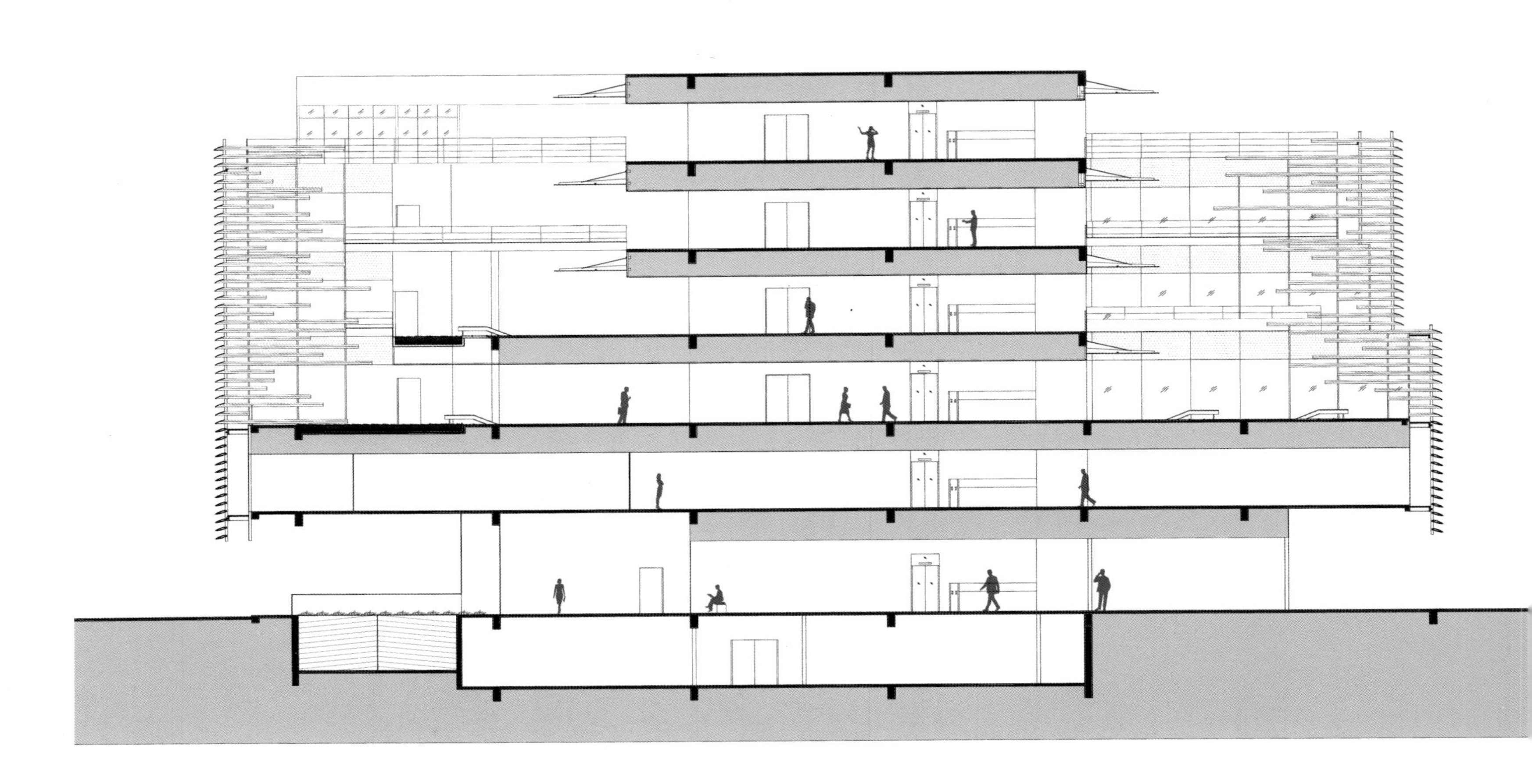

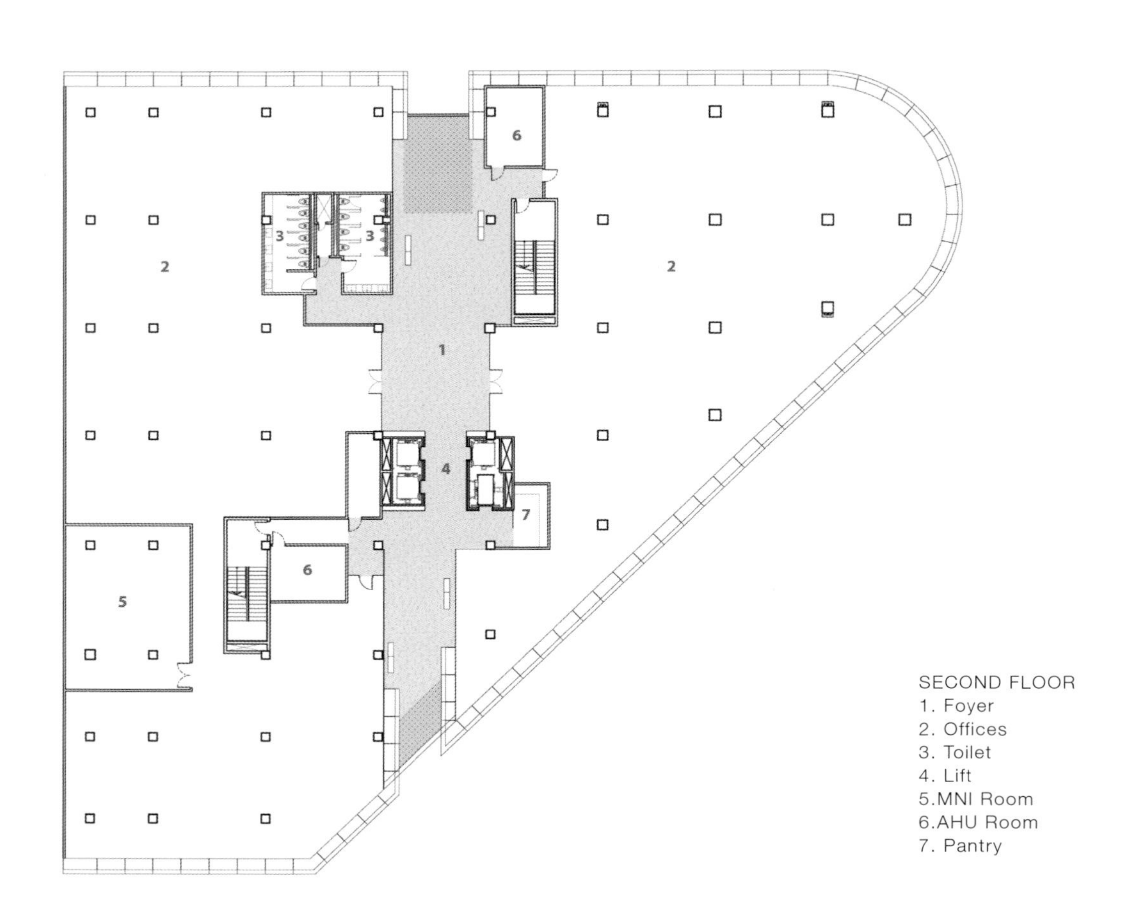
SECOND FLOOR
1. Foyer
2. Offices
3. Toilet
4. Lift
5.MNI Room
6.AHU Room
7. Pantry

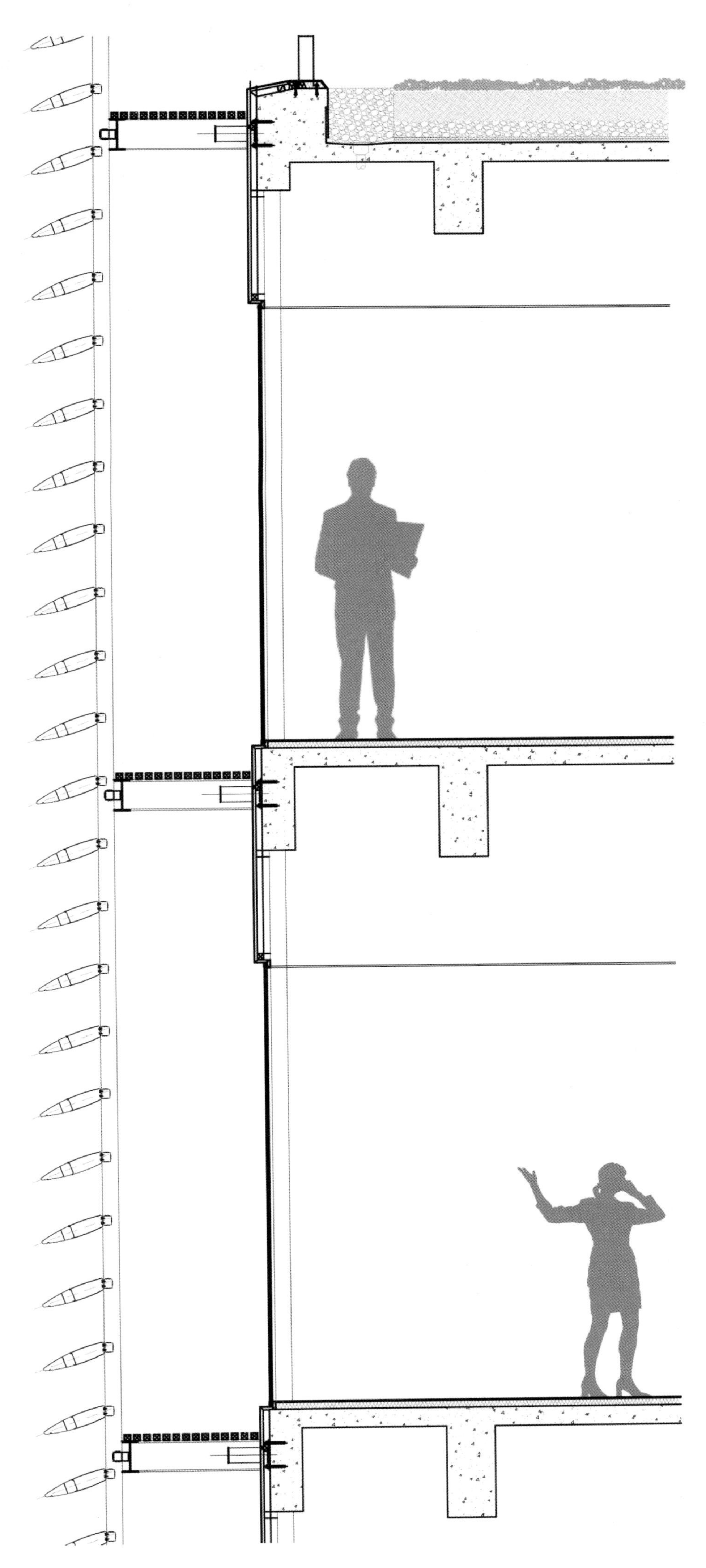

Dolce & Gabbana Offices

Location: Milan, Italy
Architect: Studio Piuarch
Client: Dolce & Gabbana S.R.L.
Built Area: 5000 m^2
Completion Date: 2006
Photographers: Ruy Teixeira, Alberto Piovano & Andrea Martiradonna

The Milan offices and showroom of Dolce & Gabbana occupies an area of 5,000 m^2 and is the result of a careful process of consolidation and remodeling of the two adjacent buildings, one dating from the 1920s and the other from the 1960s. The contrast between volumes upon which the design plays is very effective, enabling the two structures, one classical and one contemporary, to carry on a perfectly harmonious dialogue between them and with the surrounding urban context. The facade of the most recent building has been completely redone in glass, its rhythm marked out by a series of vertical blinds in opal glass. The volume is bounded by three streets to form a single glass block, simple yet of great visual impact, thanks to the play of light and shadow that enlivens the building during both day and night. The transparency and simplicity of the architecture result in a volume screened by vertical elements while remaining open to daylight and to the view of passers-by, who can admire the clothing hanging on the racks that run the entire perimeter of the facade.

With five floors above ground and two below, the complex hosts the offices and meeting rooms in the older building, while the newer structure contains the open-space showrooms, which occupy three floors, with the top floor consisting of small terraces created by the configuration of the volumes of the restaurant. The ground floor opens onto an interior courtyard paved with white stones bordered by sinuously contoured garden areas. Also facing the courtyard is the new structure that connects the two buildings, faced entirely in glass and articulated by the sheet metal of the staircases inside. The interior is characterized by the pairing of metals—polished steel for the load-bearing elements and sheet metal for the horizontal surfaces. The more private spaces are equipped with hi-tech plasma screens.

The decision to use natural materials like Namibia stone and transparent and/or reflective materials like glass and polished steel generates a play of lighting effects through the entire building, enabling to change its aspect depending on the time of day, and to be both eclectic and sophisticated at the same time, in perfect keeping with the style of the clients.

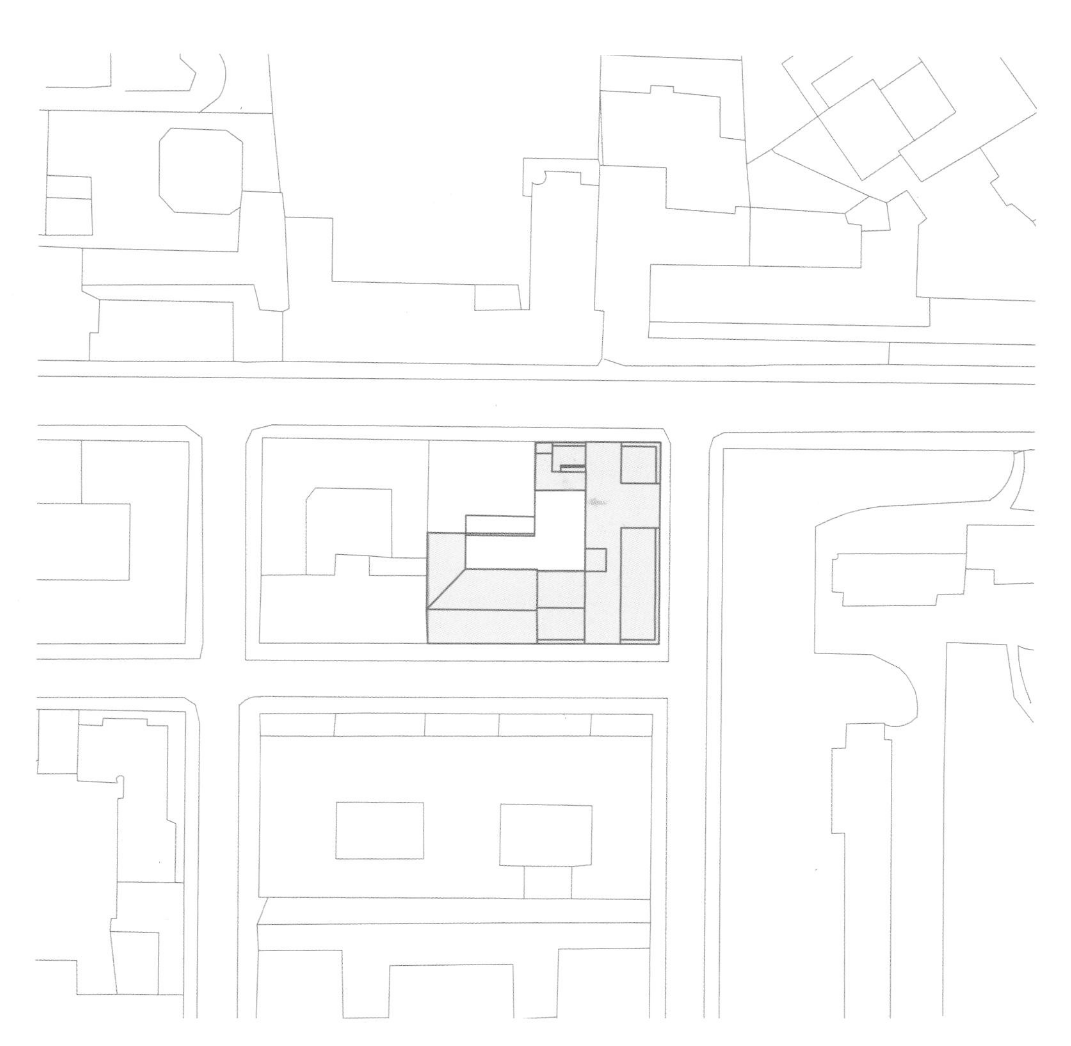

Site Plan

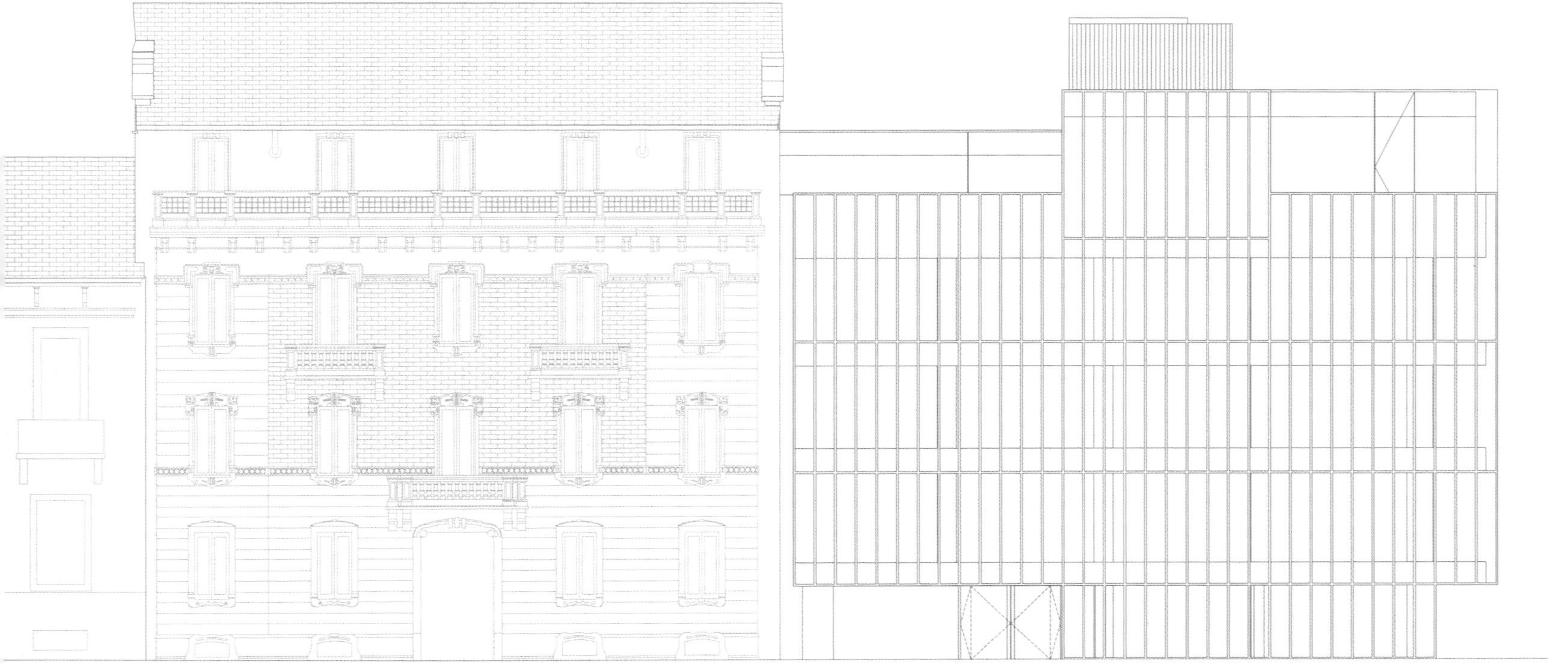

rospetto

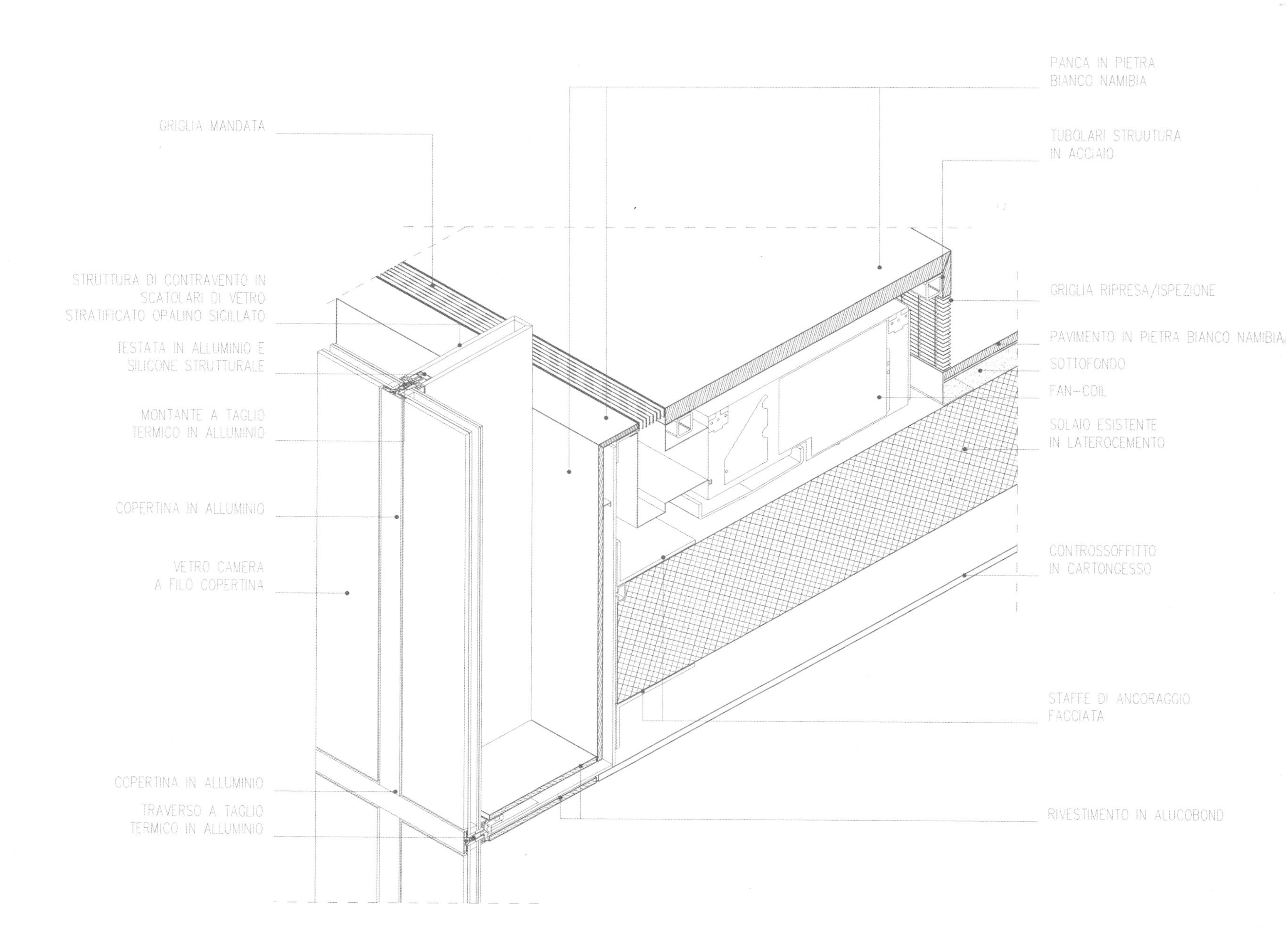

PANCA IN PIETRA
BIANCO NAMIBIA
GRIGLIA MANDATA
TUBOLARI STRUUTURA
IN ACCIAIO
STRUTTURA DI CONTRAVENTO IN
SCATOLARI DI VETRO
STRATIFICATO OPALINO SIGILLATO
GRIGLIA RIPRESA/ISPEZIONE
PAVIMENTO IN PIETRA BIANCO NAMIBIA
TESTATA IN ALLUMINIO E
SILICONE STRUTTURALE
SOTTOFONDO
FAN-COIL
MONTANTE A TAGLIO
TERMICO IN ALLUMINIO
SOLAIO ESISTENTE
IN LATEROCEMENTO
COPERTINA IN ALLUMINIO
VETRO CAMERA
A FILO COPERTINA
CONTROSSOFFITTO
IN CARTONGESSO
STAFFE DI ANCORAGGIO
FACCIATA
COPERTINA IN ALLUMINIO
TRAVERSO A TAGLIO
TERMICO IN ALLUMINIO
RIVESTIMENTO IN ALUCOBOND

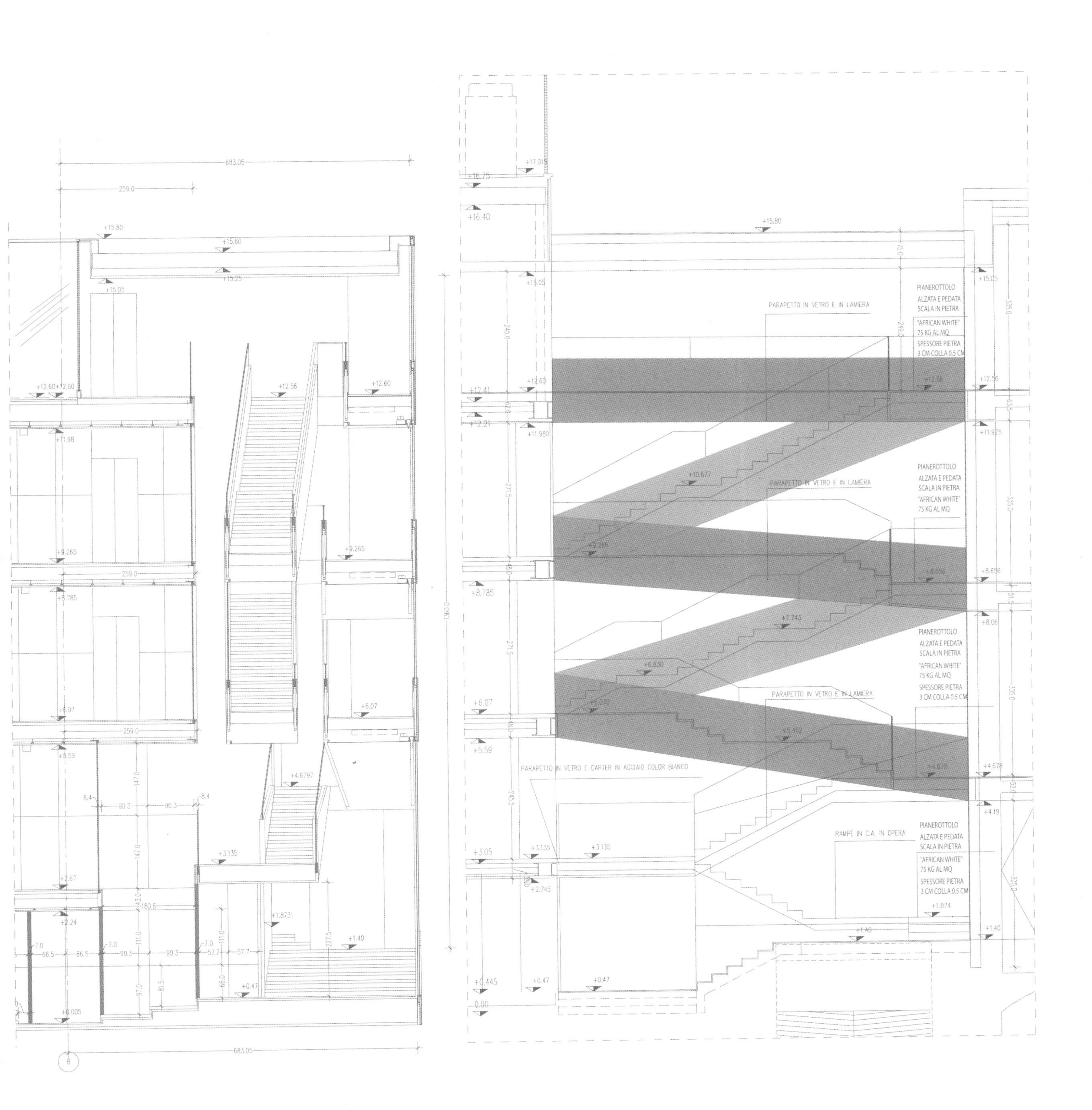
PARAPETTO IN VETRO E IN LAMIERA
PIANEROTTOLO
ALZATA E PEDATA
SCALA IN PIETRA
"AFRICAN WHITE"
75 KG AL MQ
SPESSORE PIETRA
3 CM COLLA 0.5 CM
PARAPETTO IN VETRO E CARTER IN ACCIAIO COLOR BIANCO
RAMPE IN C.A. IN OPERA

La Rioja Technology Transfer Centre

Location: Logroño, Spain
Architect: AZPML
Client: Regional Government of La Rioja
Photography: Ramon Pratt, Sérgio Padura, Jordi Todó

The complex is located in the outskirts of Logroño, the capital of La Rioja Region and is due to host three different institutions dedicated to the education, research and the nurturing of individuals and companies in the sector of world-wide-web services and technology.

The building is organized in such a way that the three institutions become part of a single organization, allowing the collective facilities to be shared between the three institutions, and to minimize the security, cleaning and maintenance costs and also aims to maximize the integration of the landscape into the building's spaces, and adopts a linear structure that maximizes the surface of contact with the outside. The classrooms and offices that constitute the majority of the functional spaces in the building are organized along a corridor space that threads through all the dependencies. The corridor space will contain the public spaces of the building, opened to the outside gardens and the rooms will be opened on the other side of the building towards the river landscape and the tree farms. This linear structure has been placed roughly on a north-south orientation, parallel to the topographical cornice that forms the western edge to the site. Such location allows the building to surround the Elms along the Camino de Los Lírios, claiming the slope as an internal garden featuring the trees. The building has been organized into a two storey bundled structure that encloses part of the site as a more internalized outside space, branching out on different levels to connect both with the surrounding urban levels and with the future River Park. On a smaller scale, the western face of the building provides terraces looking towards the Elm garden, and shredding into open-air ramps establishing topographical continuities between the building and the garden. The sectional dislocation between the two floors of the building automatically generates these terraces and a cantilever on the East facade will protect the fully glazed facade from the sun. The roof of the building becomes a kind of public belvedere over the river park, being connected to the city level on two ends through lifting bridges. A green pergola extends from this level to protect the glazed facade from direct sun exposure.

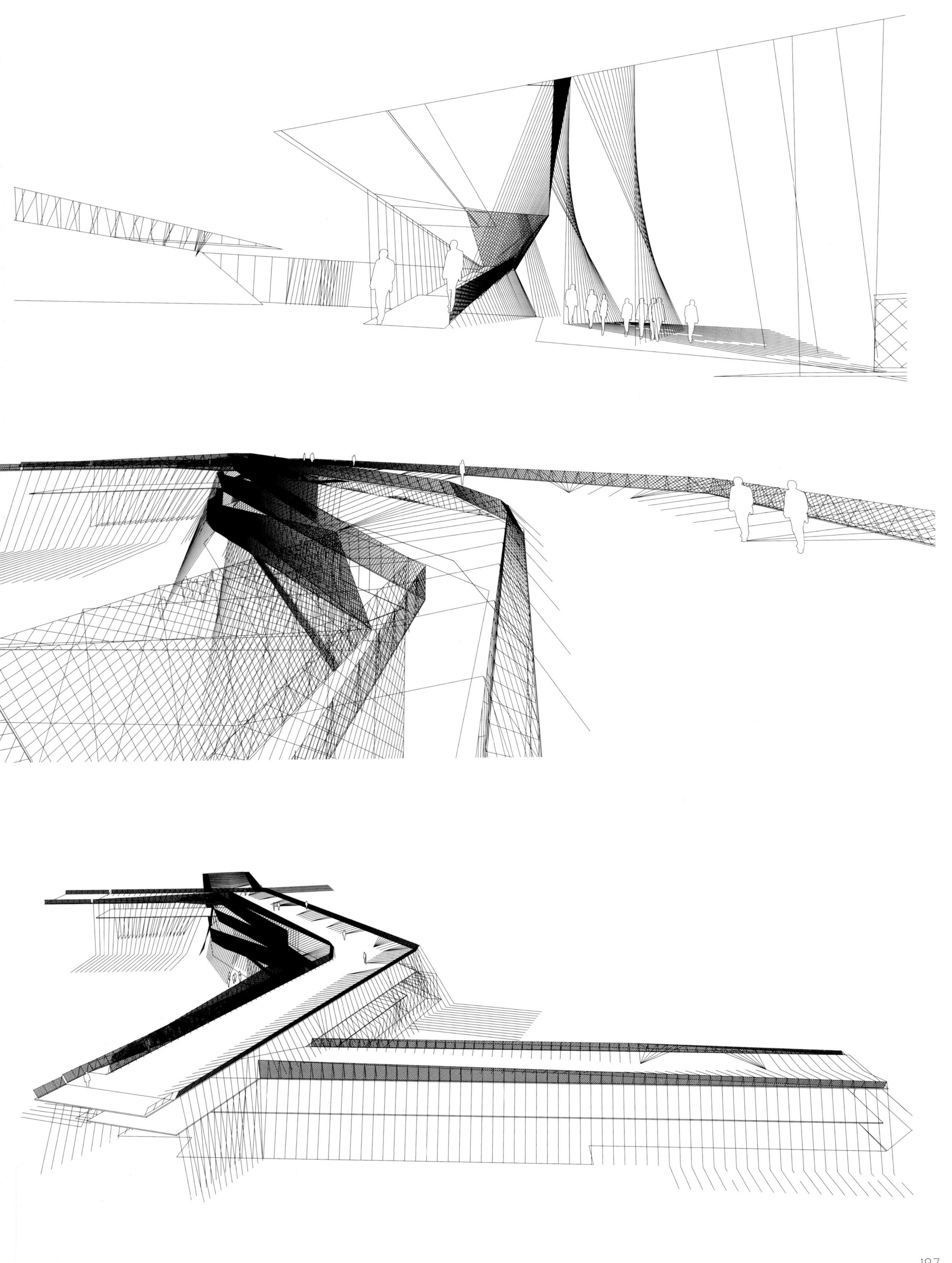

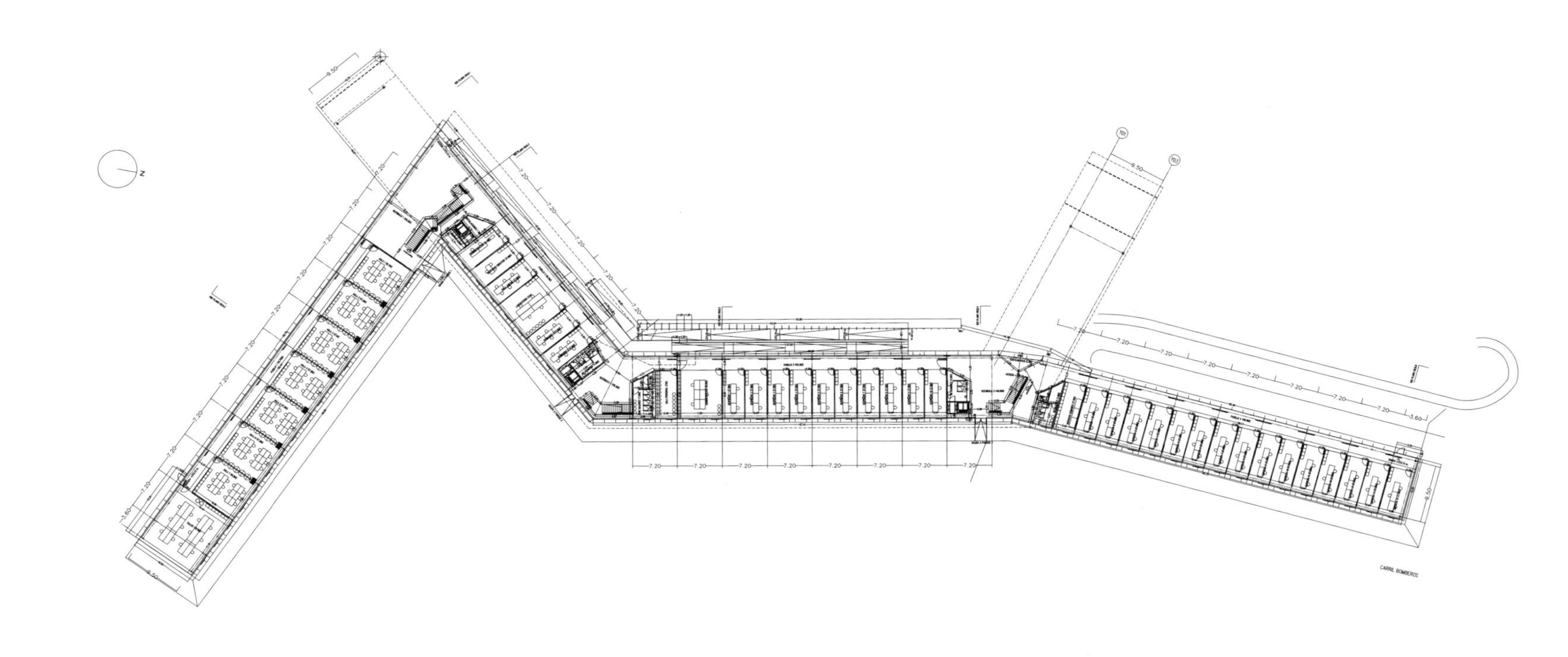

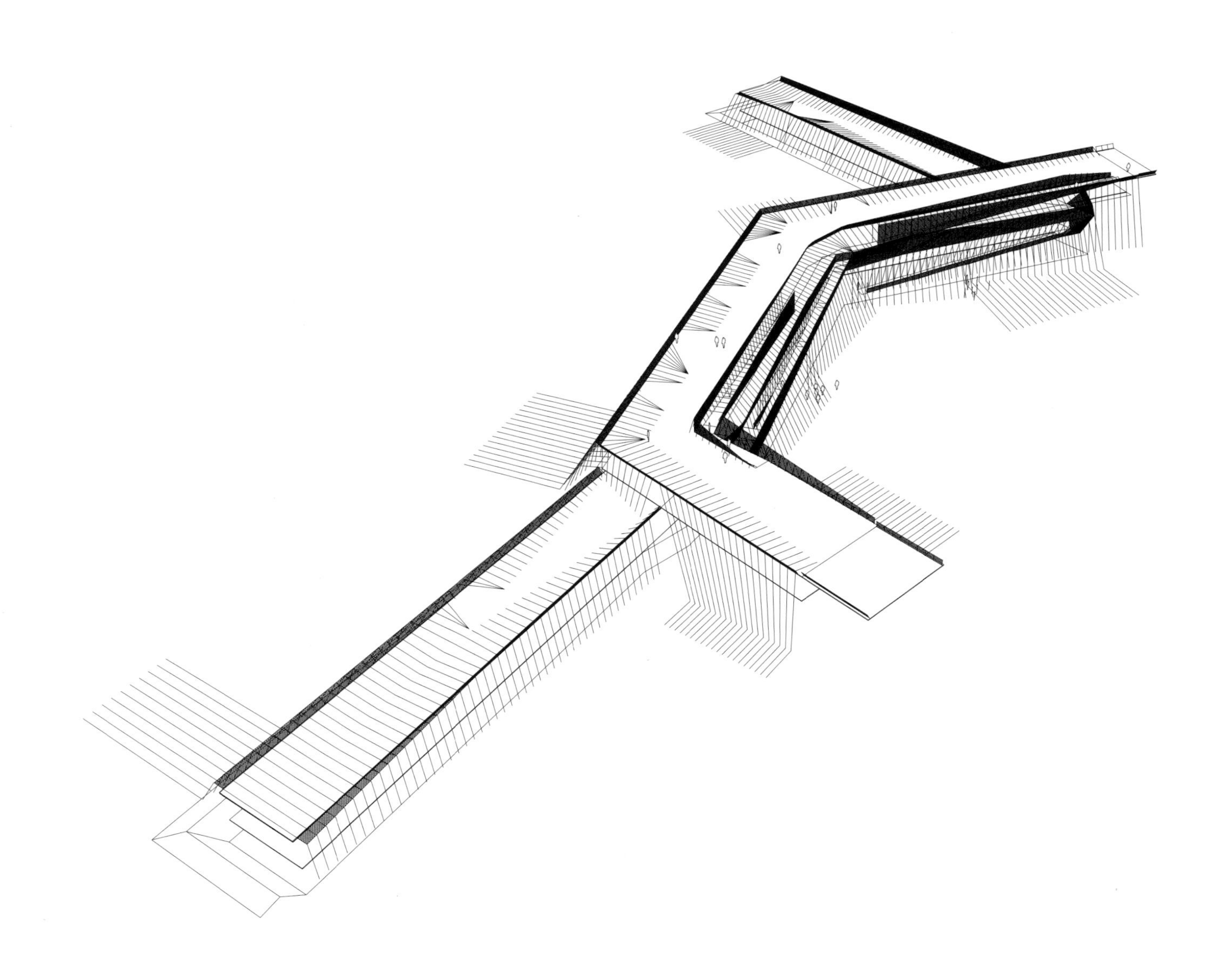

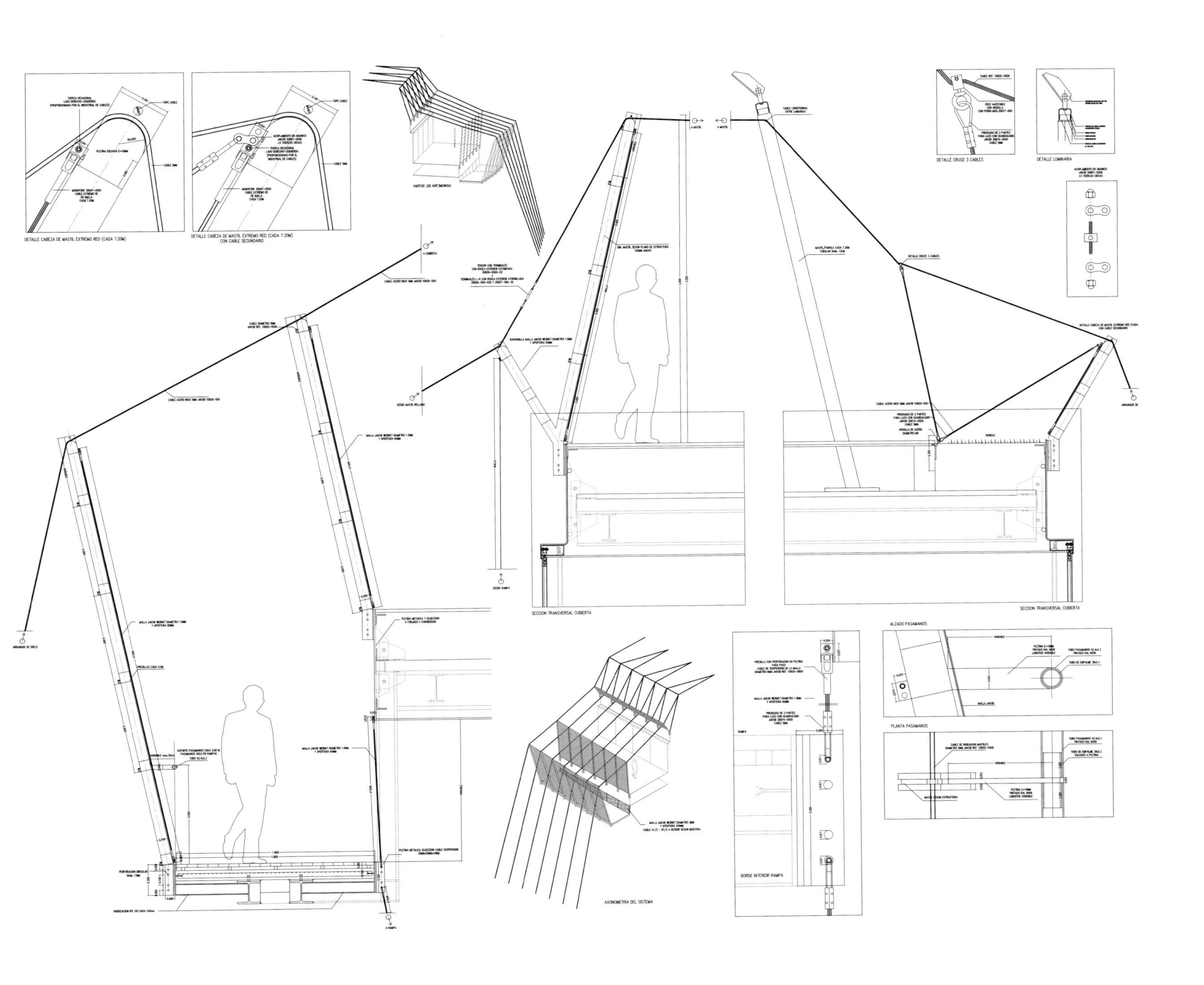

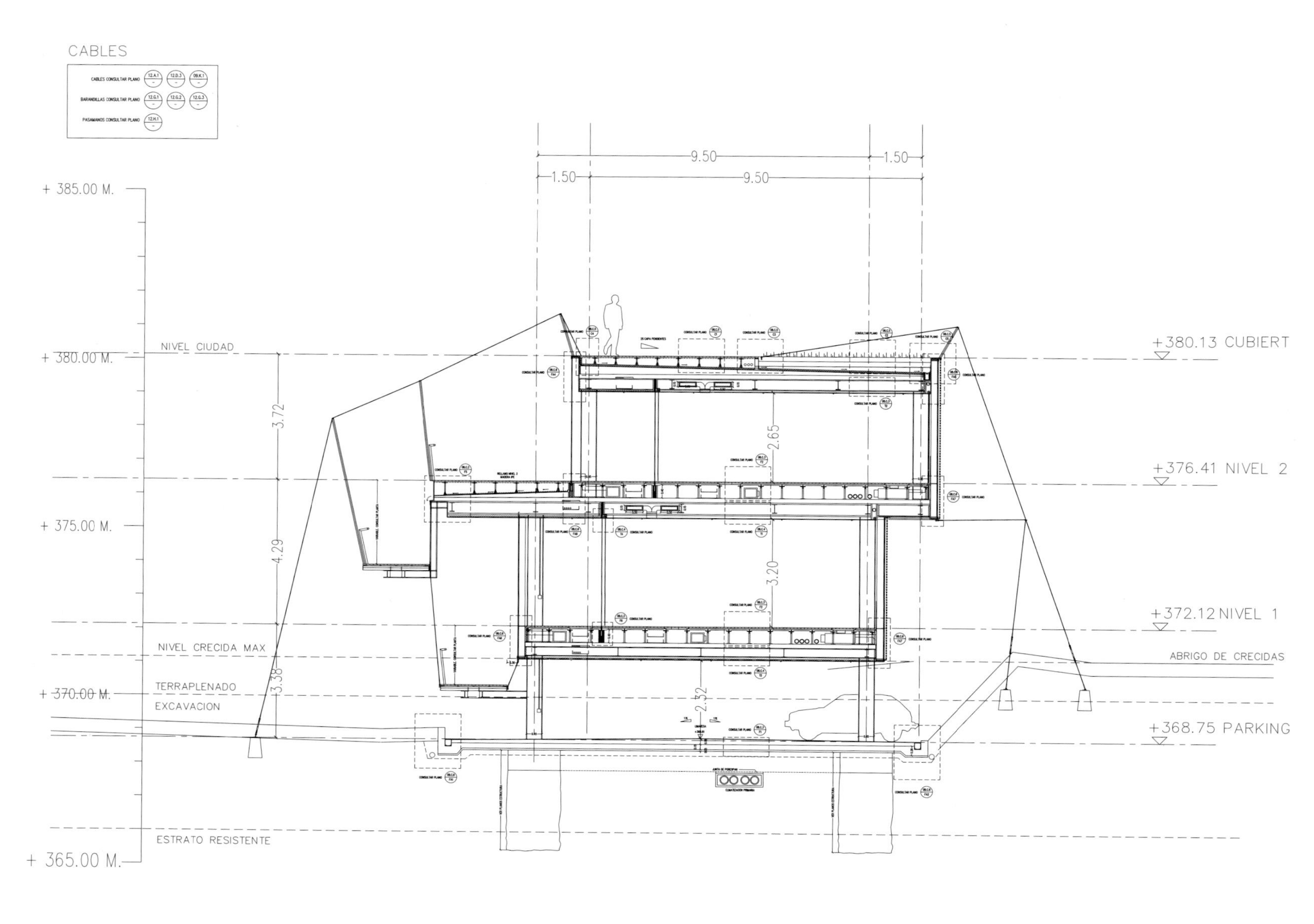
CABLES
9.50
1.50
1.50
9.50
+ 385.00 M.
NIVEL CIUDAD
+ 380.00 M.
+380.13 CUBIERT
3.72
2.65
+376.41 NIVEL 2
+ 375.00 M.
4.29
3.20
+372.12 NIVEL 1
NIVEL CRECIDA MAX
ABRIGO DE CRECIDAS
+ 370.00 M.
TERRAPLENADO
EXCAVACION
3.38
2.32
+368.75 PARKING
ESTRATO RESISTENTE
+ 365.00 M.

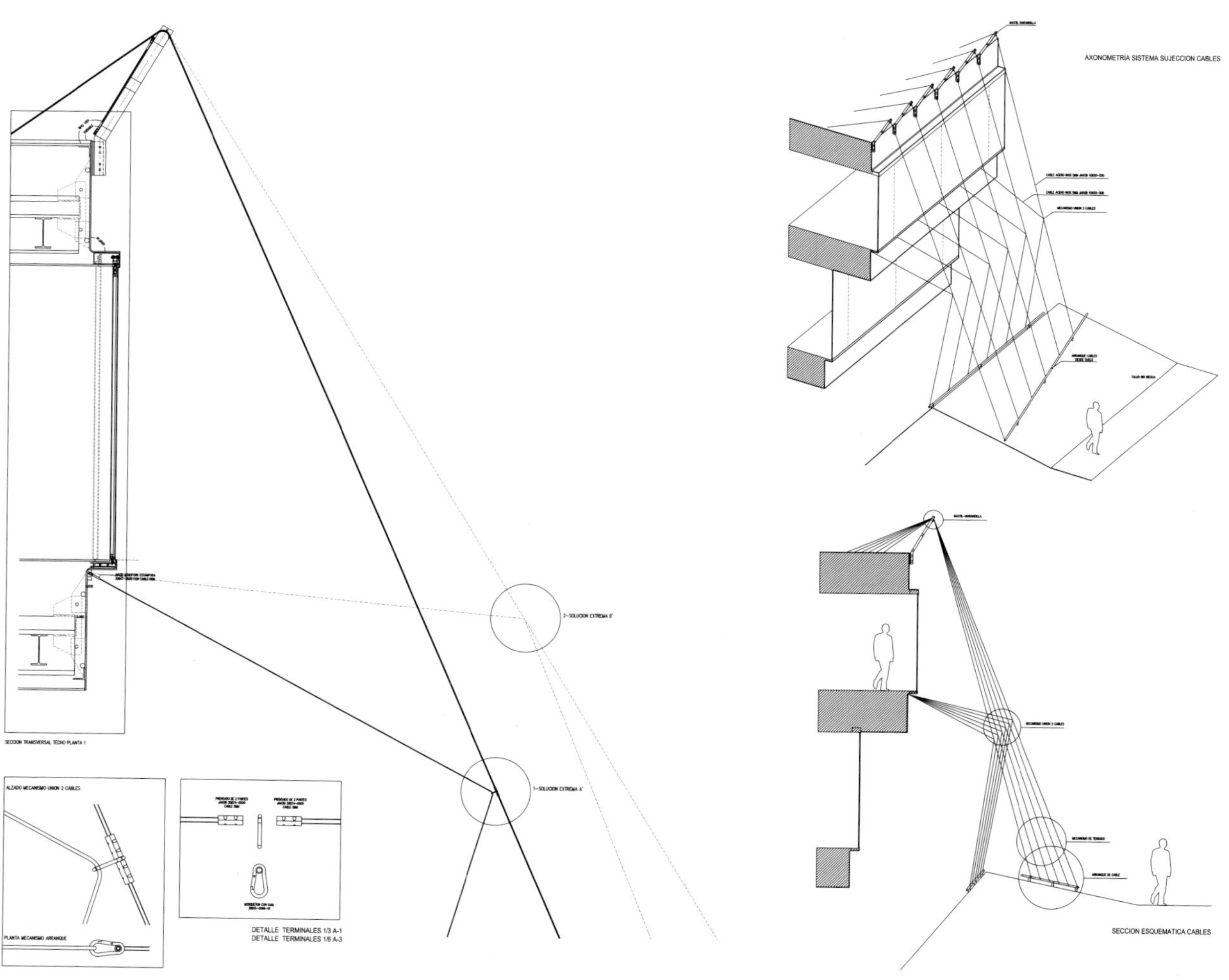
AXONOMETRIA SISTEMA SUJECCION CABLES
DETALLE TERMINALES 1/3 A-1
DETALLE TERMINALES 1/6 A-3
SECCION ESQUEMATICA CABLES

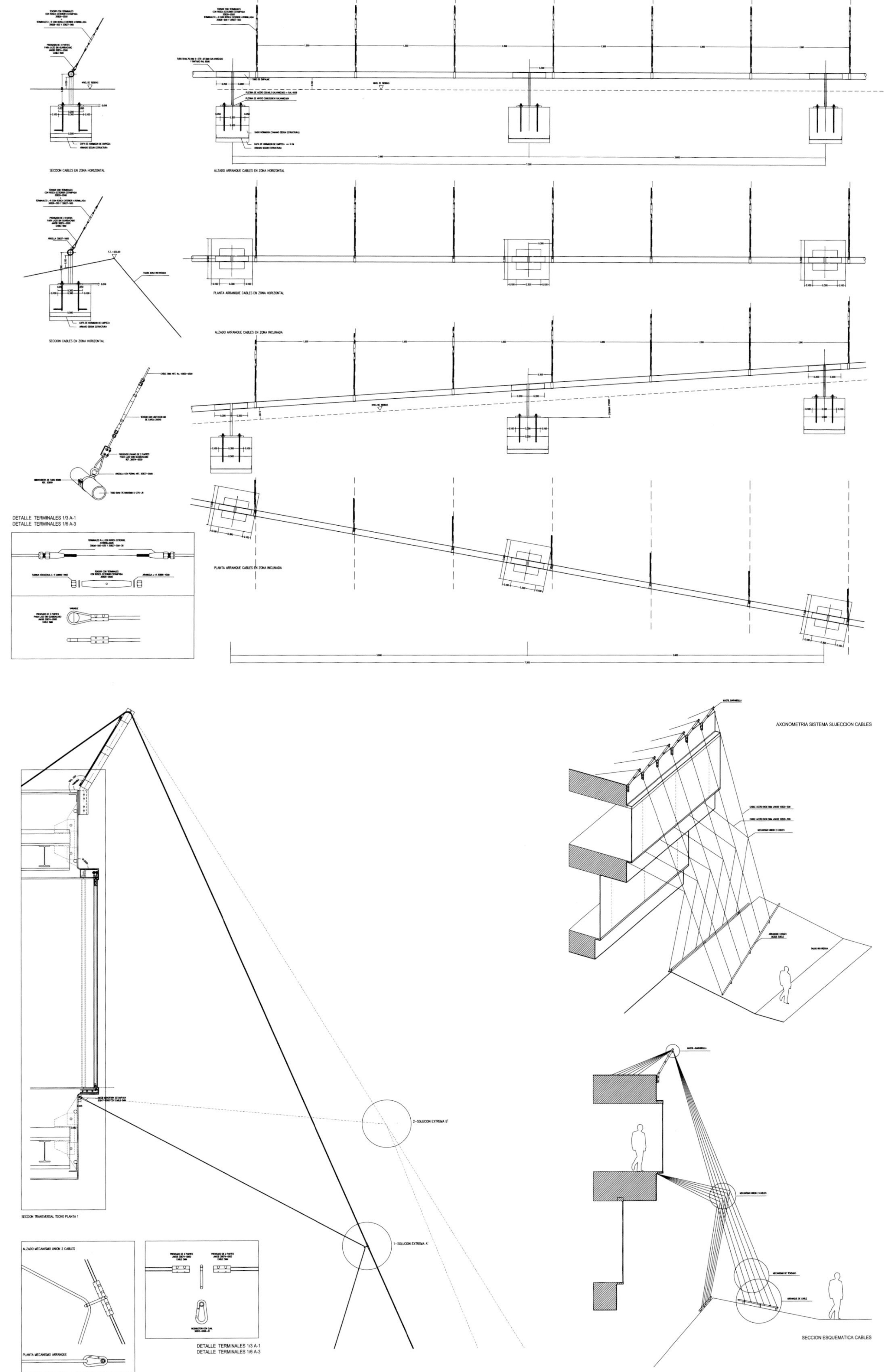
SECCION CABLES EN ZONA HORIZONTAL
ALZADO ARRANQUE CABLES EN ZONA HORIZONTAL
PLANTA ARRANQUE CABLES EN ZONA HORIZONTAL
ALZADO ARRANQUE CABLES EN ZONA INCLINADA
PLANTA ARRANQUE CABLES EN ZONA INCLINADA
DETALLE TERMINALES 1/3 A-1
DETALLE TERMINALES 1/6 A-3
AXONOMETRIA SISTEMA SUJECCION CABLES
SECCION TRANSVERSAL TECHO PLANTA 1
ALZADO MECANISMO UNION 2 CABLES
PLANTA MECANISMO ARRANQUE
DETALLE TERMINALES 1/3 A-1
DETALLE TERMINALES 1/6 A-3
2-SOLUCION EXTREMA 'B'
1-SOLUCION EXTREMA 'A'
SECCION ESQUEMATICA CABLES

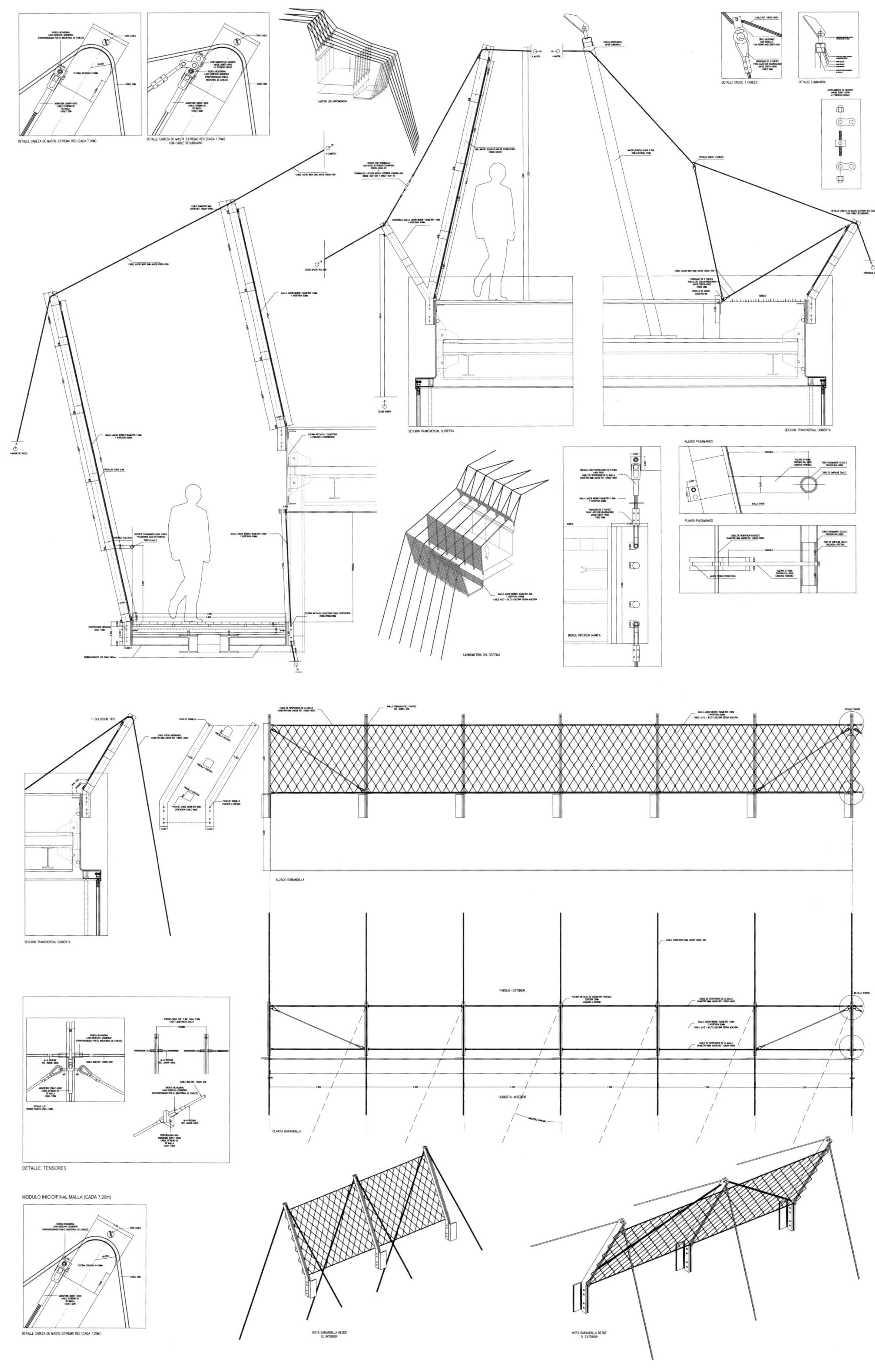
DETALLE CABEZA DE MASTIL EXTREMO RED (CADA 7,20M)
DETALLE CABEZA DE MASTIL EXTREMO RED (CADA 7,20M) CON CABLE SECUNDARIO
DETALLE CRUCE 3 CABLES
DETALLE LUMINARIA
SECCION TRANSVERSAL CUBIERTA
AXONOMETRIA DEL SISTEMA
BORDE INTERIOR RAMPA
ALZADO PASAMANOS
PLANTA PASAMANOS
SECCION TRANSVERSAL CUBIERTA
ALZADO BARANDILLA
PARQUE-EXTERIOR
CUBIERTA-INTERIOR
PLANTA BARANDILLA
DETALLE TENSORES
MODULO INICIO/FINAL MALLA (CADA 7,20m)
DETALLE CABEZA DE MASTIL EXTREMO RED (CADA 7,20M)
VISTA BARANDILLA DESDE EL INTERIOR
VISTA BARANDILLA DESDE EL EXTERIOR

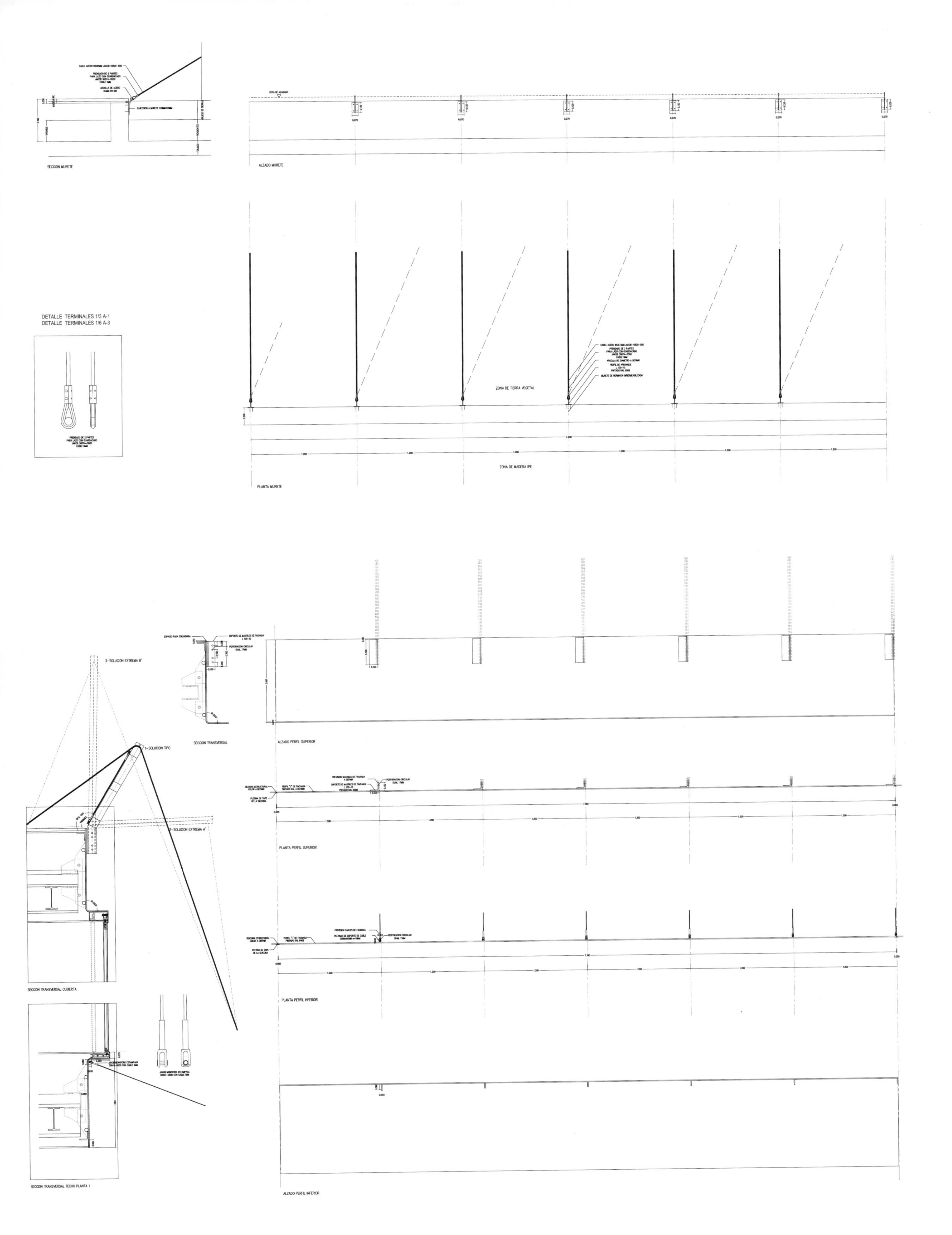

SECCION MURETE
ALZADO MURETE
DETALLE TERMINALES 1/3 A-1
DETALLE TERMINALES 1/6 A-3
ZONA DE TIERRA VEGETAL
ZONA DE MADERA IPE
PLANTA MURETE
3-SOLUCION EXTREMA B'
1-SOLUCION TIPO
SOLUCION EXTREMA A'
SECCION TRANSVERSAL
ALZADO PERFIL SUPERIOR
PLANTA PERFIL SUPERIOR
PLANTA PERFIL INFERIOR
SECCION TRANSVERSAL CUBIERTA
SECCION TRANSVERSAL TECHO PLANTA 1
ALZADO PERFIL INFERIOR

Hans Klotz Ltd.

Location: Bolzano, Italy
Architect: Monovolume Architecture + Design
Client: Hans Klotz limited company
Floor Area: 280 m²
Volume: 1,200 m³
Completion Date: 2011
Photographer: Urlich Egger

Hans Klotz Ltd. is a long-established and successful wholesale company for fruit and vegetables with headquarter in Bozen-Sigmundskron. The new offices should be integrated into an existing building on the company area. To confer an independent character, the necessary spaces were added as an autonomous, two storey volume above the existing ground floor as a kind of architectural parasite. The new building takes up the roof lines of the old, then distorting them to open the volume on the north and west side, towards huge apple plantations. Hence the new construction allows an optimal orientation of the office spaces to the north, combined with a great outlook.

The first floor contains two open space offices, the entrance and reception, as well as small secondary rooms. An ample stair along the glass façade connects the internal and the external spaces as well as the storeys with each other. The second floor is conceived as a gallery with a cafeteria opening towards a roof terrace. There are also a conference room and two private offices.

The new administrative building will be wrapped by a gleaming skin, which at the same time quotes the old roof and underlines the new building's independence.

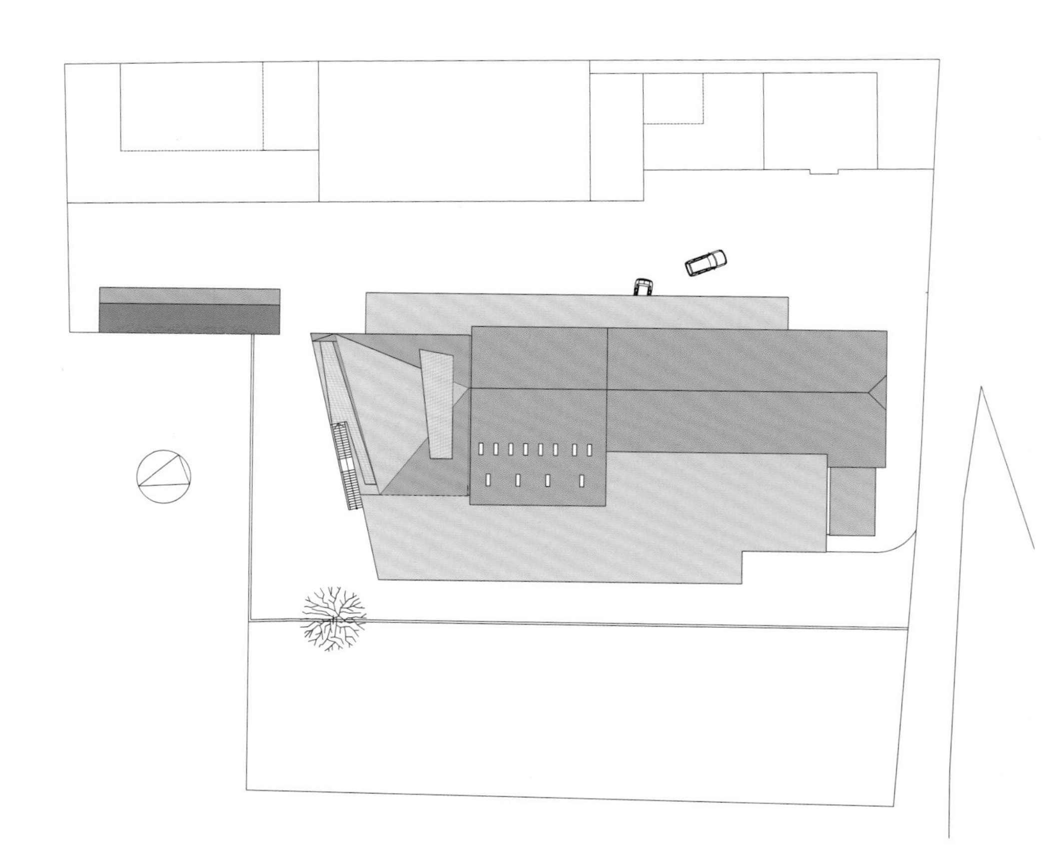

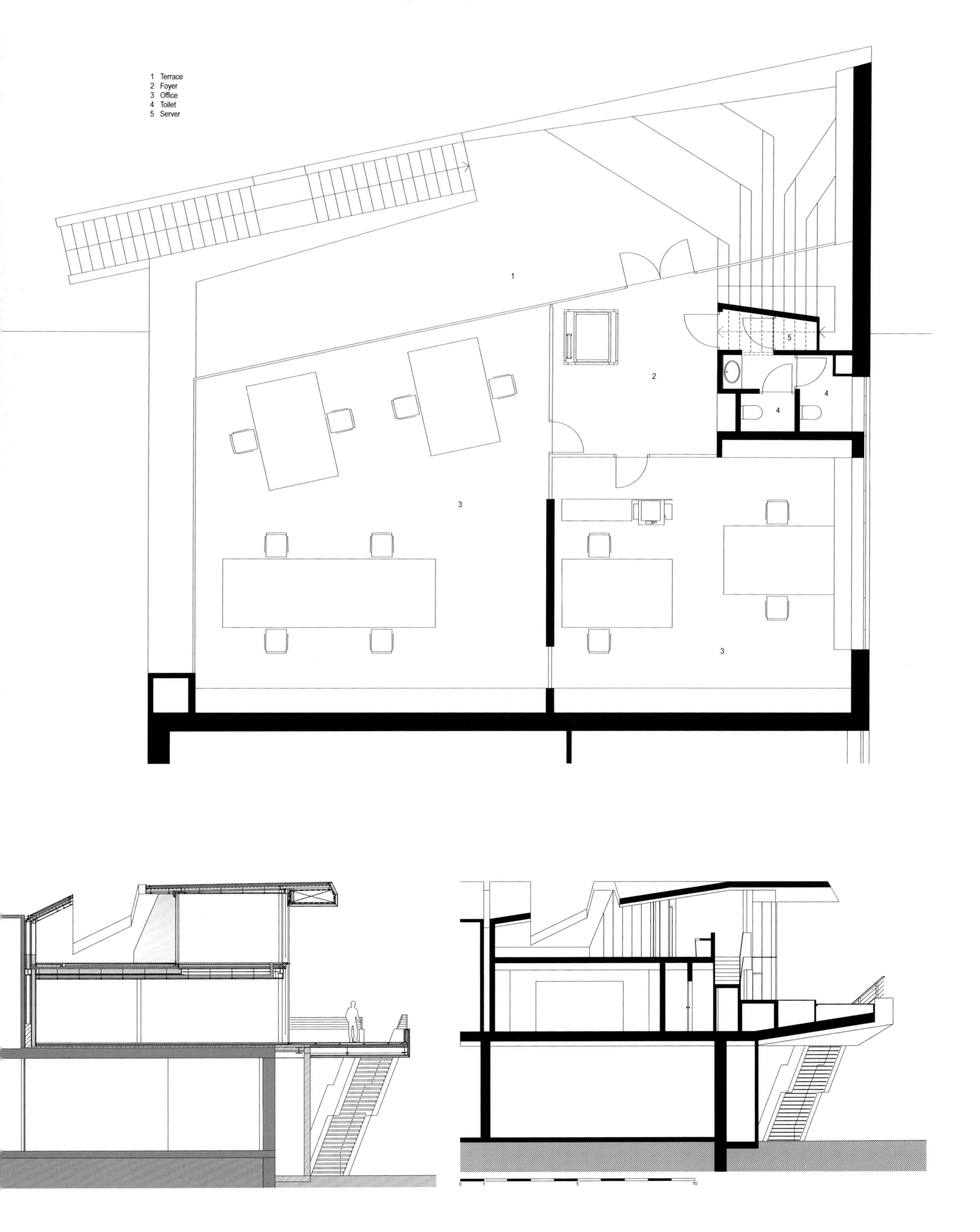
1 Terrace
2 Foyer
3 Office
4 Toilet
5 Server
1
2
3
3
4
4
5
0
1
5
10

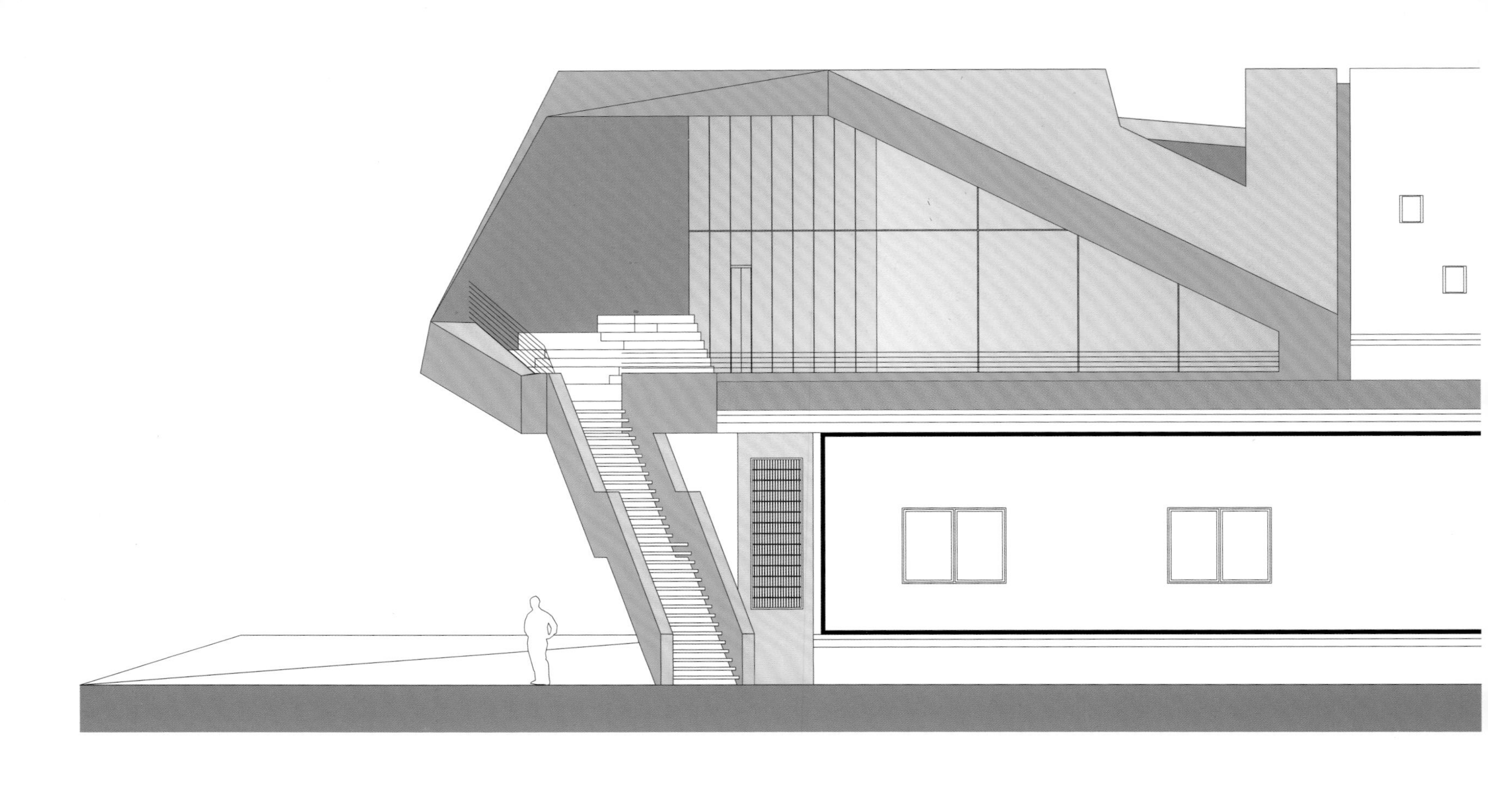

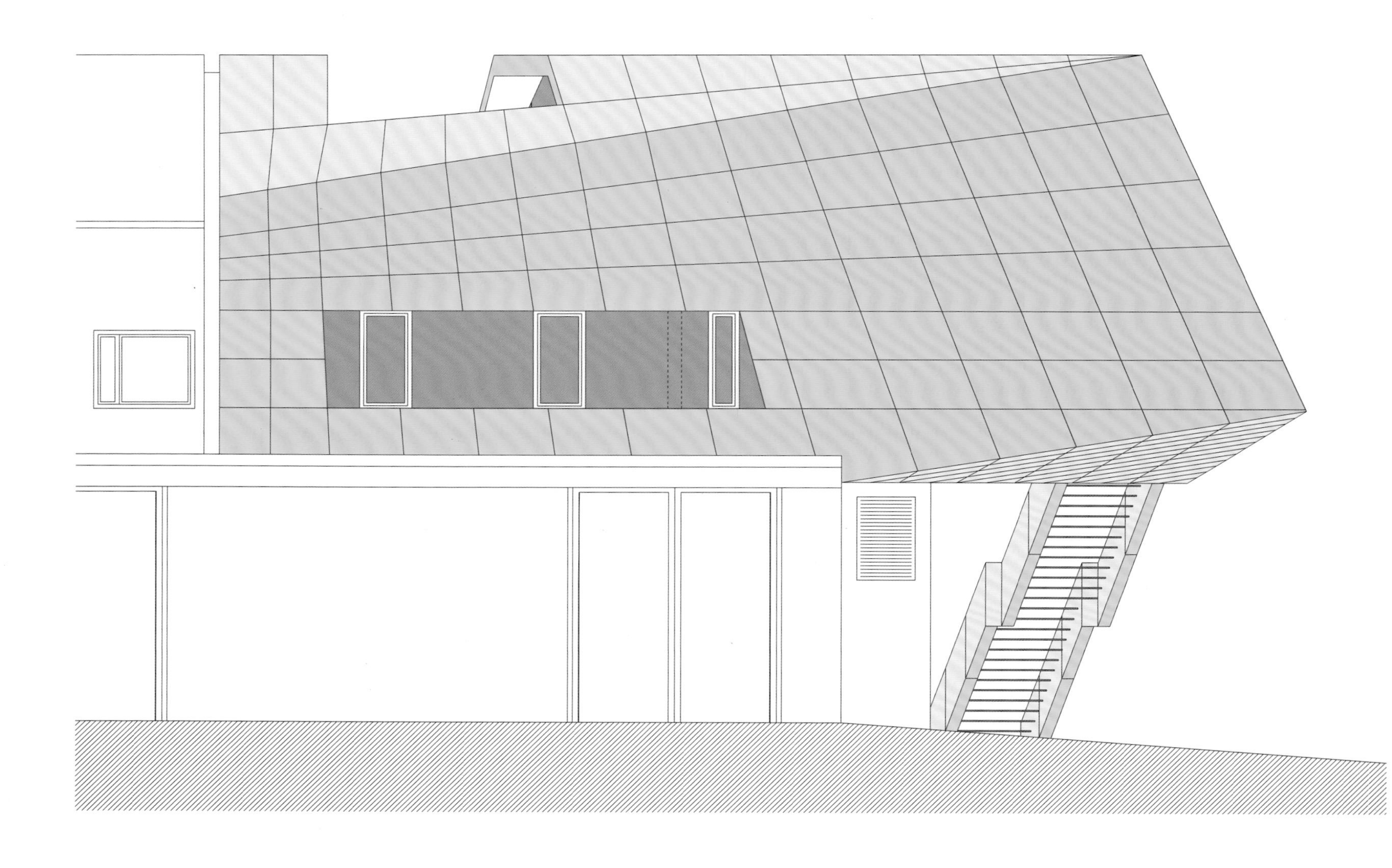

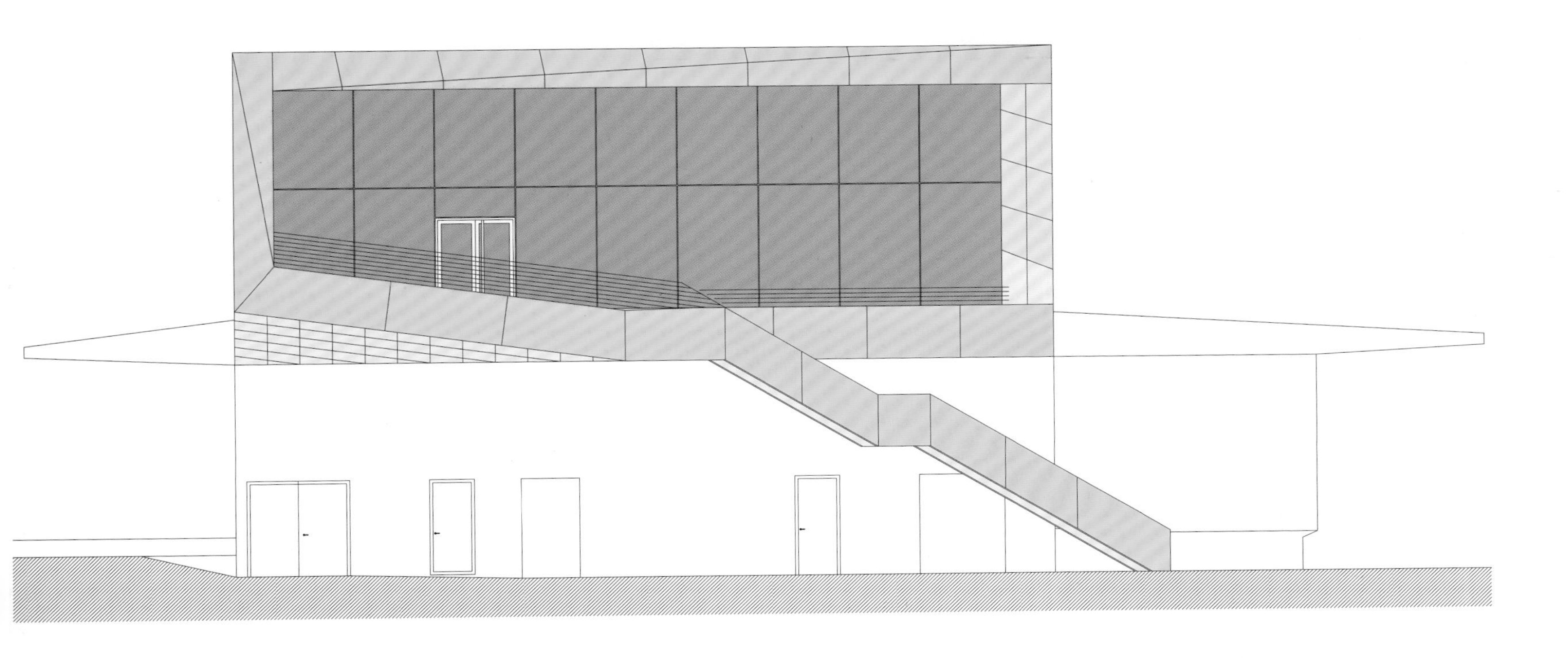

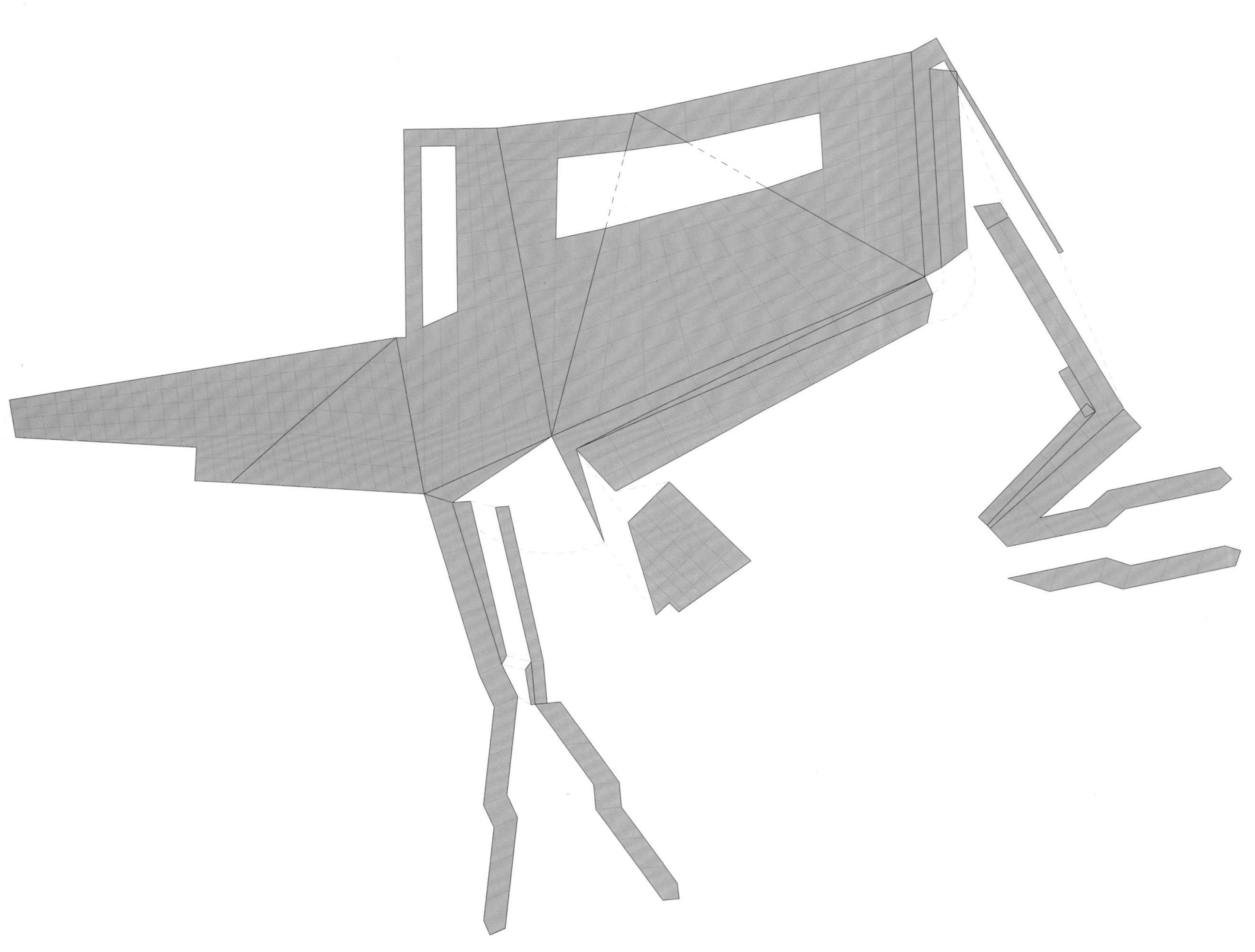

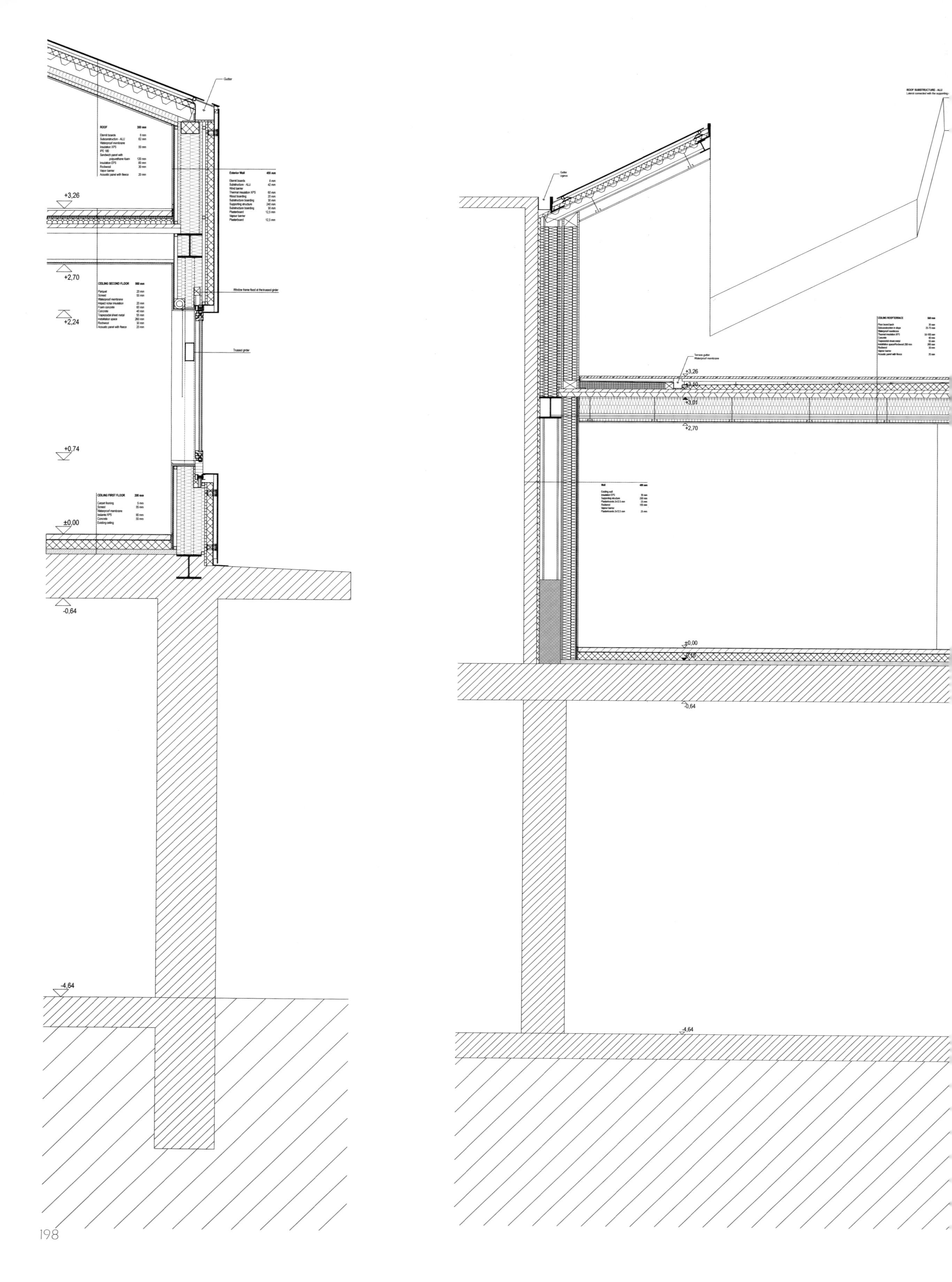
Gutter
ROOF 350 mm
Eternit boards 8 mm
Subconstruction - ALU 62 mm
Waterproof membrane
Insulation XPS 50 mm
IPE 180
Sandwich panel with polyurethane foam 120 mm
Insulation EPS 80 mm
Rockwool 30 mm
Vapor barrier
Acoustic panel with fleece 20 mm
Exterior Wall 455 mm
Eternit boards 8 mm
Substructure - ALU 42 mm
Wind barrier
Thermal insulation XPS 60 mm
Wood boarding 20 mm
Substructure boarding 30 mm
Supporting structure 240 mm
Substructure boarding 30 mm
Plasterboard 12,5 mm
Vapour barrier
Plasterboard 12,5 mm
+3,26
+2,70
CEILING SECOND FLOOR 560 mm
Parquet 20 mm
Screed 55 mm
Waterproof membrane
Impact noise insulation 20 mm
Foam concrete 60 mm
Concrete 40 mm
Trapezoidal sheet metal 55 mm
Installation space 260 mm
Rockwool 30 mm
Acoustic panel with fleece 20 mm
Window frame fixed at the trussed girder
+2,24
Trussed girder
+0,74
CEILING FIRST FLOOR 200 mm
Carpet flooring 5 mm
Screed 55 mm
Waterproof membrane
Isolante XPS 90 mm
Concrete 50 mm
Existing ceiling
±0,00
-0,64
-4,64
ROOF SUBSTRUCTURE - ALU
Gutter
Uginox
CEILING ROOFTERRACE
Floor board larch
Subconstruction in slope 25-75 mm
Waterproof membrane
Thermal insulation XPS
Concrete
Trapezoidal sheet metal
Installation space/Rockwool 200 mm
Rockwool
Vapour barrier
Acoustic panel with fleece
Terrace gutter
Waterproof membrane
+3,26
+3,10
+3,01
+2,70
Wall
Existing wall
Insulation EPS
Supporting structure
Plasterboards 2x12.5 mm
Rockwool
Vapour barrier
Plasterboards 2x12.5 mm
±0,00
-0,64
-4,64

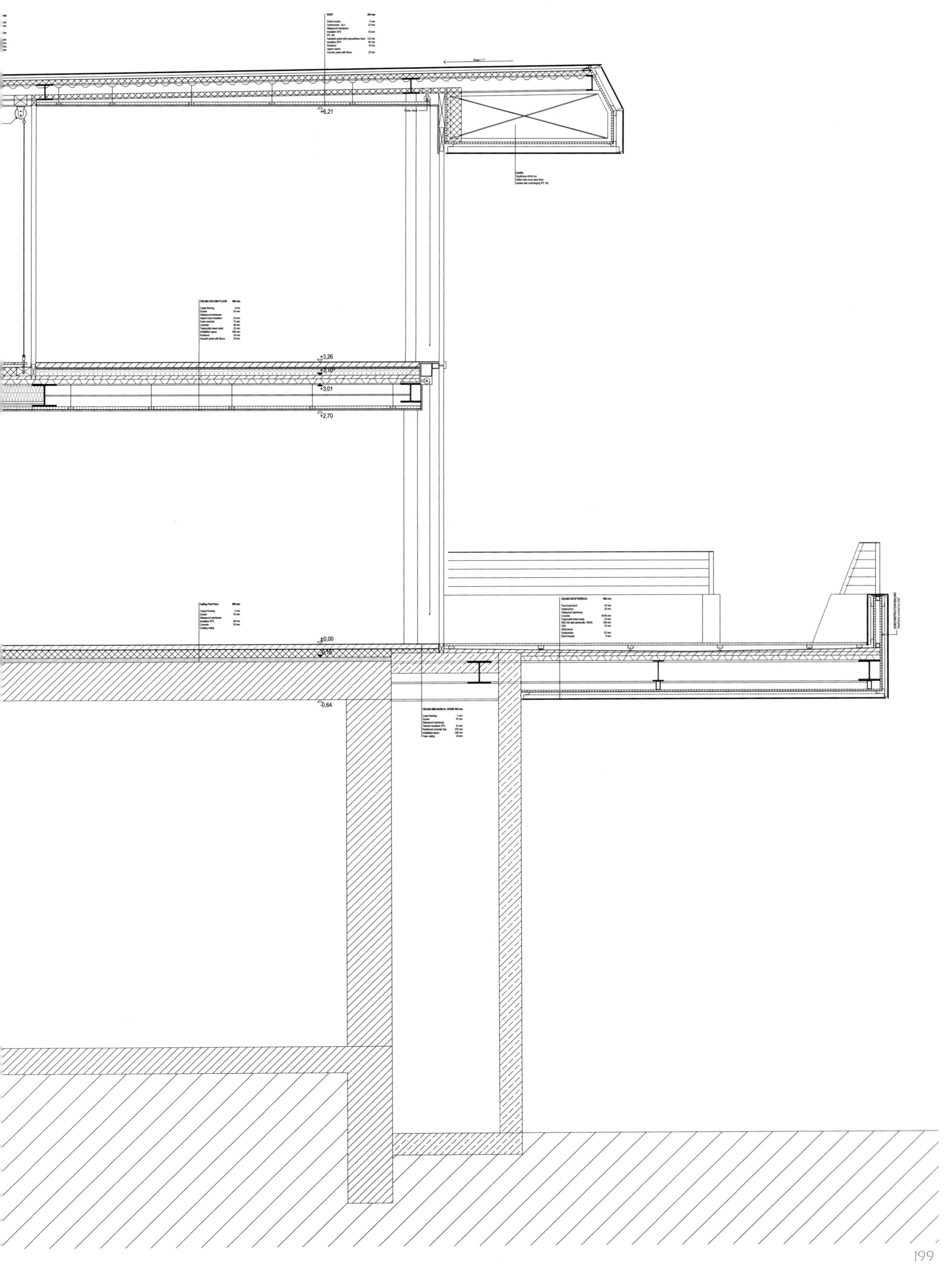
ROOF
Insulation XPS 50 mm
IPE 180
Sandwich panel with polyurethane foam 120 mm
Insulation EPS 60 mm
Rockwool 30 mm
Vapour barrier
Acoustic panel with fleece 20 mm
+6,21
EAVES
CEILING SECOND FLOOR
Carpet flooring 5 mm
Screed 55 mm
Impact noise insulation 20 mm
Foam concrete 75 mm
Concrete 40 mm
Trapezoidal sheet metal 55 mm
Installation space 290 mm
Rockwool 30 mm
Acoustic panel with fleece 20 mm
+3,26
+3,10
+3,01
+2,70
Ceiling First Floor
Carpet Flooring 5 mm
Screed 55 mm
Waterproof membrane
Insulation XPS 90 mm
Concrete 50 mm
Existing ceiling
CEILING ROOFTERRACE
Floor board larch 30 mm
Substructure 30 mm
Waterproof membrane
Concrete 40-60 mm
HEB 260 with steelprofile 100/50 340 mm
OSB 22 mm
Substructure 82 mm
Eternit boards 8 mm
±0,00
-0,64
CEILING MECHANICAL ROOM
Carpet flooring 5 mm
Screed 55 mm
Waterproof membrane
Thermal insulation XPS 45 mm
Installation space 300 mm
False ceiling 20 mm

La Forgiatura

Location: Milan, Italy
Architect: Giuseppe Tortato & Partners Architects
Client: La Forgiatura S.r.l.
Floor Area: 24,000,00 m^2
Photography: Stefano Topuntoli per GDM Costruzioni + Andrea Puggiotto

Giuseppe Tortato & Partners Architects signs La Forgiatura, the redevelopment of an industrial area of Milan that demonstrates how it is possible to work on an urban area in a nonviolent way. The new multi-purpose complex is characterized by a distinctive sign that arises from the culture and history of the original place marrying technology and natural environment, with a strong integration between architecture and landscape.

Located in the north west area of Milan, La Forgiatura rappresented for many years the excellence of Italian and international metallurgical industry. Its regeneration has allowed to upgrade more than 14.000 m^2 of existing buildings to which were added 10.000 m^2 of new construction , for a total of 7 buildings for office and showroom space. This is a complex of properties for different uses, certified in B class, all served by a common conditioning air system managed by Building Management System, that allows the optimization of different applications at the same time.

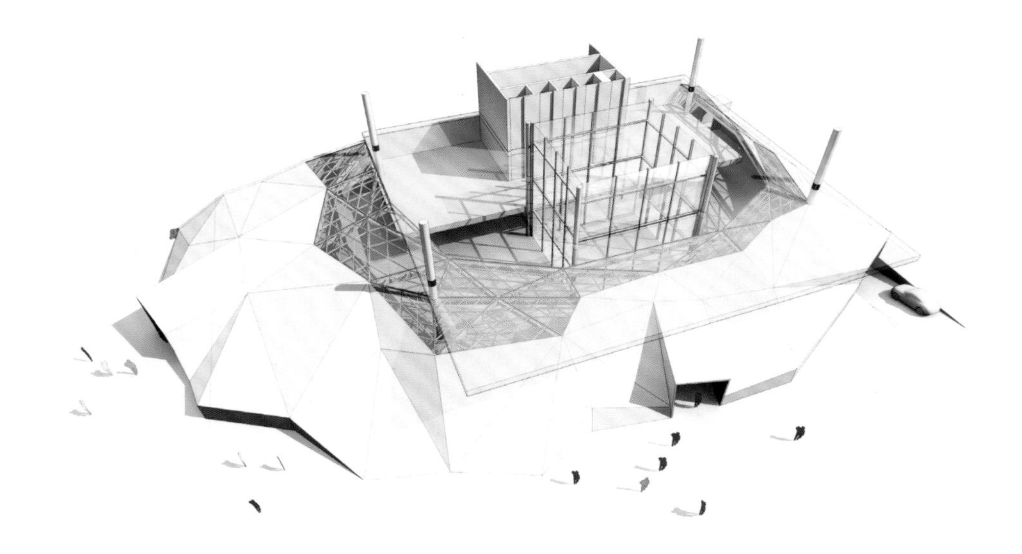

A real urban oasis, characterized by a three-dimensional development of the green landscape that incorporates old industrial structures joined smoothly to the new buildings. Thanks to the movement of the ground, in some cases real artificial hills eight meters high, you can access the buildings from various levels, including the roofs, enjoying a unique sensory relationship, given by the alternation of emotions: thanks to green patios, great heights, natural light, a unique relationship between interior and exterior greenery , old and new structures.

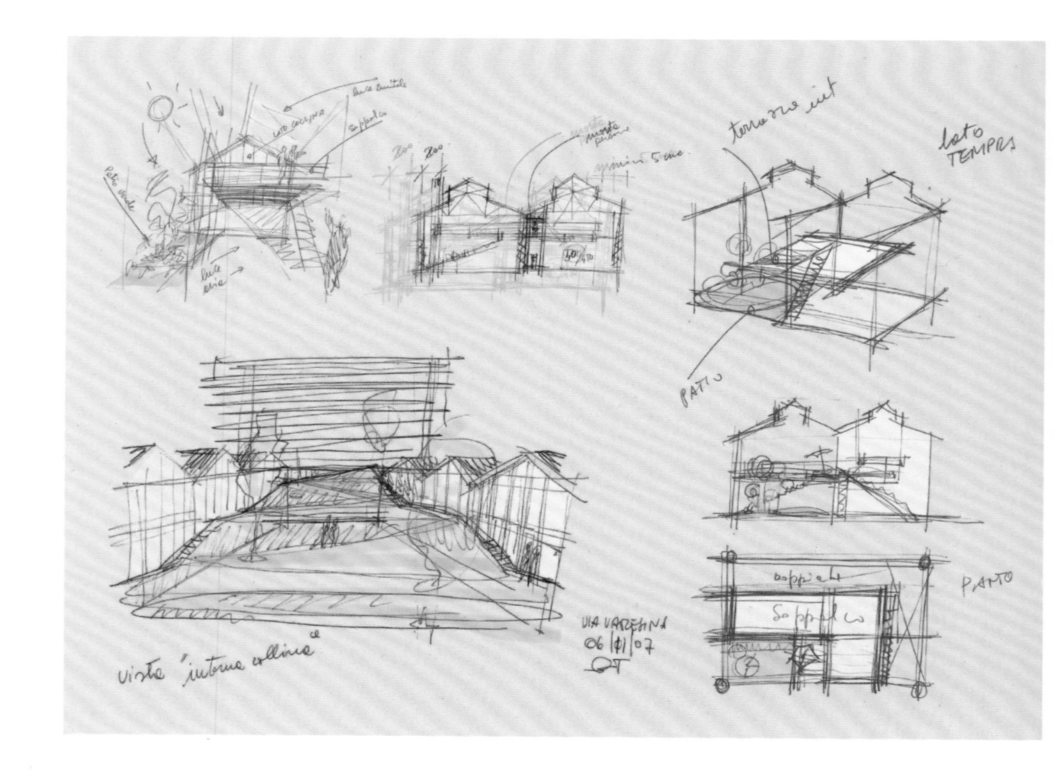

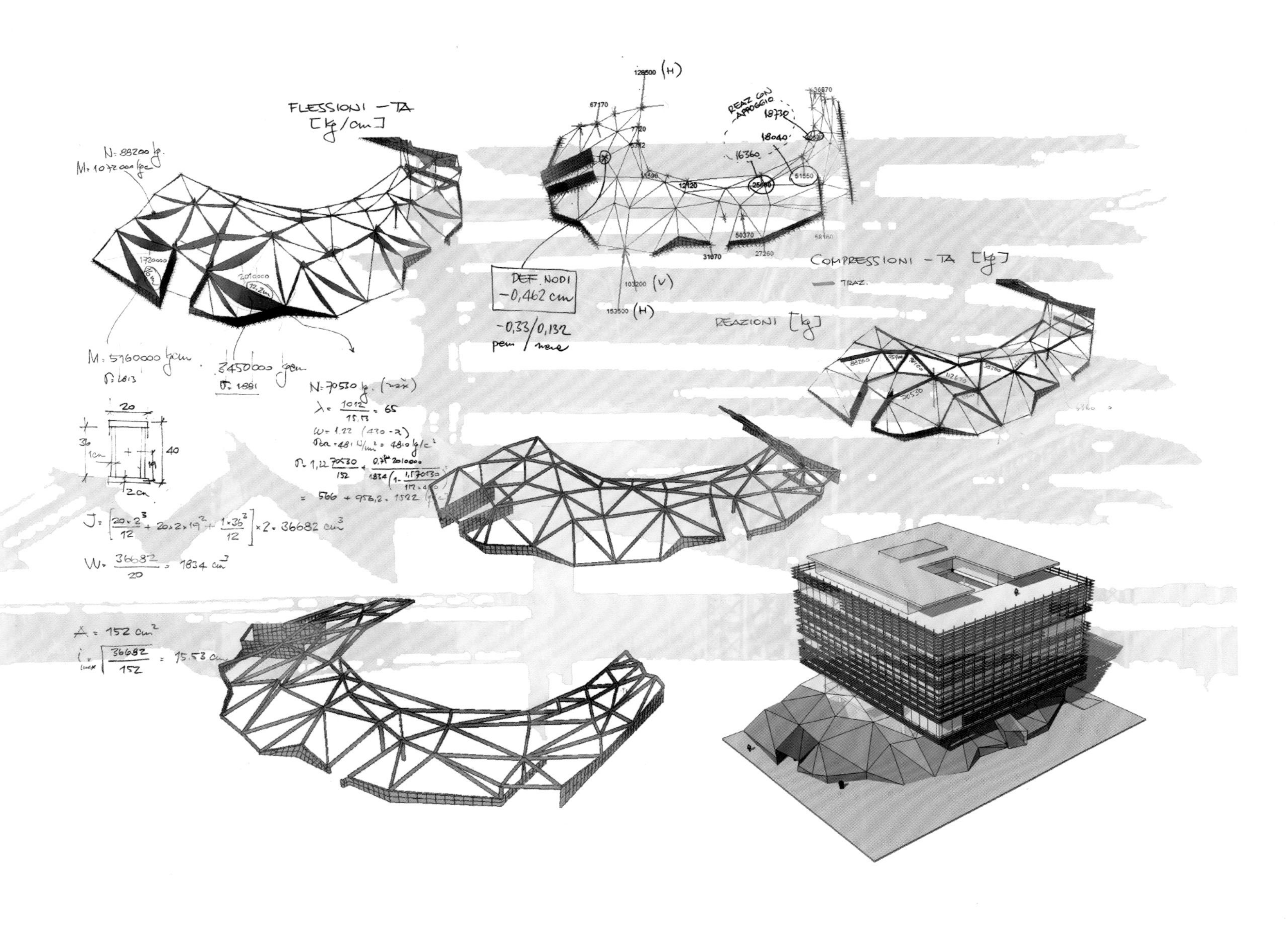
FLESSIONI – TA
[kg/cm]
COMPRESSIONI – TA [kg]
REAZIONI [kg]
DEF. NODI
-0,462 cm
REAZ. CON APPOGGIO

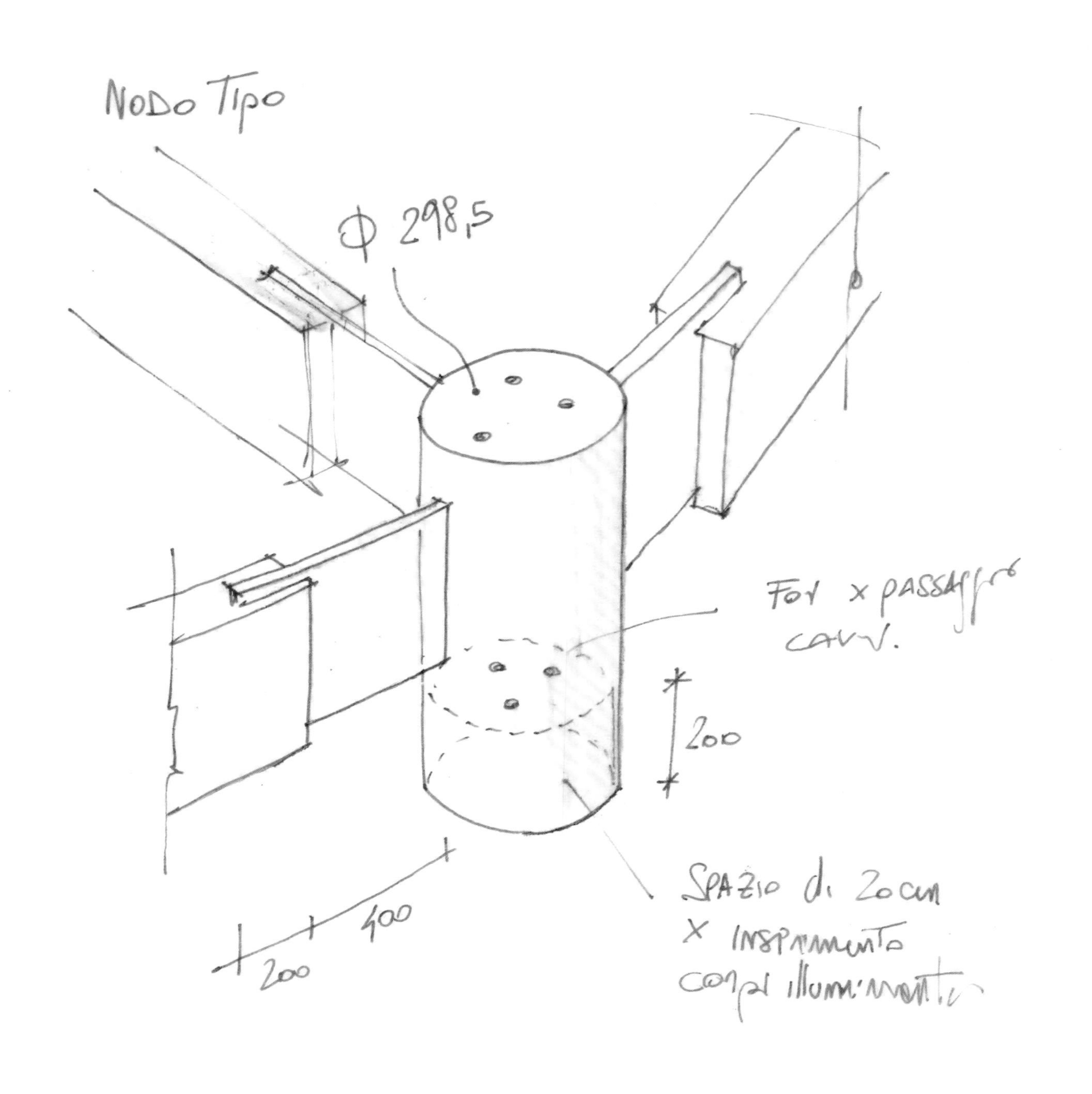
NODO TIPO
Φ 298,5
200
400
200
Spazio di 20 cm

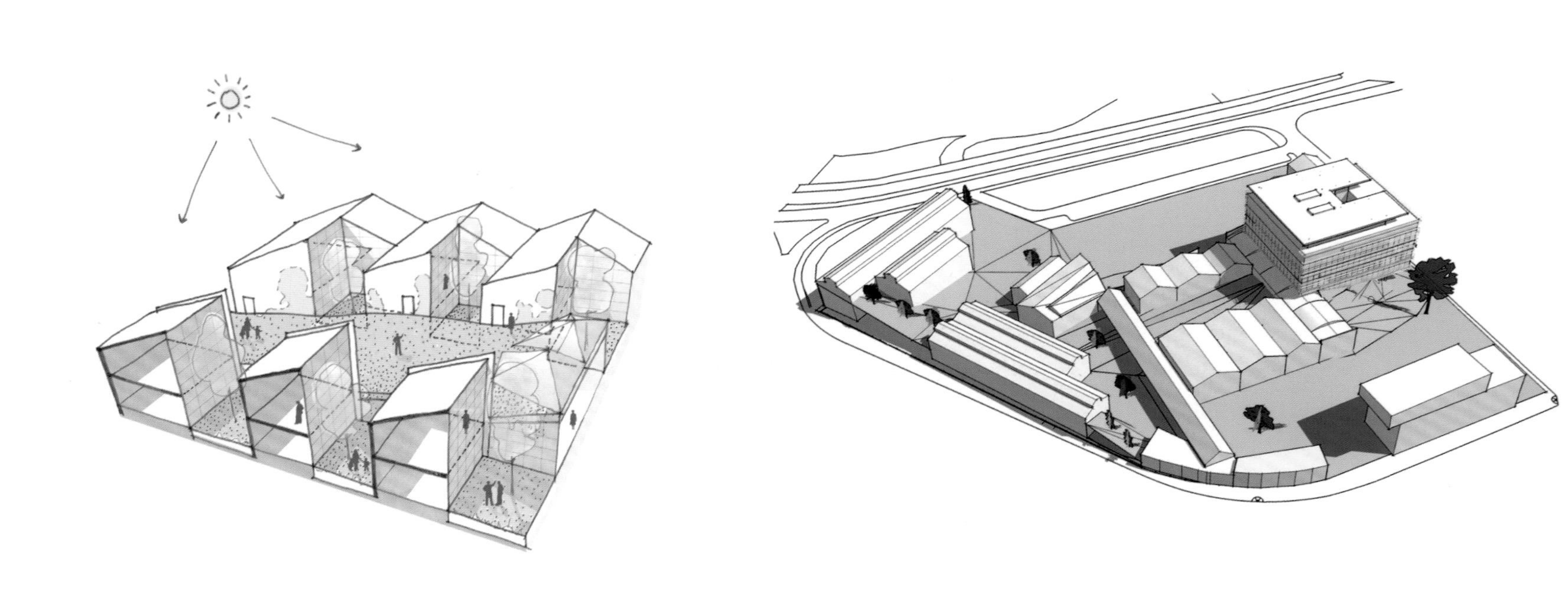

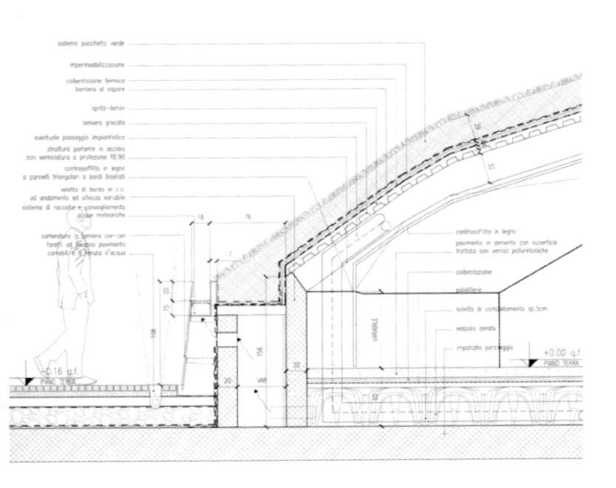

sistema pacchetto verde
impermeabilizzazione
coibentazione termica
barriera al vapore
spritz-beton
lamiera grecata
eventuale passaggio impiantistico
struttura portante in acciaio
con verniciatura a protezione RE90
controsoffitto in legno
a pannelli triangolari a bordi bisellati
veletta di bordo in c.a.
ad andamento ed altezza variabile
sistema di raccolta e convogliamento
acque meteoriche
controsoffitto in legno
pavimento in cemento con superficie
trattata con vernici poliuretaniche
coibentazione
polietilene
soletta di completamento sp.5cm
vespaio aerato
impalcato parcheggio
-0.16 q.f.
+0.00 q.f.

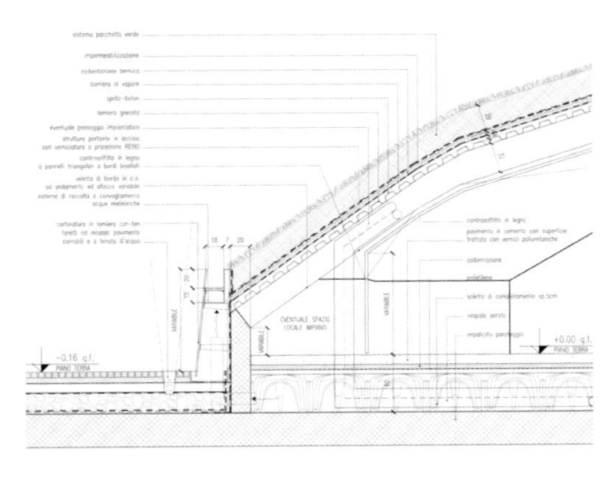

EVENTUALE SPAZIO
LOCALE IMPIANTI
-0.16 q.f.
PIANO TERRA
+0.00 q.f.
PIANO TERRA

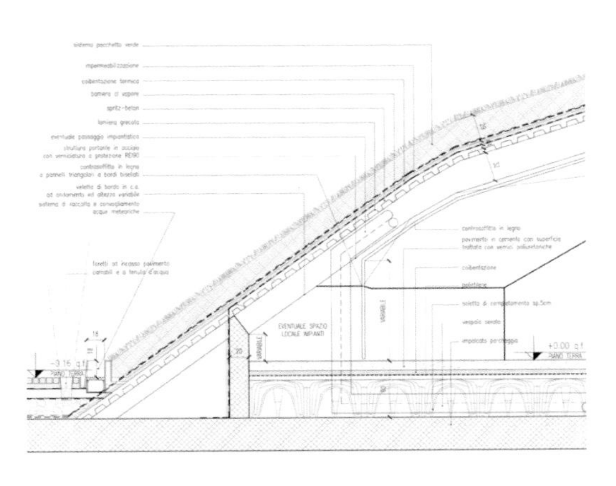

EVENTUALE SPAZIO
LOCALE IMPIANTI
-3.16 q.f.
PIANO TERRA
+0.00 q.f.
PIANO TERRA

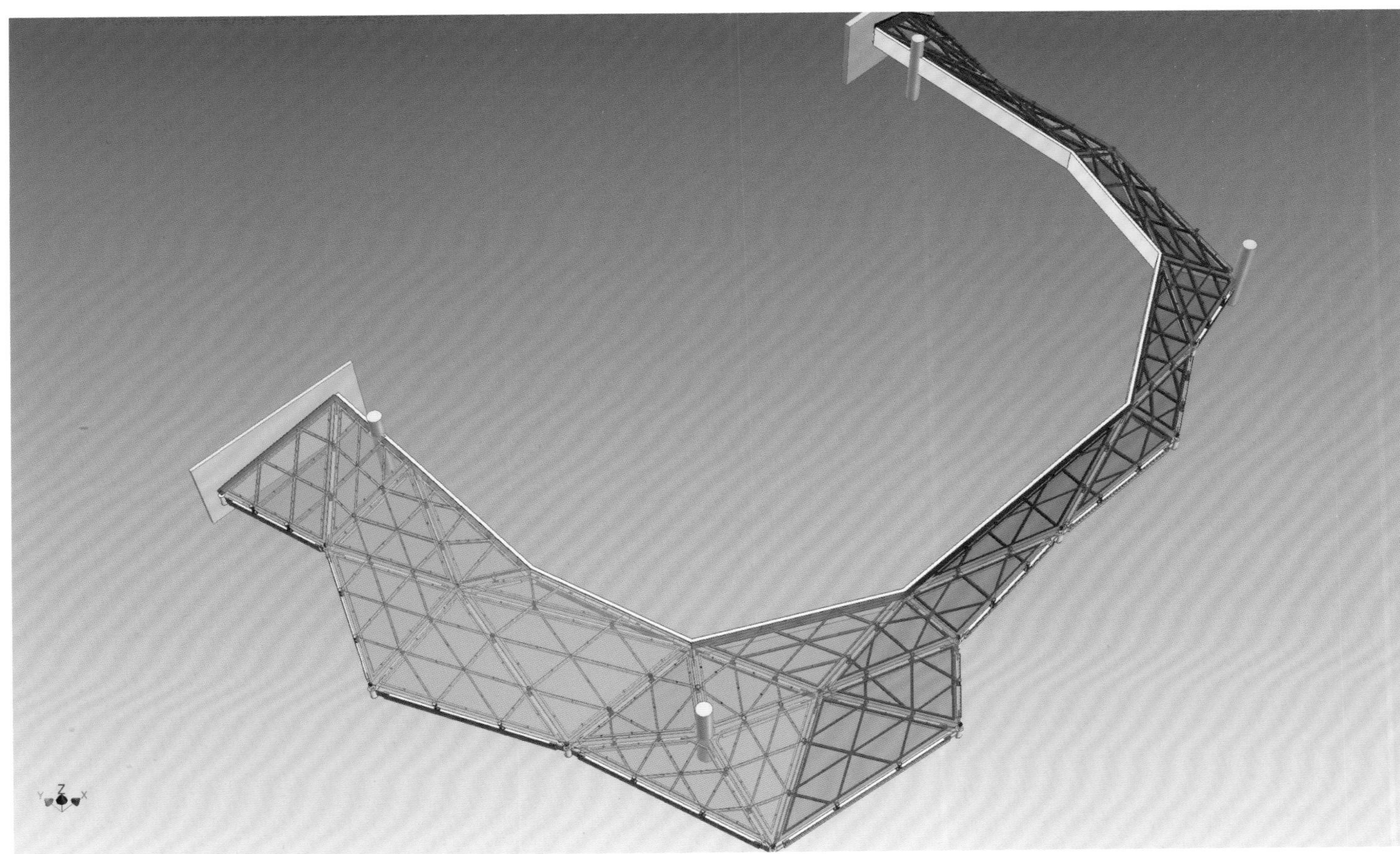

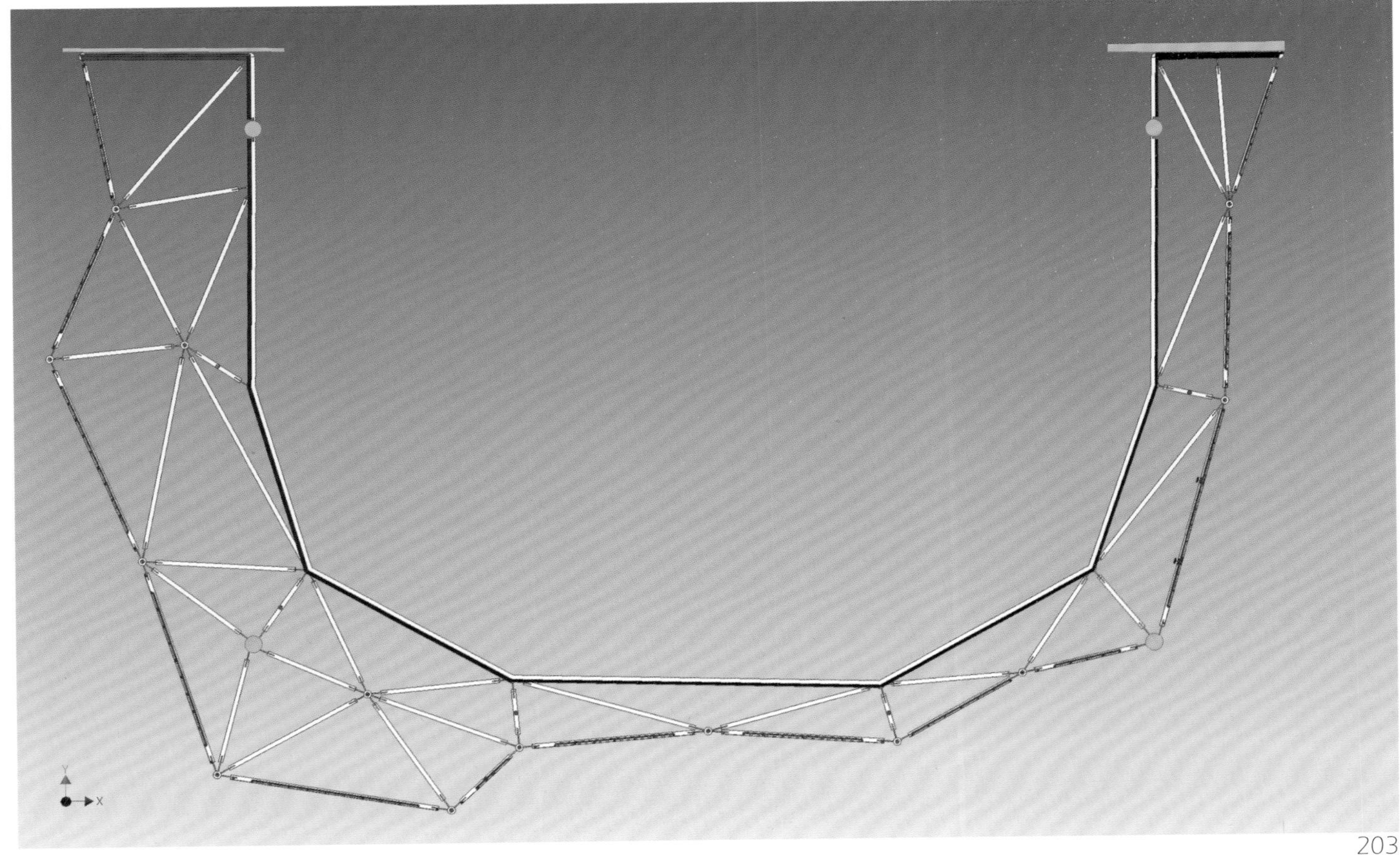

Tahiliani Design Headquarters

Location: New Delhi, India
Architect: S.P.A Design
Built Up Area: 4,000 m²
Completion Date: 2008
Photographer: Sanjit Wahi, Stephane Paumier, Shaily

The project aims at creating inward looking spaces around a garden, like a traditional cloister, a calm environment ideal for long hours of work. The studios and offices are suspended above the buzz of production located on the lower floors, getting zenithal light adapted for precision work. The construction is rectangular in shape with numerous skylights giving subdued light to the working spaces, adapted to the harsh climate of Delhi. Assembly production and embroidery are located in the basement and on the ground floor. The first floor is assigned to the design studios and the sampling. The trial rooms, showroom, personal design studio and offices for the executives are located on the top floor. The production area on the ground floor is the heart of the building with its vaulted skylight and magnificent double height.

The building is an exposed reinforced cement concrete structure supported on large mushroom columns and flat slabs, allowing beam free office spaces with generous height. The central grid of mushroom columns in the production hall transforms the flat ceiling into a vault with skylight in between. The vault forms the planters for the upper floors. The main design studio areas get a double height space with one side of the hall looking from below, at the hanging gardens on the either side of the courtyard. On the upper floor, there is a terrace garden above the skylight - a garden enclosed in a noisy and dusty area of the industries. The Main atelier of the fashion designer is placed in another vaulted space on the second floor looking on to the garden.

The project has used exposed ordinary bricks for external facade giving the precinct a warm character. The arched shape of the structure is inspired by the Islamic monuments of Delhi, reinterpreted into a contemporary design. The entrance is an entre-deux created out of a vault open in the center between the 2 T shaped pillars, the logo of Tarun Tahiliani design firm. The entrance faces east, and it is almost surreal when people enter the factory in the morning with the morning sun beaming through the building in foreground.

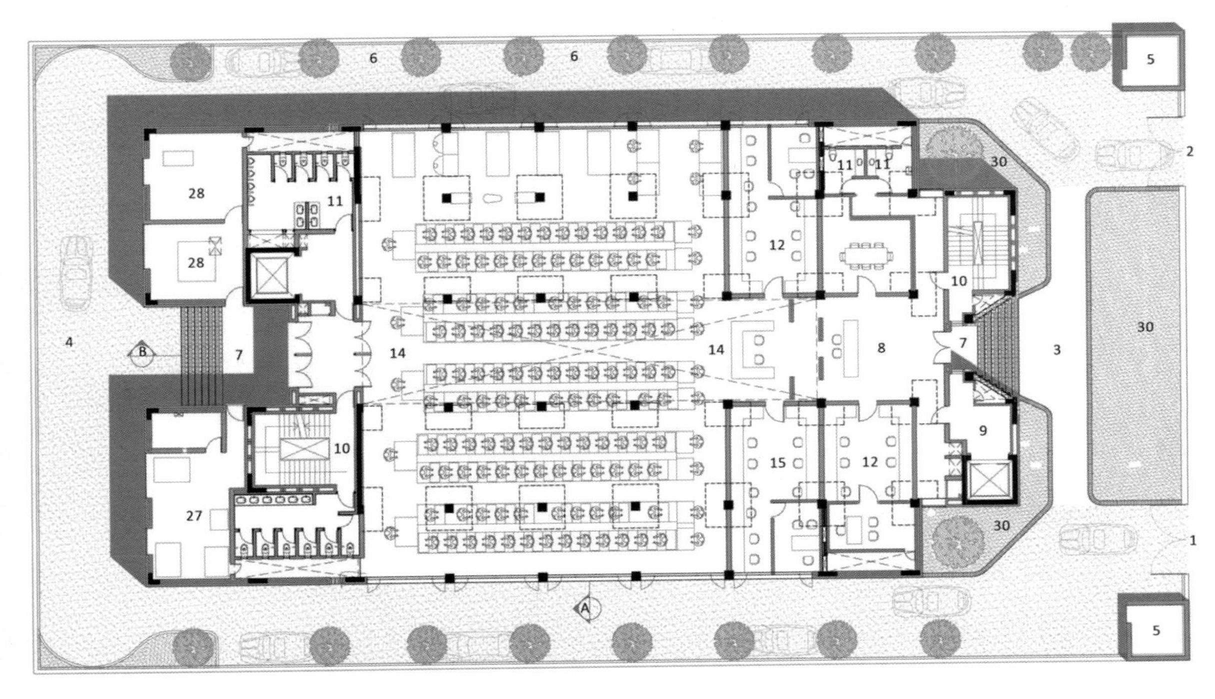

Site Plan

Legend
1. Entry
2. Exit
3. Drop off
4. Vehicular road
5. Guard room
6. Parking
7. Entrance
8. Reception/entrance lobby
9. Lift lobby
10. Staircase
11. Toilets
12. Administration
13. Conference room
14. Production area
15. Finance dept.
16. Couture design sampling hall
17. Pret-a-porter design sam pling hall
18. Designer studio
19. Hod office
20. Showroom
21. Computer lab
22. Office of tarun tahiliani
23. Secretary office
24. Executive offices
25. Archives
26. Pantry
27. Washing room
28. Electrical room
29. DG set
30. Landscaped area
31. Storage
32. Terrace garden

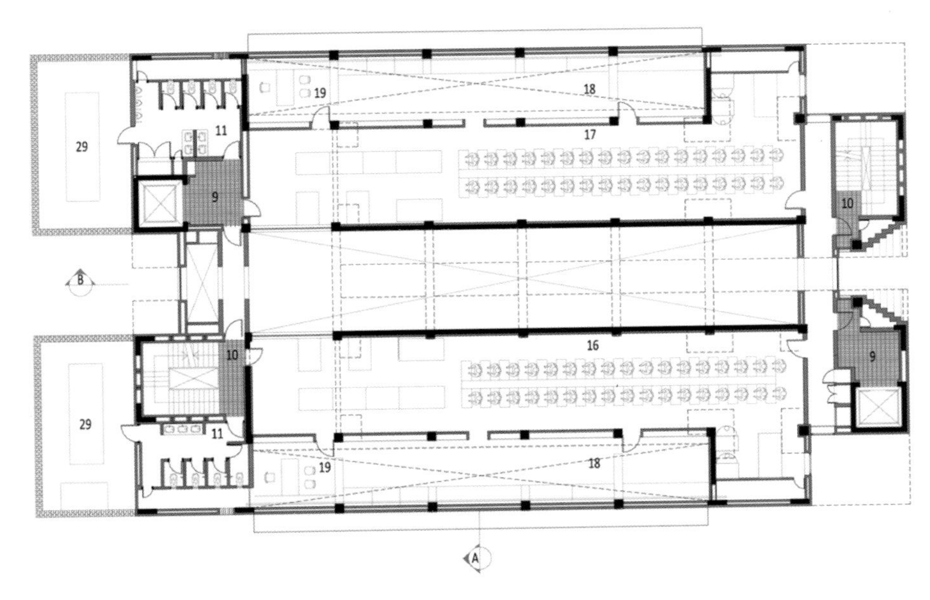

First Floor Plan

Legend
1. Entry
2. Exit
3. Dropoff
4. Vehicular road
5. Guard room
6. Parking
7. Entrance
8. Reception/entrance lobby
9. Liftlobby
10. Staircase
11. Toilets
12. Administration
13. Conference room
14. Production area
15. Finance dept.
16. Couture design sampling hall
17. Pret-a-porter design sampling hall
18. Designer studio
19. Hod office
20. Showroom
21. Computer lab
22. Office of tarun tahiliani
23. Secretary office
24. Executive offices
25. Archives
26. Pantry
27. Washing room
28. Electrical room
29. DG set
30. Landscaped area
31. Storage
32. Terrace garden

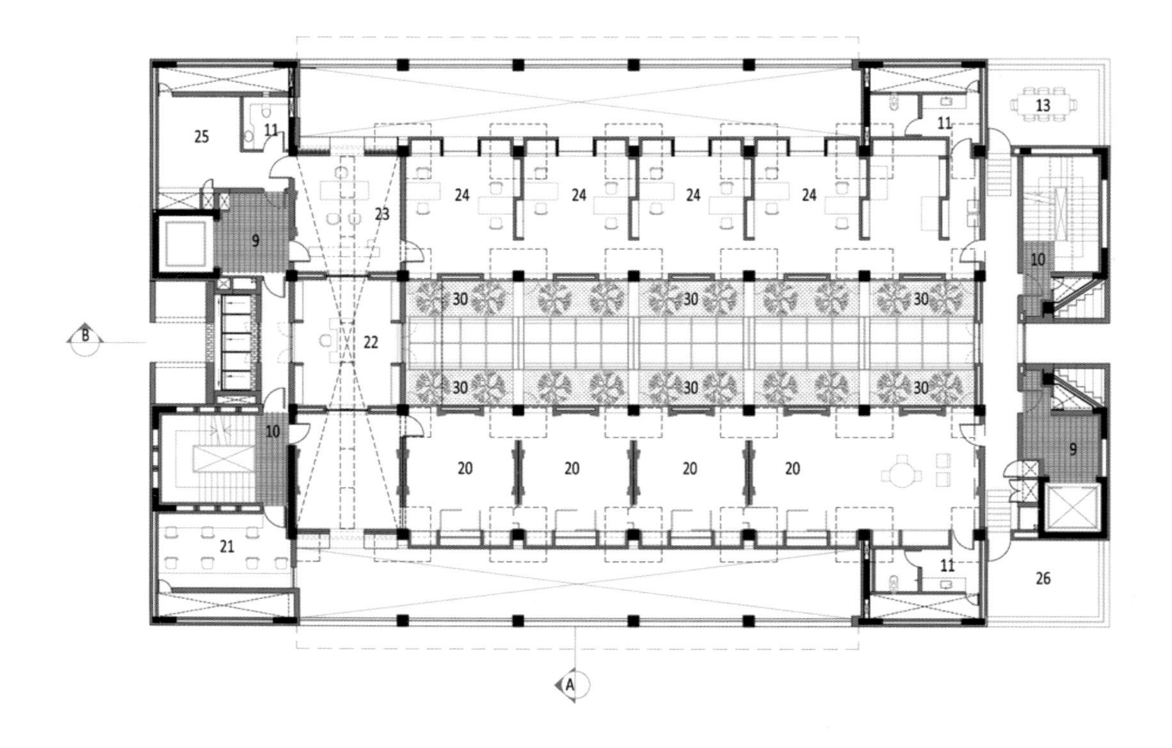

Legend
1. Entry
2. Exit
3. Drop off
4. Vehicular road
S. Guard room
6. Parking
7. Entrance
8. Reception/entrance lobby
9. Lift lobby
10. Staircase
11. Toilets
12. Administration
13. Conference room
14. Production area
15. Finance dept.
16. Couture design sampling hall
17. Pret-a-porter design sampling hall
18. Designer studio
19. Hod office
20. Showroom
21. Computer lab
22. Office of tarun tahiliani
23. Secretary office
24. Executive offices
25. Archives
26. Pantry
27. Washing room
28. Electrical room
29. DG set
30. Landscaped area
31. Storage
32. Terrace garden

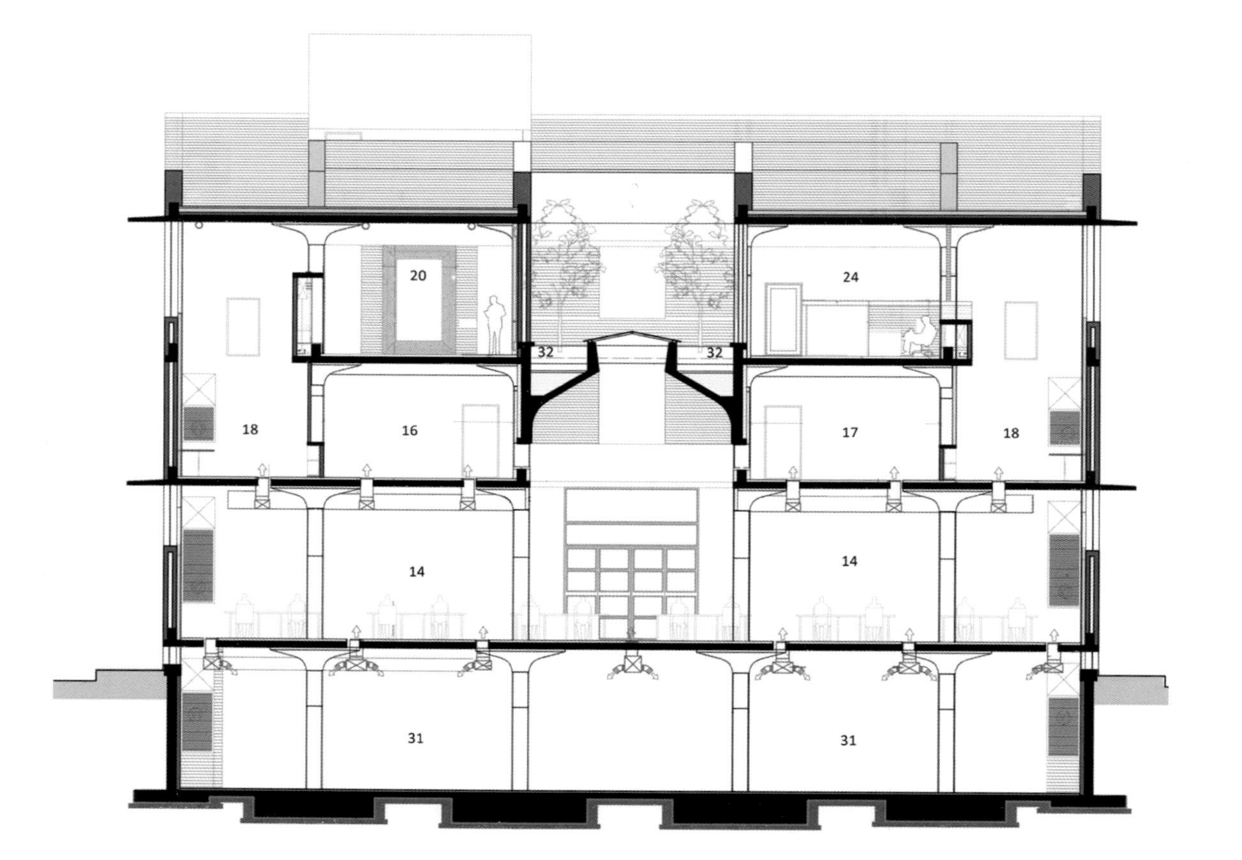

Legend
1. Entry
2. Exit
3. Drop off
4. Vehicular road
5. Guard room
6. Parking
7. Entrance
8. Reception/entrance lobby
9. Lift lobby
10. Staircase
11. Toilets
12. Administration
13. Conference room
14. Production area
15. Finance dept.
16. Couture design sampling hall
17. Pret-a-porter design sampling hall
18. Designer studio
19. Hod office
20. Showroom
21. Computer lab
22. Office of tarun tahiliani
23. Secretary office
24. Executive offices
25. Archives
26. Pantry
27. Washing room
28. Electrical room
29. DG set
30. Landscaped area
31. Storage
32. Terrace garden

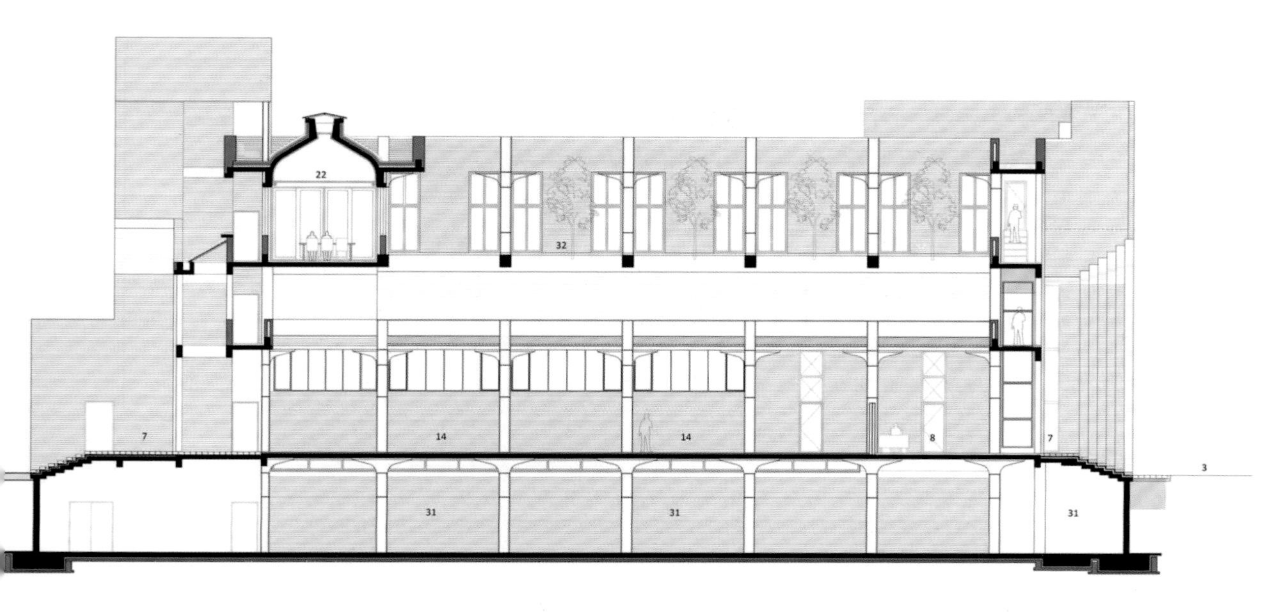

Legend
1. Entry
2. Exit
3. Drop off
4. Vehicular road
S. Guard room
6. Parking
7. Entrance
8. Reception/entrance lobby
9. Lift lobby
10. Staircase
11. Toilets
12. Administration
13. Conference room
14. Production area
15. Finance dept.
16. Couture design sampling hall
17. Pret·a·porter design sampling hall
18. Designer studio
19. Hod office
20. Showroom
21. Computer lab
22. Office of tarun tahiuani
23. Secretary office
24. Executive offices
2S. Archives
26. Pantry
27. Washing room
28. Electrical room
29. DG set
30. Landscaped area
31. Storage
32. Terrace garden

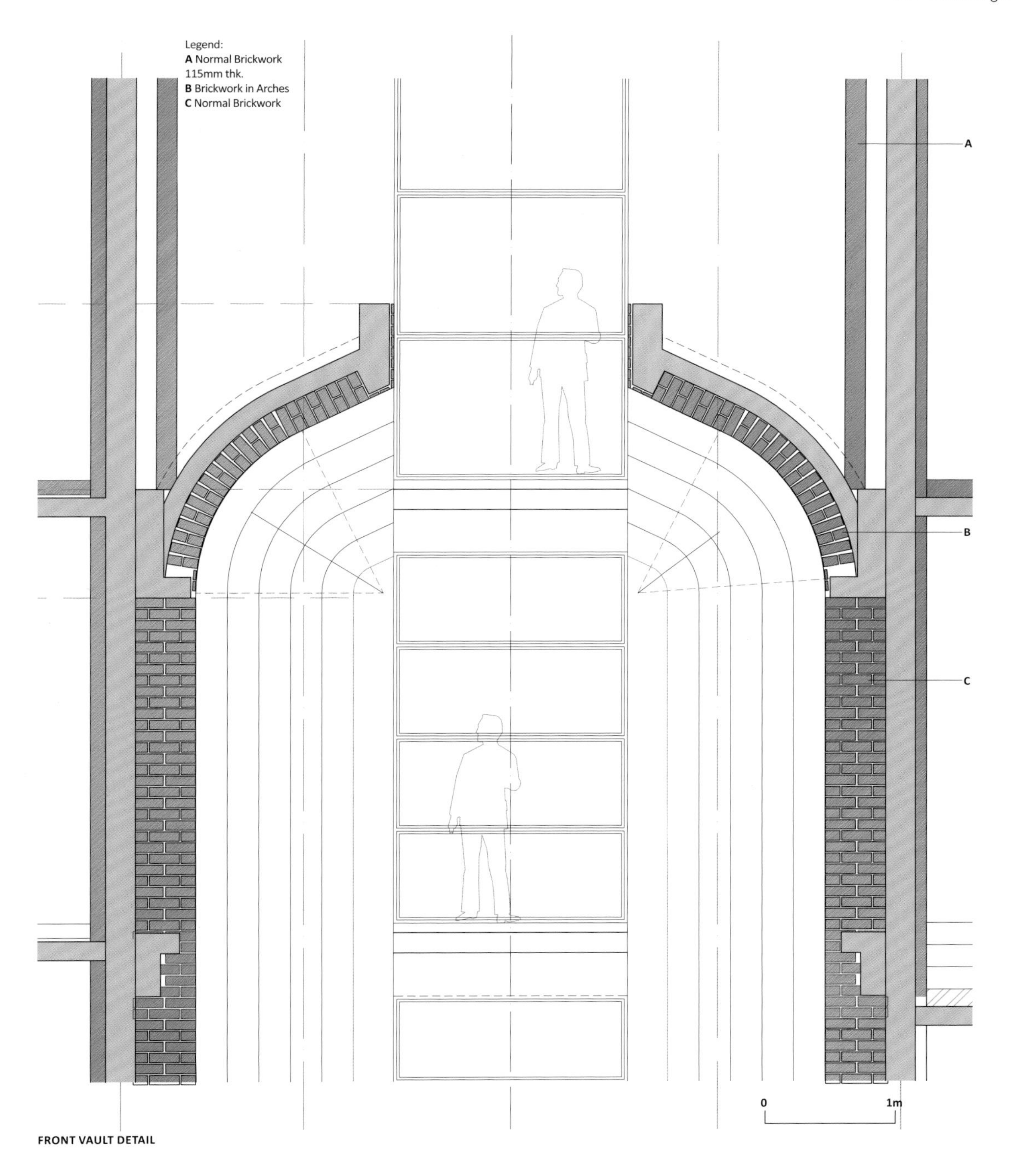

FRONT VAULT DETAIL

Building A – Sub-district 8

Location: La Spezia, Italy
Client: Subotto Scarl
Architects: MMAA (Studio Manfroni & Associati srl)
Gross Floor Area: 6,750 m²
Completion Date: 2012
Photographer: Roberto Buratta

The project involves the construction of a new tertiary pole through the urban reorganization of sub-district 8 in the former IP refinery in the north-east of La Spezia.

The main project target is to reuse the non-permeable area, preserving the green frame behind it: the new building is localized in the existing parking area.

The four-storey building is characterized by a double-facing and takes a soft and tapered shape emphasized by the evident bevel angles. It consists of a basement and ground floor where you can watch free pillars supporting the upper floors closed in a more compact volume.

On the ground floor there are areas with independent access, while the areas for tertiary use are on the upper floors.

Within the string courses, which correspond to the three upper floors of the building, the ratio between full and empty (cladding panels and windows) can change over time according to the needs of interior spaces, without affecting the general balance of the whole facade surface.

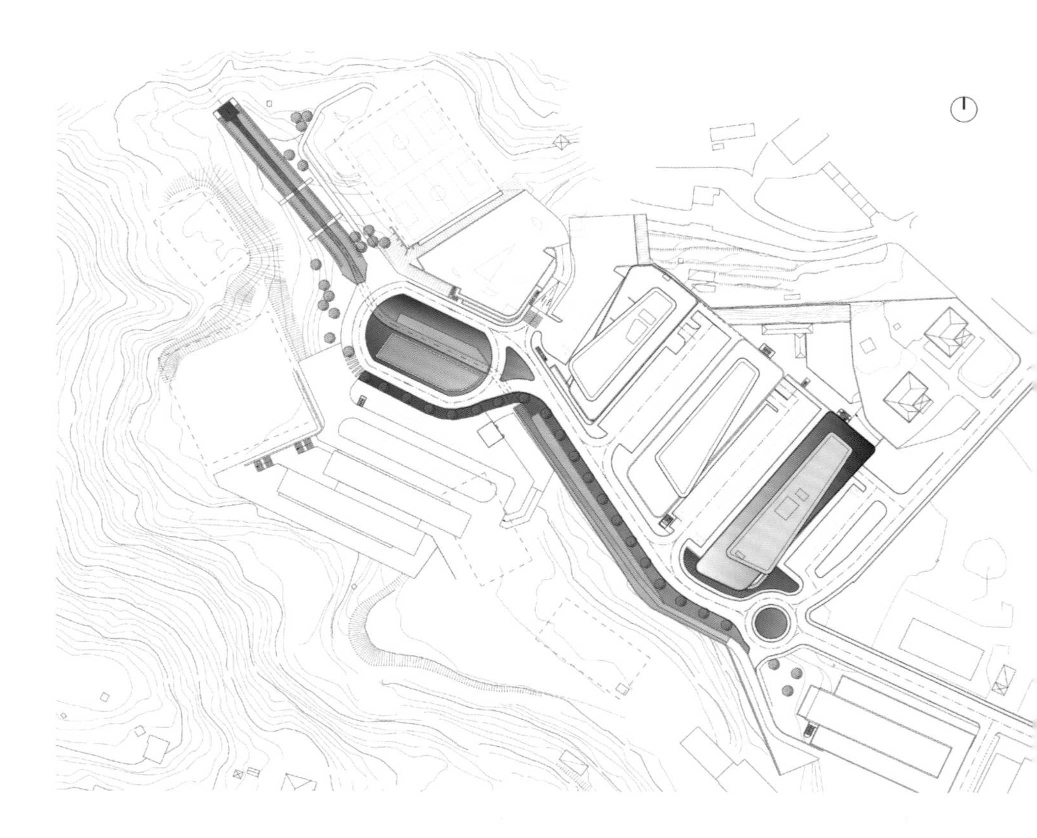

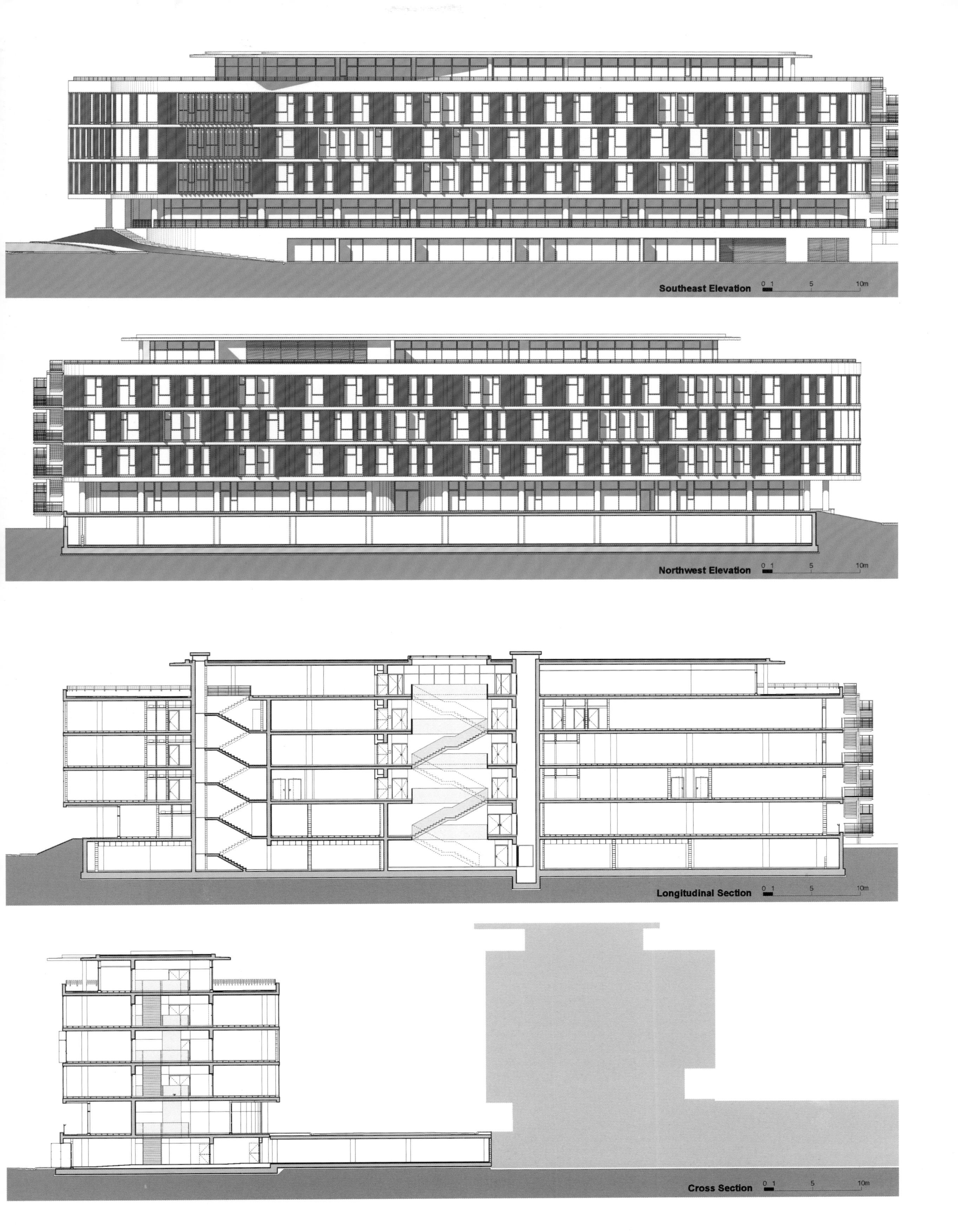
Southeast Elevation
0 1
5
10m
Northwest Elevation
0 1
5
10m
Longitudinal Section
0 1
5
10m
Cross Section
0 1
5
10m

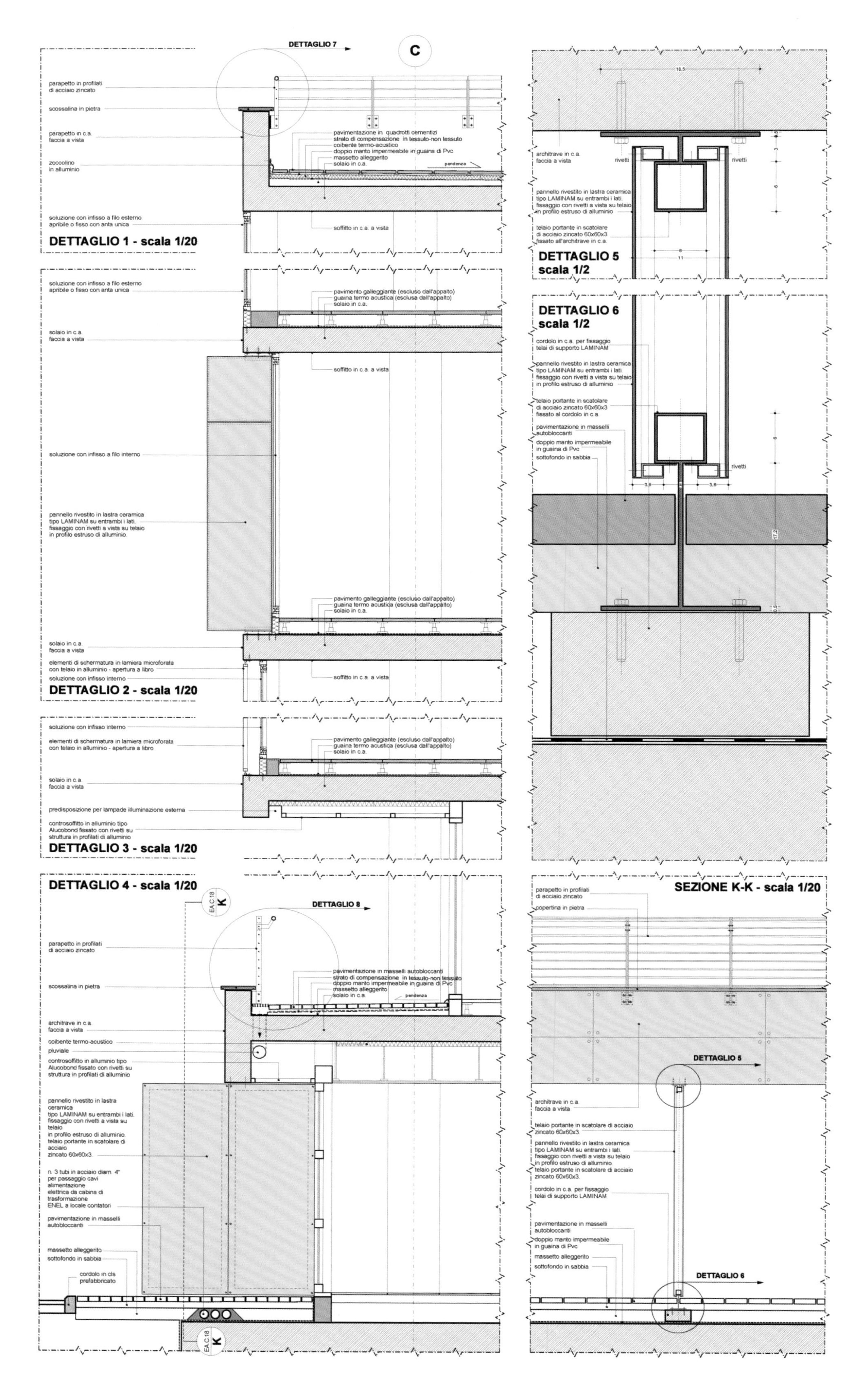
DETTAGLIO 7
C
parapetto in profilati di acciaio zincato
scossalina in pietra
parapetto in c.a. faccia a vista
zoccolino in alluminio
pavimentazione in quadrotti cementizi
strato di compensazione in tessuto-non tessuto
coibente termo-acustico
doppio manto impermeabile in guaina di Pvc
massetto alleggerito
solaio in c.a.
pendenza
soluzione con infisso a filo esterno apribile o fisso con anta unica
soffitto in c.a. a vista
DETTAGLIO 1 - scala 1/20
soluzione con infisso a filo esterno apribile o fisso con anta unica
pavimento galleggiante (escluso dall'appalto)
guaina termo acustica (esclusa dall'appalto)
solaio in c.a.
solaio in c.a. faccia a vista
soffitto in c.a. a vista
soluzione con infisso a filo interno
pannello rivestito in lastra ceramica tipo LAMINAM su entrambi i lati. fissaggio con rivetti a vista su telaio in profilo estruso di alluminio.
pavimento galleggiante (escluso dall'appalto)
guaina termo acustica (esclusa dall'appalto)
solaio in c.a.
solaio in c.a. faccia a vista
elementi di schermatura in lamiera microforata con telaio in alluminio - apertura a libro
soluzione con infisso interno
soffitto in c.a. a vista
DETTAGLIO 2 - scala 1/20
soluzione con infisso interno
elementi di schermatura in lamiera microforata con telaio in alluminio - apertura a libro
pavimento galleggiante (escluso dall'appalto)
guaina termo acustica (esclusa dall'appalto)
solaio in c.a.
solaio in c.a. faccia a vista
predisposizione per lampade illuminazione esterna
controsoffitto in alluminio tipo Alucobond fissato con rivetti su struttura in profilati di alluminio
DETTAGLIO 3 - scala 1/20
DETTAGLIO 4 - scala 1/20
EA.C.18
K
DETTAGLIO 8
parapetto in profilati di acciaio zincato
scossalina in pietra
pavimentazione in masselli autobloccanti
strato di compensazione in tessuto-non tessuto
doppio manto impermeabile in guaina di Pvc
massetto alleggerito
solaio in c.a.
pendenza
architrave in c.a. faccia a vista
coibente termo-acustico
pluviale
controsoffitto in alluminio tipo Alucobond fissato con rivetti su struttura in profilati di alluminio
pannello rivestito in lastra ceramica tipo LAMINAM su entrambi i lati. fissaggio con rivetti a vista su telaio in profilo estruso di alluminio. telaio portante in scatolare di acciaio zincato 60x60x3.
n. 3 tubi in acciaio diam. 4" per passaggio cavi alimentazione elettrica da cabina di trasformazione ENEL a locale contatori
pavimentazione in masselli autobloccanti
massetto alleggerito
sottofondo in sabbia
cordolo in cls prefabbricato
EA.C.18
K
18,5
architrave in c.a. faccia a vista
rivetti
rivetti
pannello rivestito in lastra ceramica tipo LAMINAM su entrambi i lati. fissaggio con rivetti a vista su telaio in profilo estruso di alluminio
telaio portante in scatolare di acciaio zincato 60x60x3 fissato all'architrave in c.a.
6
11
DETTAGLIO 5 scala 1/2
DETTAGLIO 6 scala 1/2
cordolo in c.a. per fissaggio telai di supporto LAMINAM
pannello rivestito in lastra ceramica tipo LAMINAM su entrambi i lati. fissaggio con rivetti a vista su telaio in profilo estruso di alluminio
telaio portante in scatolare 60x60x3 di acciaio zincato 60x60x3 fissato al cordolo in c.a.
pavimentazione in masselli autobloccanti
doppio manto impermeabile in guaina di Pvc
sottofondo in sabbia
rivetti
3,6
3,6
17,2
SEZIONE K-K - scala 1/20
parapetto in profilati di acciaio zincato
copertina in pietra
DETTAGLIO 5
architrave in c.a. faccia a vista
telaio portante in scatolare di acciaio zincato 60x60x3.
pannello rivestito in lastra ceramica tipo LAMINAM su entrambi i lati. fissaggio con rivetti a vista su telaio in profilo estruso di alluminio. telaio portante in scatolare di acciaio zincato 60x60x3.
cordolo in c.a. per fissaggio telai di supporto LAMINAM
pavimentazione in masselli autobloccanti
doppio manto impermeabile in guaina di Pvc
massetto alleggerito
sottofondo in sabbia
DETTAGLIO 6

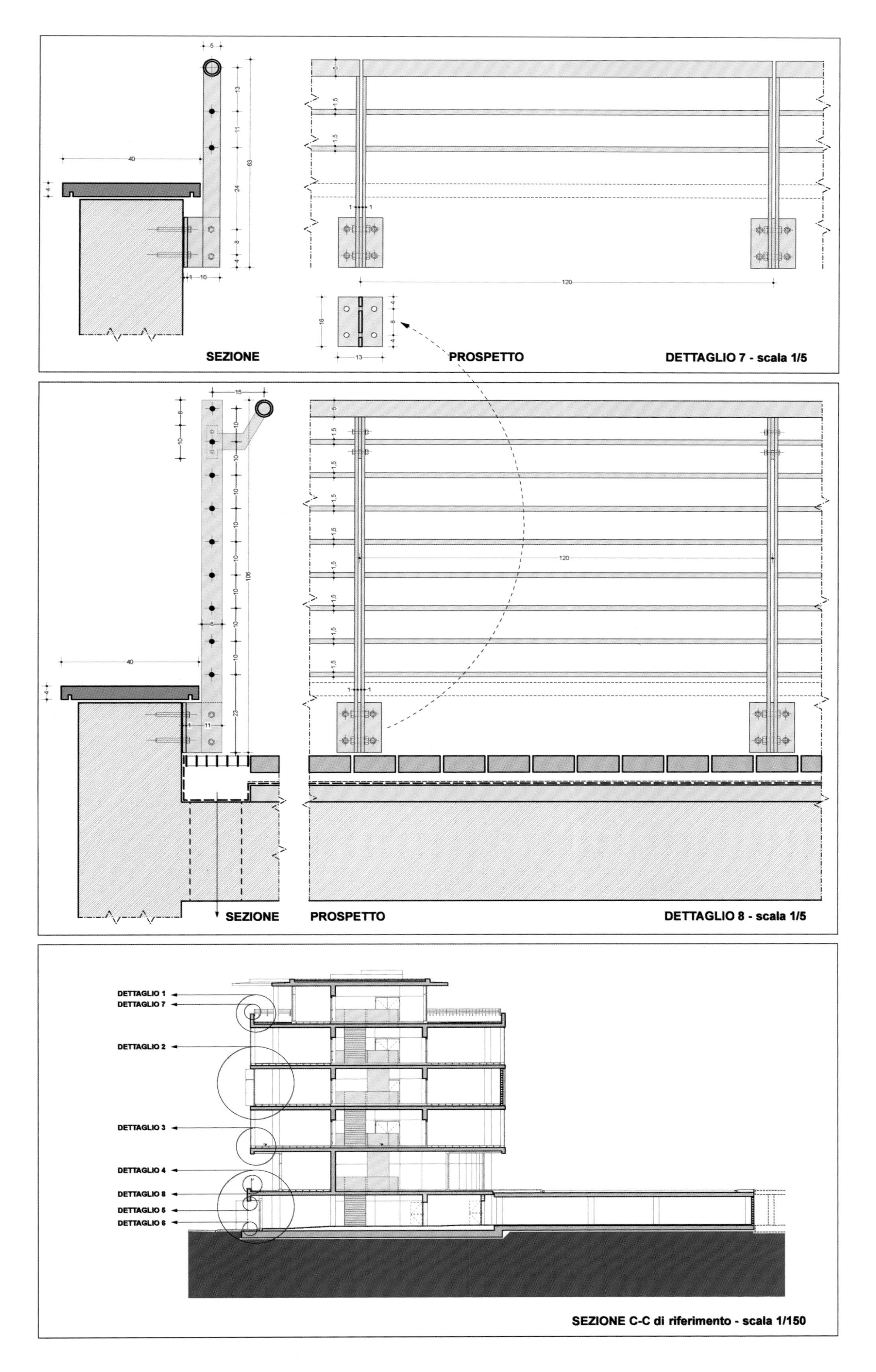
SEZIONE
PROSPETTO
DETTAGLIO 7 - scala 1/5
SEZIONE
PROSPETTO
DETTAGLIO 8 - scala 1/5
DETTAGLIO 1
DETTAGLIO 7
DETTAGLIO 2
DETTAGLIO 3
DETTAGLIO 4
DETTAGLIO 8
DETTAGLIO 5
DETTAGLIO 6
SEZIONE C-C di riferimento - scala 1/150

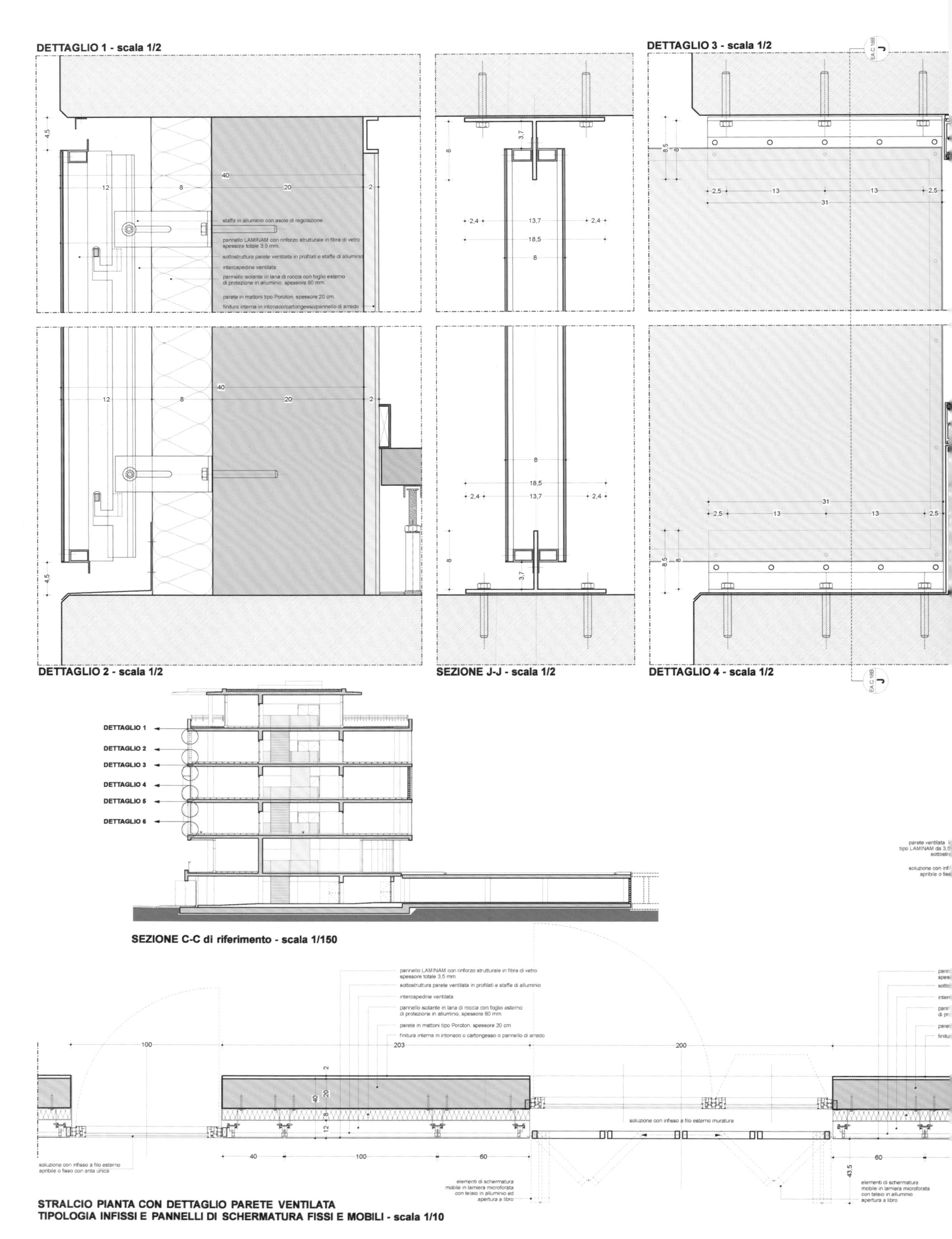

STRALCIO PIANTA CON DETTAGLIO PARETE VENTILATA
TIPOLOGIA INFISSI E PANNELLI DI SCHERMATURA FISSI E MOBILI - scala 1/10

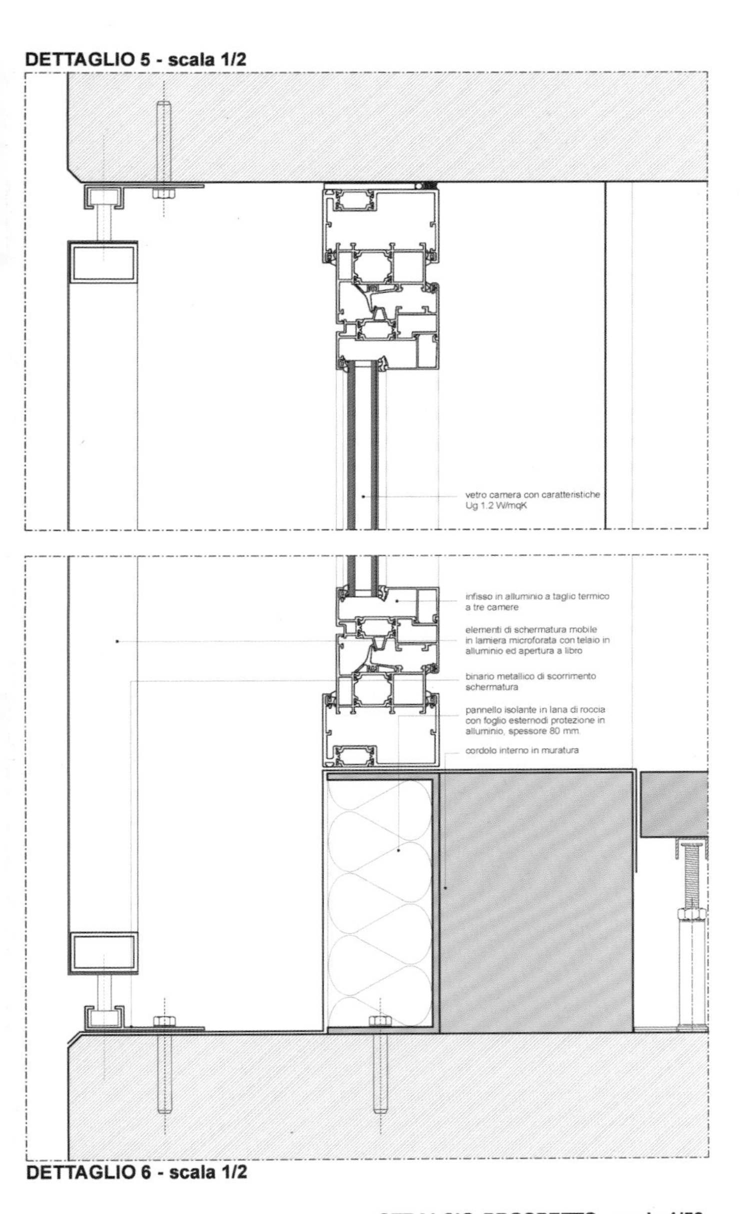

DETTAGLIO 5 - scala 1/2
vetro camera con caratteristiche Ug 1.2 W/mqK
infisso in alluminio a taglio termico a tre camere
elementi di schermatura mobile in lamiera microforata con telaio in alluminio ed apertura a libro
binario metallico di scorrimento schermatura
pannello isolante in lana di roccia con foglio esternodi protezione in alluminio, spessore 80 mm.
cordolo interno in muratura
DETTAGLIO 6 - scala 1/2

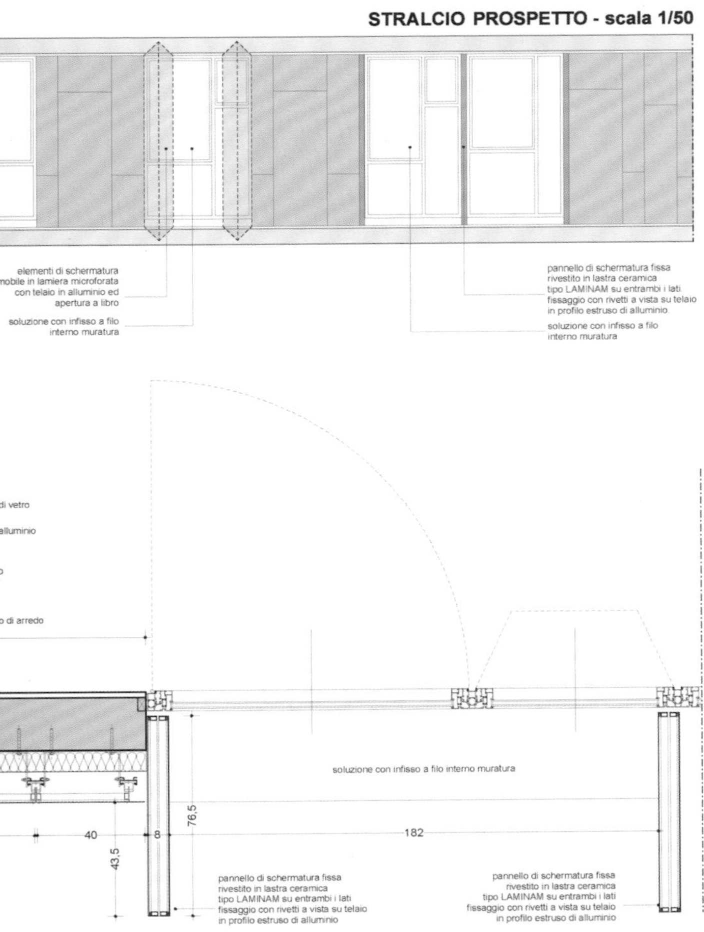

STRALCIO PROSPETTO - scala 1/50
elementi di schermatura mobile in lamiera microforata con telaio in alluminio ed apertura a libro
soluzione con infisso a filo interno muratura
pannello di schermatura fissa rivestito in lastra ceramica tipo LAMINAM su entrambi i lati fissaggio con rivetti a vista su telaio in profilo estruso di alluminio
soluzione con infisso a filo interno muratura
di vetro
alluminio
lo di arredo
soluzione con infisso a filo interno muratura
76,5
40
8
182
43,5
pannello di schermatura fissa rivestito in lastra ceramica tipo LAMINAM su entrambi i lati fissaggio con rivetti a vista su telaio in profilo estruso di alluminio
pannello di schermatura fissa rivestito in lastra ceramica tipo LAMINAM su entrambi i lati fissaggio con rivetti a vista su telaio in profilo estruso di alluminio

Nestlé Headquarters

Location: Vevey, Switzerland
Architect: Richter Dahl Rocha & Associés architectes
Client: Nestec SA
Floor Area: 60,000 m²
Completion Date: 2000
Photography: Mario Carrieri, Yves Andre, Nestec SA

The Nestlé Headquarters in Vevey is one of the most remarkable administrative buildings in the French-speaking part of Switzerland. Completed in 1960, this major work of the architect Jean Tschumi was extended in 1975 and has been the object of diverse transformations or additions. The obsolescence of the technical installations and the evolution of functional demands clearly mandated an integral renovation of Tschumi's building. The magnitude of this operation allowed for a wider revision of the project, which in turn resulted in the proposal for a new element to link the two buildings and thus enhance the unity and coherence of the whole. This intervention on a modern building (one listed in the Historical Monument inventory) poses new questions, particularly about the strategies of approach and the choice of means for intervention, in a complex situation combining different degrees of action such as the restoration, the transformation, and the creation of new elements.

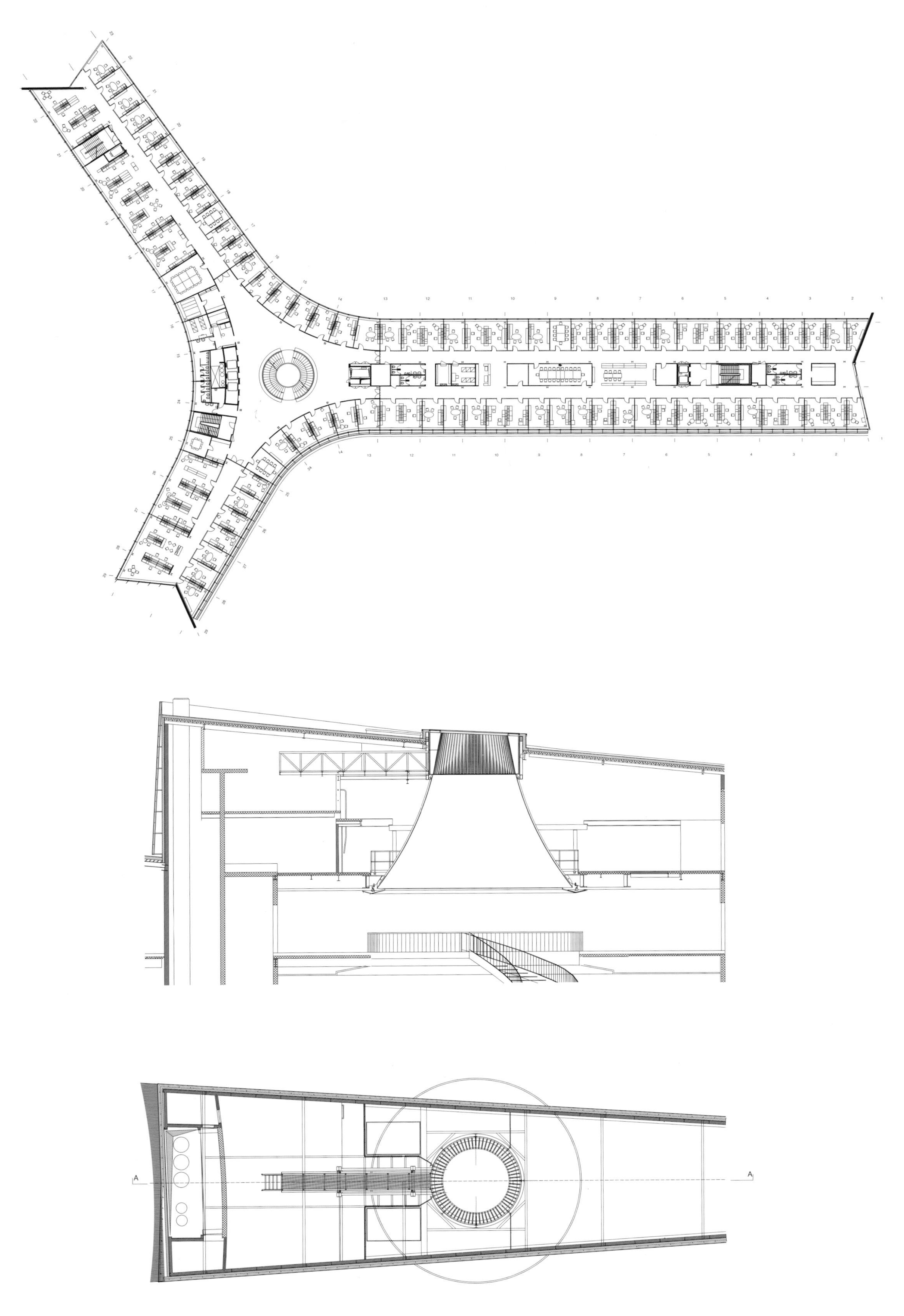

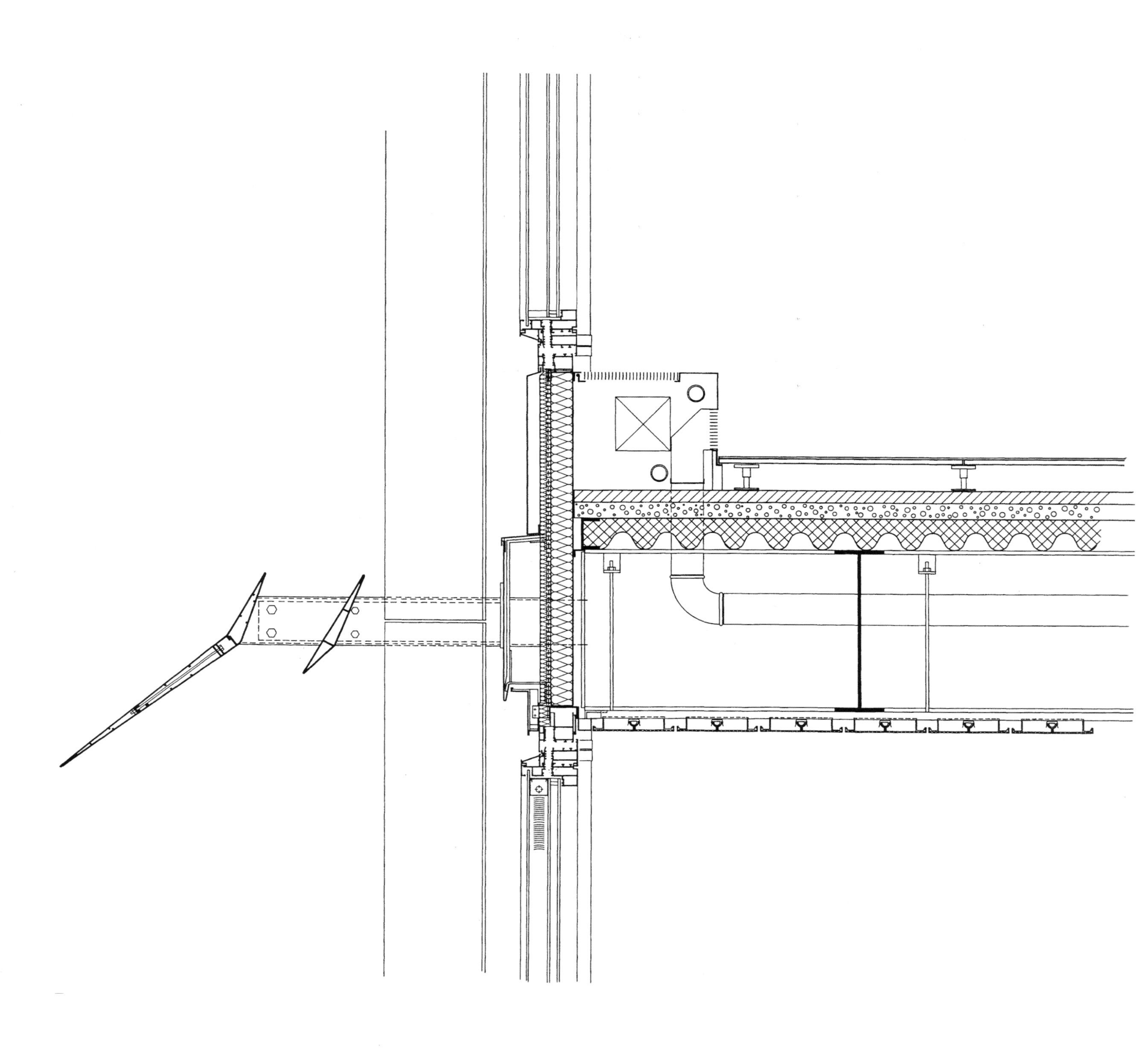

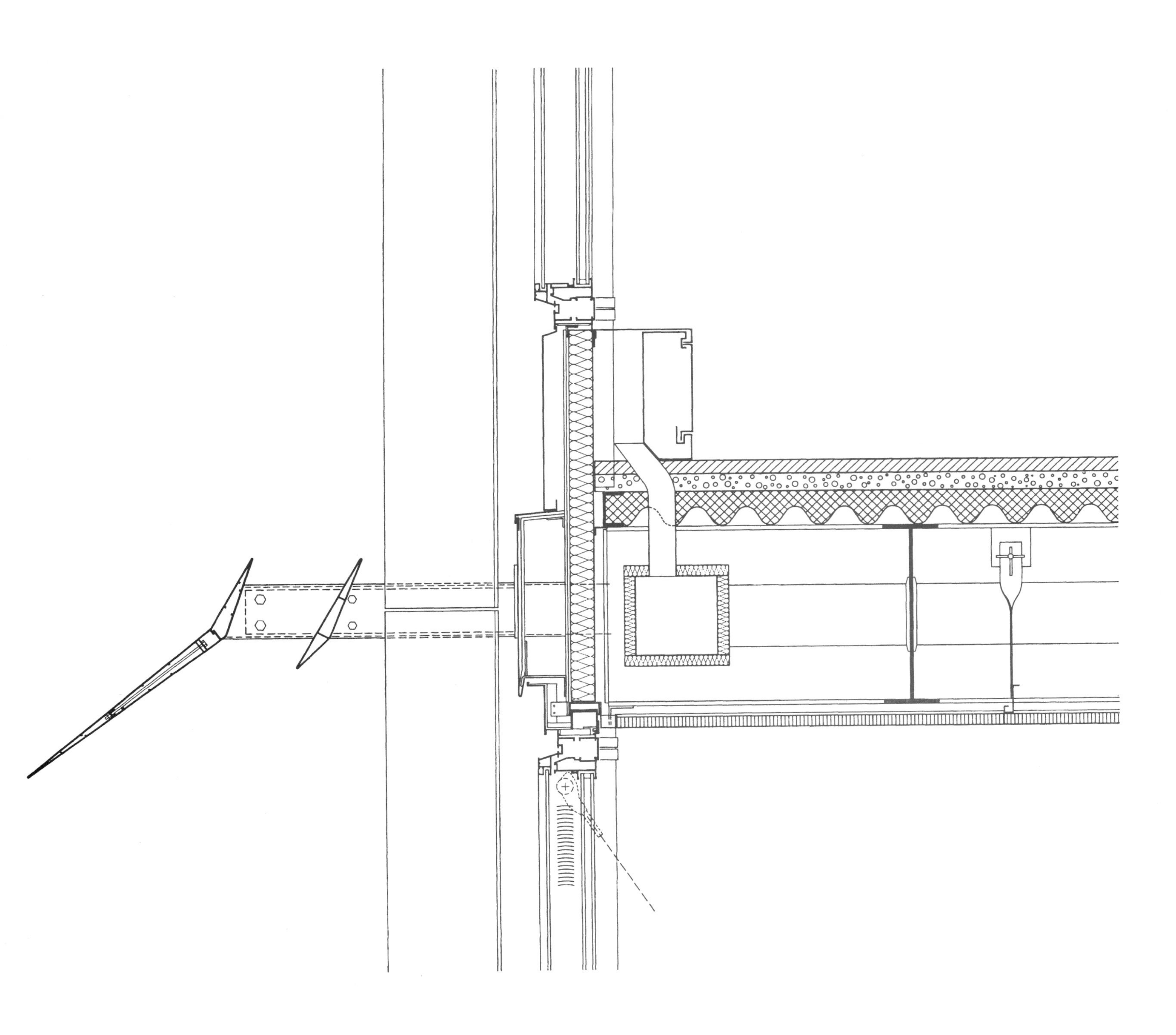

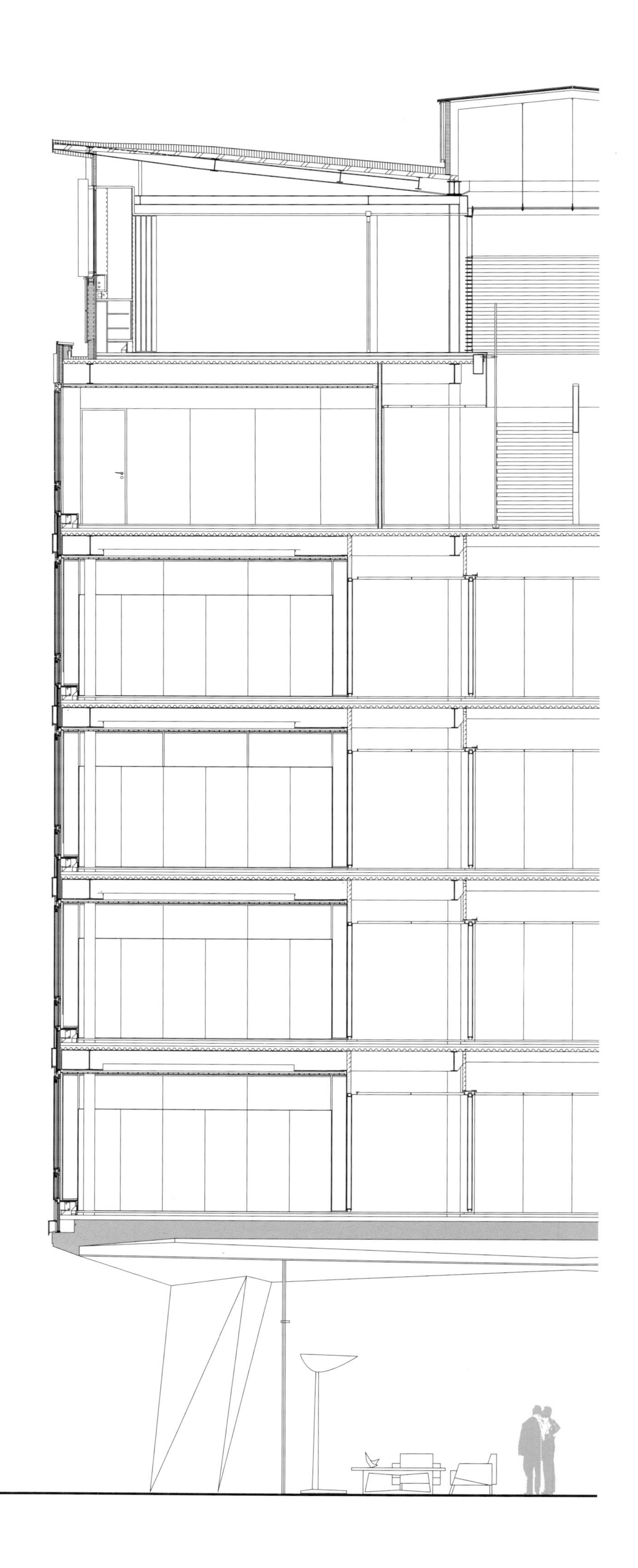

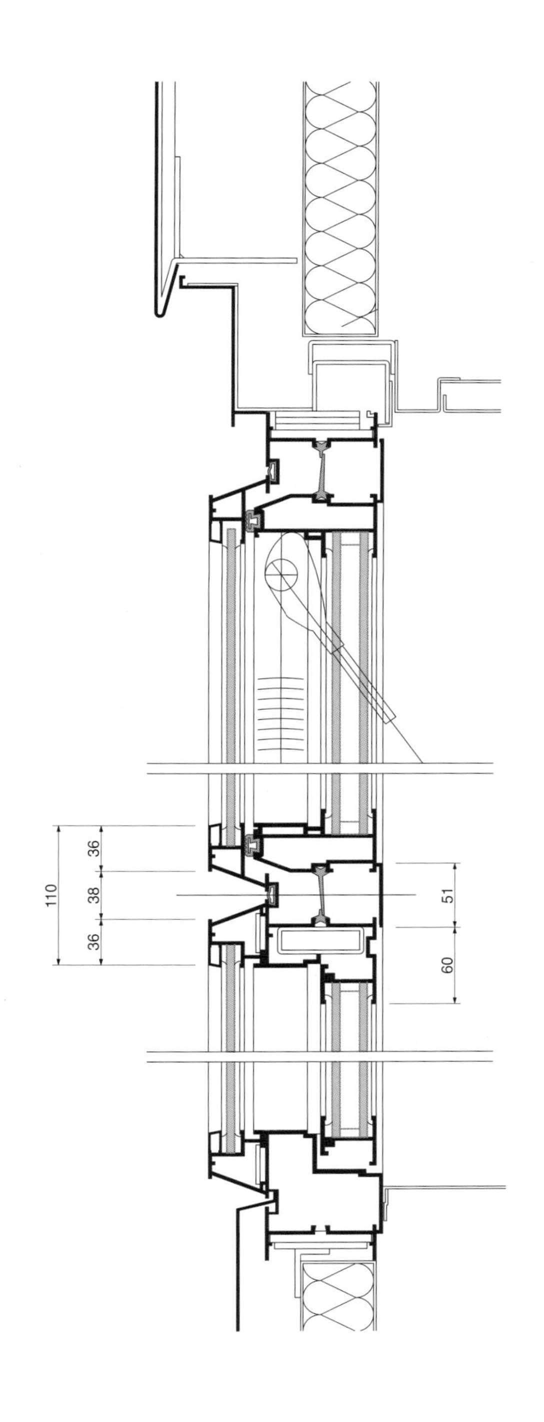
110
36
38
36
51
60

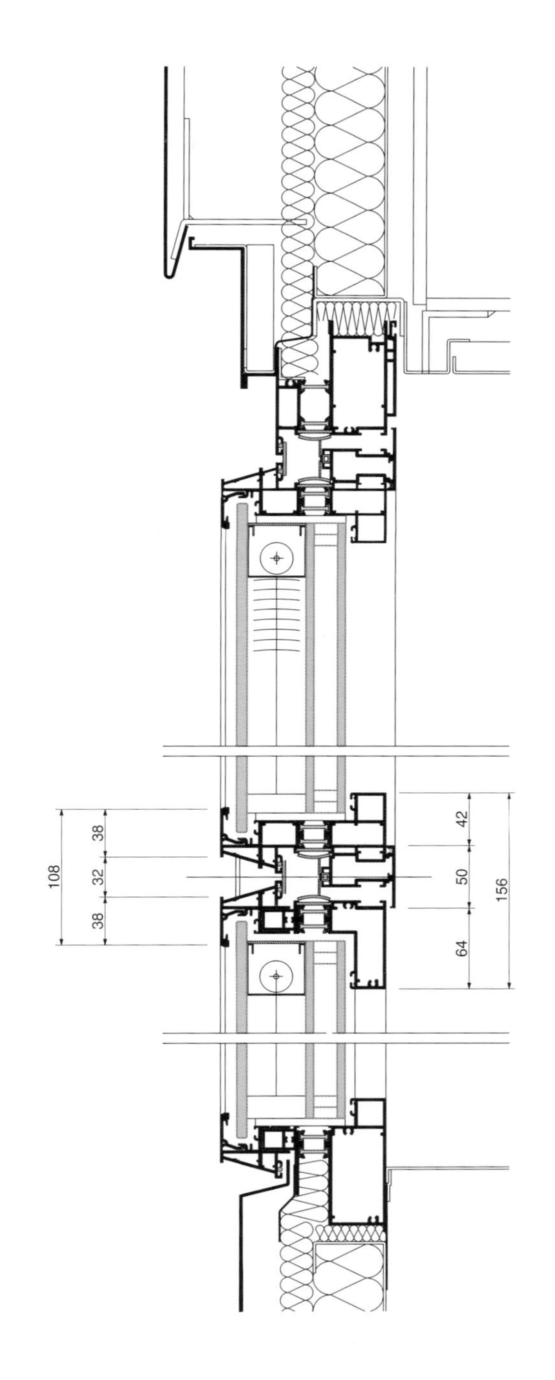
108
38
32
38
42
50
64
156

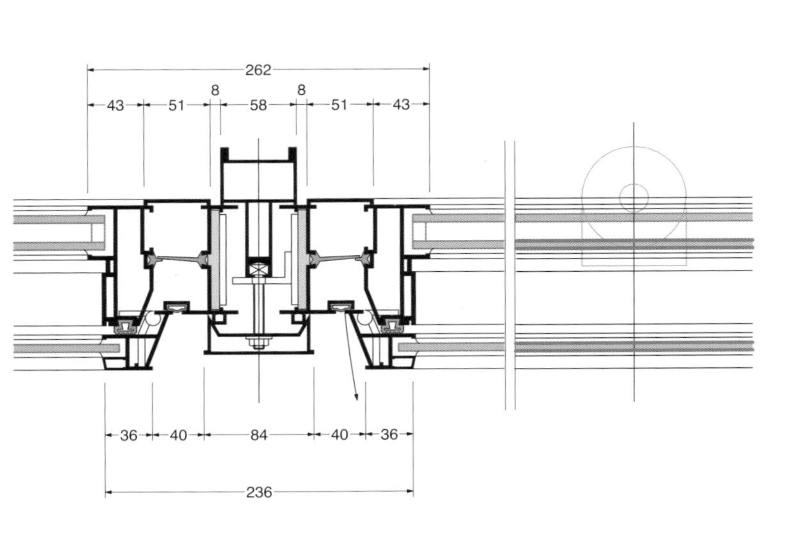
262
43
51
8
58
8
51
43
36
40
84
40
36
236

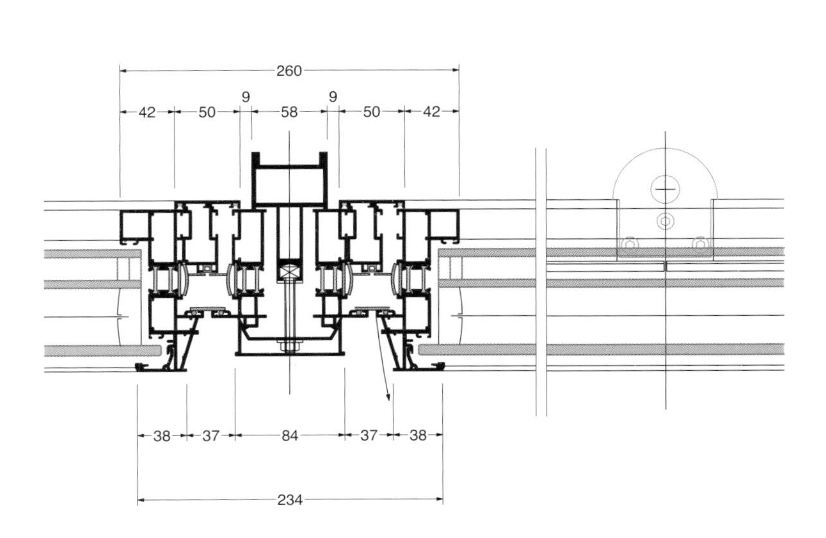
260
42
50
9
58
9
50
42
38
37
84
37
38
234

Porta Nuova

Location: Milan, Italy
Architect: Studio Piuarch
Client: Hines Italia SGR S.P.A.
Floor Area: 22,500 m^2
Completion Date: 2013
Photographer: Andrea Martiradonna

This building for showrooms and offices is part of the Porta Nuova urban renewal project, which comprises the Garibaldi, Varesine and Isola areas in the heart of Milan. The area occupies approximately 300,000 m^2 and functions as a juncture between the three different urban contexts that surround it. The goal of its requalification is to give this vital area back to the city and make it an integral part of the future urban dynamic. Within this multifaceted and complex plan, the building seeks to establish a dialogue with the public part of the area as a whole, configuring itself as an access point to the park, which is situated at a higher elevation with respect to street level—thus the decision to position the building on a sort of podium that connects the two spaces. Bounded on one side by the central square, adjacent to Cesar Pelli's three towers, the building has an area of about 22,500 m^2 distributed over five stories and a ground floor for an overall height of 30 m. This choice was conditioned by the need to respect the heights of the existing residential buildings as well as by the desire to make the new structure a strong and recognizable landmark within the requalified area, in contrast with the general plan which calls for much taller buildings.

The simple and sinuous form of the building articulates and integrates the two volumes in a single element distinguished by a deep central fissure. A projecting roof on the southern side runs the entire perimeter of the building, resolving at the base of the first floor to form a sort of container open on the long sides. The two façades are treated differently: the one facing north and overlooking the central square is characterized by large windows that act as a backdrop for the pedestrian area and the Porta Nuova park, while the curved southern façade is protected by a system of sun blinds that mark out its rhythm. The building, which is about 140 m long, incorporates a system of internal courtyards with coloured walls. The ground floor is wrapped by an ample portico that opens directly onto the main square and is covered by the projecting volume above.

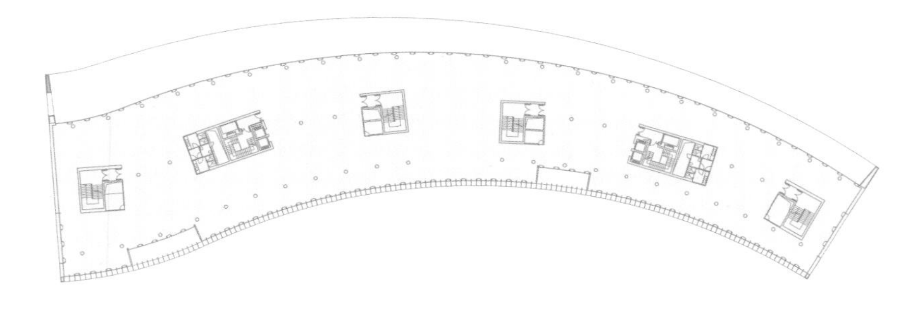

PIANTA PIANO TERZO
THIRD FLOOR PLAN

0 10M

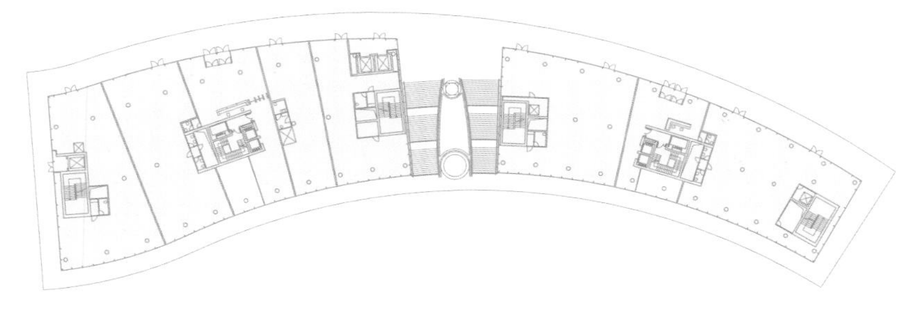

PIANTA PIANO TERRA
GROUND FLOOR PLAN

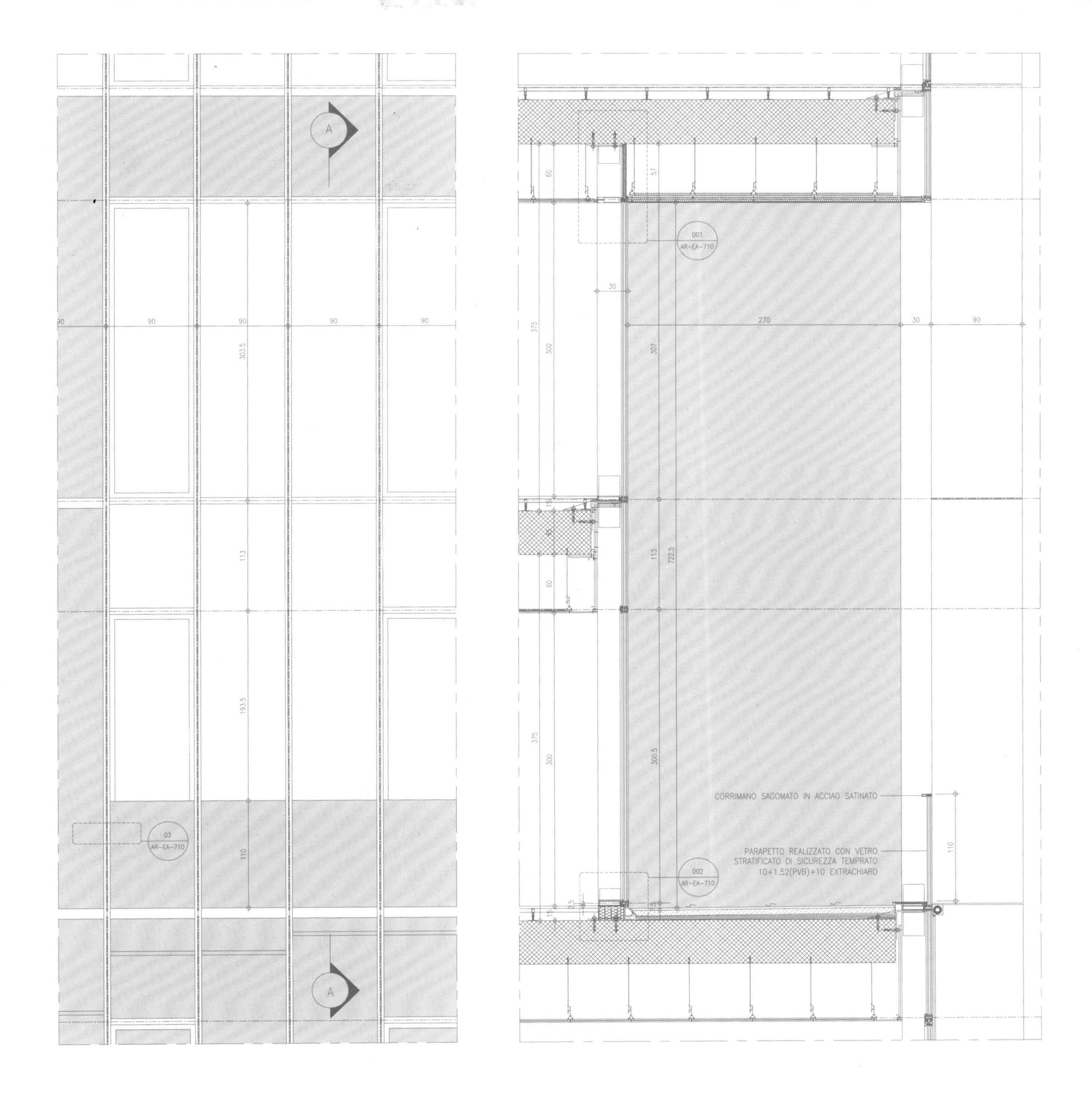
A
001
AR-EA-710
CORRIMANO SAGOMATO IN ACCIAO SATINATO
PARAPETTO REALIZZATO CON VETRO
STRATIFICATO DI SICUREZZA TEMPRATO
10+1.52(PVB)+10 EXTRACHIARO
002
AR-EA-710
03
AR-EA-710
A

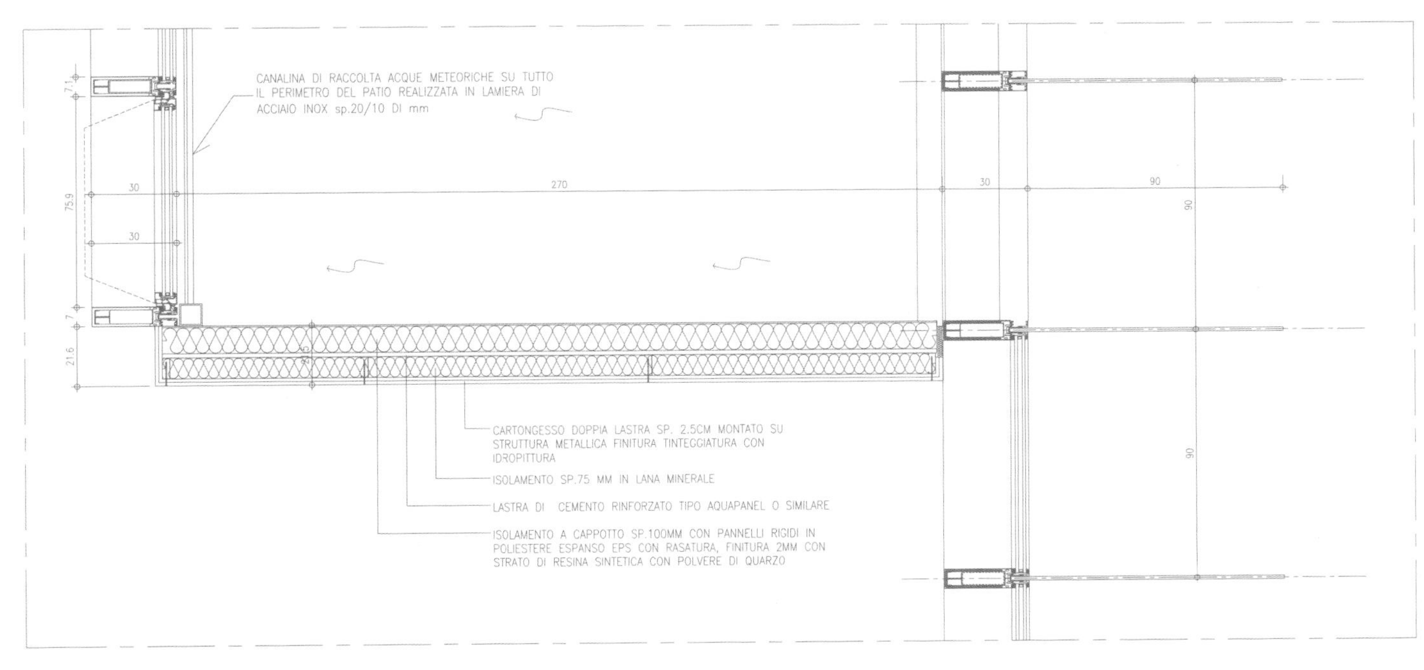
CANALINA DI RACCOLTA ACQUE METEORICHE SU TUTTO
IL PERIMETRO DEL PATIO REALIZZATA IN LAMIERA DI
ACCIAIO INOX sp.20/10 DI mm
CARTONGESSO DOPPIA LASTRA SP. 2.5CM MONTATO SU
STRUTTURA METALLICA FINITURA TINTEGGIATURA CON
IDROPITTURA
ISOLAMENTO SP.75 MM IN LANA MINERALE
LASTRA DI CEMENTO RINFORZATO TIPO AQUAPANEL O SIMILARE
ISOLAMENTO A CAPPOTTO SP.100MM CON PANNELLI RIGIDI IN
POLIESTERE ESPANSO EPS CON RASATURA, FINITURA 2MM CON
STRATO DI RESINA SINTETICA CON POLVERE DI QUARZO

Transportation

Novoperedelkino Metro Station

Location: Moscow, Russia
Architect: U-R-A | United Riga Architects
Client: MosInzhProekt
Size: 7,800 m²

The main idea of the concept was to combine the archetypal motifs of the Moscow architecture with modern ways of interior decorations and, moreover, make the station a truly unique and dynamic place.

Metal panels, perforated in the Russian-style pattern, comprise light boxes and disperse light. The patterns create a lasting and expressive experience.

The prime feature of the station is a variety of experiences created. Being equipped with RGB LED emitters, light boxes – either flat or domed – may change their colour scheme. During city holidays and other events, such patterns may become blue, red, or multi-colour or even broadcast videos.

The Novoperedelkino district is renowned for its unique forests and such identity was expressed in the pavilion decorations – laser engravings in the decorative glasses are reminiscent of forests made-up of pines and birches. LED backlighting makes the engravings fluoresce, creating an almost 3D image of a forest around.

Thus, the architects aimed at supporting time-proven trends of artistic architecture at the Moscow subway, through creating innovative ideas rather than simple copying of old concepts.

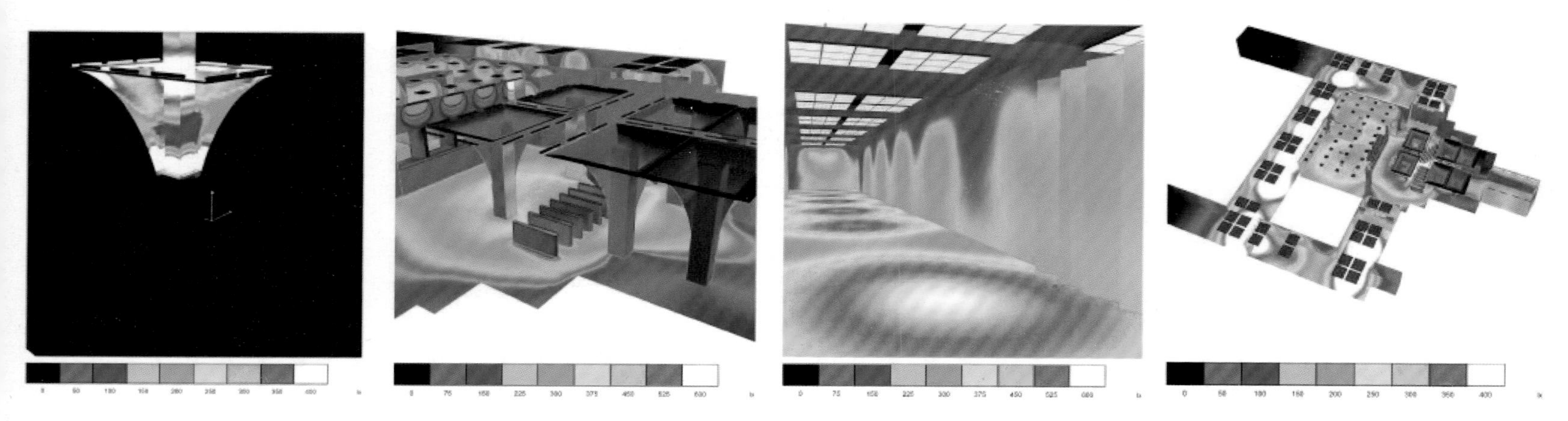

Detali

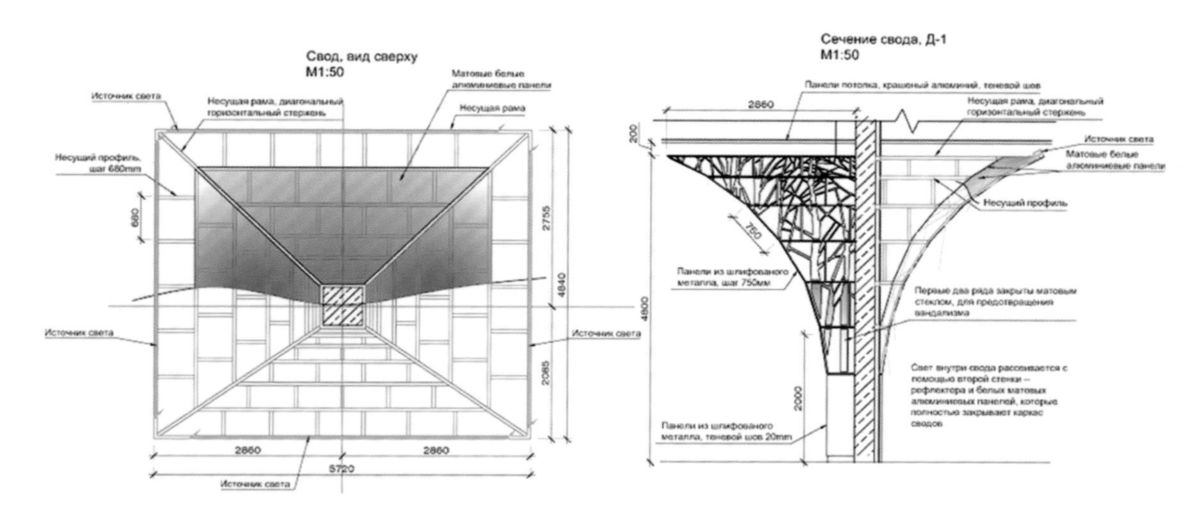

Detail

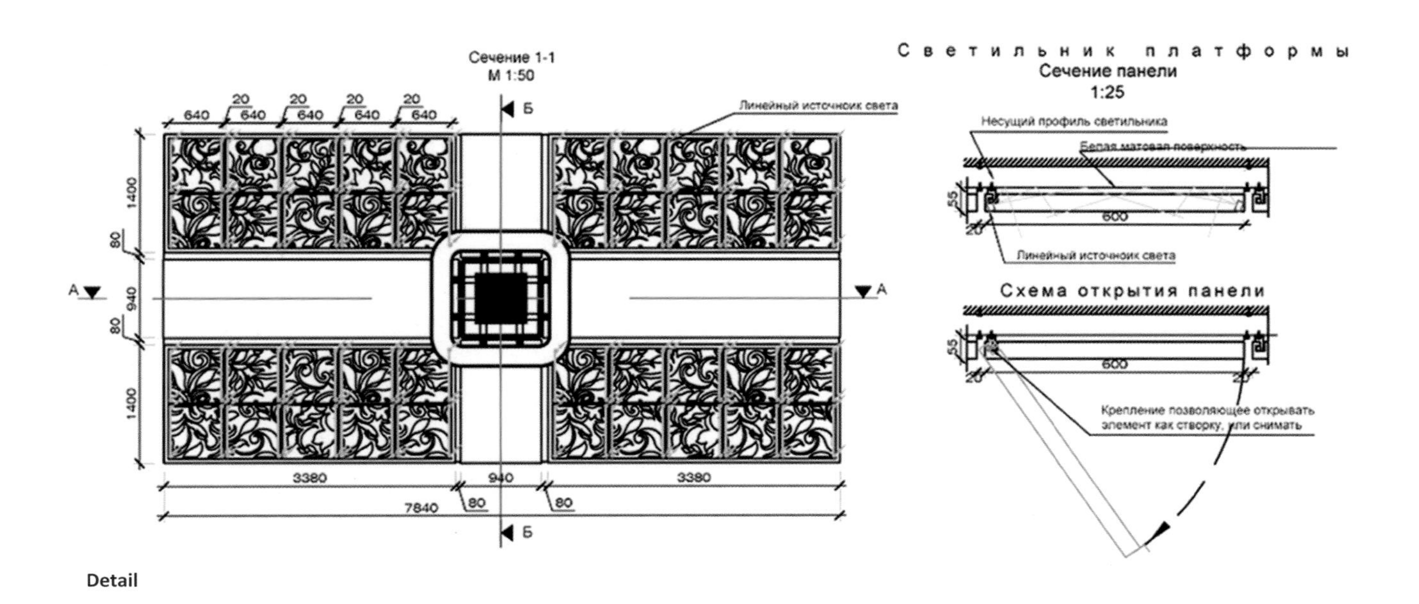

Detail

СВЕТОВЫЕ КОРОБА В ВИДЕ СВОДОВ
ПРЕДЛАГАЕМОЕ РЕШЕНИЕ: ОТРАЖЕННЫЙ СВЕТ - 3D ЭФФЕКТ

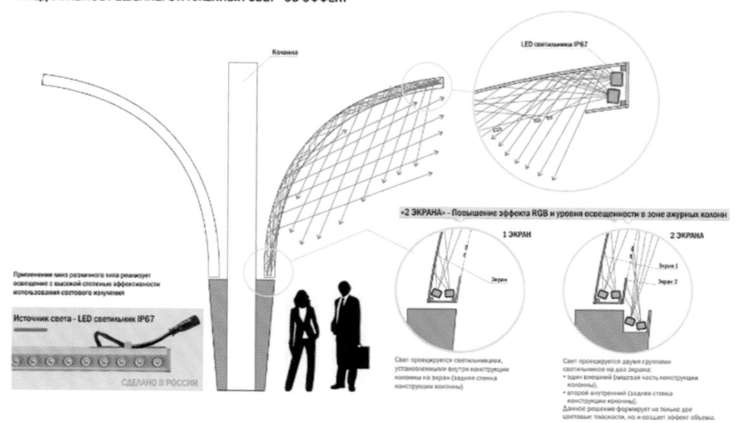

СВЕТОВЫЕ КОРОБА В ВИДЕ СВОДОВ
ПРЕДЛАГАЕМОЕ РЕШЕНИЕ: ОТРАЖЕННЫЙ СВЕТ

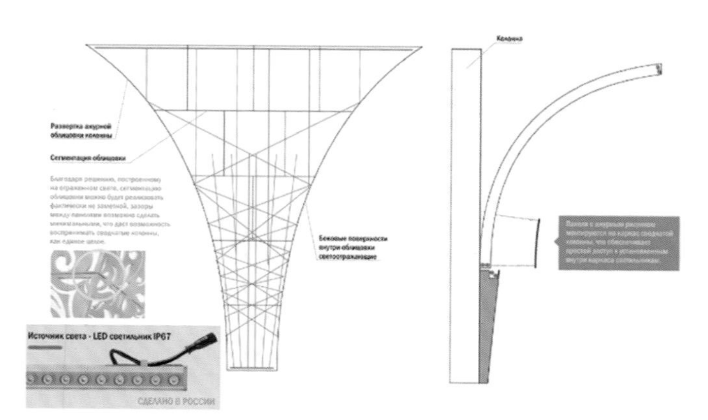

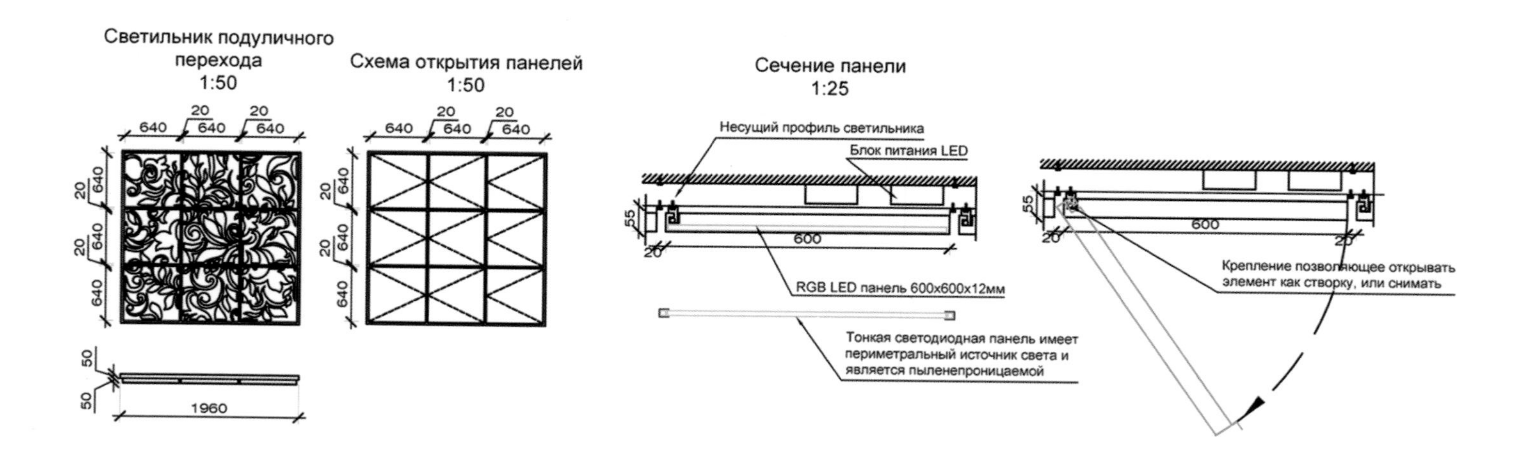

Детали платформенных светильников

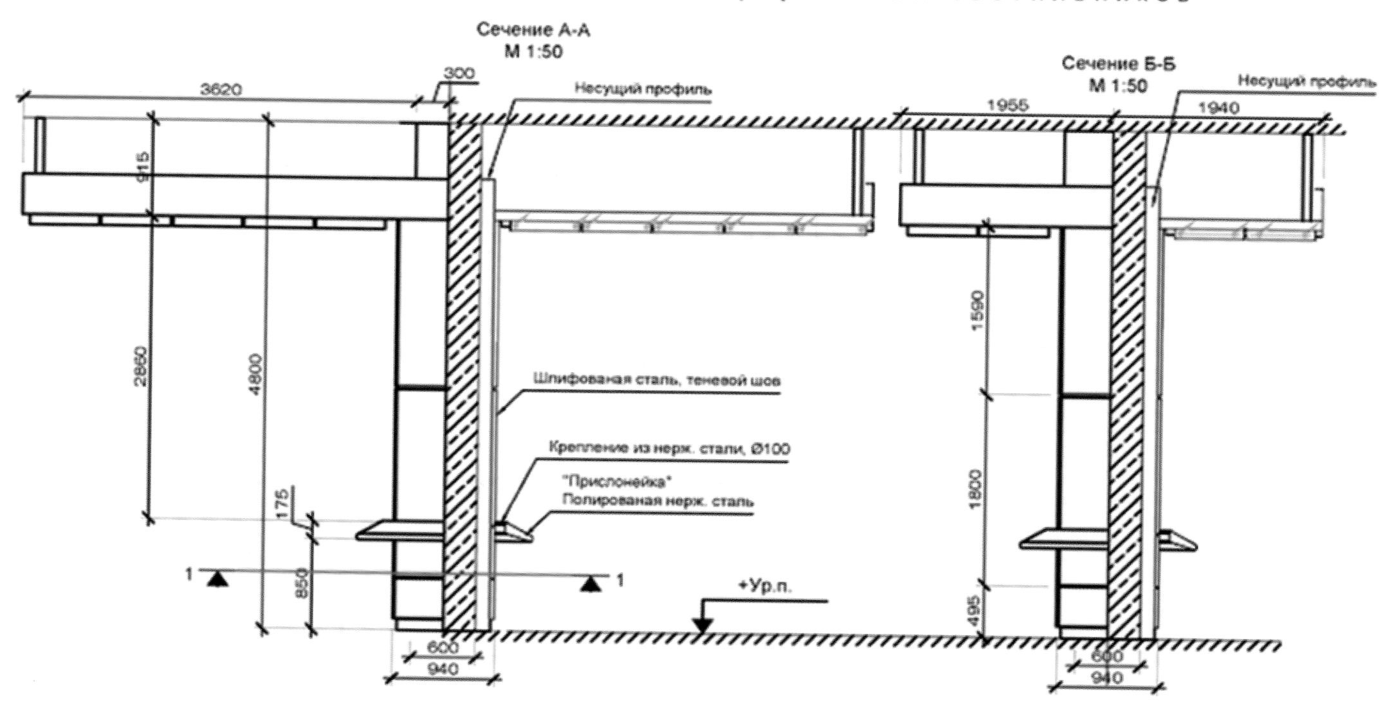

ПЛАН ПОЛА ПЛАТФОРМЕННОЙ ЧАСТИ
М 1:200

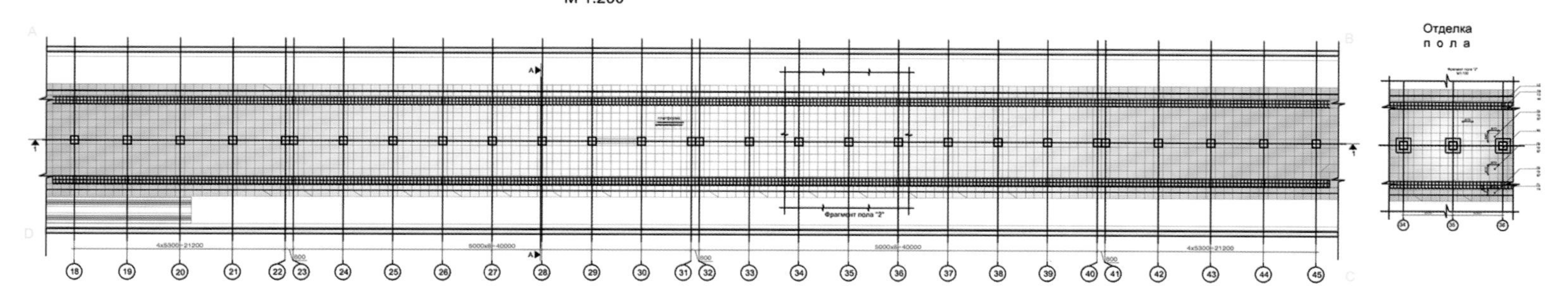

Генеральный план

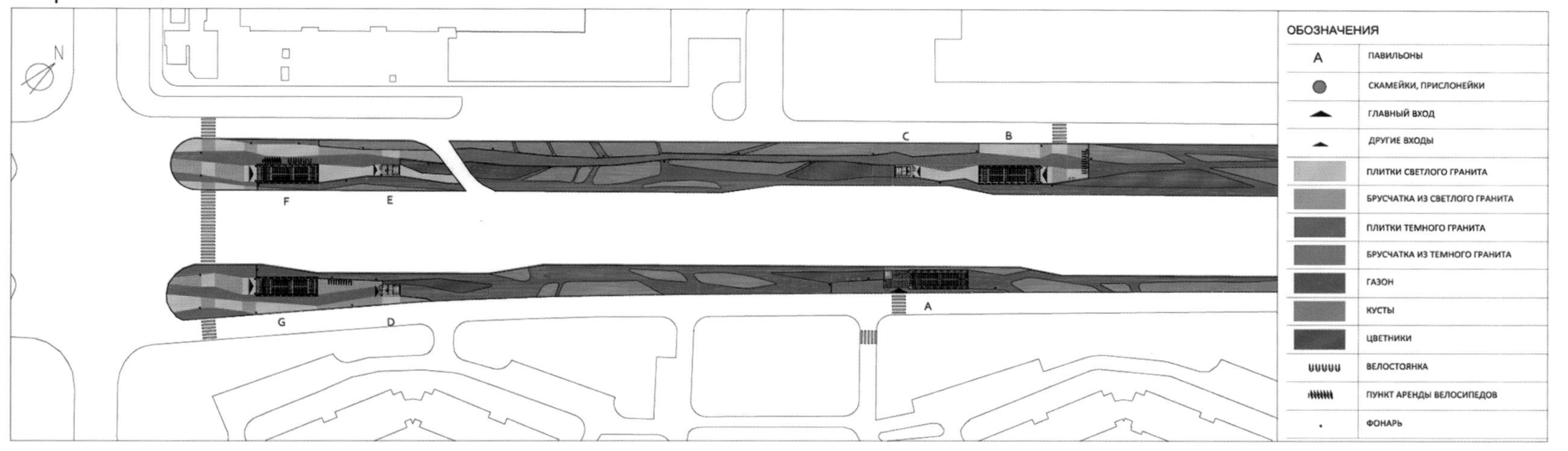

КАССОВЫЙ ЗАЛ N1,N2 .РАЗРЕЗ А-А,В-В. М 1:100

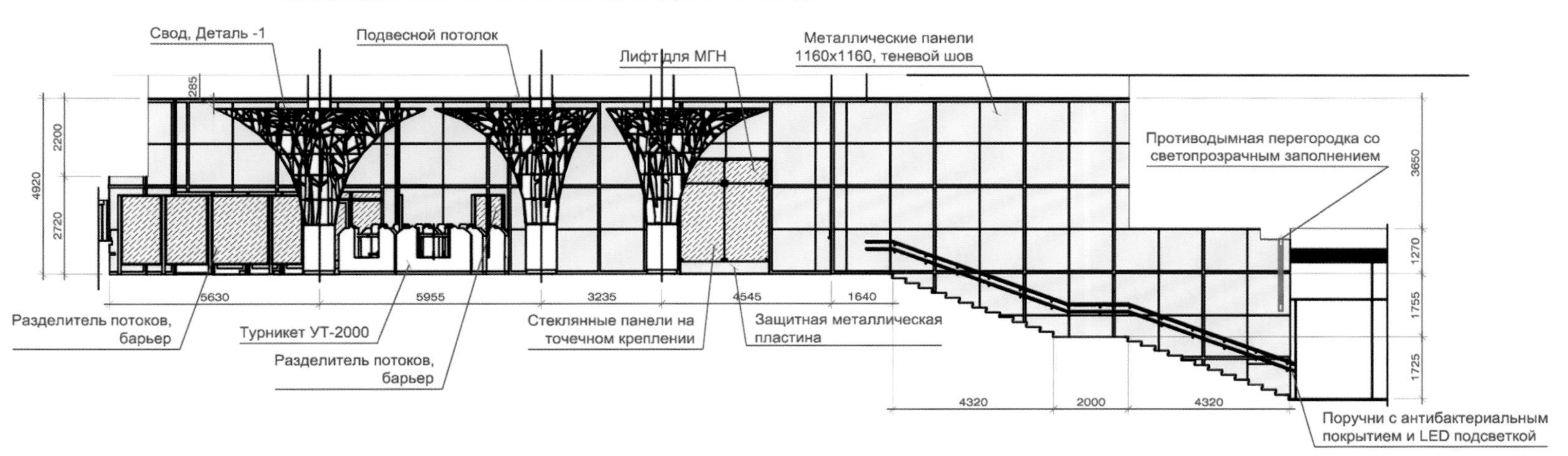

КАССОВЫЙ ЗАЛ N1,N2 .РАЗРЕЗ 1-1,2-2. М 1:100

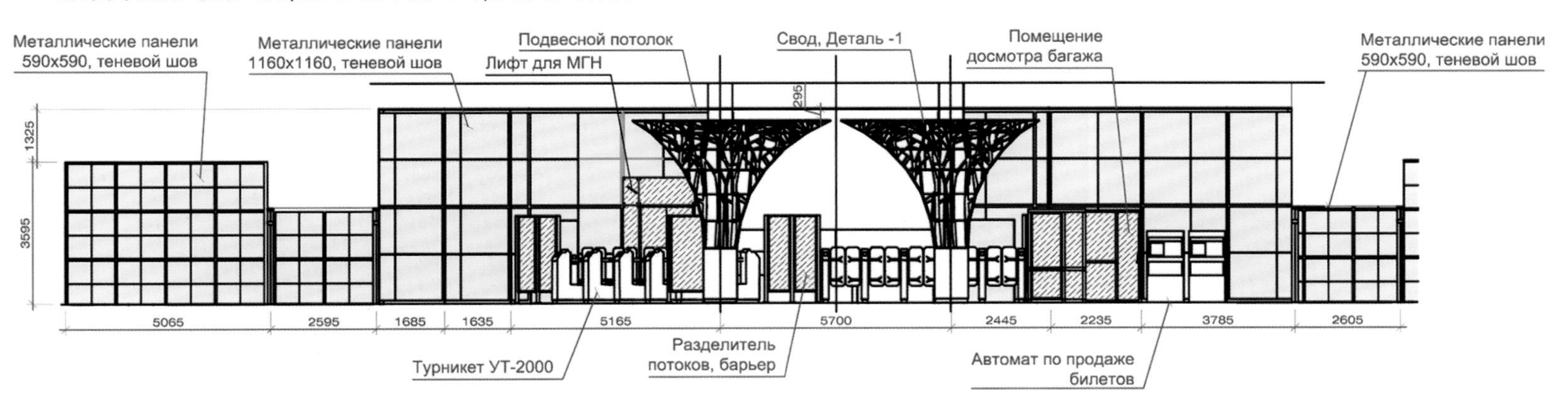

Павильоны F-E,G-D,C-B
продольный разрез
М1:200

Профнастил, скат 10%
Облицовочные металлические панели
+6.860
+5.860
Логотип Московскоо метро
Несущая стойка, шаг 2310mm
Стекло с декоративной обработкой
Бетонный цоколь
+0.000(+13.057)
Панели отделки потолка, шлифованный металл, теневой шов
Конструктивная связь, металлический трос
Стекло крепится на точечные крепления
Цветовая и тактильная маркировка первых и последних ступеней марша
Встроенные в стены светильники располагаются над площадками между маршами
Поручень с антибактериальным покрытием и LED подсветкой
Стекло с декоративной обработкой
Профнастил, скат 10%
Облицовочные металлические панели
+3.900
Входной светильник
Несущая стойка, шаг 2310mm
Пандус 1:12
+0.000(+13.057)

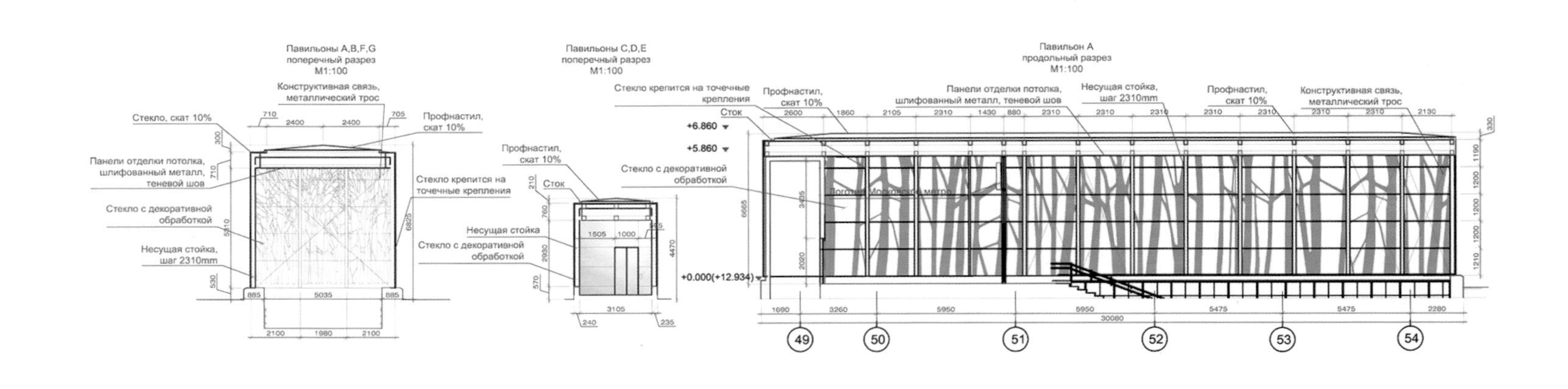

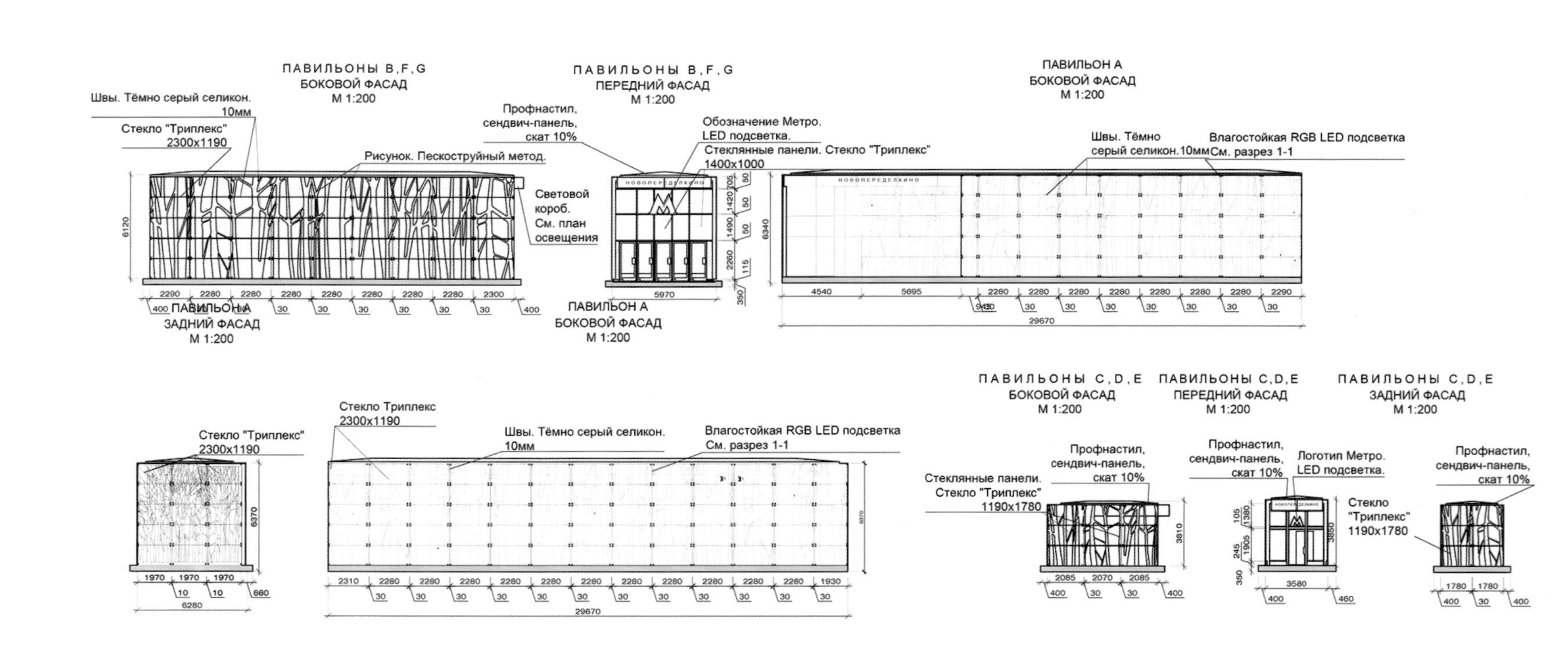

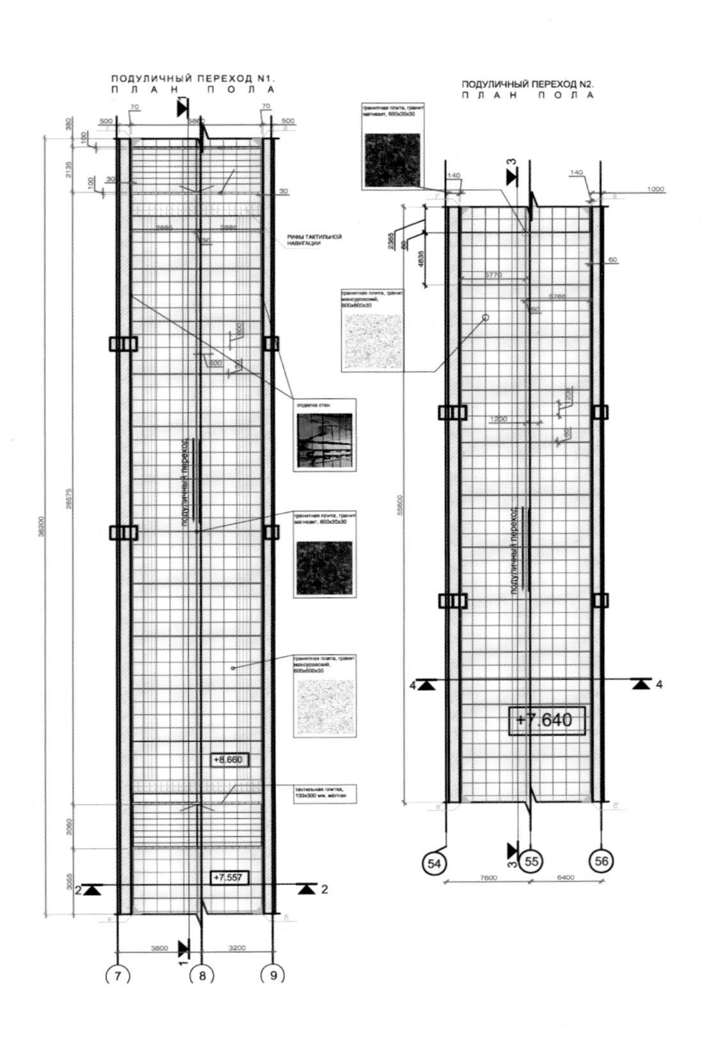
ПОДУЛИЧНЫЙ ПЕРЕХОД N1.
ПЛАН ПОЛА
ПОДУЛИЧНЫЙ ПЕРЕХОД N2.
ПЛАН ПОЛА
+8.660
+7.557
+7.640

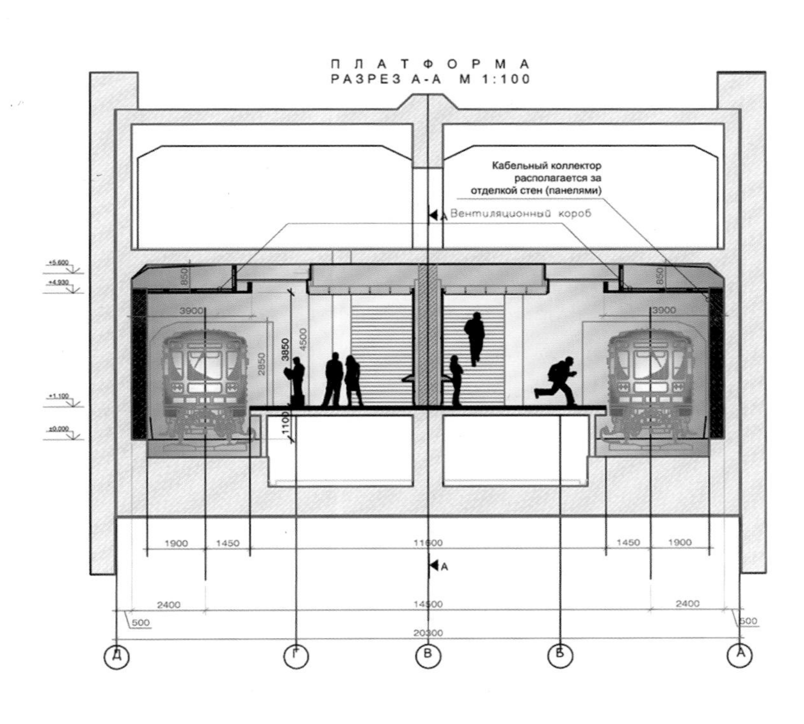
ПЛАТФОРМА
РАЗРЕЗ А-А М 1:100
Кабельный коллектор располагается за отделкой стен (панелями)
Вентиляционный короб

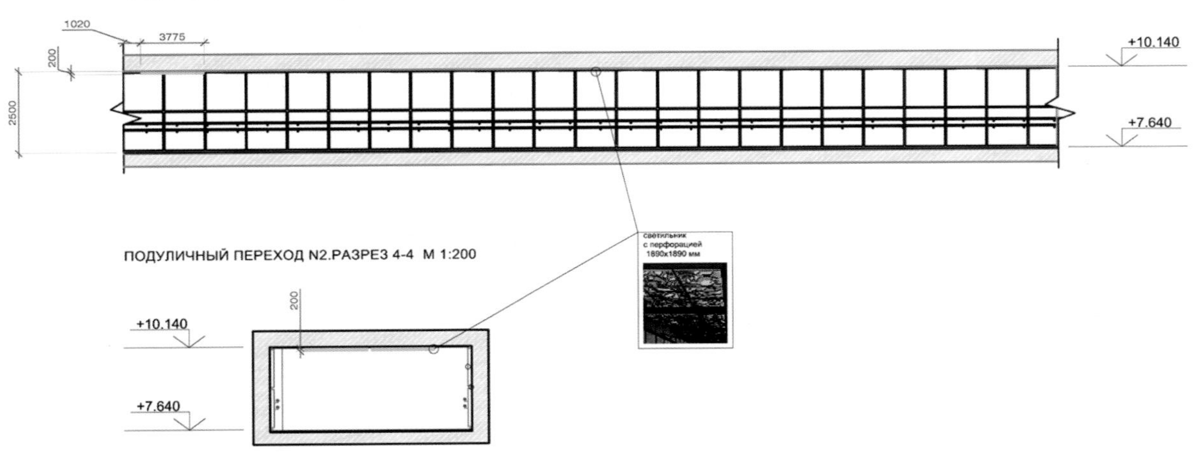
ПОДУЛИЧНЫЙ ПЕРЕХОД N2.РАЗРЕЗ 3-3 М 1:200
+10.140
+7.640
ПОДУЛИЧНЫЙ ПЕРЕХОД N2.РАЗРЕЗ 4-4 М 1:200
+10.140
+7.640
светильник с перфорацией 1890x1890 мм

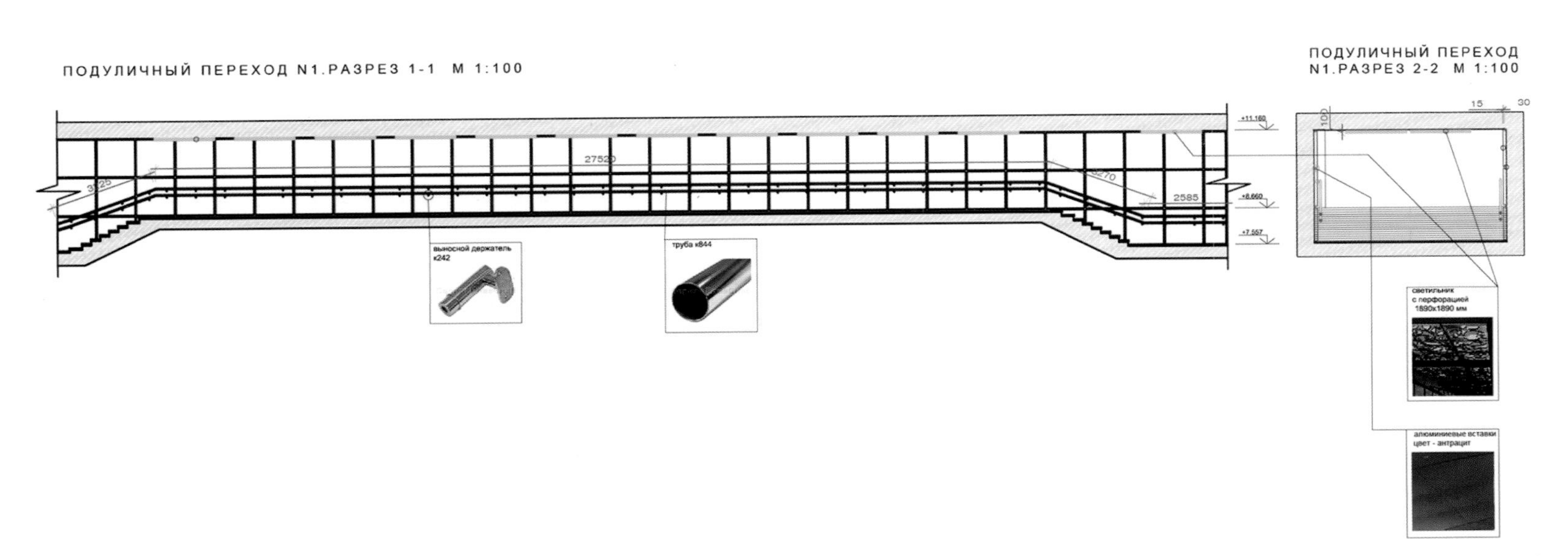
ПОДУЛИЧНЫЙ ПЕРЕХОД N1.РАЗРЕЗ 1-1 М 1:100
ПОДУЛИЧНЫЙ ПЕРЕХОД N1.РАЗРЕЗ 2-2 М 1:100
выносной держатель x242
труба x844
светильник с перфорацией 1890x1890 мм
алюминиевые вставки цвет - антрацит

Herma Parking Building

Location: Gyong Gi-Do, Korea
Architect: JOHO Architecture
Site Area: 853.7 m²
Gross Floor Area: 2,554.29 m²
Completion Date: 2010
Photographer: Sun Namgoong

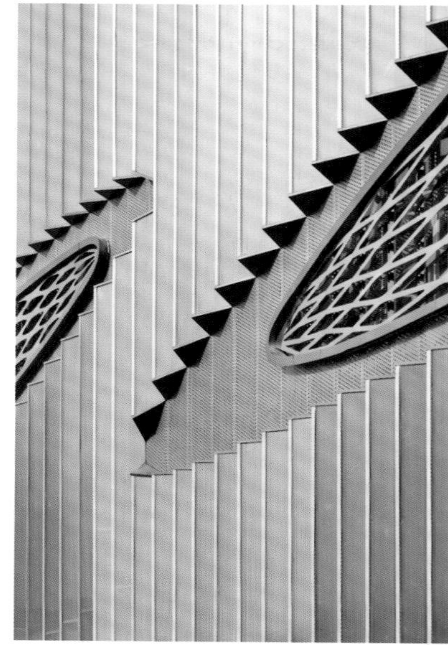

Parking lot either frankly reveals itself to be a parking lot without any consideration for design or surroundings, or disguises like a commercial facility. However, both cases are not welcomed in a city and ruin the city landscape.

Development districts of New Downtown in Korea that are fully filled with gigantic real estate goods only emphasize the legal maximum floor area ratio. The city identity is represented by the wall-covering signs not the presence of space nor void. The massive box lumps clutter the city with the logic of capitalism that is composed of investment, lease and presale, rather than the respective regional characteristics. This project intended to change the urban landscape through a proposition of a certain symbolically designed icon on a dry city.

In Korea for the development of a new town, developers parcel out the land in lots according to the usage such as a business area and a residence area, based on the Building codes. Parking lot zone also gets placed in major site following the business and residence area zoning program. A plot for a parking lot can be planned with cheaper land price and eased regulations than a business area.

However, it has a restriction that the commercial facility-to-parking facility ratio has to be 2 to 8. After all, deriving the land cost and construction cost with very 20% commercial area is the poor reality of Parking lot construction in Korea. This substandard business value causes the parking buildings to be built without any design concerns for lower construction price. Furthermore, a vicious circle of gradually deteriorating surrounding urban environment continues because of the box-shaped boring parking buildings. The Herma parking building was started from a fundamental consideration of current parking lots in Korea.

Construction Process

CIP Piles + Mat Foundation

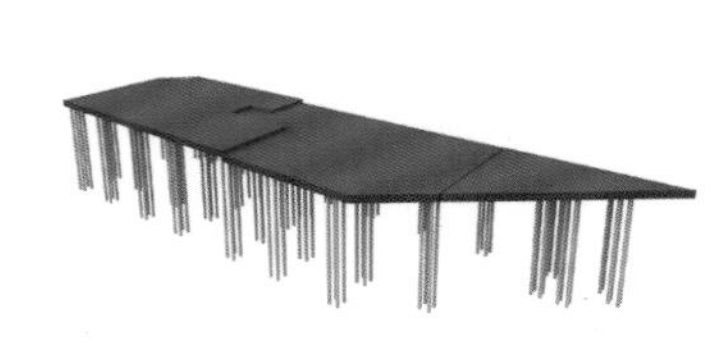

Core

Rooftop Plant

Steel Stairs

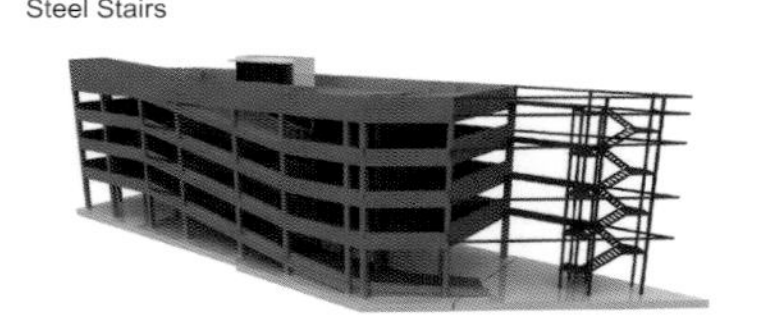

Column

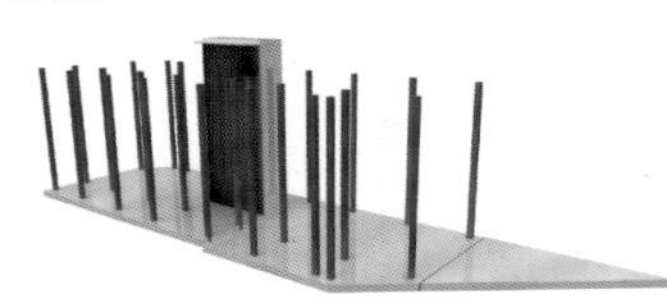

Girder

Galvanized Pipe Frame

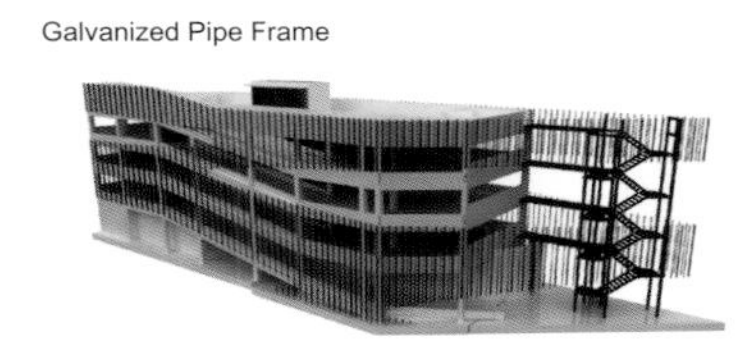

Aluminium Perforated Panel Frame

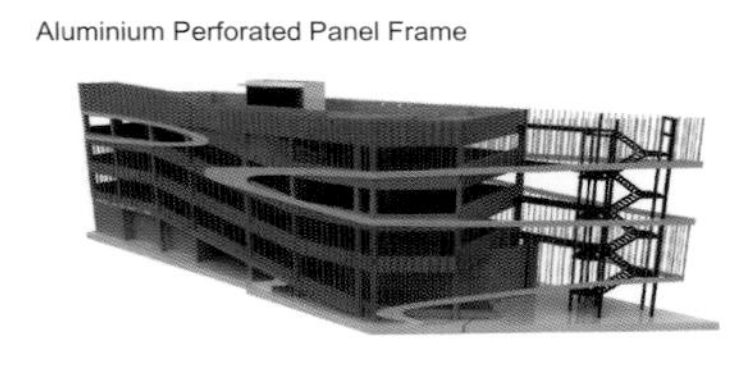

Ramp

Slab

Polycarbonate Panel

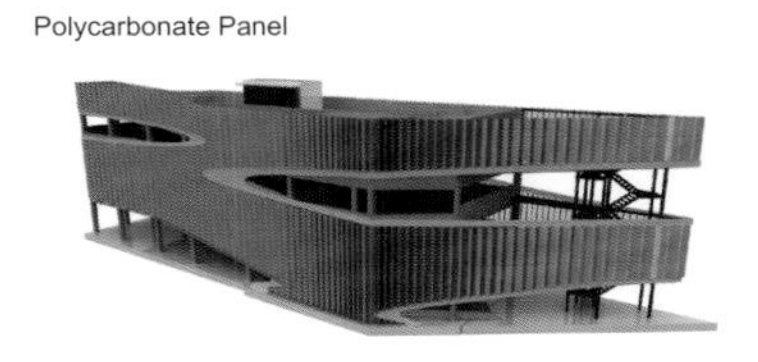

Galvanized Pipe + Paint

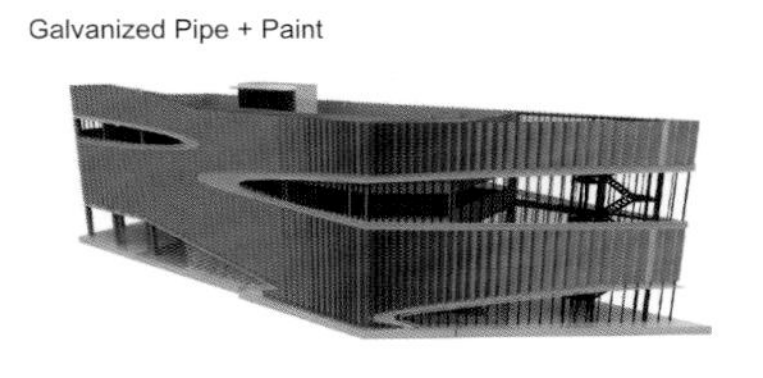

Parapet

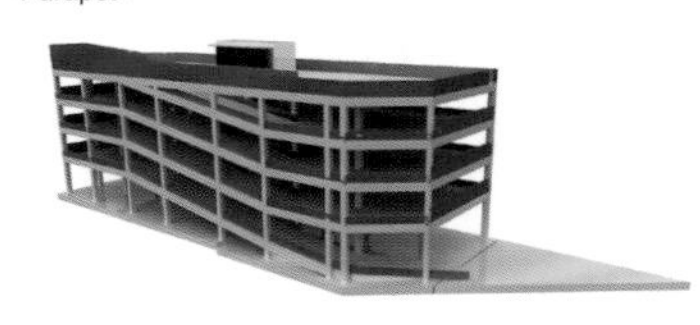

Wall + Glass

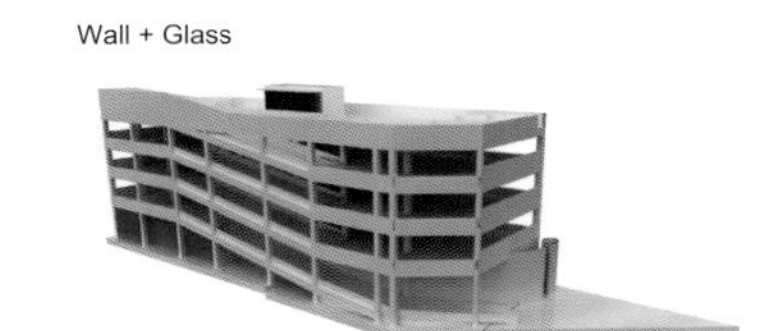

Stainless Louver Frame

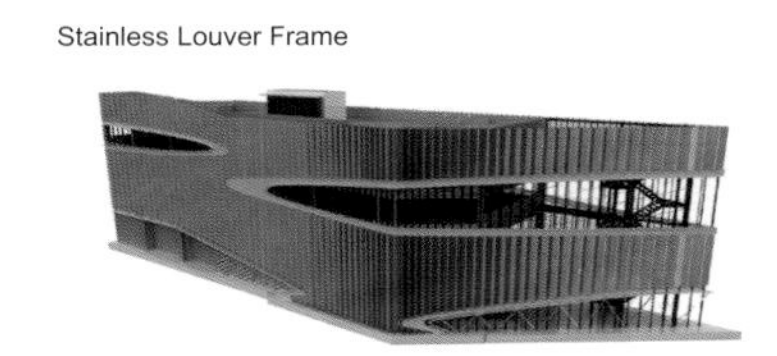

Stainless Pattern Frame

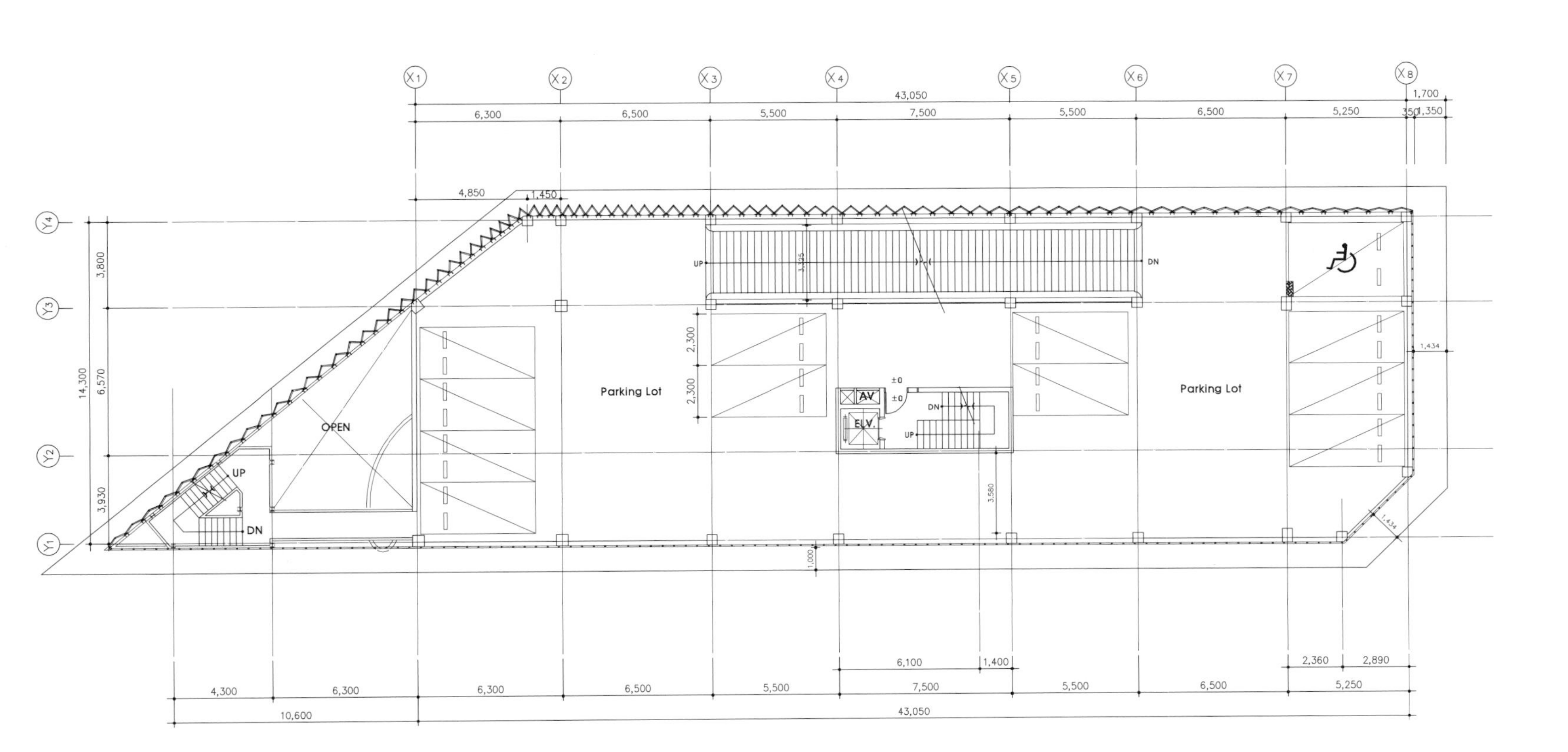

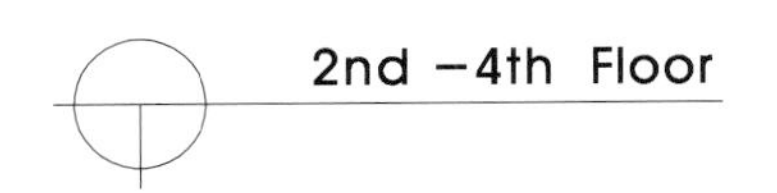
2nd –4th Floor

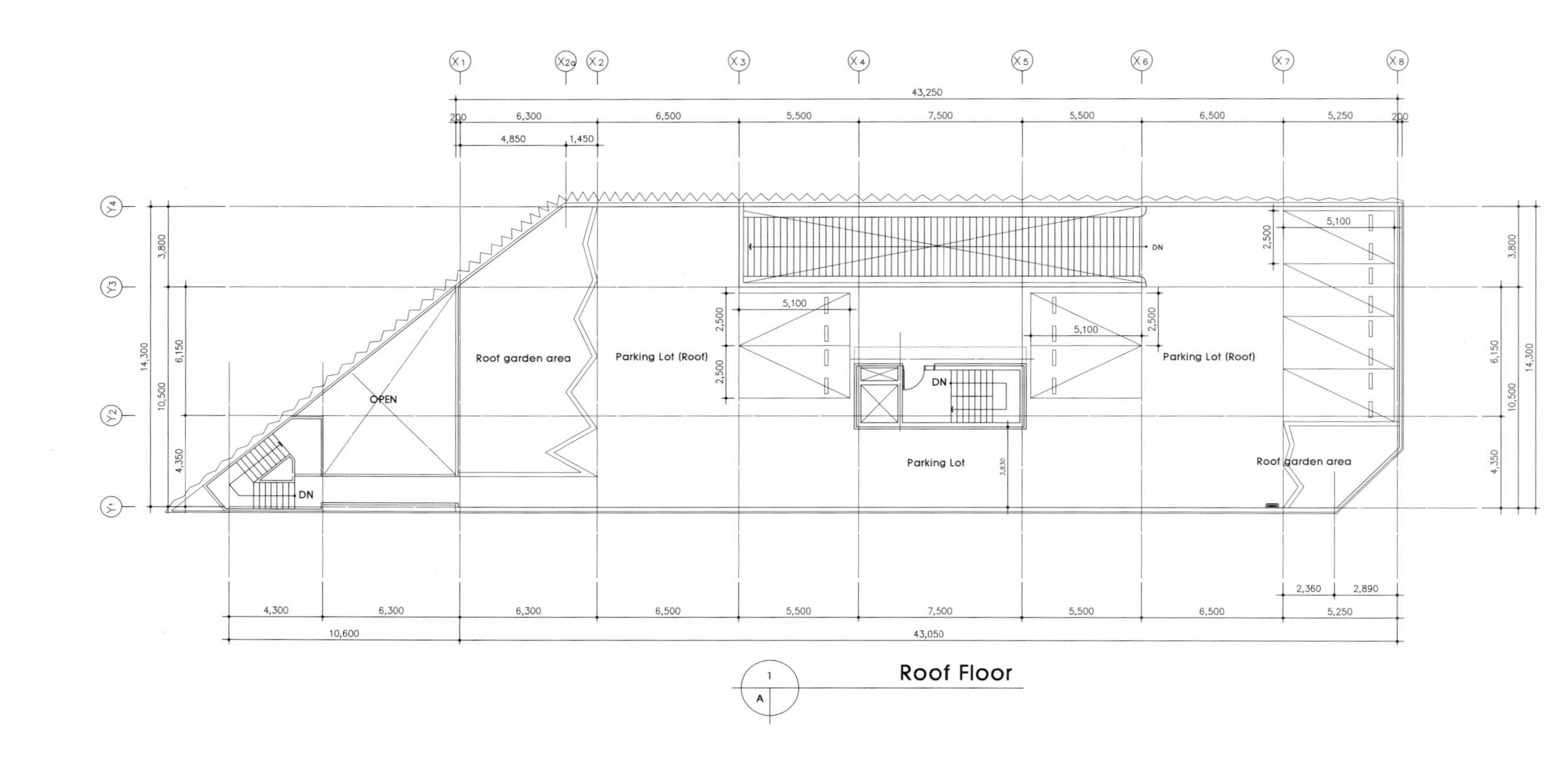
Roof garden area
Parking Lot (Roof)
OPEN
DN
Parking Lot
Parking Lot (Roof)
Roof garden area
Roof Floor

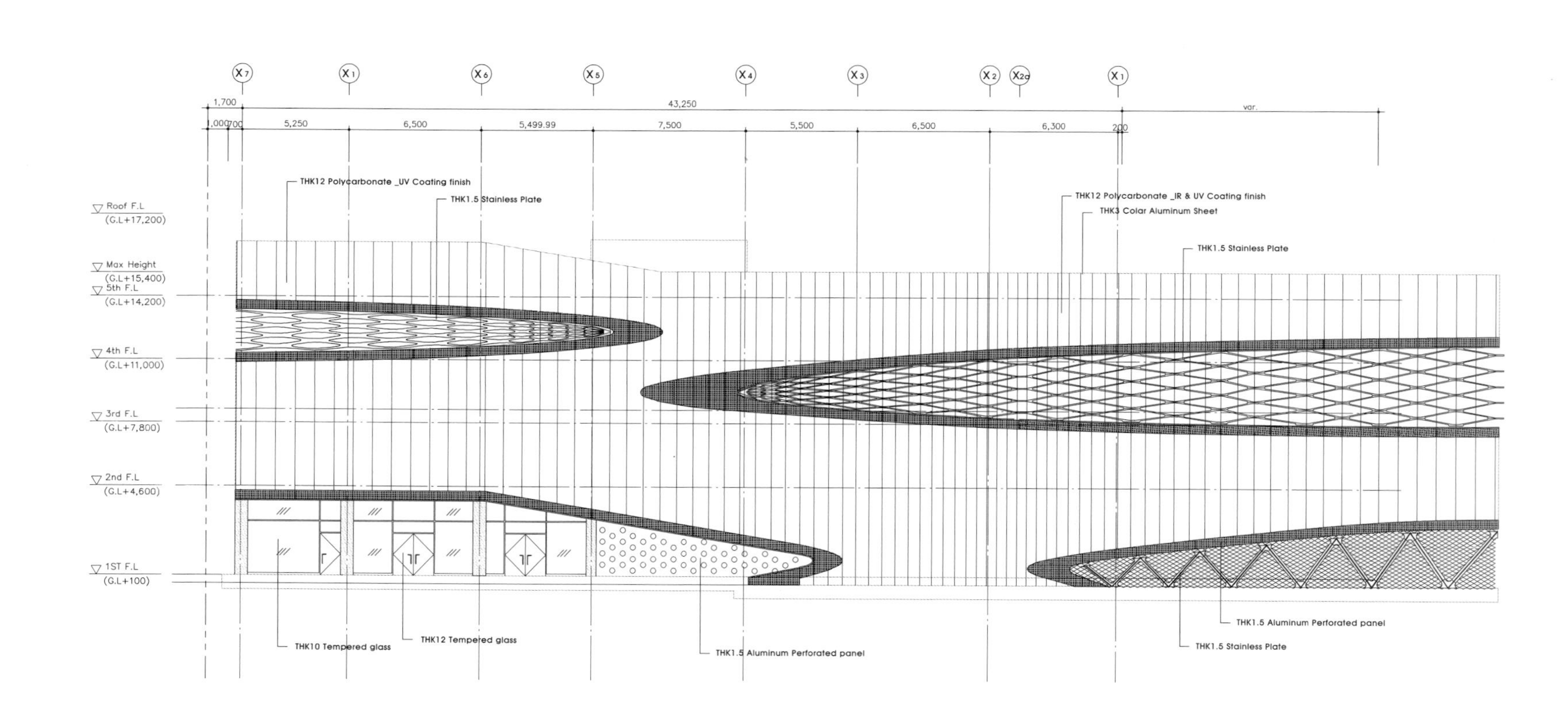
Roof F.L (G.L+17,200)
Max Height (G.L+15,400)
5th F.L (G.L+14,200)
4th F.L (G.L+11,000)
3rd F.L (G.L+7,800)
2nd F.L (G.L+4,600)
1ST F.L (G.L+100)
THK12 Polycarbonate _UV Coating finish
THK1.5 Stainless Plate
THK12 Polycarbonate _IR & UV Coating finish
THK3 Color Aluminum Sheet
THK1.5 Stainless Plate
THK10 Tempered glass
THK12 Tempered glass
THK1.5 Aluminum Perforated panel
THK1.5 Aluminum Perforated panel
THK1.5 Stainless Plate

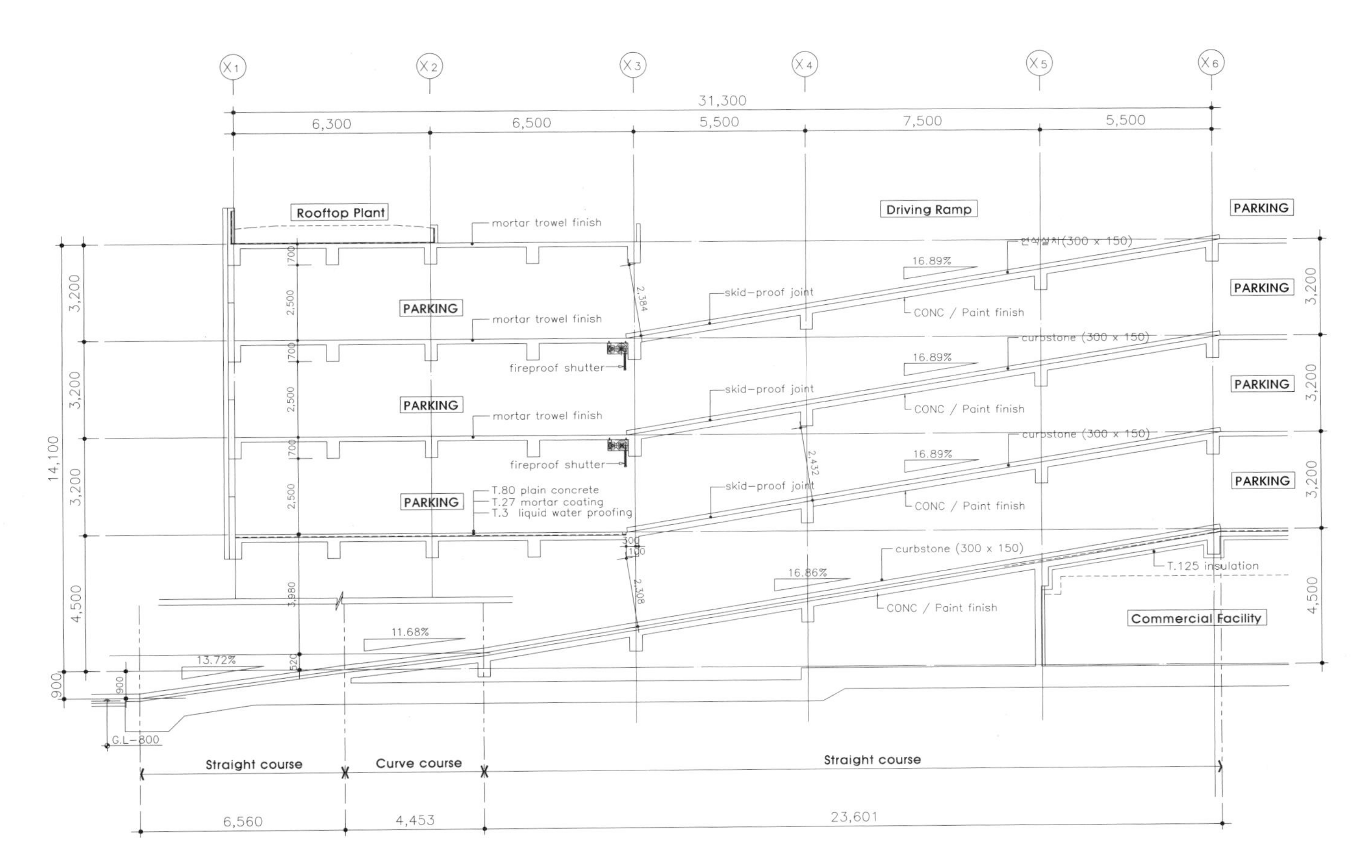

Ramp Section

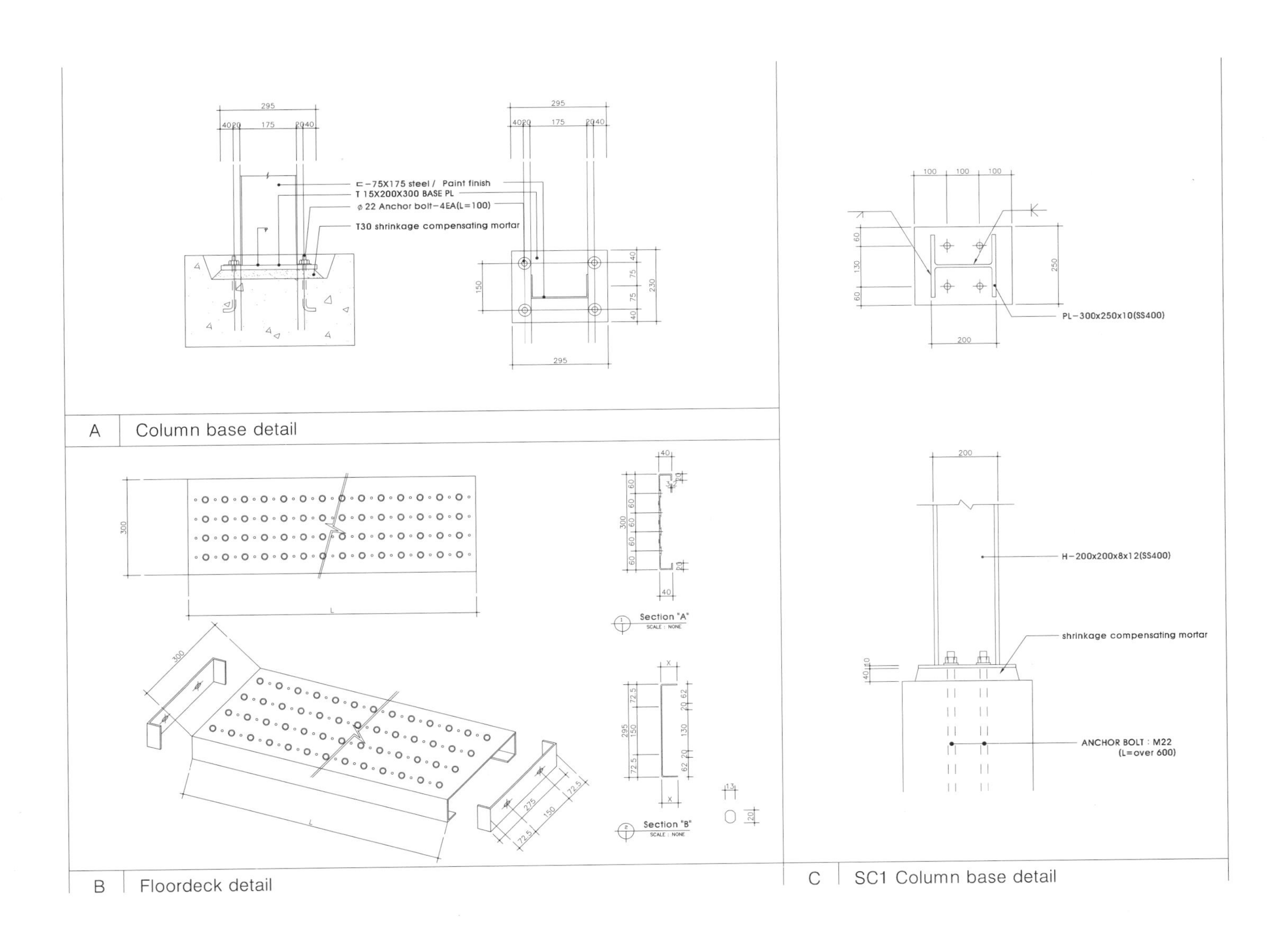

A Column base detail

B Floordeck detail

C SC1 Column base detail

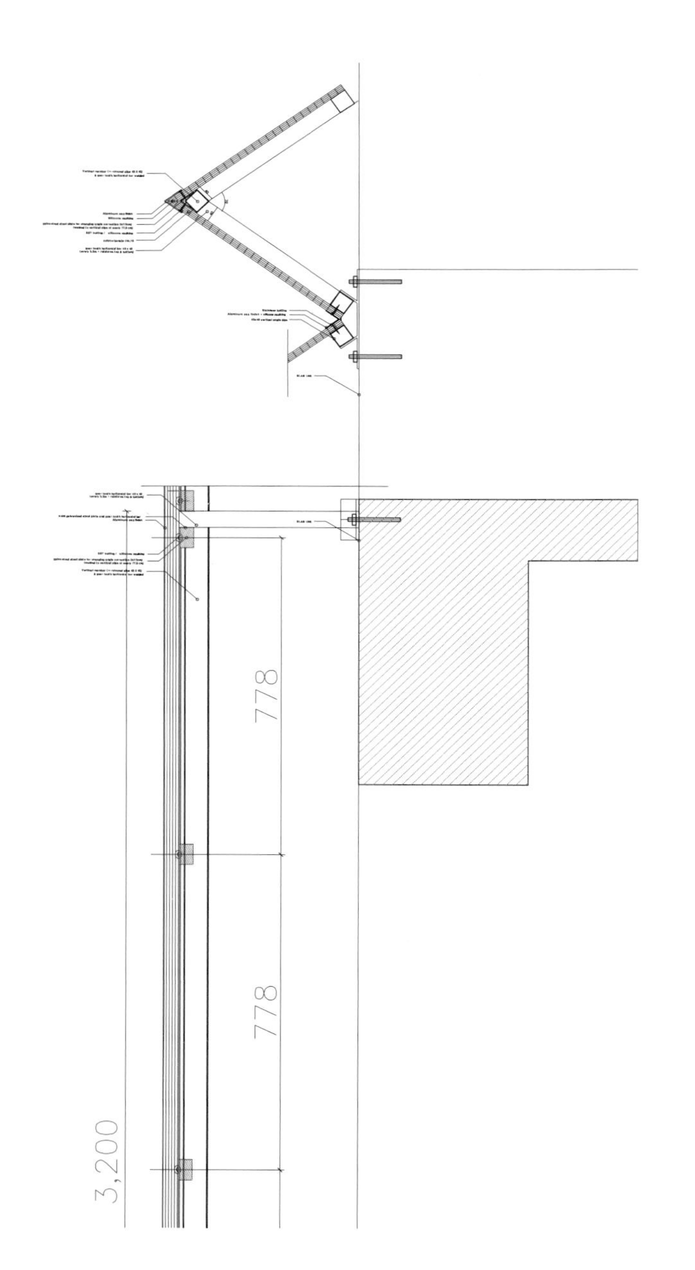

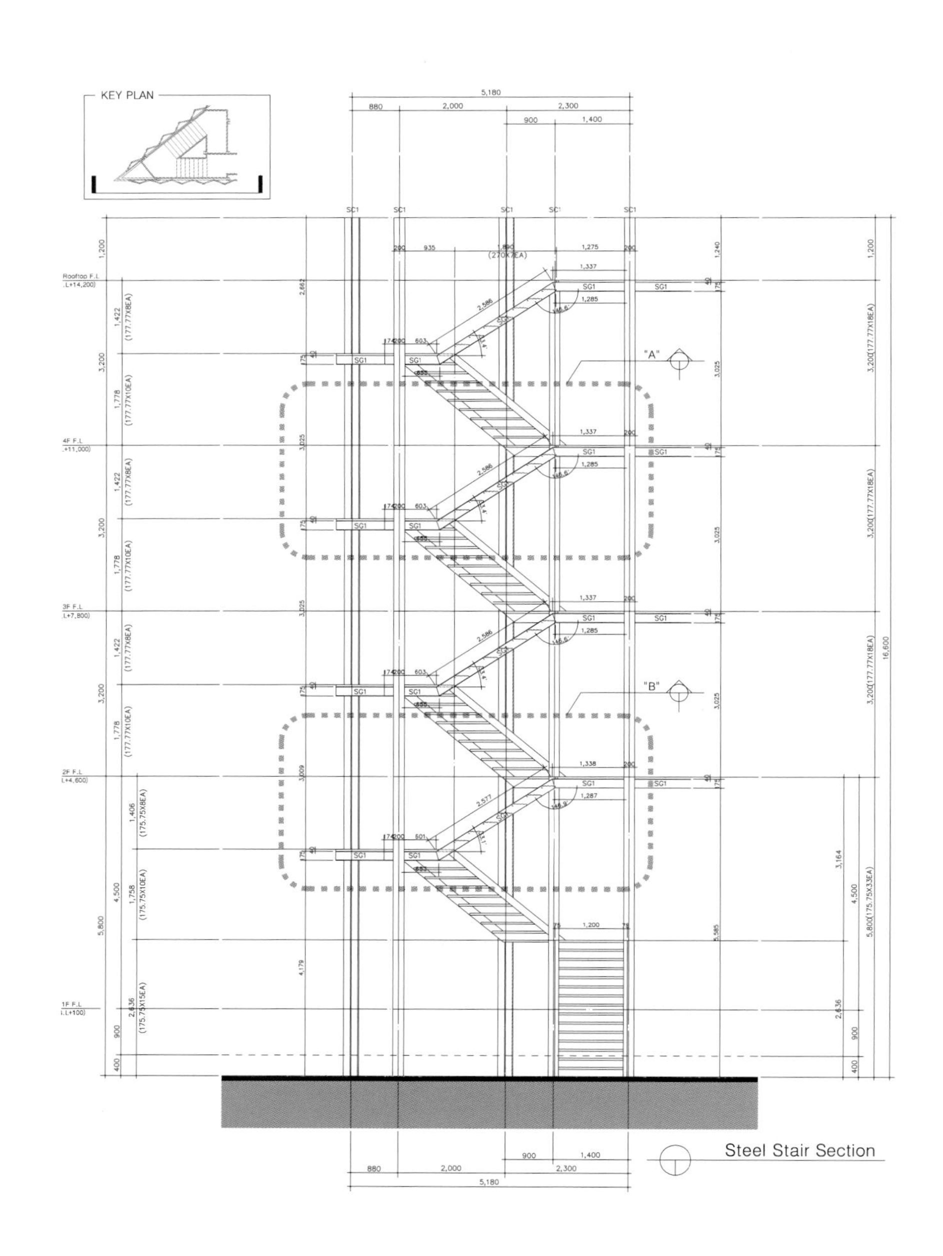

Steel Stair Section

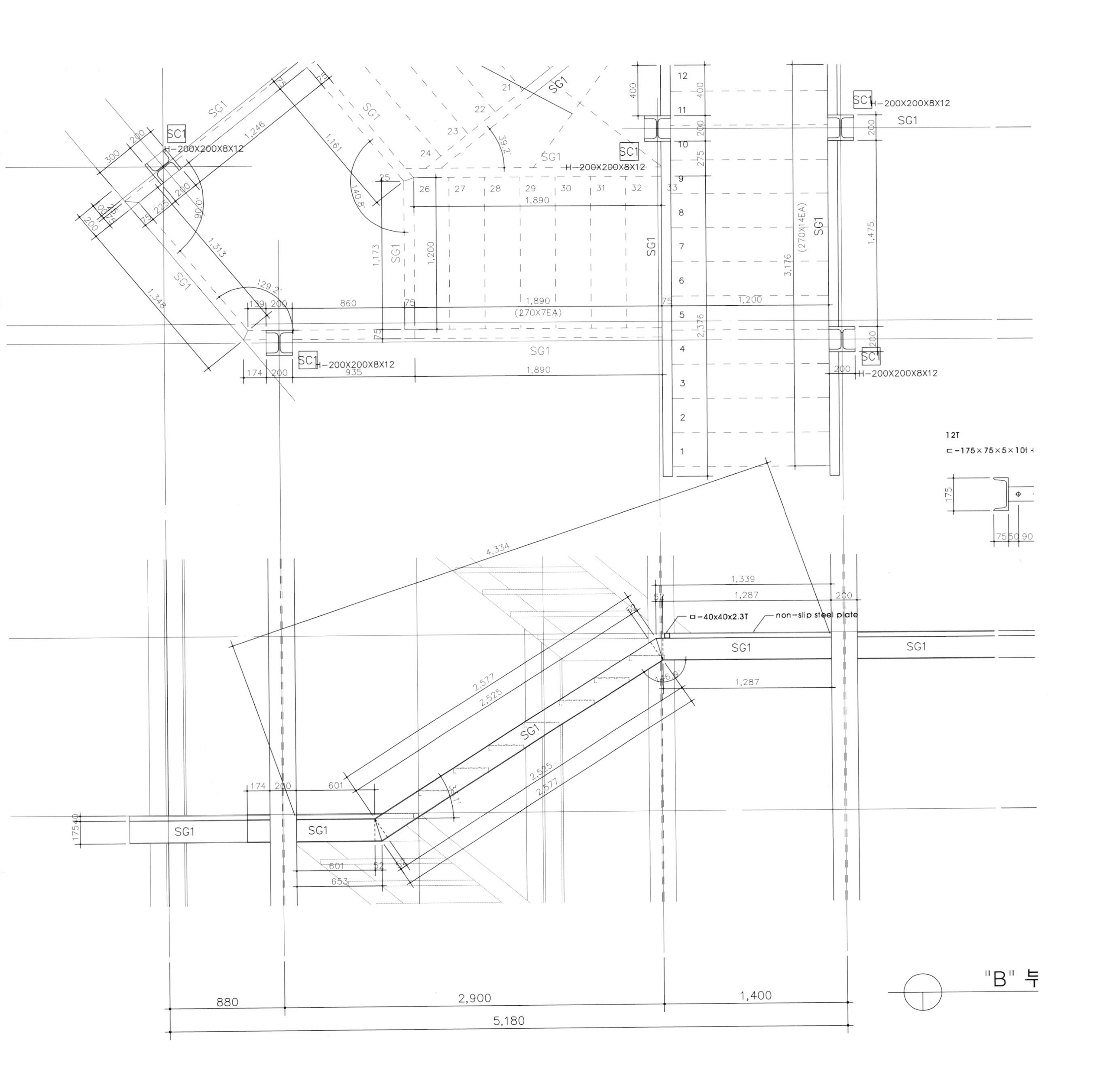

SC1
H-200X200X8X12
SG1
1,890
(270X7EA)
(270X14EA)
3,176
1,475
1,200
12T
4,334
2,577
2,525
1,339
1,287
□-40x40x2.3T
non-slip steel plate
880
2,900
1,400
5,180
"B" 부

Pulkovo Airport

Location: St Petersburg, Russia
Architect: Grimshaw Architects
Client: Pulkovo Airport
Size: 110,000 m^2
Completion Date: 2014

The brief asked for a new, second terminal to be constructed adjacent to the existing 1950s Terminal 1, and to be organized around a new airport city master plan. Our approach proposes a new axial boulevard, which forms the Airport Gateway and is designed to be reminiscent of St Petersburg's wide avenues. The boulevard terminates in a large square, which fronts the old terminal to create a new 'heritage quarter'. The new terminal is also highlighted within the master plan, sited to offer passengers long, framed views of the front facade and roof on approach.

Externally, our main consideration was to design a roof that would stop snow from drifting, which can result in uneven loading of the structure. The roof is therefore predominantly flat, ensuring that snow accumulates as evenly as possible across its 50,000 sq m surface. Internally, the faceted forms of the soffit are evocative of the pitched roof forms of the city, and their surface textures are reminiscent of the crystal patterns of ice and snowflakes.

The roof is also designed to deal with meltwater. Subtle falls in the flat roof are arranged as inverted prisms, with their lowest points above the centreline of each column. This allows the deepest build-up of snow to be above the point of greatest structural support and meltwater to be collected at the appropriate point for drainage.

To welcome natural daylight deep into the building, roof lights are arranged between the roof ribbons. In order for them to remain effective throughout the year, even during winter when sunlight is at a premium and snow lies thick on the roof, the prism-like roof lights project above the flat roof to a height of over 4m at the centre, then taper away sharply along their length. The soffit is colored gold to gild the light entering the building. This effect is characteristic of the reflections from St Petersburg's gilded domes and spires.

The facades are designed to respond to specific functional and climatic demands. The east- and west-facing facades are predominantly solid with only small apertures relating to internal planning requirements. As well as allowing the walls to be finished with highly insulated cladding, this also provides the internal flexibility to build against the

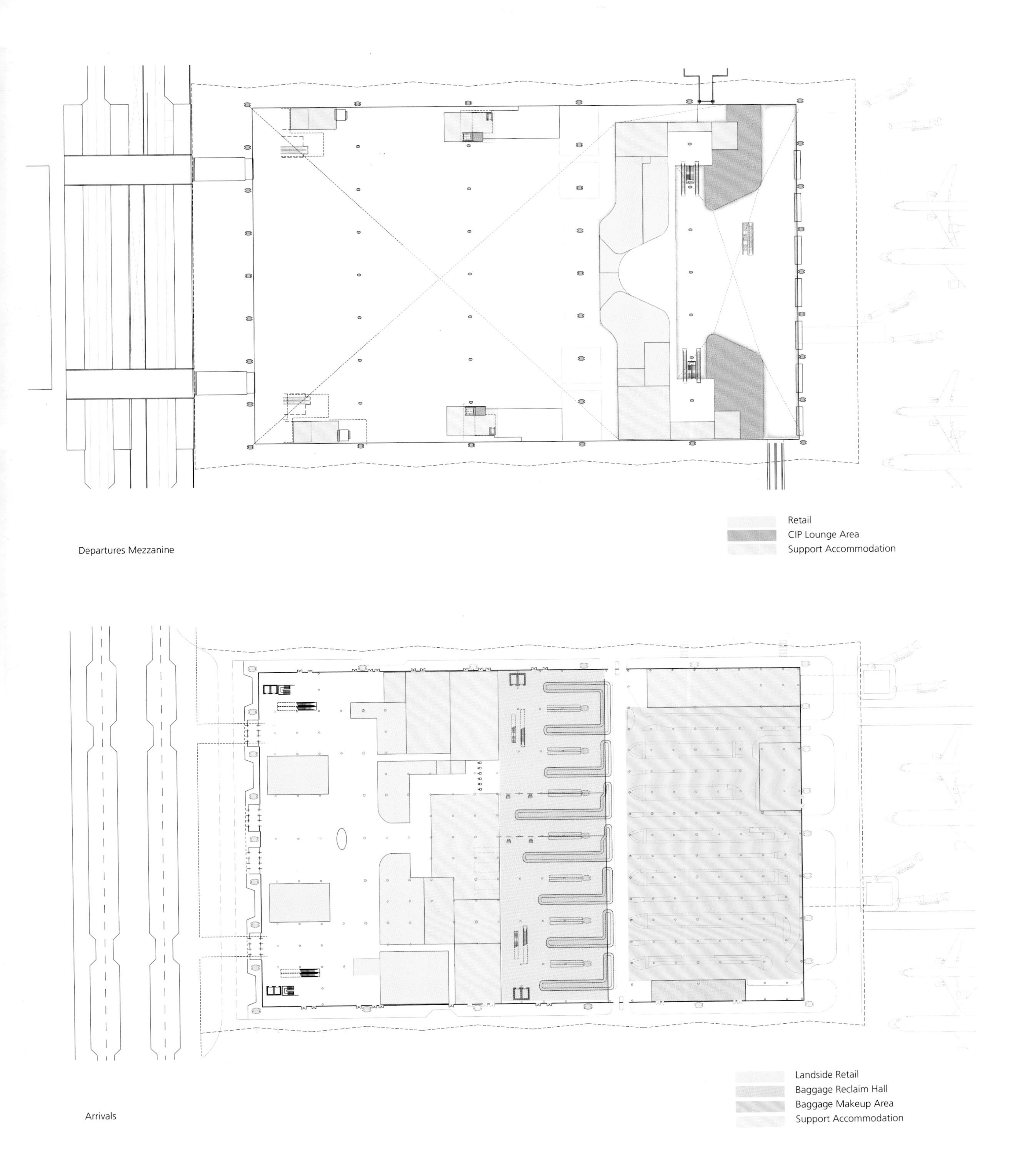

Departures Mezzanine

Arrivals

walls without compromising its external appearance.

In contrast, the north and south facades – landside and airside – are largely glazed, as they present the airport to arriving and departing passengers. They are designed as large civic windows, echoing those found on some of St Petersburg's most significant buildings. Organised between the primary columns of the terminal, they frame passenger views into and out of the airport.

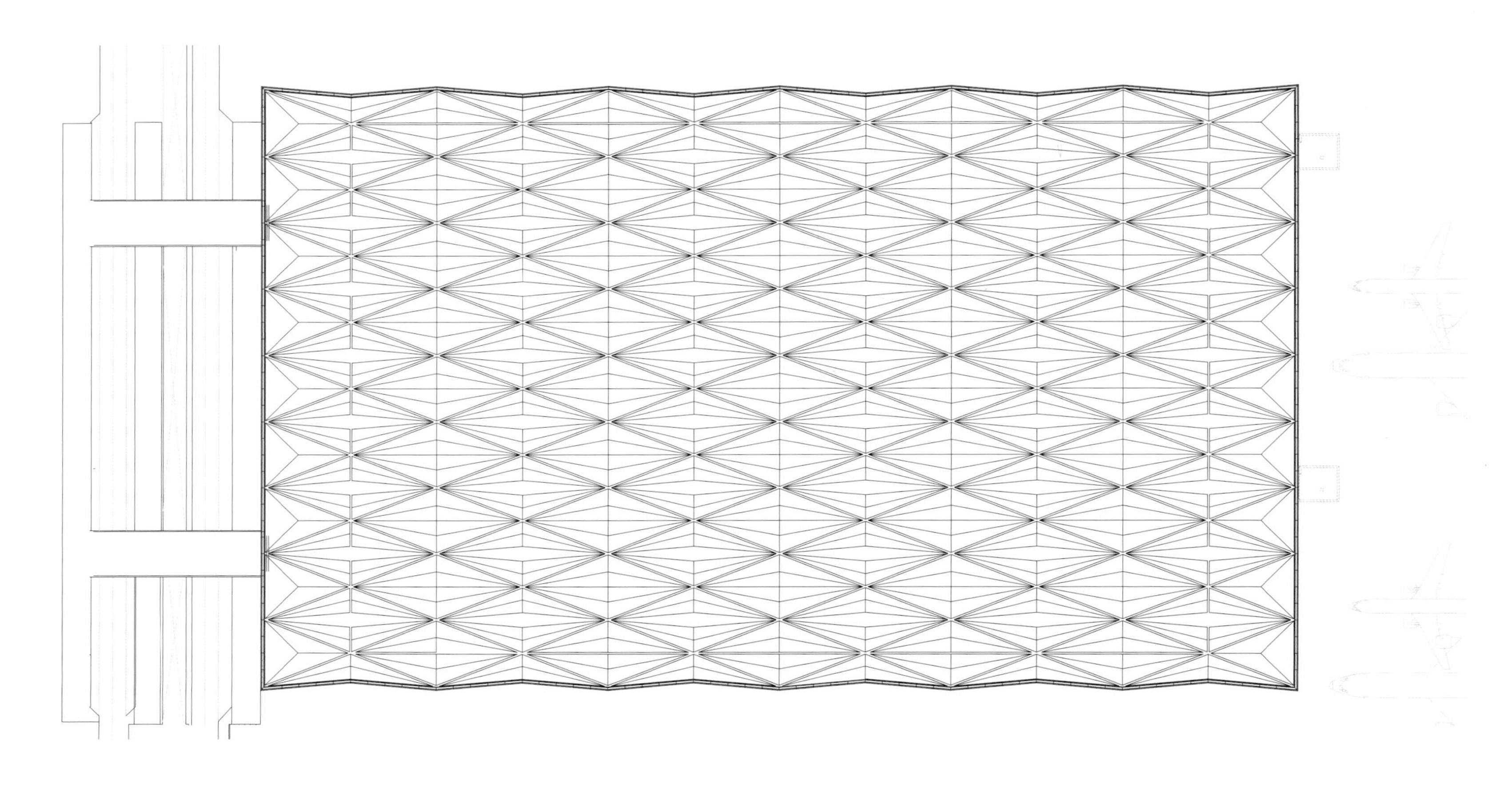

Ceiling Geometry

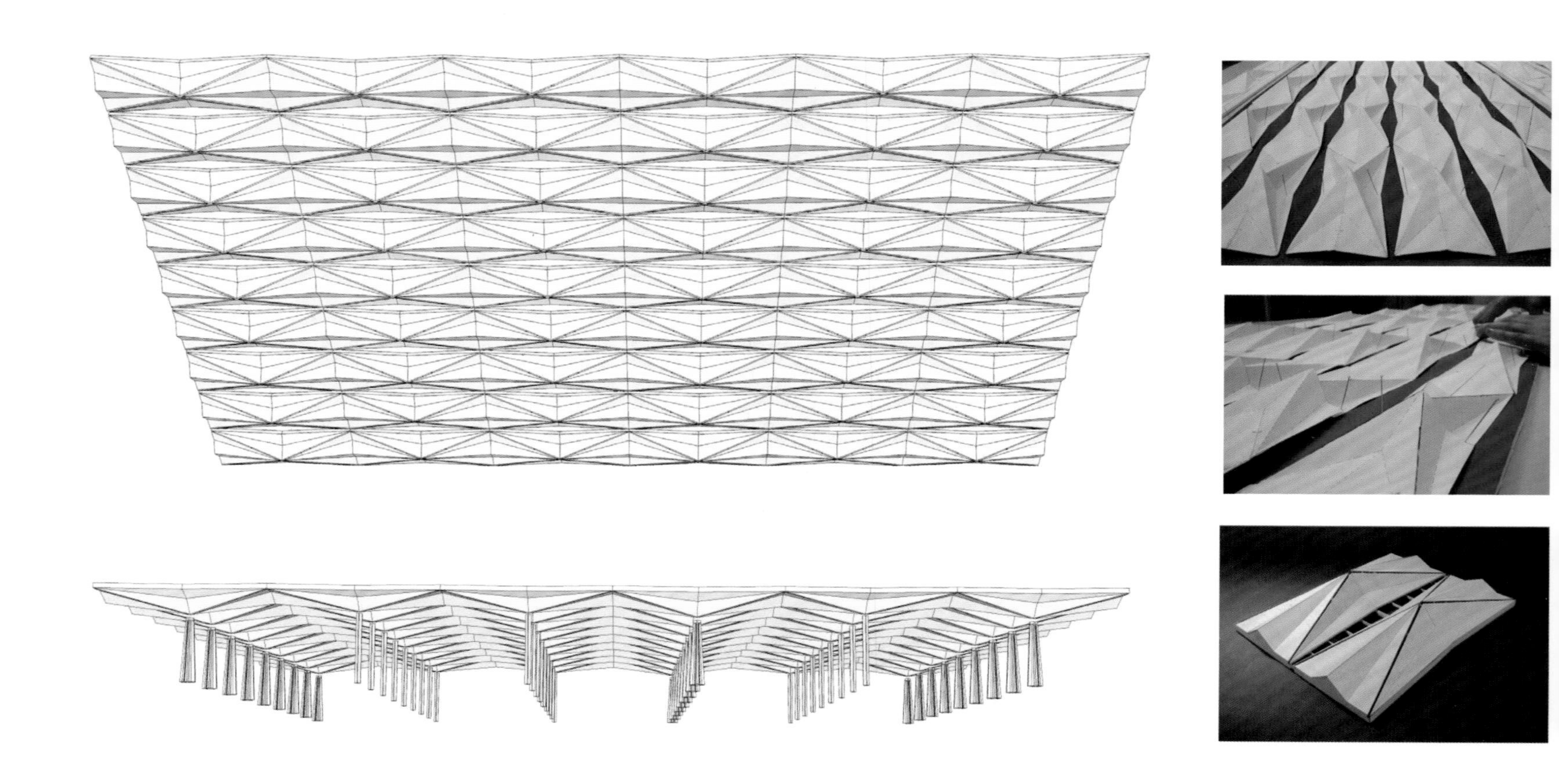

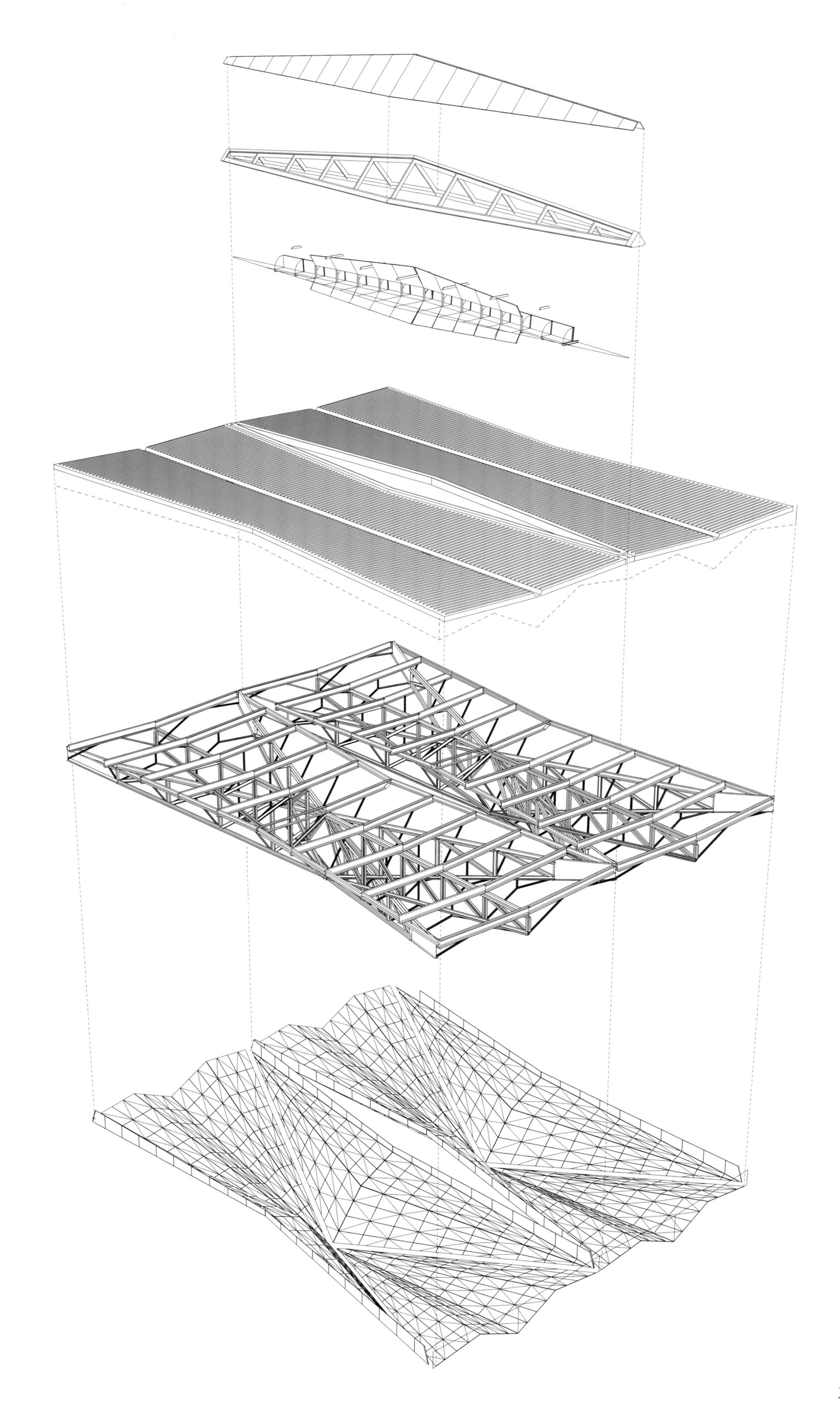

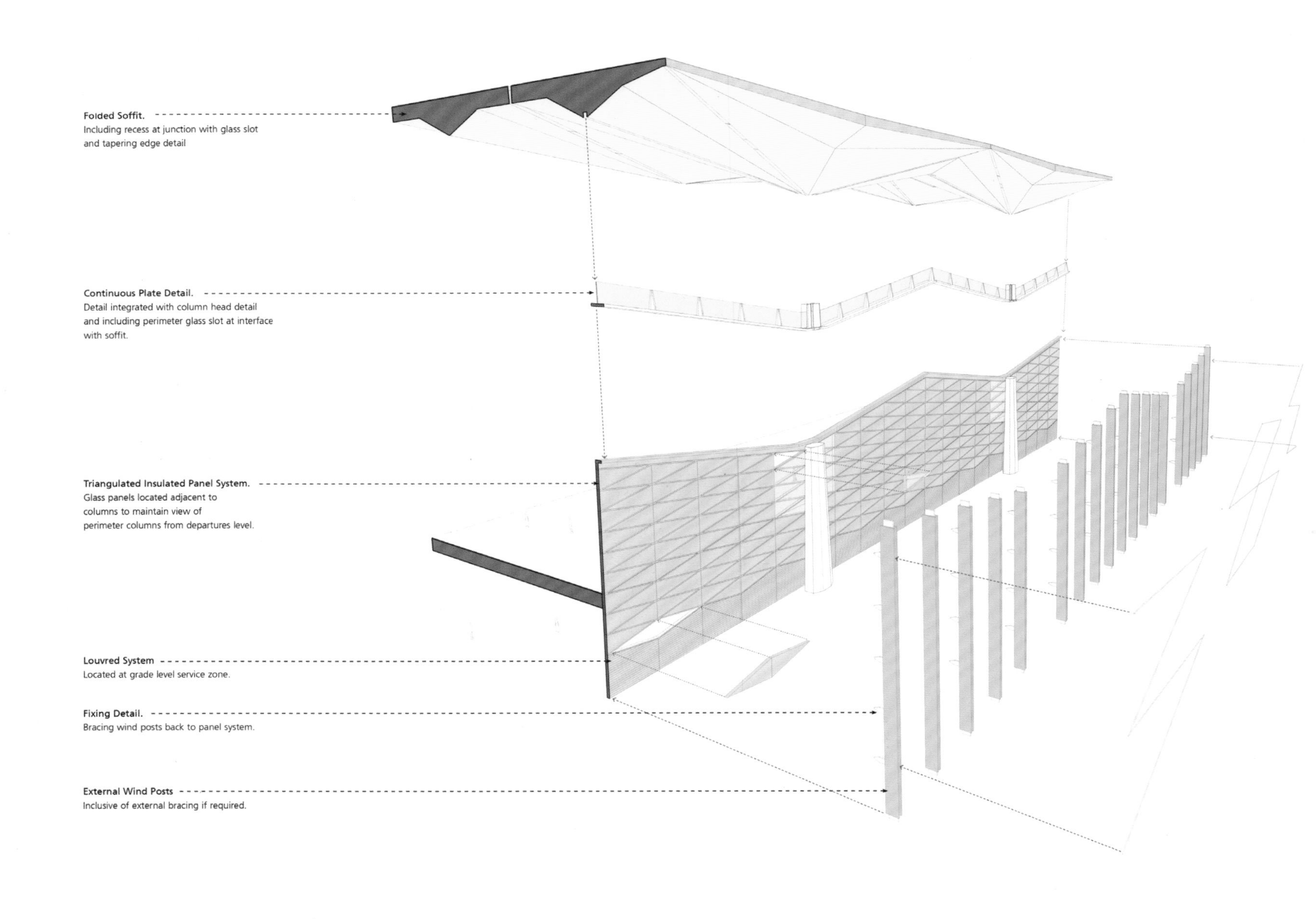
Folded Soffit.
Including recess at junction with glass slot
and tapering edge detail
Continuous Plate Detail.
Detail integrated with column head detail
and including perimeter glass slot at interface
with soffit.
Triangulated Insulated Panel System.
Glass panels located adjacent to
columns to maintain view of
perimeter columns from departures level.
Louvred System
Located at grade level service zone.
Fixing Detail.
Bracing wind posts back to panel system.
External Wind Posts
Inclusive of external bracing if required.

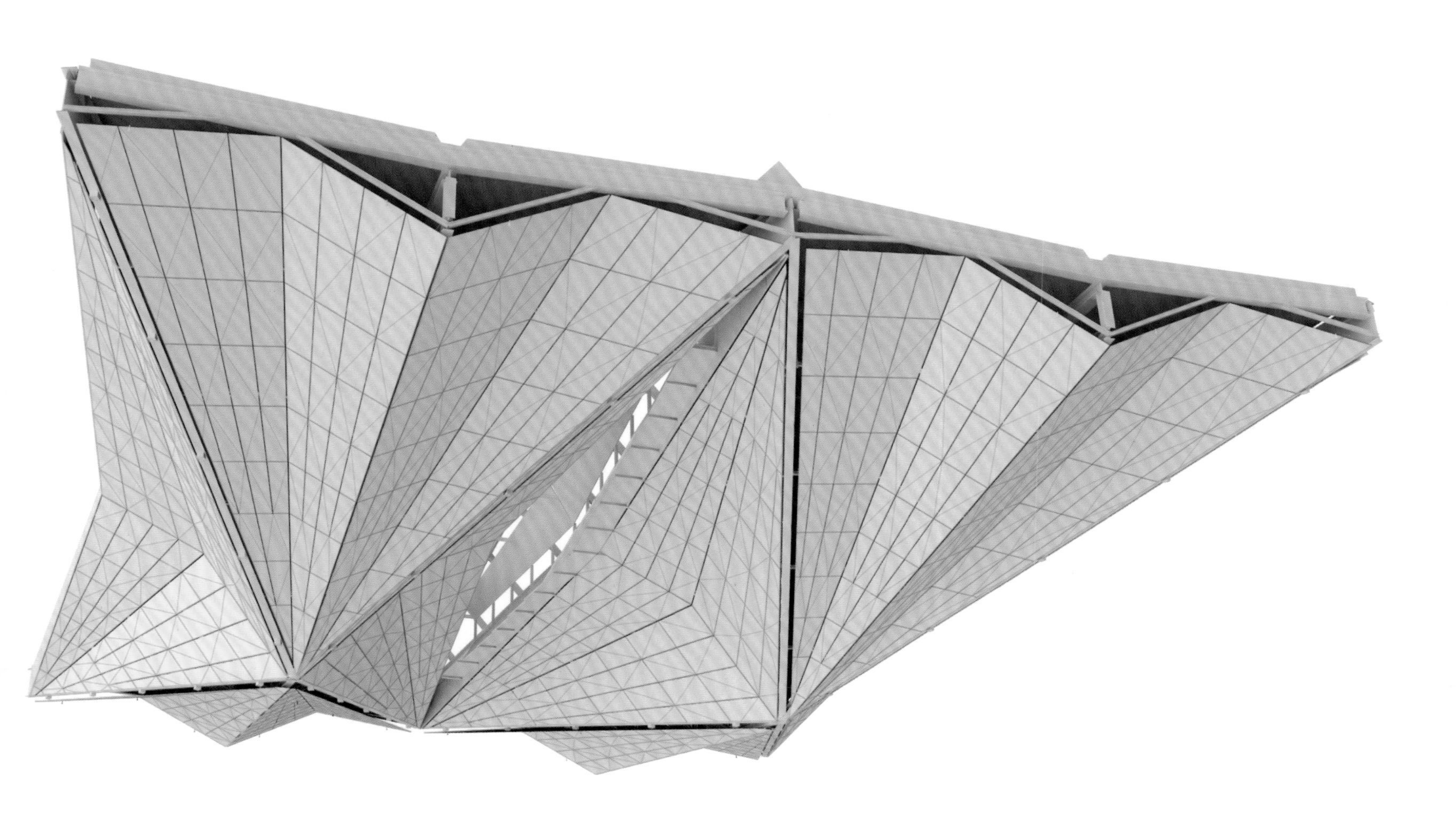

Sports, Health & Others

Care Campus

Location: Schiedam, Netherlands
Architect: Möhn + Bouman BV / René Bouman
Client: Ipse de Bruggen foundation
Size: 1,460 m²
Completion Date: 2013
Photographer: Sarah Blee

The project is part of a larger renewal of a small campus area for a community of people with disabilities. On the border Schiedam and its rural surroundings the campus comprises this newly renovated daycare centre along with a housing facility currently under design.

The centre has been stripped of unnecessary details and has been painted dark brown, and provided with high quality wooden terraces on all sides, which in turn are wrapped by a wooden screen. In keeping with the local environment, the screen is constructed from chestnut wood and blends into the rural atmosphere. At entrance areas, the screen's shape adapts, and frames with printed glass cut into the wood, creating a sleek, modern contrast with the rougher, more agricultural material.

The glass prints, made by artist Carol Fulton, continue throughout the interior of the centre as they surround the new lobby in the heart of the building. They form a panoramic view of the coppiced chestnut grove, forming a visual association with the screens outside.

In the heart of the lobby a polyurethane drop rises from the floor. This contains the main kitchen and the sanitary units, and the material creates another tactile contrast. Due to the cognitive constraints of the primary users of the building, the sensory experience of their environment becomes exceptionally important. Incorporating this aspect of the client's need, along with staging spatial experiences, was the key theme in the design process, resulting in what maybe could be coined 'Performance Architecture.'

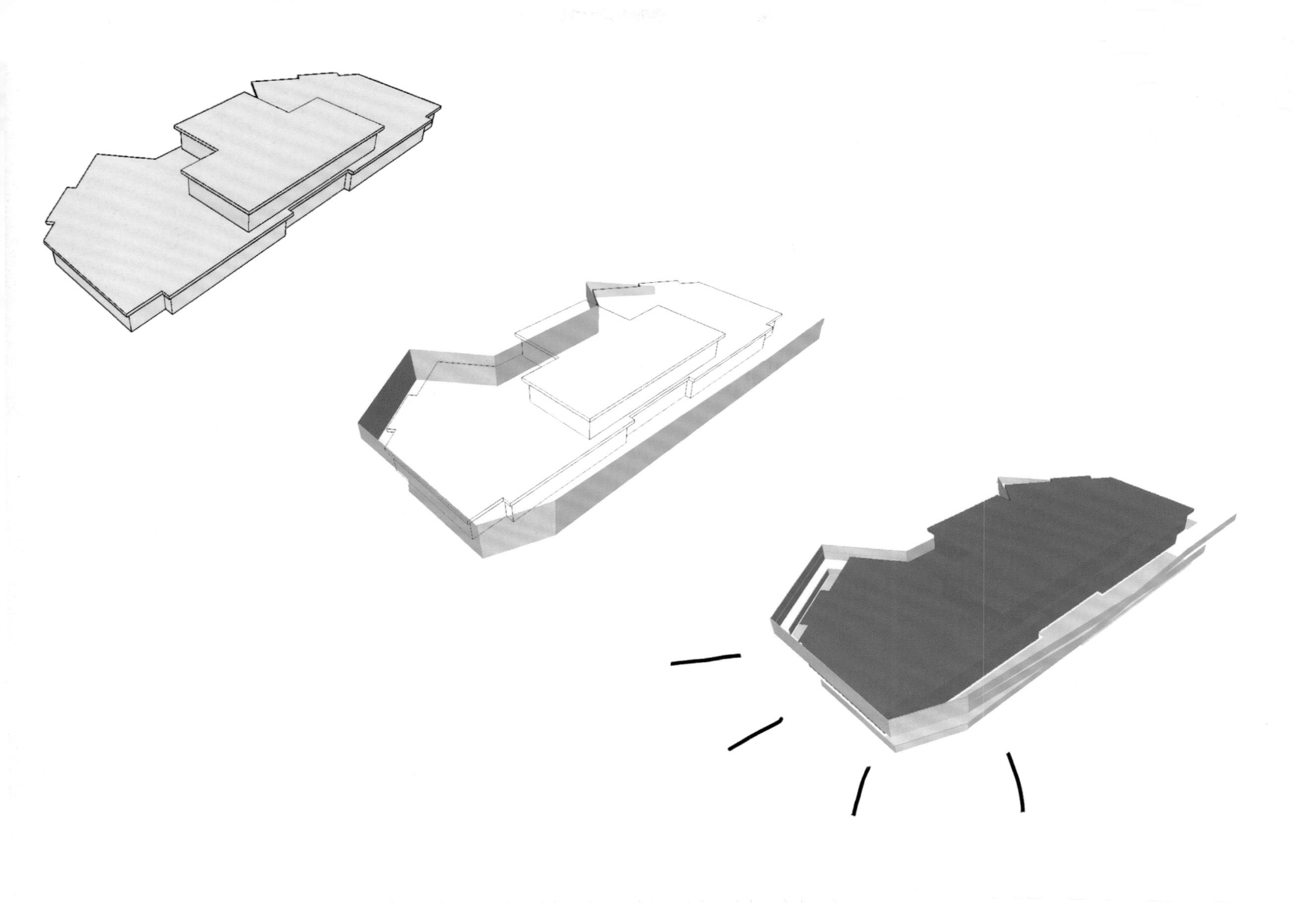

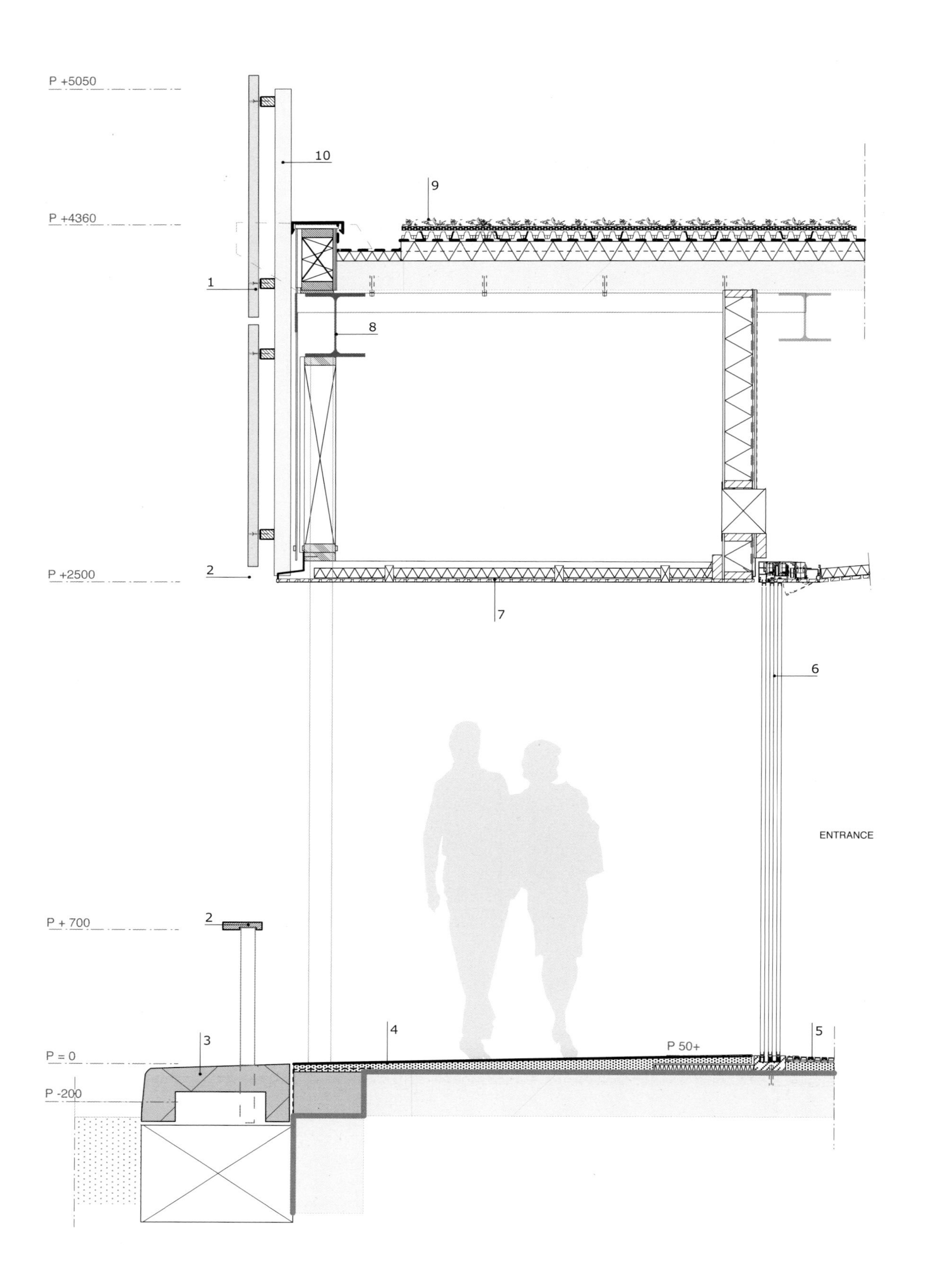

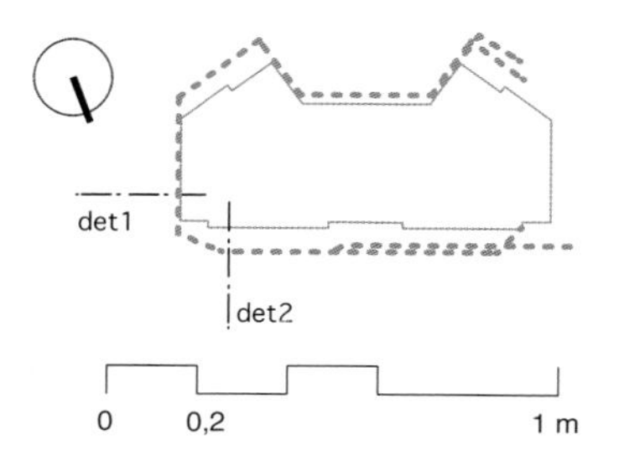

det. 1

1. prefabricated chestnut paling fences
2. accoya frame coated with polyurethane
3. prefab concrete element
4. power-floated concrete floor coated with polyurethane
5. entrance mat
6. sliding door
7. wooden ceiling with led strip light
8. new steel beam HEB 340
9. green roof
10. channel profiel UNISTRUT P5000

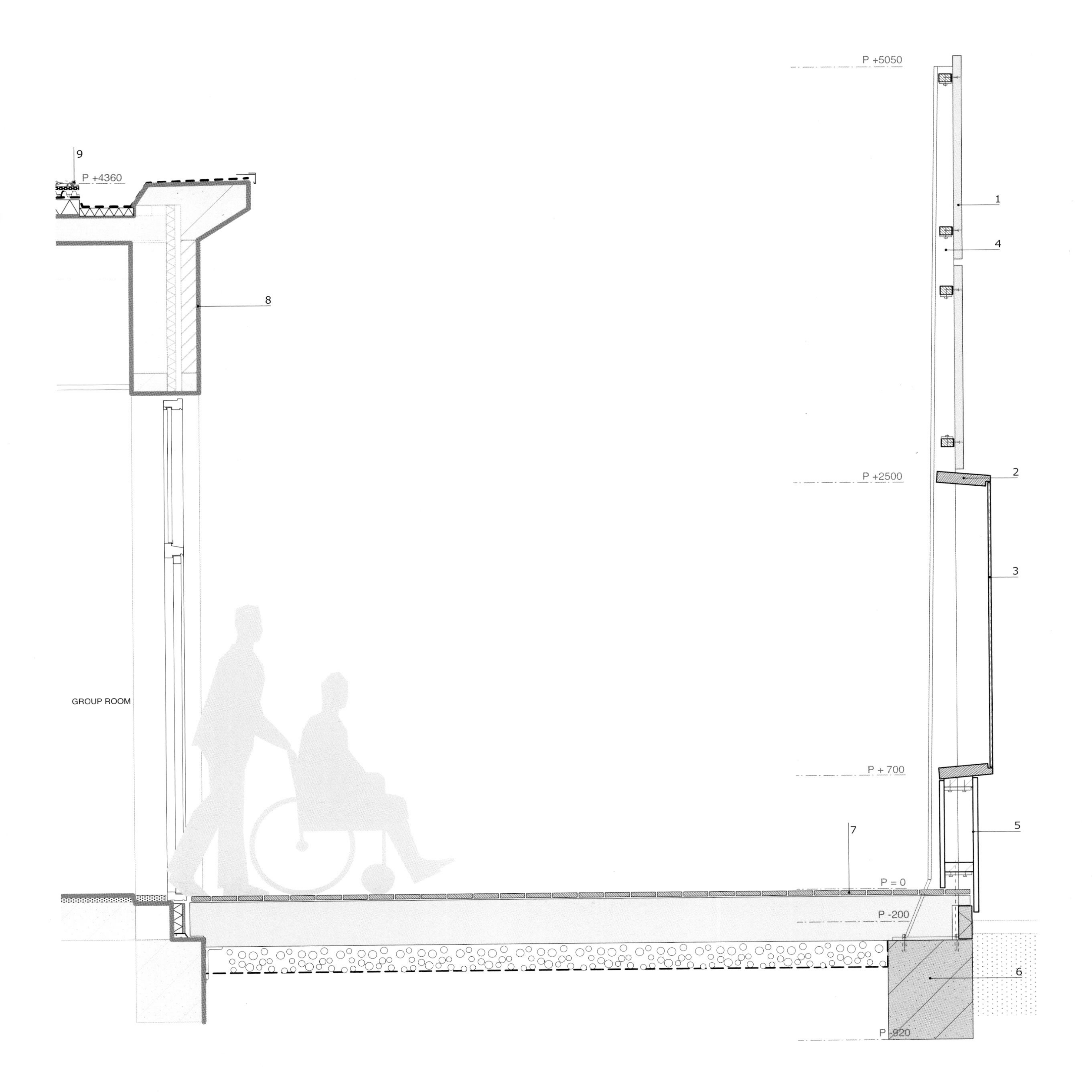

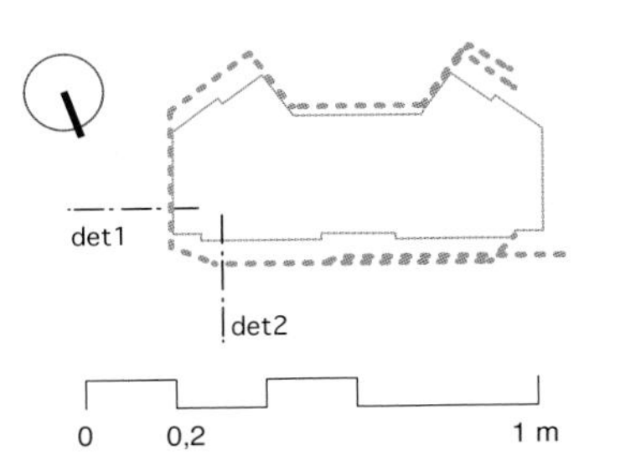

det. 2

1. prefabricated chestnut paling fences
2. accoya frame coated with polyurethane
3. printed glass
4. accoya elements
5. steel stip
6. foundation beam
7. wooden terrace
8. existing building
9. green roof

Azur Arena

Location: Antibes, France
Architect: Auer+Weber+Assoziierte
Client: Ville d'Antibes Juan-les-Pins
Gross Floor Area: 12,120 m^2
Completion Date: 2013
Photographer: Aldo Amoretti

The multifunctional hall is located in "Trois Moulins", the industrial area of Antibes. The site is split into the existing business park in the West and a planned park on the South of the lot. In the North-East you will find a landform shaped by sport infrastructure and rich vegetation. The new Azur Arena becomes the figurehead which works as a regulatory and identity-defining element.

Horizontally mounted bands of expanded metal structure the surfaces. The lines which emerge from these bands are the main organizational principle of the facade and support the central theme of soft, vivid transitions all the way to the roof. At night, the ribbon windows intersect as beams of light on the facade, which animate and represent the dynamism of sport.

The development utilizes the various heights in topography to straighten walkways and builds a connected ring of all floors and access points. The main access is via the upper floor on the East: Spectators make their way to the main courtyard from the parking in the North-East, and from there gain access to the transparent entrance hall using the directly connected, encircling foyer-ring.

This inner section allows entry to all services and leads to the main arena through "vomitories". The athletes enter using an entrance separated from the public in the West on a lower floor. The South-Westerly access and delivery point is also situated on this level; this is part of an encompassing buckle which surrounds the building and offers bypass to vehicles.

Due to the variety of entry-points, large flexibility is obtained which enables the hosting of concurrent events, such as sporting activities in the trampoline and warm-up halls while simultaneously hosting sporting or cultural events in the arena.

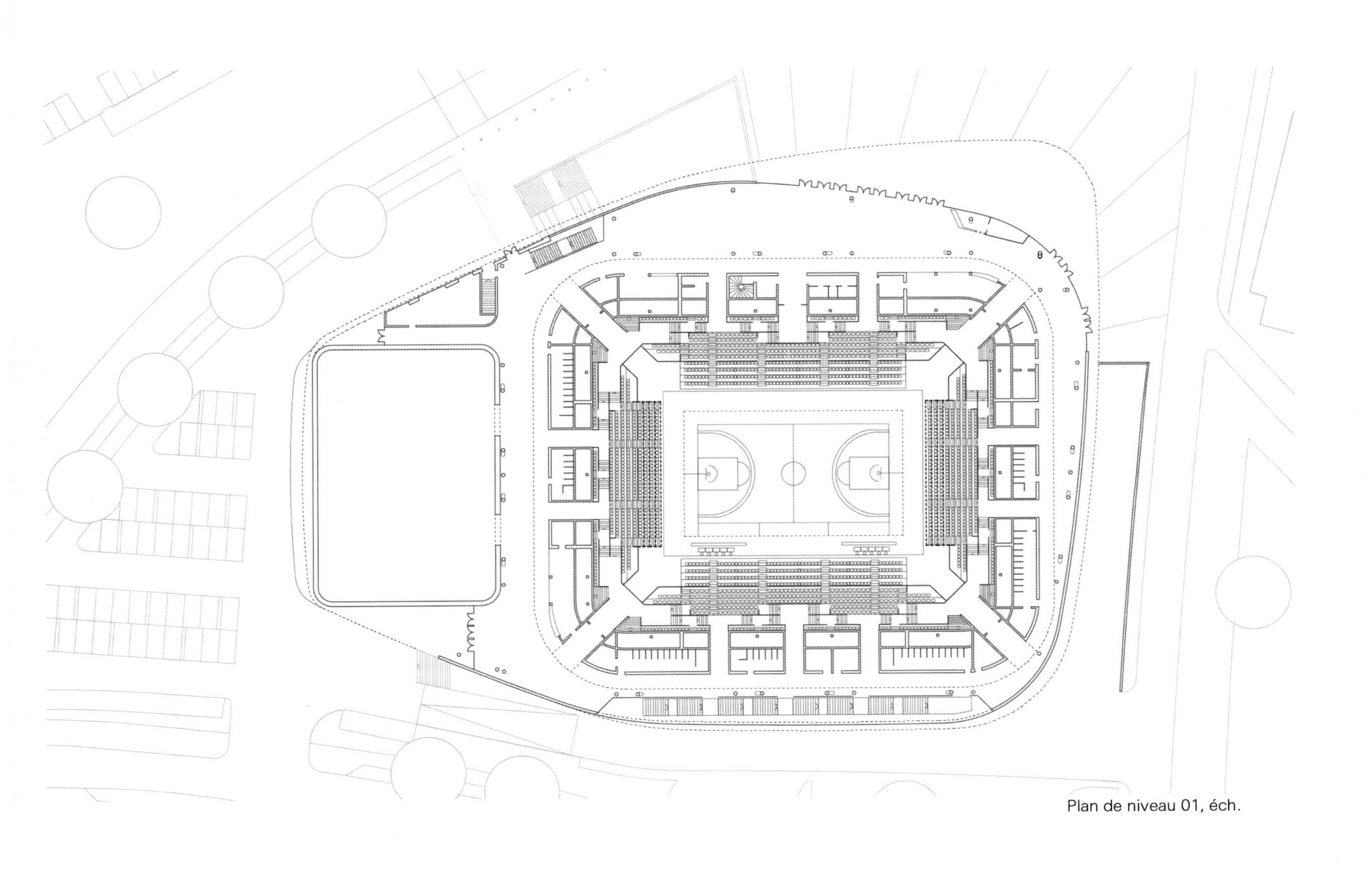

Plan de niveau 01, éch.

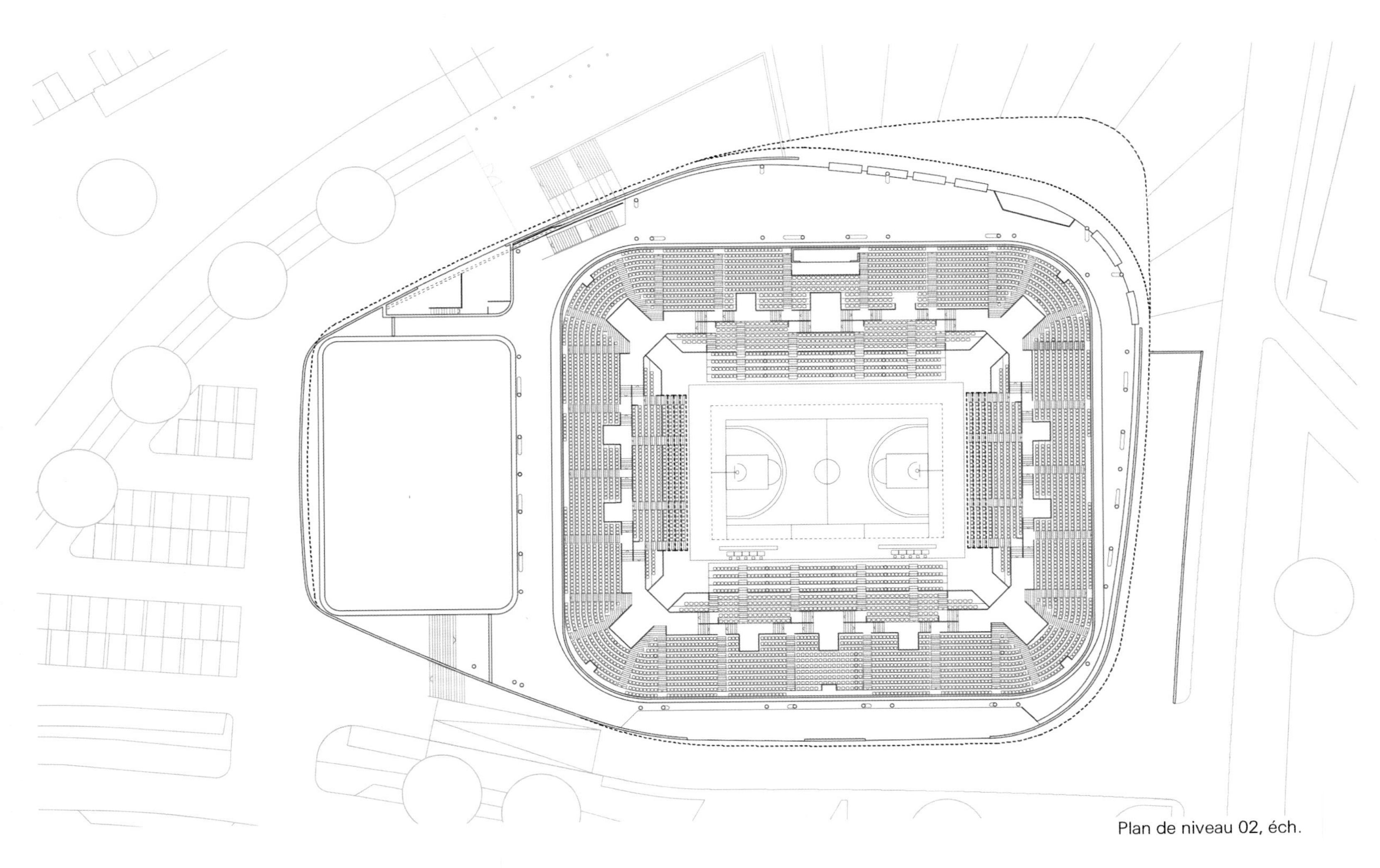

Plan de niveau 02, éch.

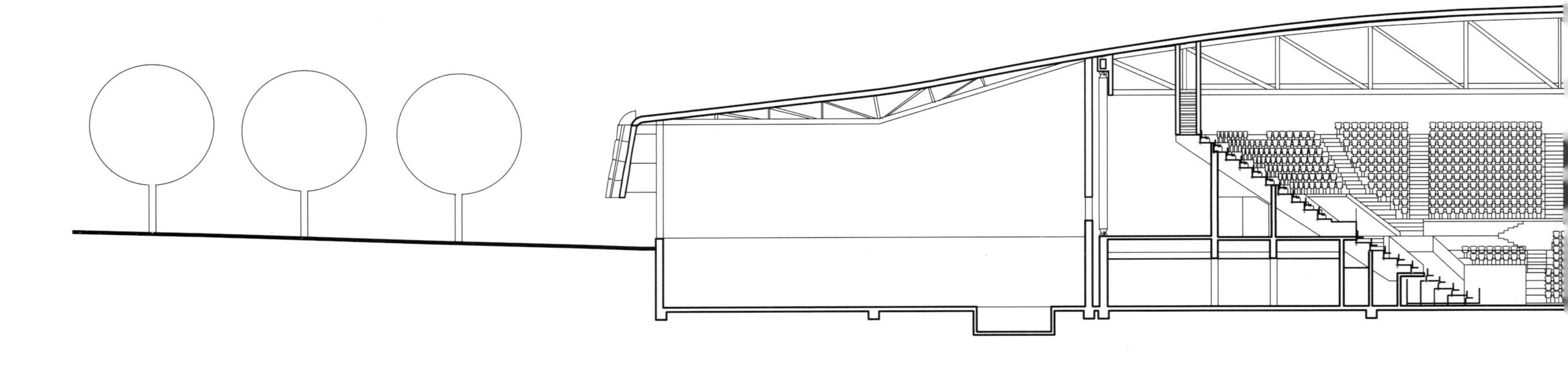

Section

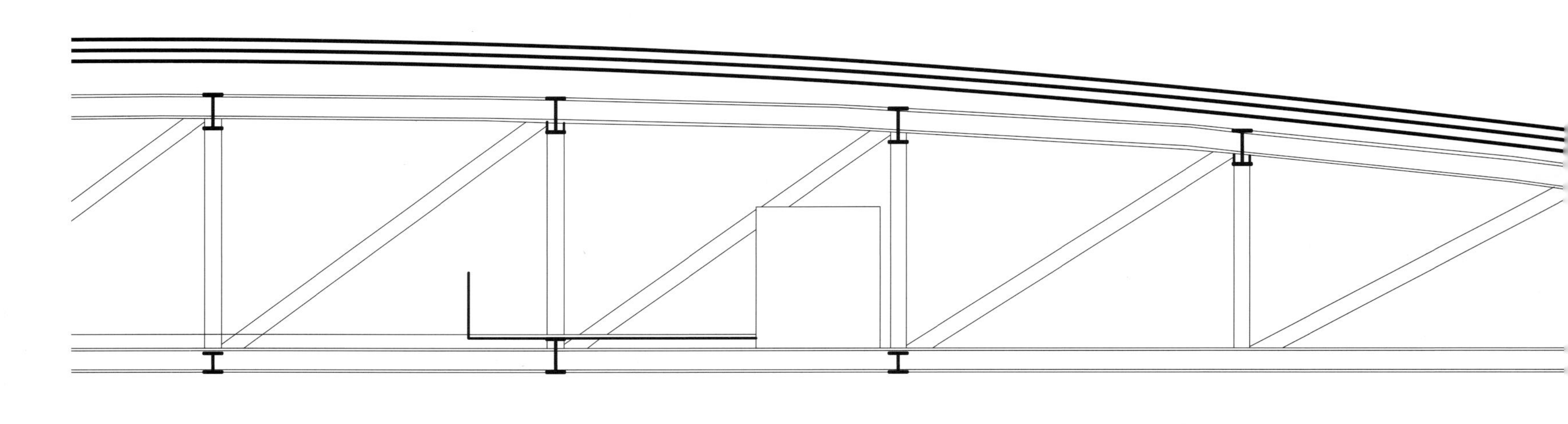

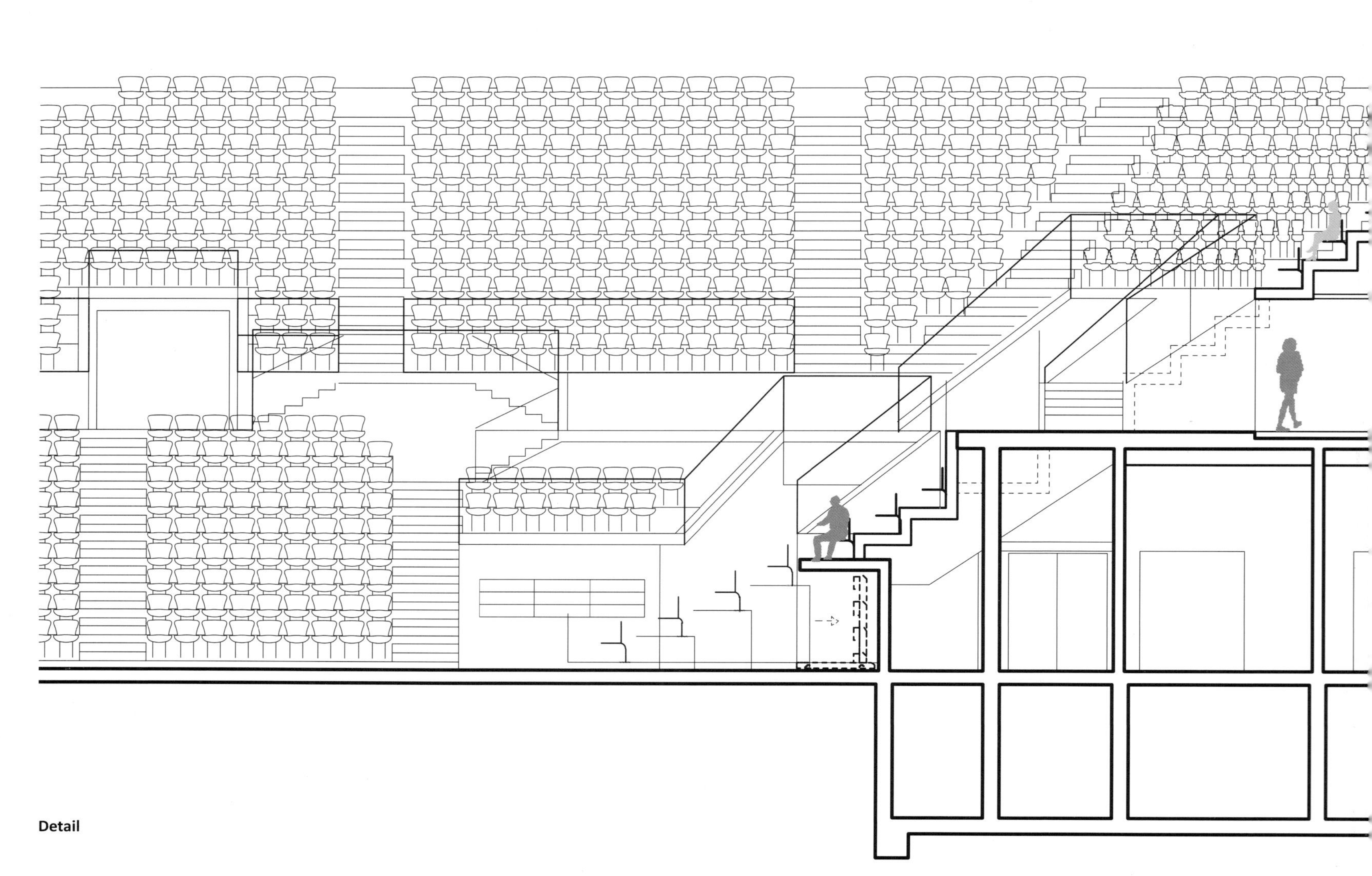

Detail

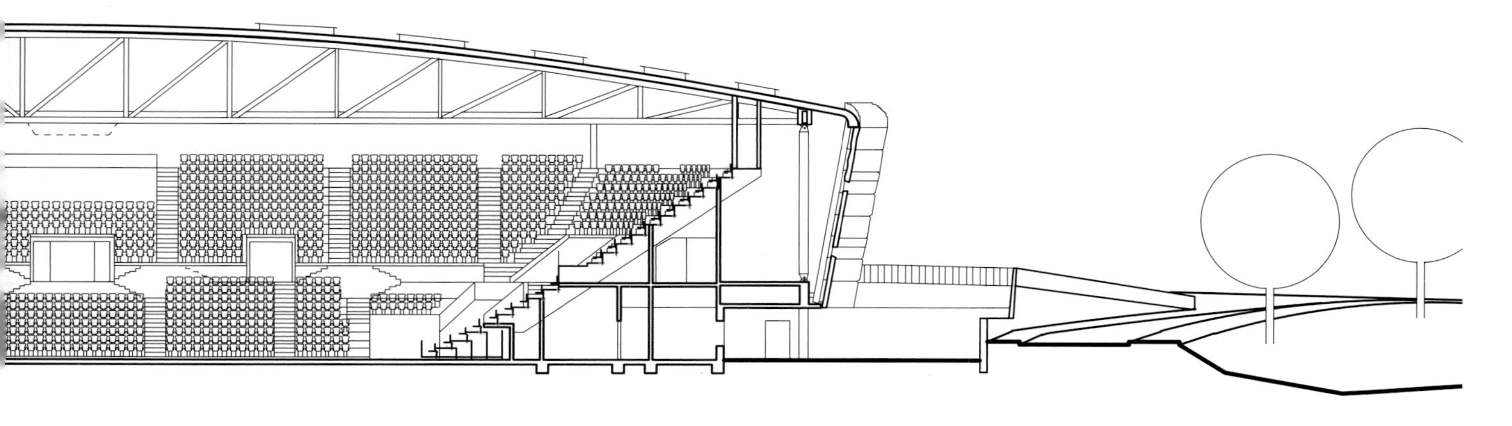

Coupe longitudinale, éch.

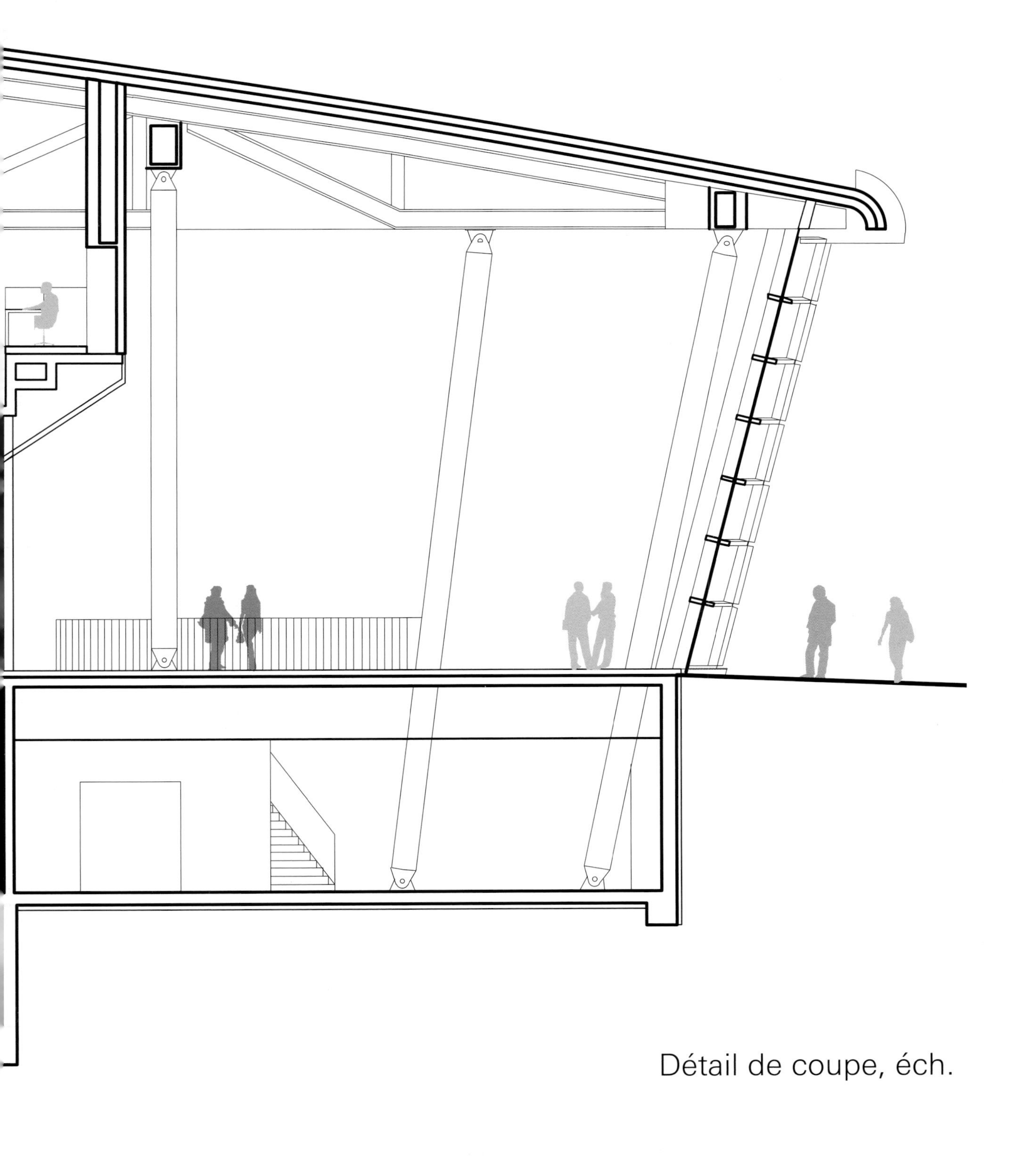

Détail de coupe, éch.

Emergencies and Infections Clinic SUS

Location: Malmö, Sweden
Architect: C.F. Moller Architects & Link Arkitektur
Client: Regionservice Södra Skåne
Size: 26,000 m²
Completion Date: 2011
Photographer: Jorgen True

The circular building housing the casualty and isolation departments at Skåne University Hospital (SUS) in Malmö, Sweden – designed to offer the best possible prevention against the spread of infection – is set to become the hospital's striking new landmark.

Patients enter the isolation ward via a special airlock leading from the top floor corridor that encircles the building. External lifts are provided exclusively for infectious patients and removal of hospital waste – with internal lifts for staff movements and incoming supplies. In addition, individual floors can be subdivided into smaller isolation units in the event of an epidemic.

The new casualty department is responsible for injured and acutely ill, somatic patients, and includes a Children's Emergency Ward, Community Health Care and a Referral Clinic. Treatments are carried out efficiently and safely, conforming to the highest national and international standards. With its facilities for treating all types of acute patients, the casualty department at Skåne University Hospital is the largest and busiest accident and emergency facility in Sweden. The entire casualty department has been planned with optimal logistics in mind – focusing on the patient and on the potential for staff teamwork. This results in fast and functional treatment and care of the highest quality.

The isolation clinic houses a new infectious-diseases department with a total of 51 single-bed rooms and a modern reception facility. This makes the Malmö isolation department Sweden's largest and most modern infectious-diseases facility.

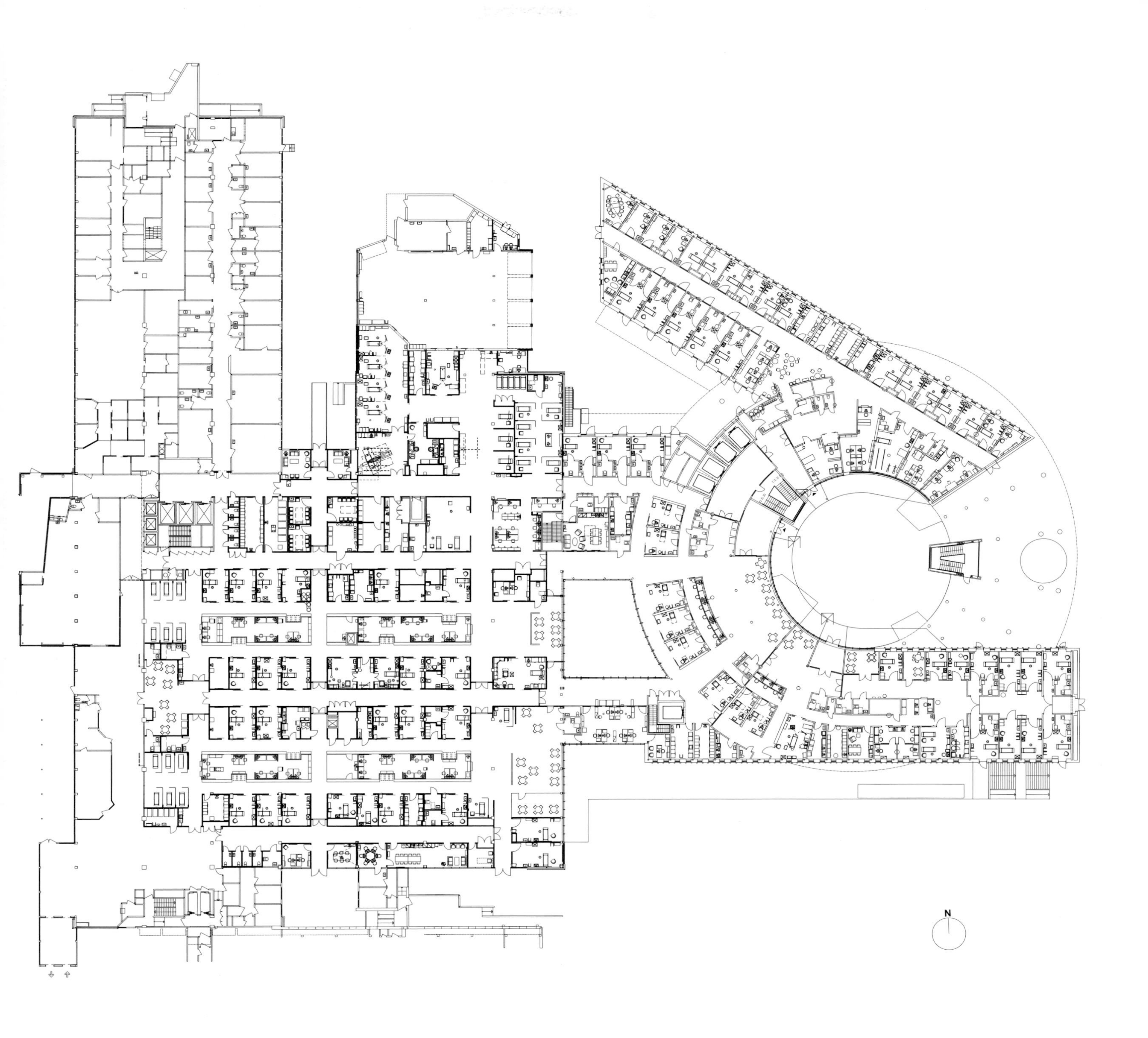

In spite of the need for isolation, the building invites in natural daylight. The wards – which have been laid out as "slices of pie" around the periphery of the building – benefit from direct, natural light. Because the cylindrical building has a central, open atrium, the "inner-circle" rooms also benefit from natural light, giving all rooms in the building direct daylight except the corridors, which – through the glazed wall sections – benefit from indirect natural light from the rooms.

Patients can be wheeled directly into the isolation wards from the outside via a glazed top-floor corridor. Between the top-floor corridor and the ward there is an airlock which maintains the partial vacuum necessary to control ventilation and thereby the risk of the spread of infection. Staffs enter the wards from their work areas in the core of the building through corresponding airlocks.

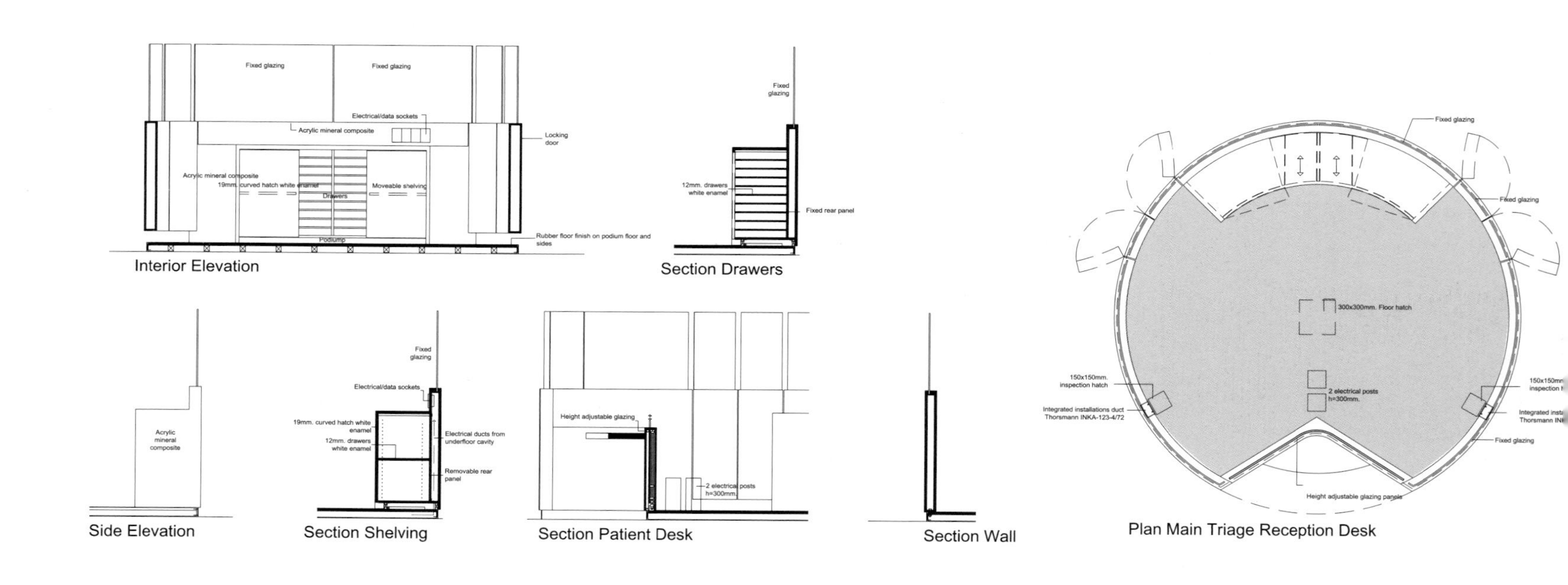

Fixed glazing
Fixed glazing
Electrical/data sockets
Acrylic mineral composite
Locking door
Acrylic mineral composite
19mm. curved hatch white enamel
Drawers
Moveable shelving
Podium
Rubber floor finish on podium floor and sides
Interior Elevation
Fixed glazing
12mm. drawers white enamel
Fixed rear panel
Section Drawers
Fixed glazing
Fixed glazing
300x300mm. Floor hatch
150x150mm. inspection hatch
Integrated installations duct Thorsmann INKA-123-4/72
2 electrical posts h=300mm.
Height adjustable glazing panels
Plan Main Triage Reception Desk
Acrylic mineral composite
Side Elevation
Fixed glazing
Electrical/data sockets
19mm. curved hatch white enamel
12mm. drawers white enamel
Electrical ducts from underfloor cavity
Removable rear panel
Section Shelving
Height adjustable glazing
2 electrical posts h=300mm.
Section Patient Desk
Section Wall

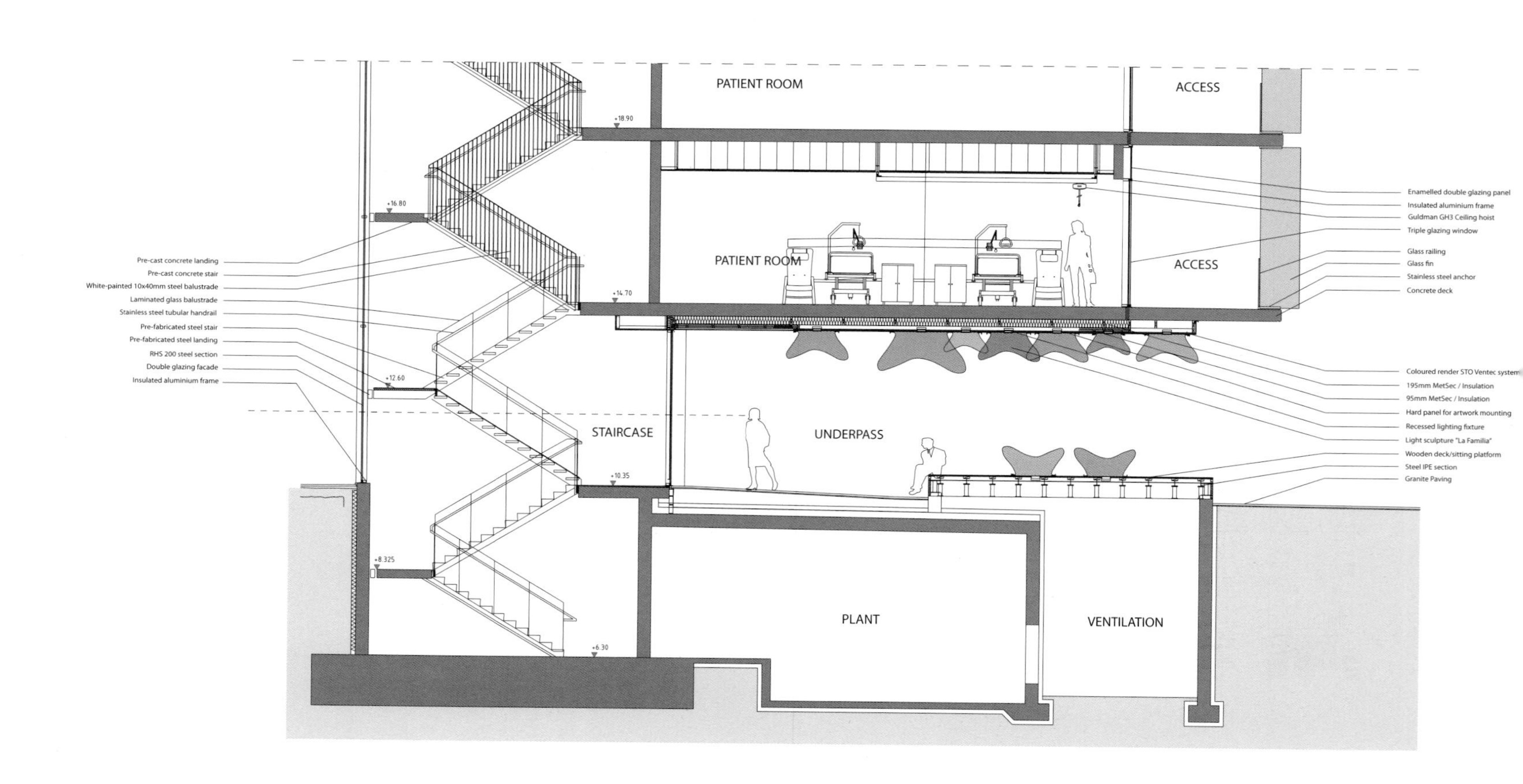

PATIENT ROOM
ACCESS
+18.90
+16.80
PATIENT ROOM
ACCESS
+14.70
+12.60
STAIRCASE
UNDERPASS
+10.35
+8.325
PLANT
VENTILATION
+6.30
Pre-cast concrete landing
Pre-cast concrete stair
White-painted 10x40mm steel balustrade
Laminated glass balustrade
Stainless steel tubular handrail
Pre-fabricated steel stair
Pre-fabricated steel landing
RHS 200 steel section
Double glazing facade
Insulated aluminium frame
Enamelled double glazing panel
Insulated aluminium frame
Guldman GH3 Ceiling hoist
Triple glazing window
Glass railing
Glass fin
Stainless steel anchor
Concrete deck
Coloured render STO Ventec system
195mm MetSec / Insulation
95mm MetSec / Insulation
Hard panel for artwork mounting
Recessed lighting fixture
Light sculpture "La Familia"
Wooden deck/sitting platform
Steel IPE section
Granite Paving

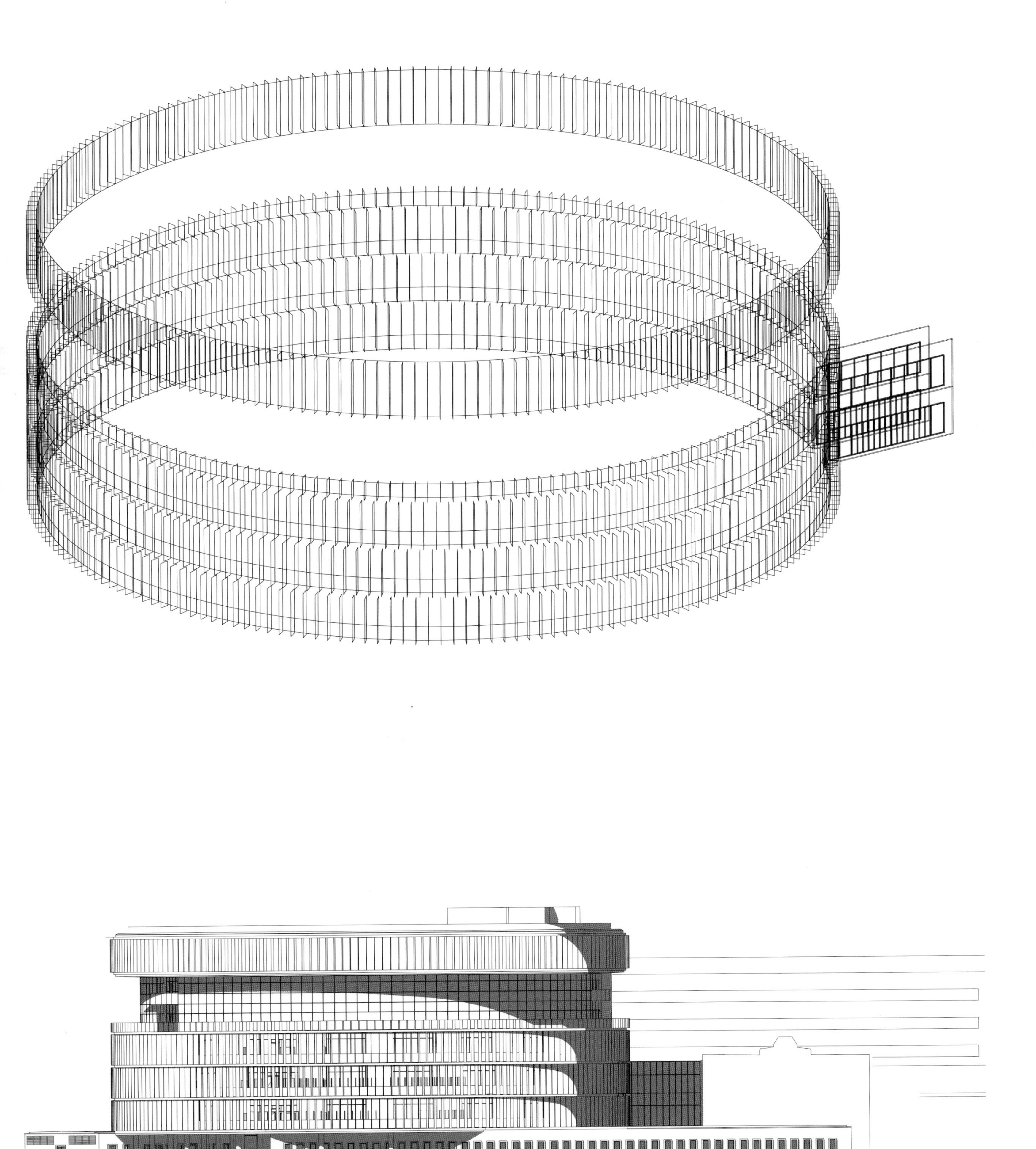

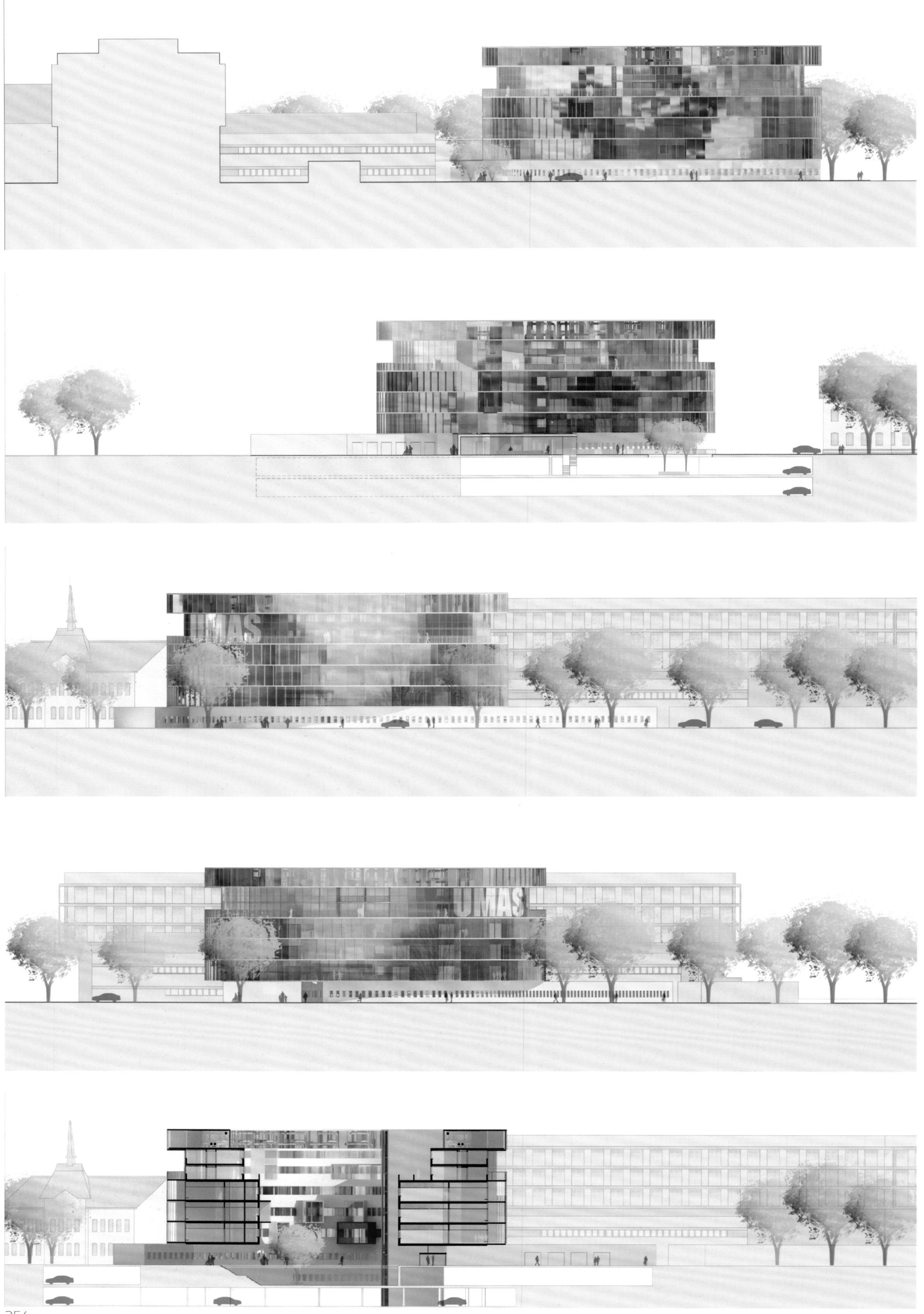
UMAS
UMAS

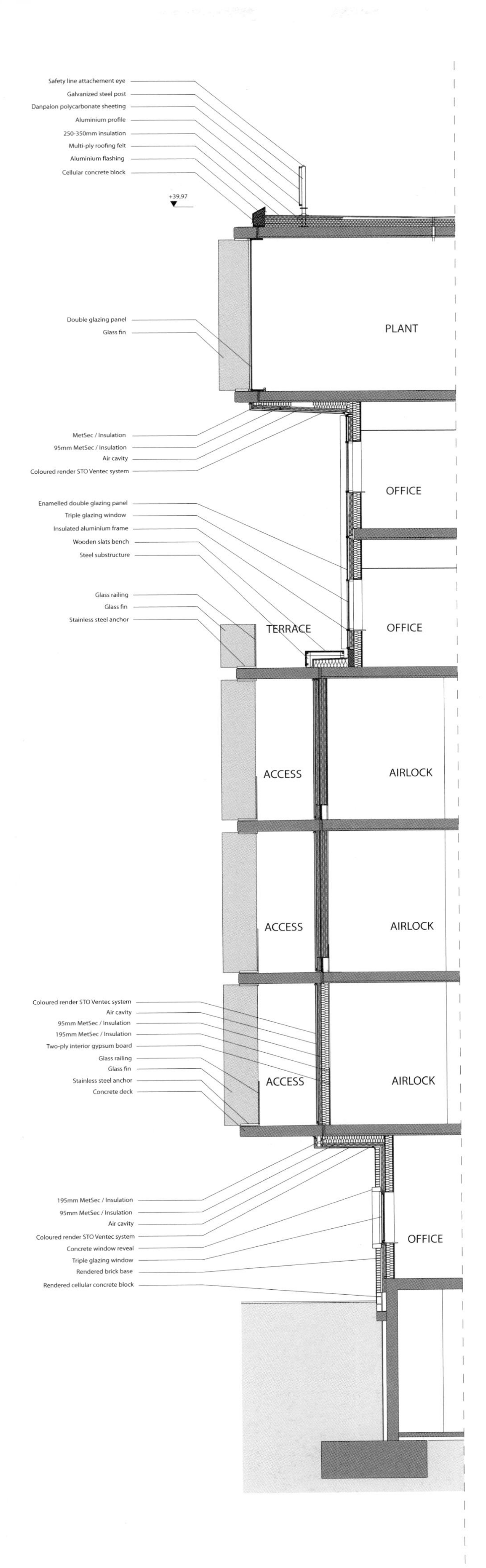

TYPICAL FACADE SECTION 1:100 @ A3

San Mames Stadium

Location: Bilbao, Spain
Architect: César A. Azcárate Gómez (ACXT-IDOM)
Client: San Mames Barria
Capacity: 53,500 Spectators
Completion Date: 2014
Photographer: Aitor Ortiz

One of the main challenges in the design of the New San Mames was maintaining the intense and magical football atmosphere of the old Cathedral. This effect has not only been sustained but increased, thoroughly satisfying the demands of one of the best fan bases in the world.

It was intended for those stadium areas that are traditionally worthless to become valuable. These are located between the stadium's perimeter and the rear part of the stands and constitute the circulation areas through which you can access and exit the stands, which are, after all, the main part of the whole football stadium. In order to give these areas an added value, the strategy of the project consisted of, not only giving them spatial features, but also making sure that they had a very intense connection with the city and the surroundings. For this purpose, a basic element that will surely give character to the New San Mames stadium is put into play on the façade. This is, the repetition of a twisted ETFE element, giving the elevation energy and unity. This element will be illuminated at night, thus creating an urban landmark over the estuary, projecting a new image of Bilbao from within, thanks to one of the most advanced dynamic lighting systems in the world. The roof, formed by powerful radial metal trusses orientated towards the centre of the pitch, is covered with white ETFE cushions, covering the entire stands. The set-up of the stands is totally focused on the field, maximizing the pressure that the fans exert on the game, just like in the old San Mames, known the world over for being like a pressure cooker where the public would be on top of the players.

The stadium has ample hospitality areas, with VIP boxes, premium seating and its leisure and meeting areas, restaurants, cafes, the Club's Museum, the Official Shop and areas for meetings, as well as a sports centre open to the general public under one of its stands. Its capacity will exceed 53,000 spectators.

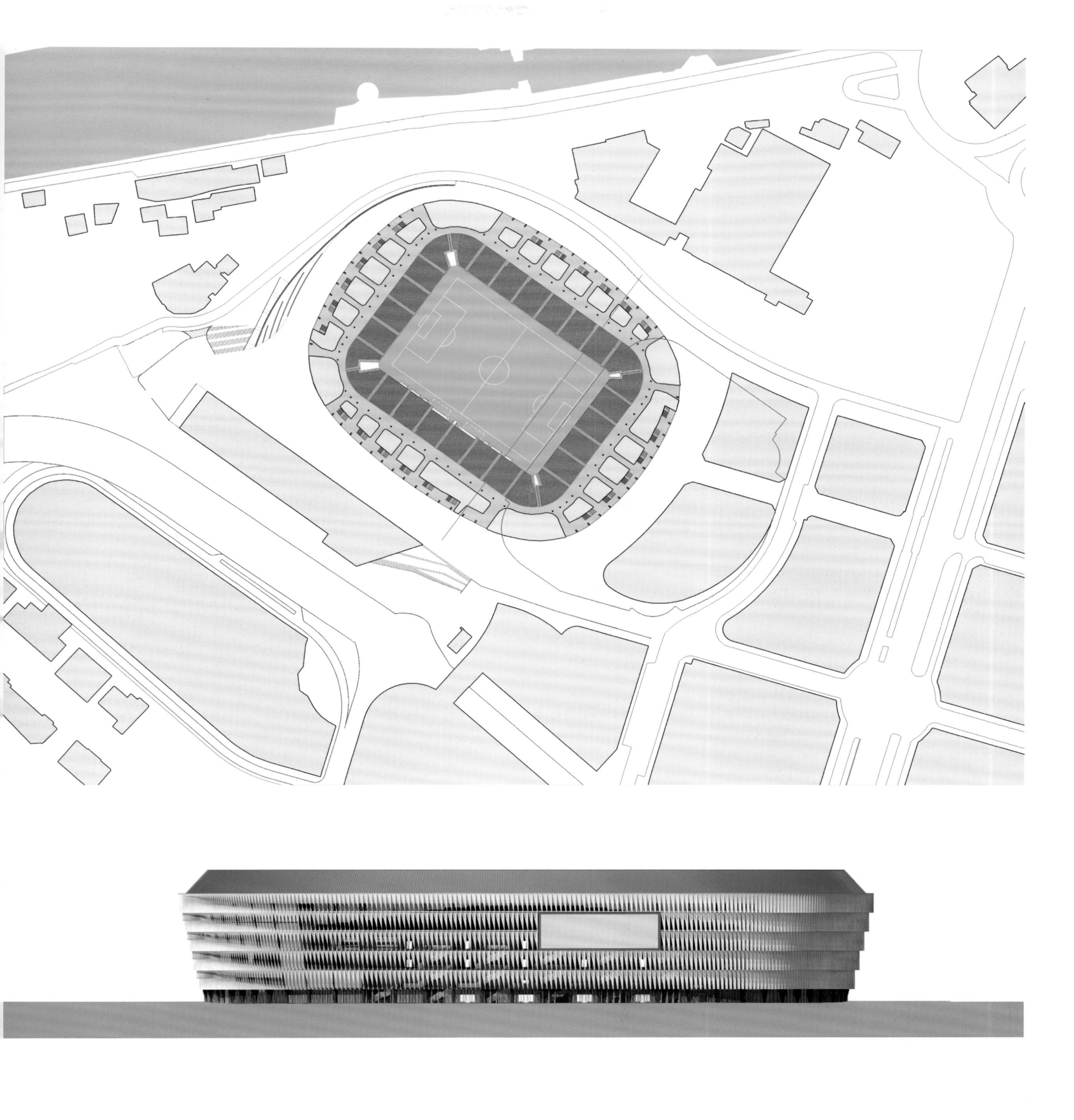

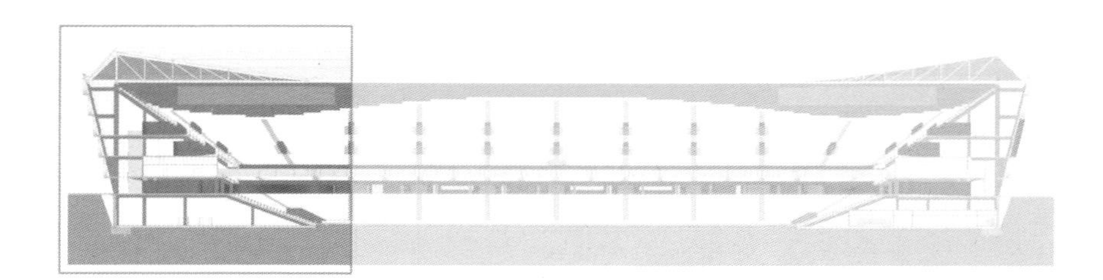

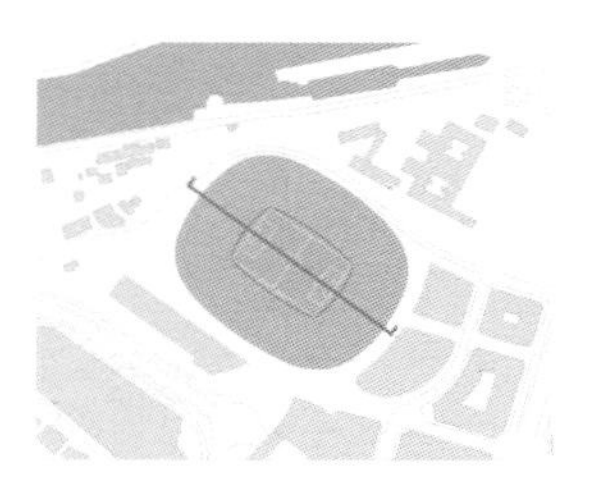

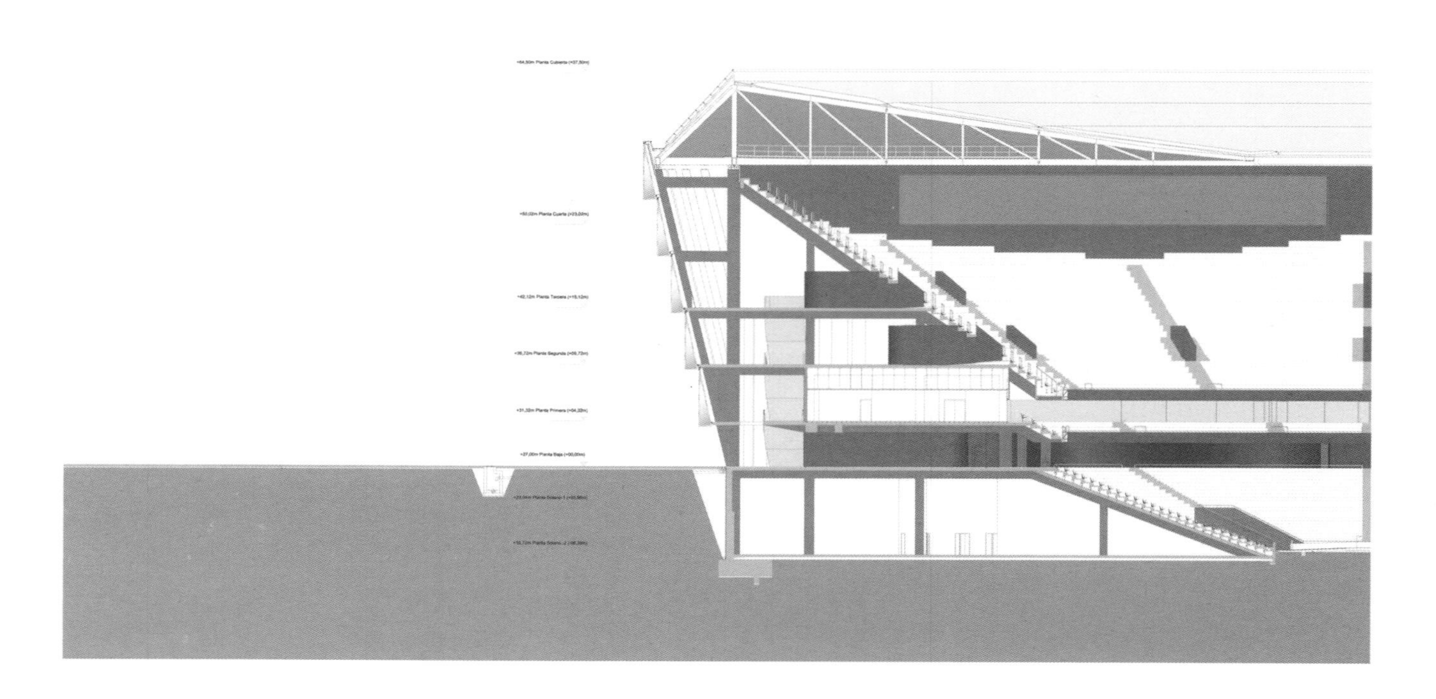

+64,50m Planta Cubierta (+37,50m)
+50,02m Planta Cuarta (+23,02m)
+42,12m Planta Tercera (+15,12m)
+36,72m Planta Segunda (+09,72m)
+31,32m Planta Primera (+04,32m)
+27,00m Planta Baja (+00,00m)

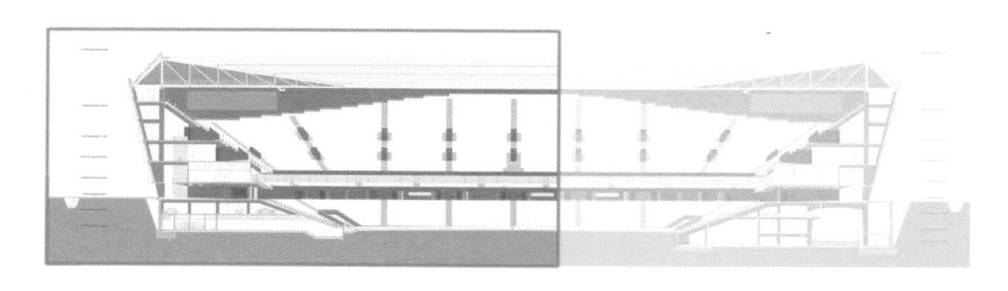

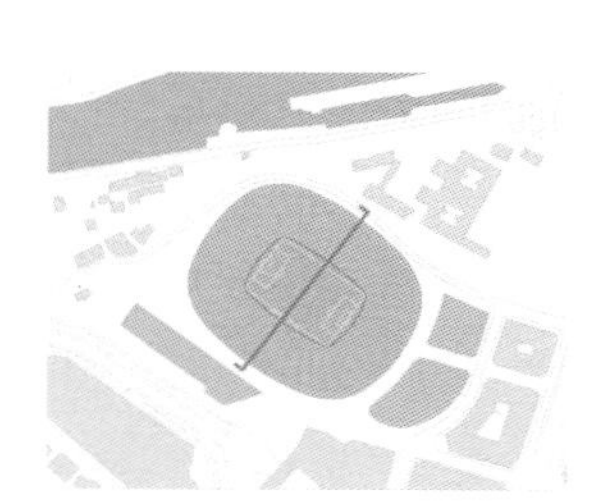

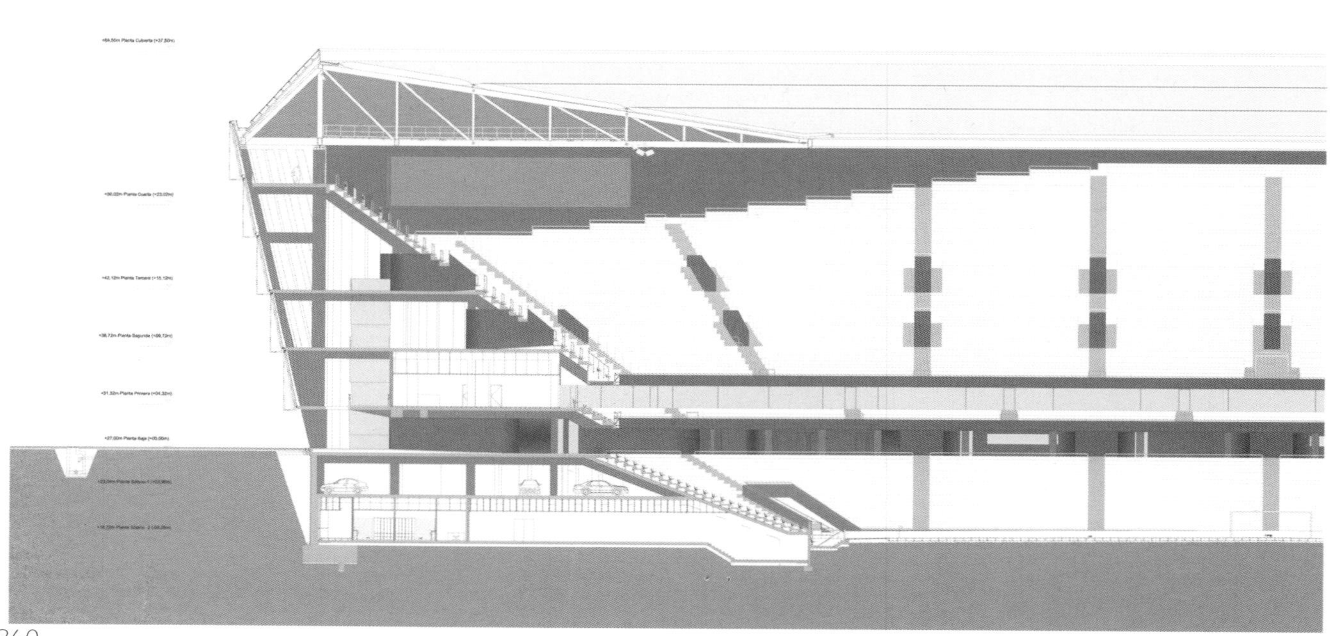

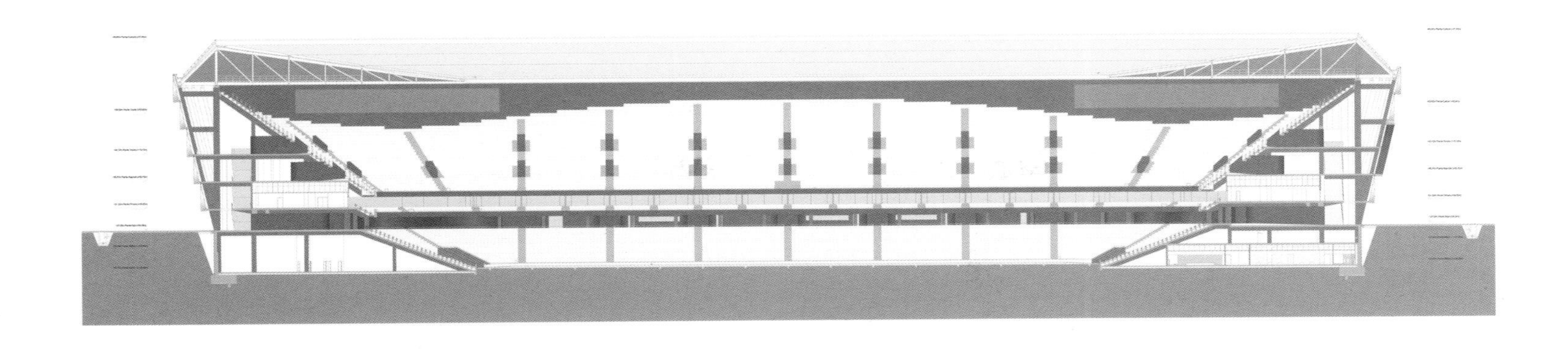

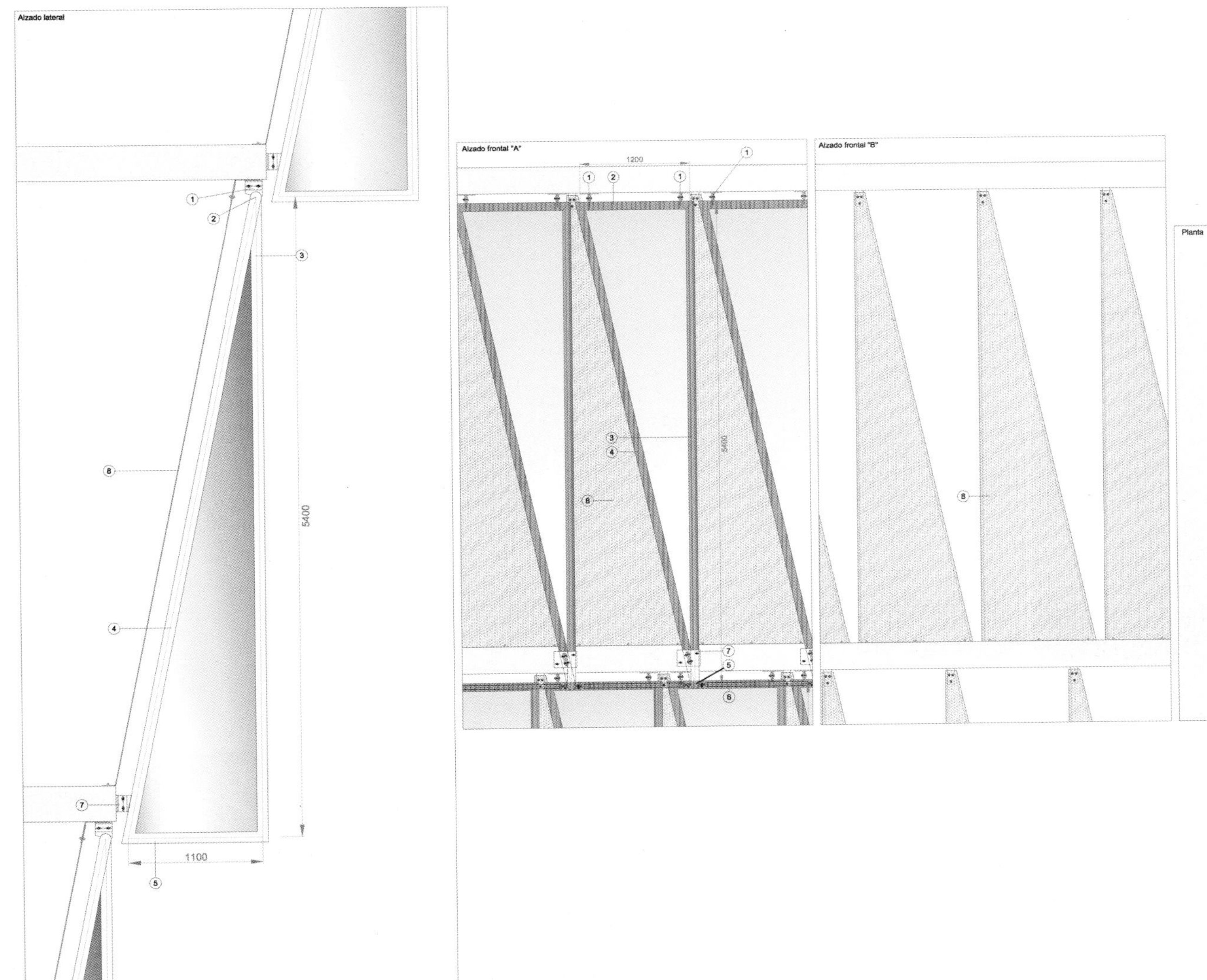

1- Sujeción superior.
2- Tubo Ø90x5 (lado superior)
3- Tubo Ø90x5 (lado largo aprox. vertical).
4- Tubo Ø90x5 (lado largo diagonal).
5- Tubo Ø90x5 (lado inferior).
6- Tirante tracción-compresión.
7- Sujeción inferior.
8- Chapa perforada e=2mm

Sustainable Hothouse

Location: Botanical Gardens, Aarhus, Denmark
Architect: C.F. Moller Architects
Client: Aarhus University and Danish University & Property Agency
Size: 3,300 m^2
Completion Date: 2013
Photographer: Julian Weyer; Quintin Lake

The snail-shaped hothouse in the Botanic Garden in Aarhus is a national icon in hothouse architecture. It was designed in 1969 by C.F. Moller Architects, and is well adapted to its surroundings. Accordingly, it was important to bear the existing architectural values in mind when designing a new hothouse to replace the former palm house which had been literally outgrown.

The design of the new hothouse is based on energy-conserving design solutions and on knowledge of materials, indoor climate and technology.

Using advanced calculations, the architects and engineers have optimized their way to the building's structure, ensuring that its form and energy consumption interact in the best possible manner and make optimal use of sunlight. The domed shape and the building's orientation in relation to the points of the compass have been chosen because this precise format gives the smallest surface area coupled with the largest volume, as well as the best possible sunlight incidence in winter, and the least possible in summer.

The transparent dome is clad with ETFE foil cushions with an interior pneumatic shading system. The support structure consists of 10 steel arches, which fan out around a longitudinal and a transverse axis, creating a net of rectangles of varying sizes. On the south-facing side, the cushions used were made with three layers, two of which were printed. Through changes in pressure, the relative positions of these printed foils can be adjusted. This can reduce or increase, as desired, the translucence of the cushions, changing the light and heat input of the building.

Floor Plan

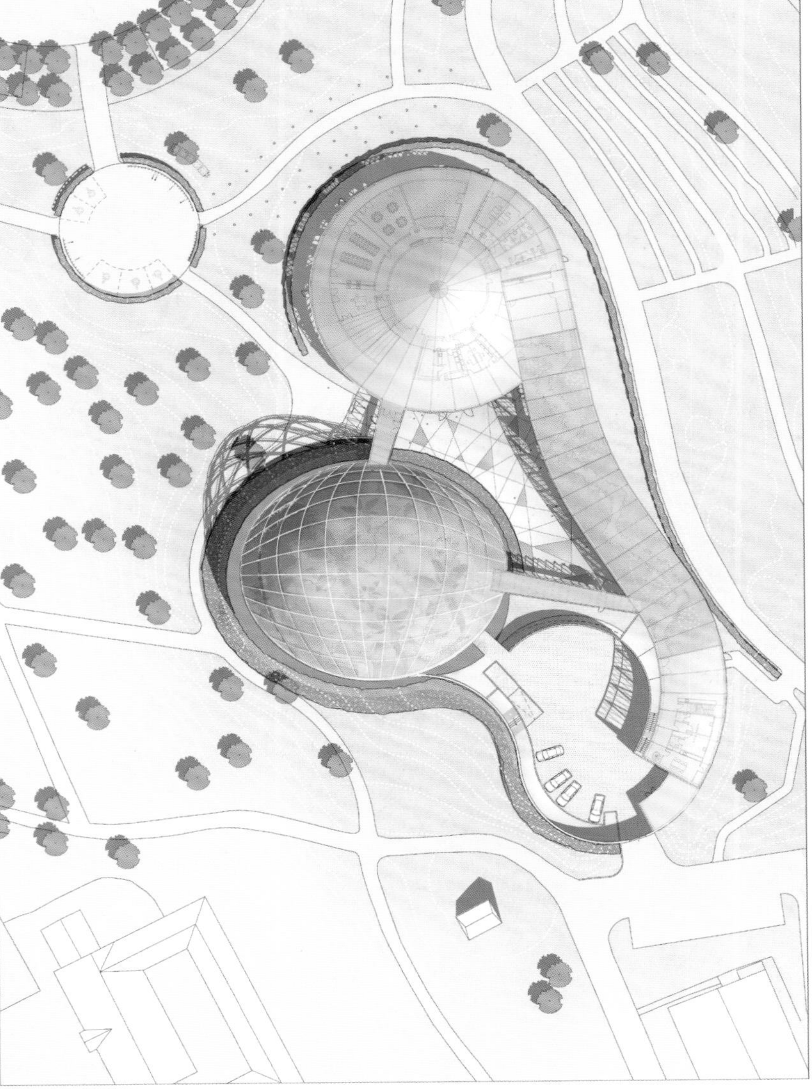

Roof Plan

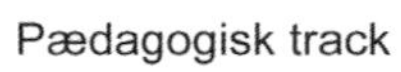
Pædagogisk track

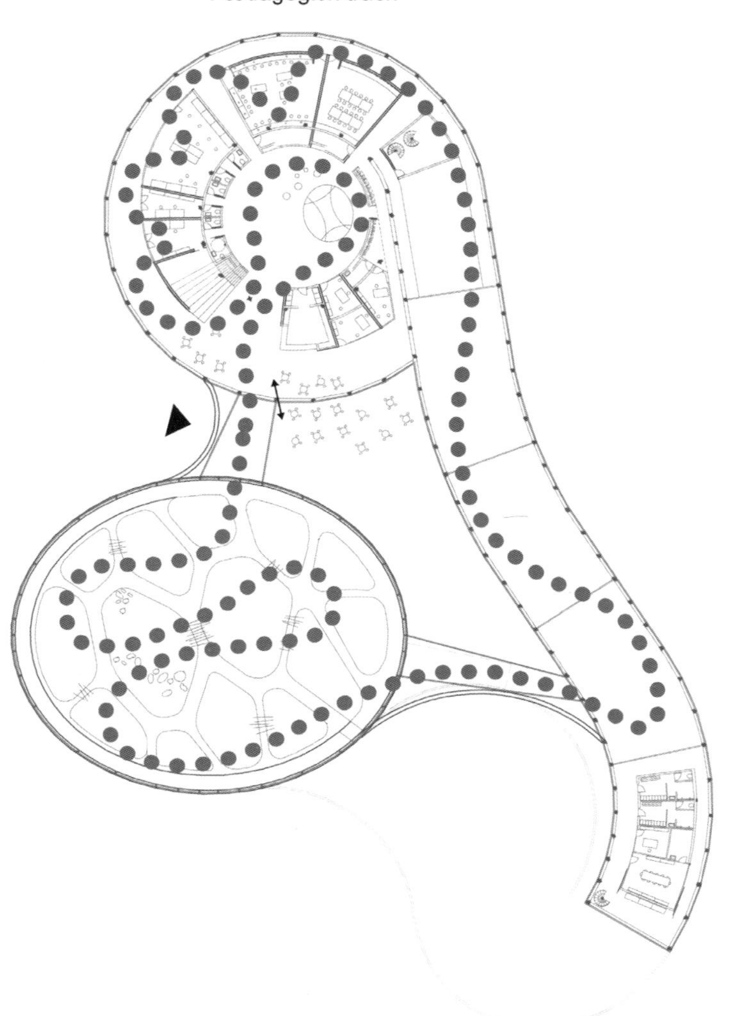

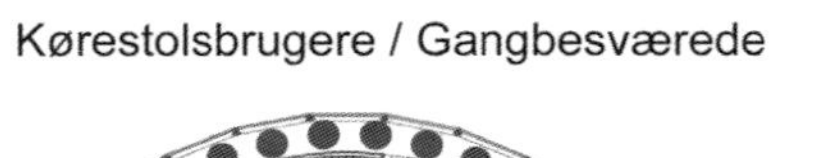
Kørestolsbrugere / Gangbesværede

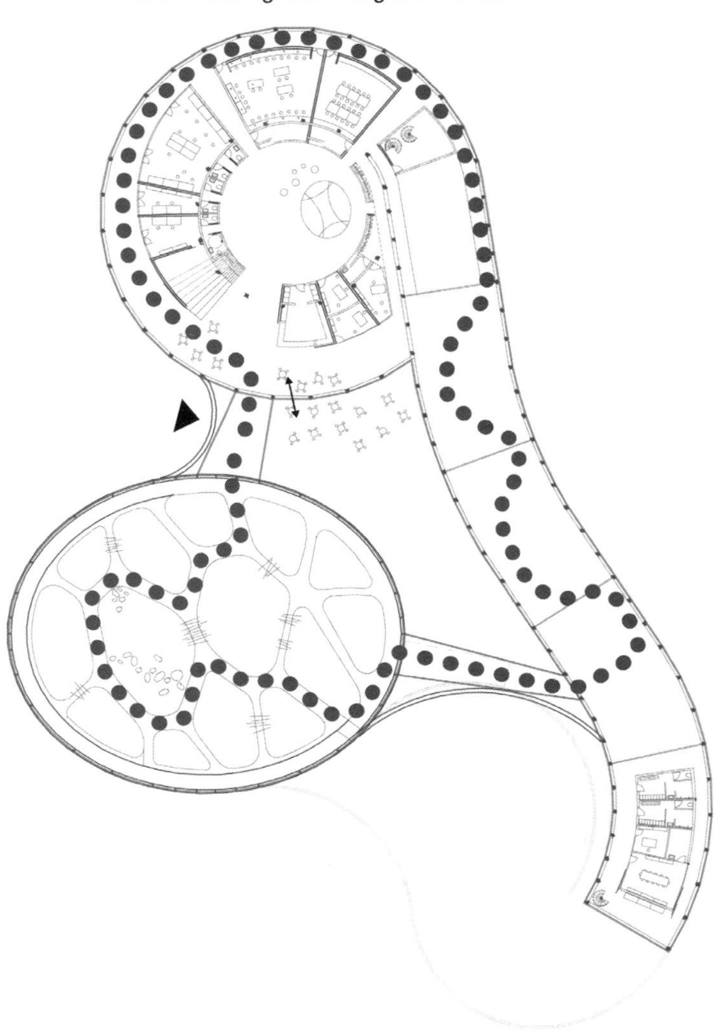

"Fast" track

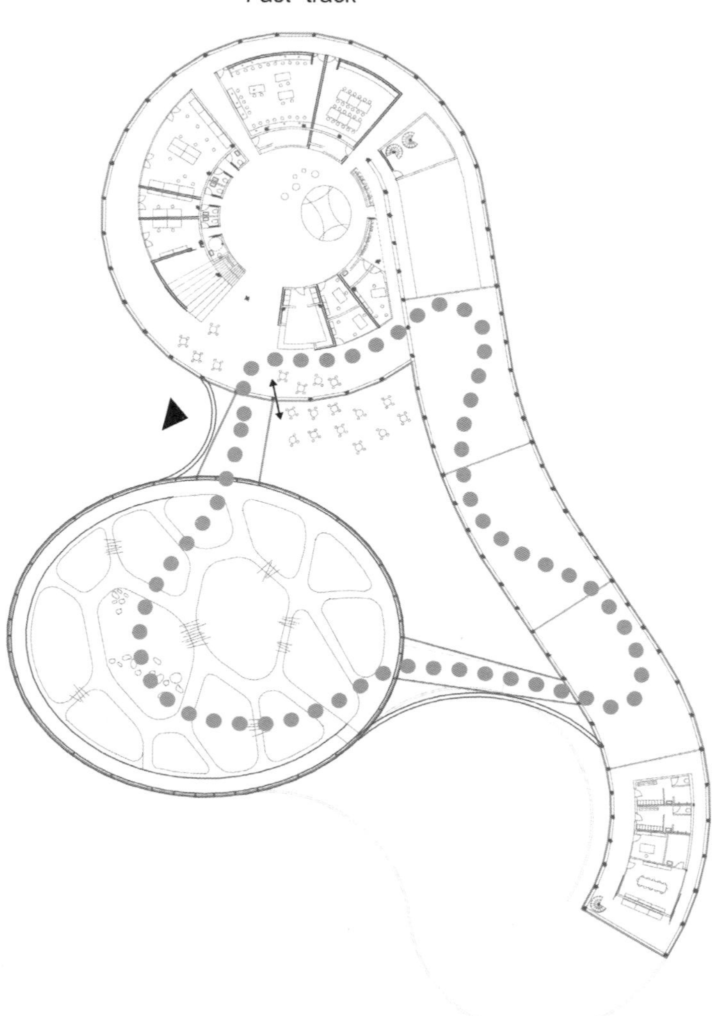

Service track

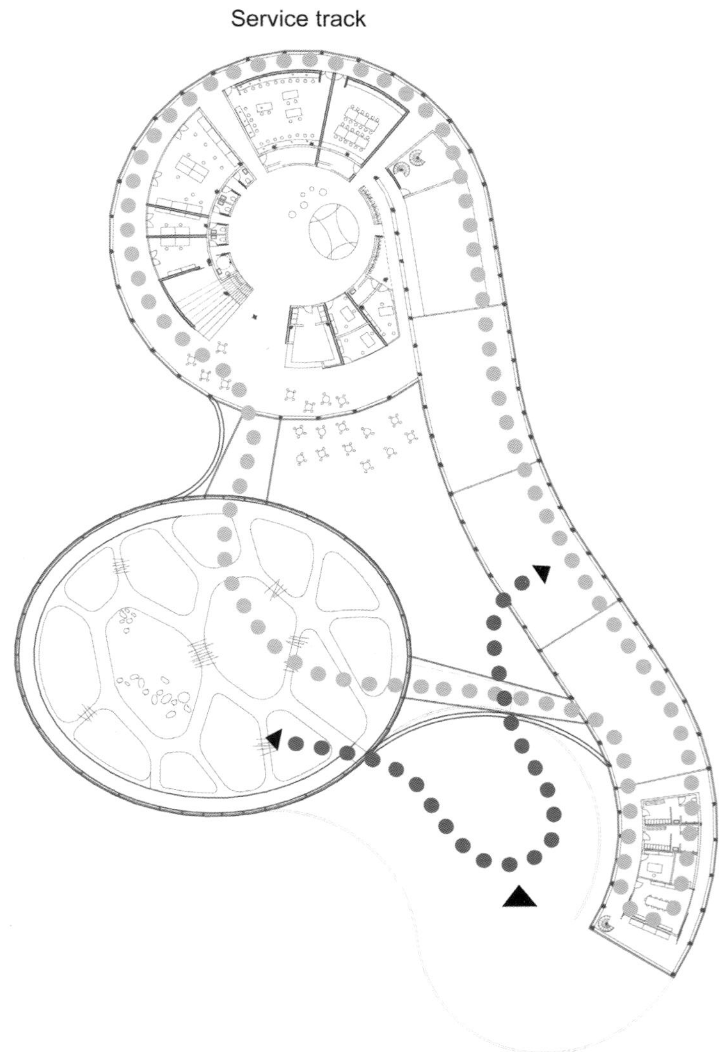

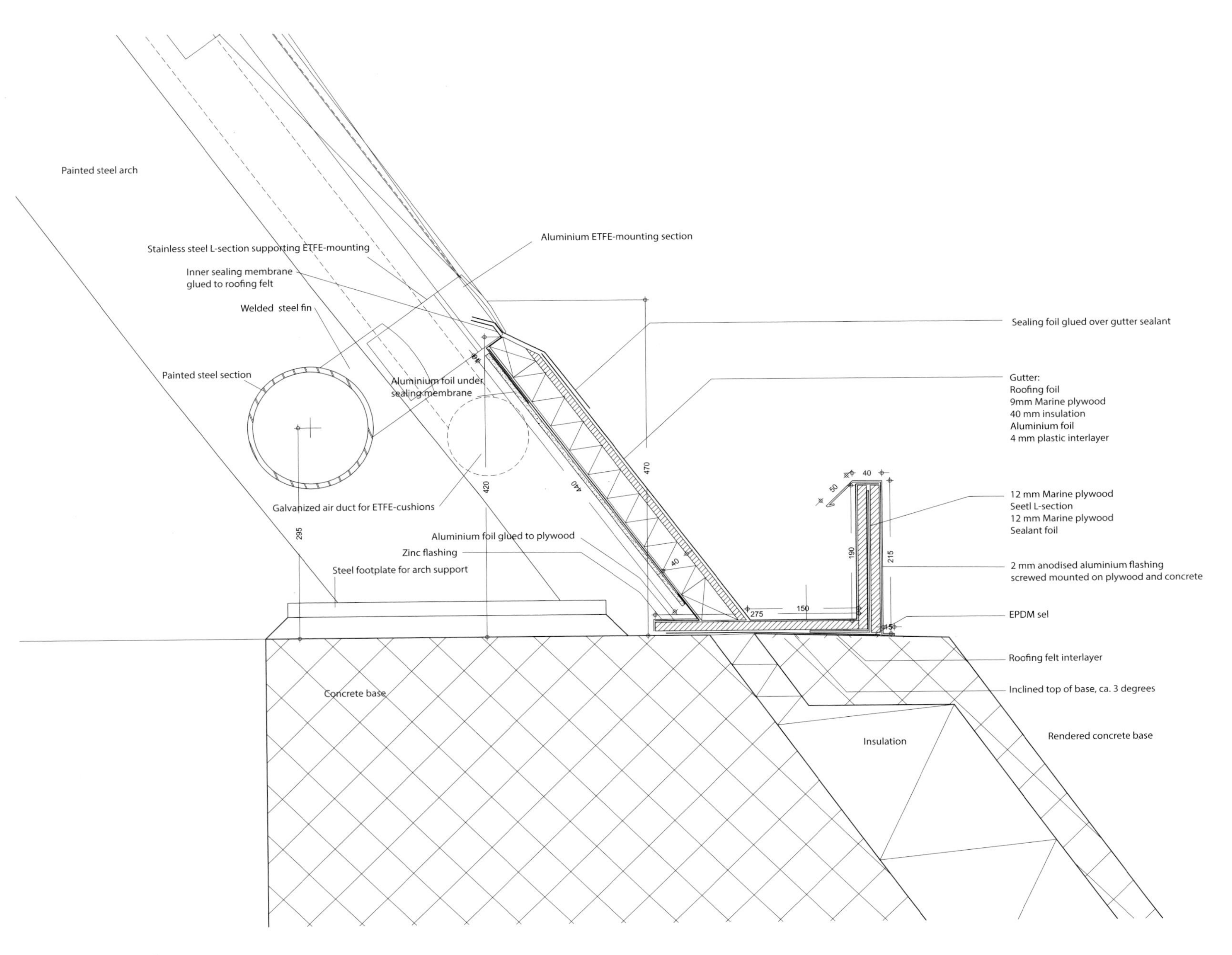

Facade Section Detail

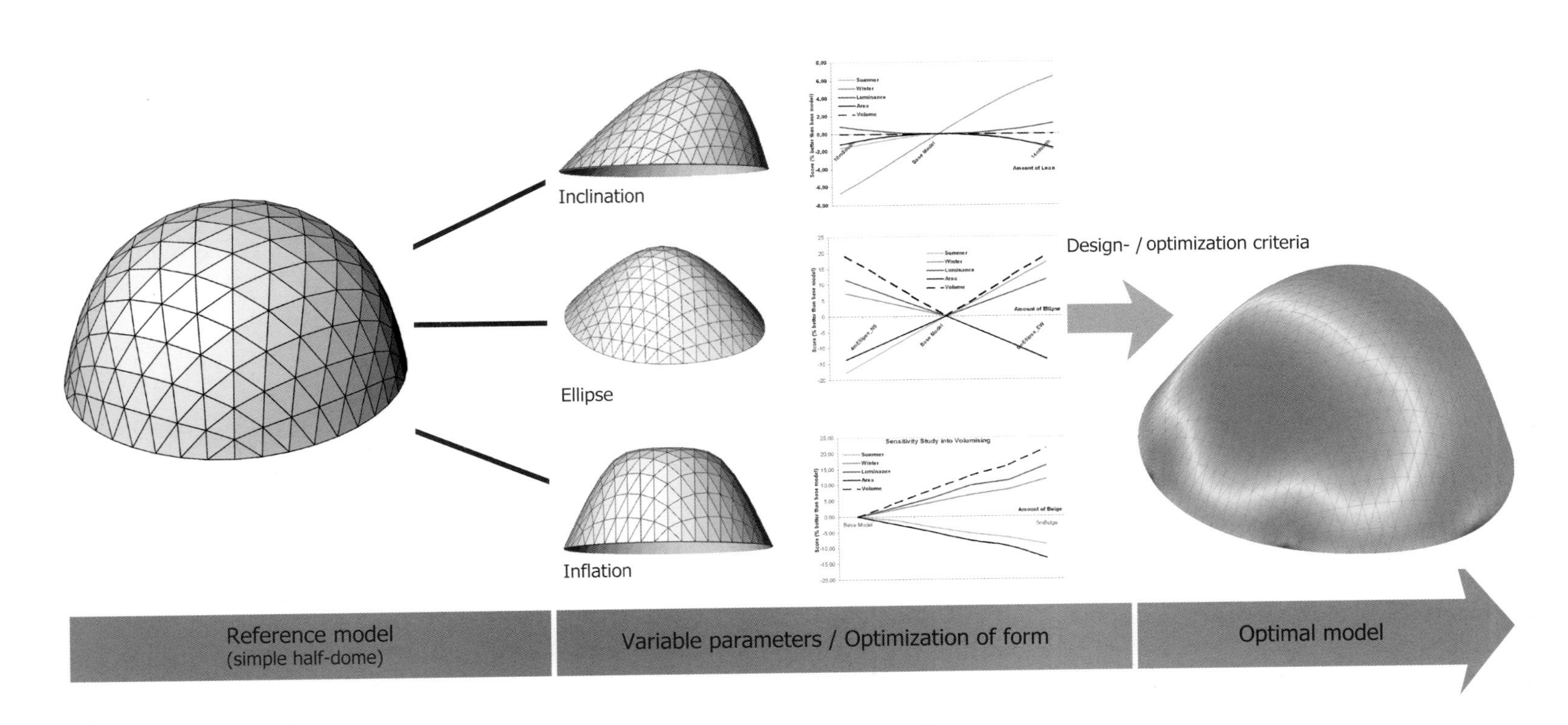

London Aquatics Centre

Location: London,United Kingdom
Architect: Zaha Hadid Architects
Client: Olympic Delivery Authority
Completion Date: 2014
Site Area: 36,875 m²
Photography: Hélène Binet, Hufton + Crow

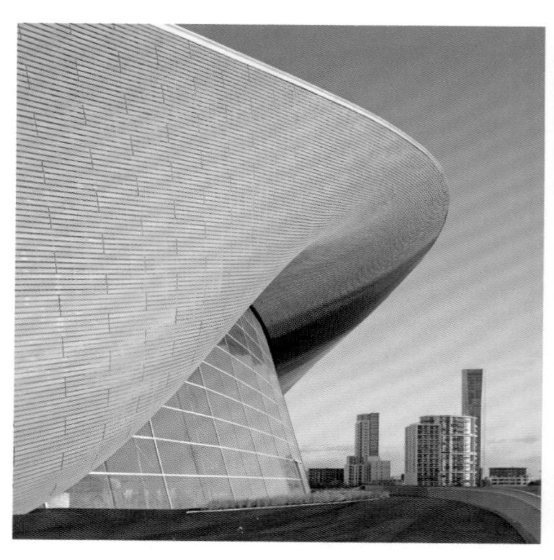

The architectural concept of the London Aquatics Centre is inspired by the fluid geometry of water in motion, creating spaces and a surrounding environment in sympathy with the river landscape of the Olympic Park. An undulating roof sweeps up from the ground as a wave – enclosing the pools of the Centre with its unifying gesture of fluidity, whilst also describing the volume of the swimming and diving pools.

The Aquatics Centre is planned on an orthogonal axis perpendicular to the Stratford City Bridge. Along this axis are laid out the three pools. The training pool is located under the bridge whilst the competition and diving pools are within a large volumetric pool hall. The overall strategy is to frame the base of the pool hall as a podium by surrounding it and connecting it into the bridge.

This podium element allows for the containment of a variety of differentiated and cellular programmatic elements into a single architectural volume which is seen to be completely assimilated with the bridge and the landscape. The podium emerges from the bridge to cascade around the pool hall to the lower level of the canal.

The pool hall is expressed above the podium level by a large roof which arches along the same axis as the pools. Its form is generated by the sightlines for the spectators during the Olympic mode. Double-curvature geometry has been used to create a structure of parabolic arches that define its form. The roof undulates to differentiate the volumes of the competition and diving pools, and extends beyond the pool hall envelope to cover the external areas of the podium and entrance on the bridge.

The roof structure is grounded at three points of the centre (two points at the northwest end on the bridge; and one single point to the south east end). This structural arrangement ensured 7,500 temporary spectator seats could be installed along either side of the pools in Olympic mode (total 15,000 temporary seats) with no structural obstructions. After the 2012 Olympic and Paralympic Games, this temporary seating has been removed and replaced with glazing panels, leaving a capacity of 2,500 seats for community use and future national/international events, with a significantly reduced pool hall volume.

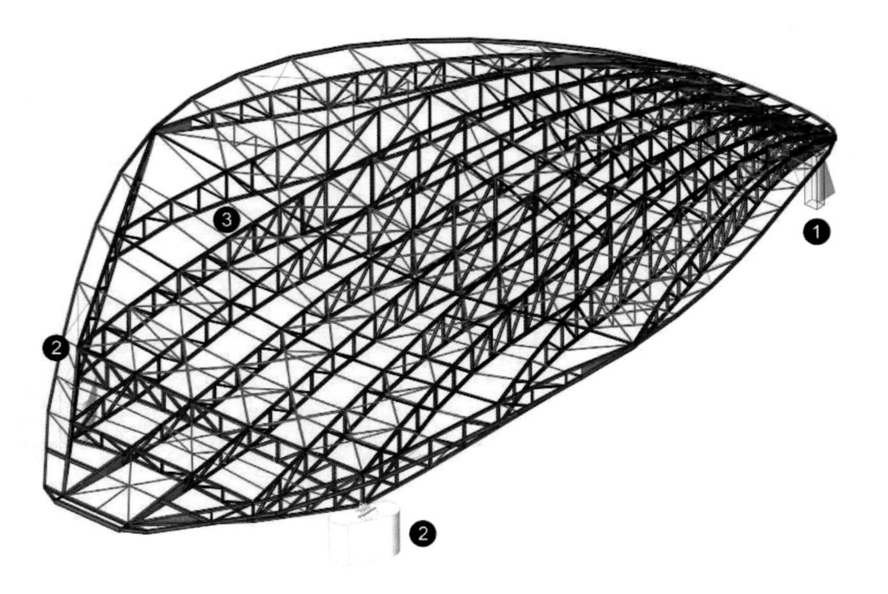

How the roof structure works

The whole roof structure is supported on just three points: a wall at the south end **(1)** and two concrete cores to the north **(2)**.

Despite its complex shape the roof is made up from relatively simple two dimensional elements.

The fan trusses **(3)** run in a north-south direction and are shaped to clear the diving and competition pools.

The trusses incline outwards from the centre like a fan, the two outer trusses **(4)** act as inclined tied arches which create two cantilevered wings on either side of the building for the temporary seating.

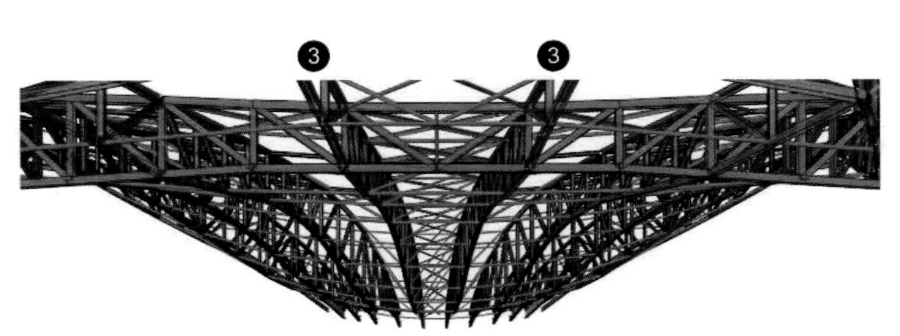

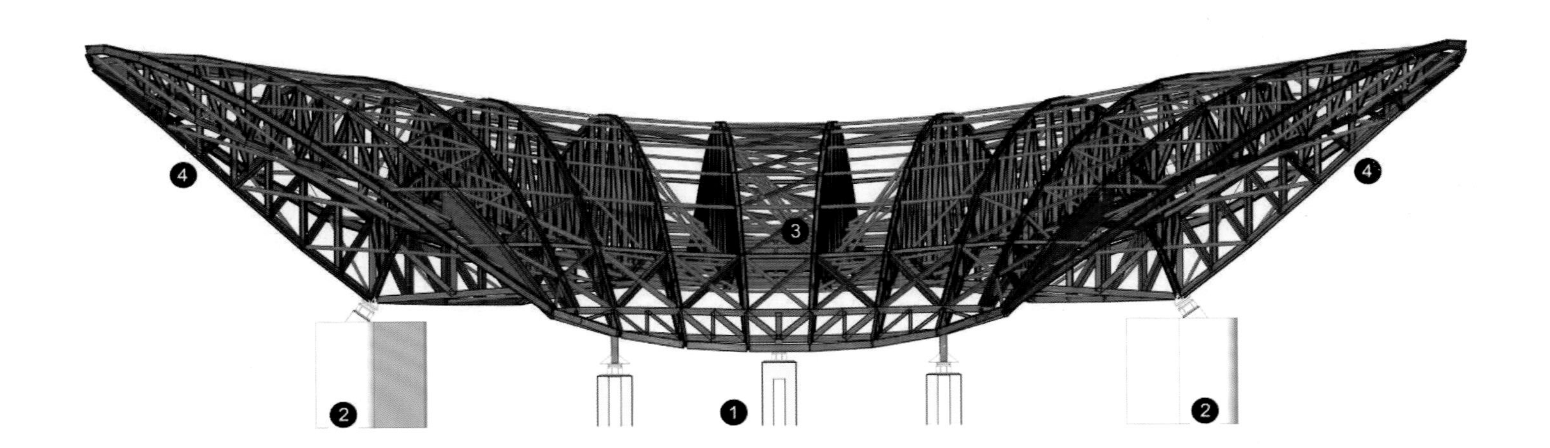

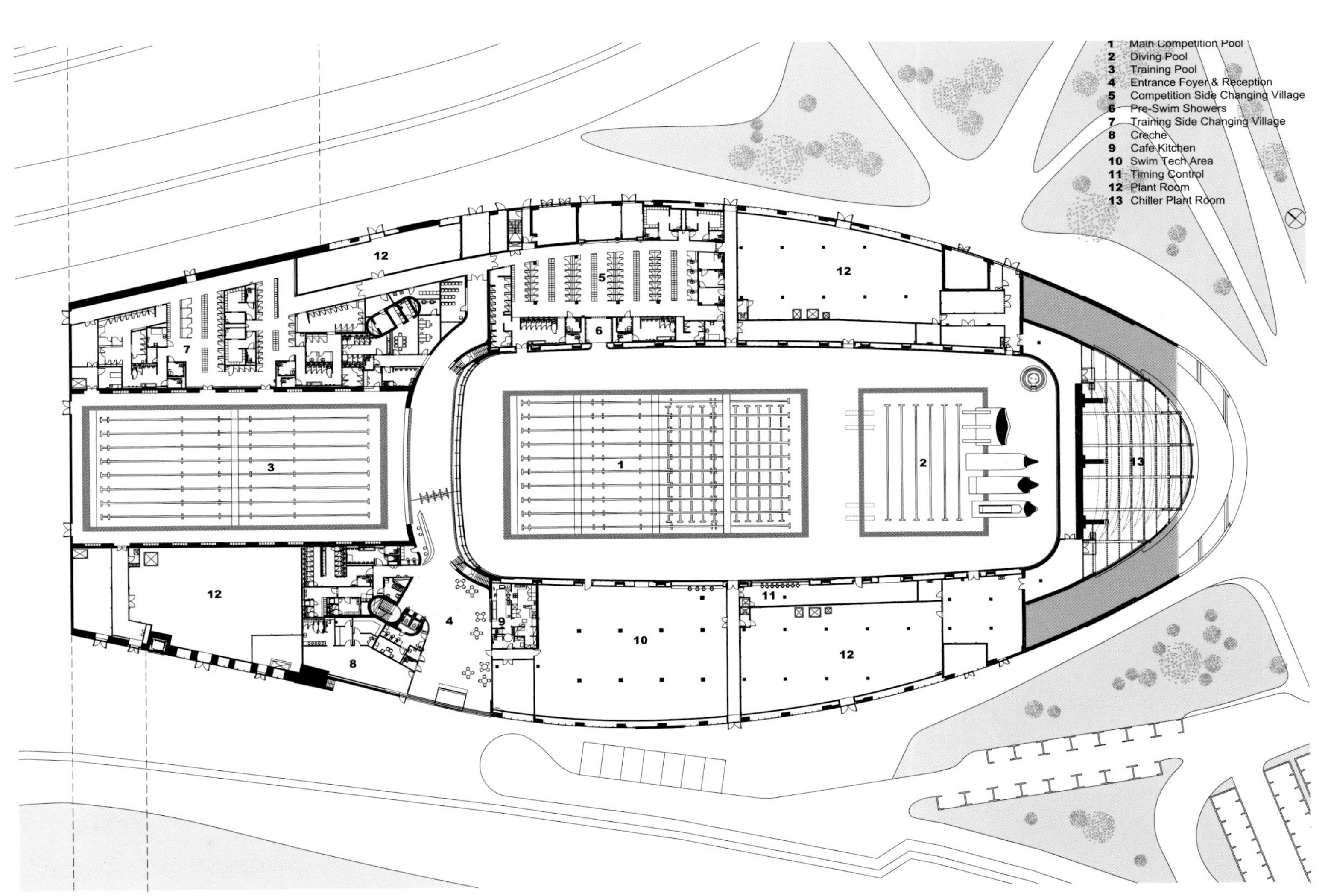

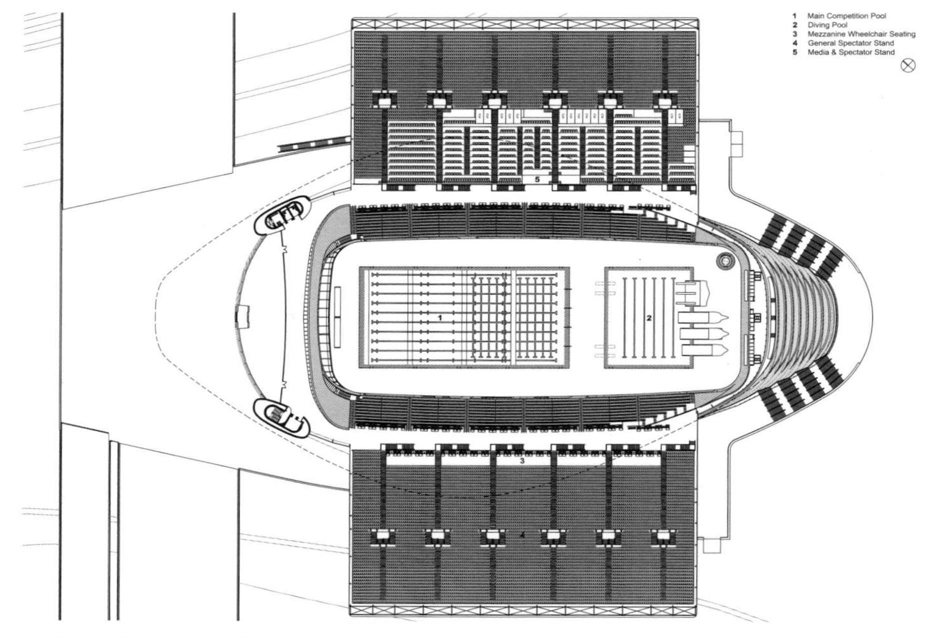

Second Floor Plan (Olympic Mode)

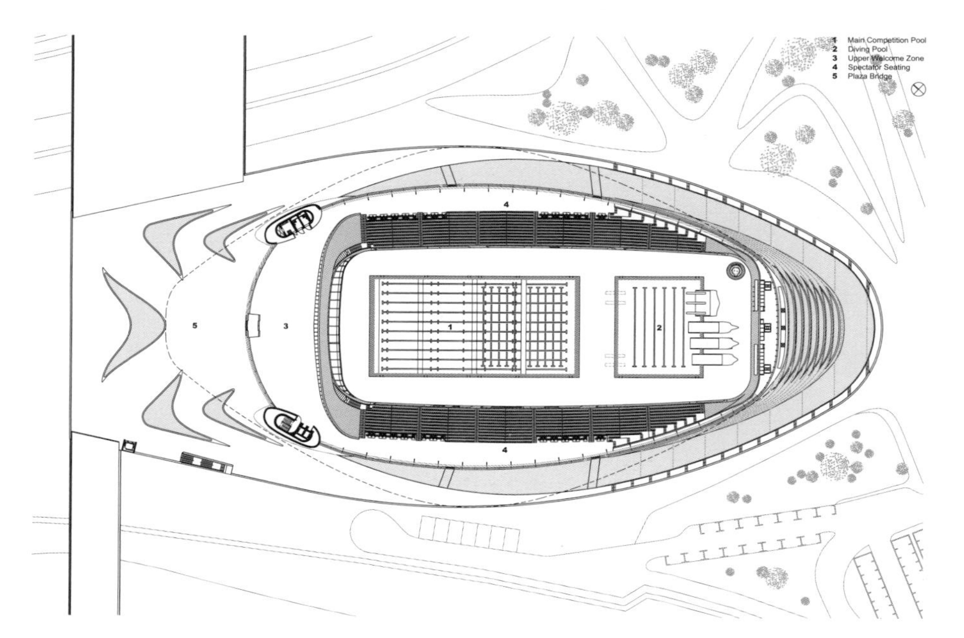

First Floor Plan (Legacy Mode)

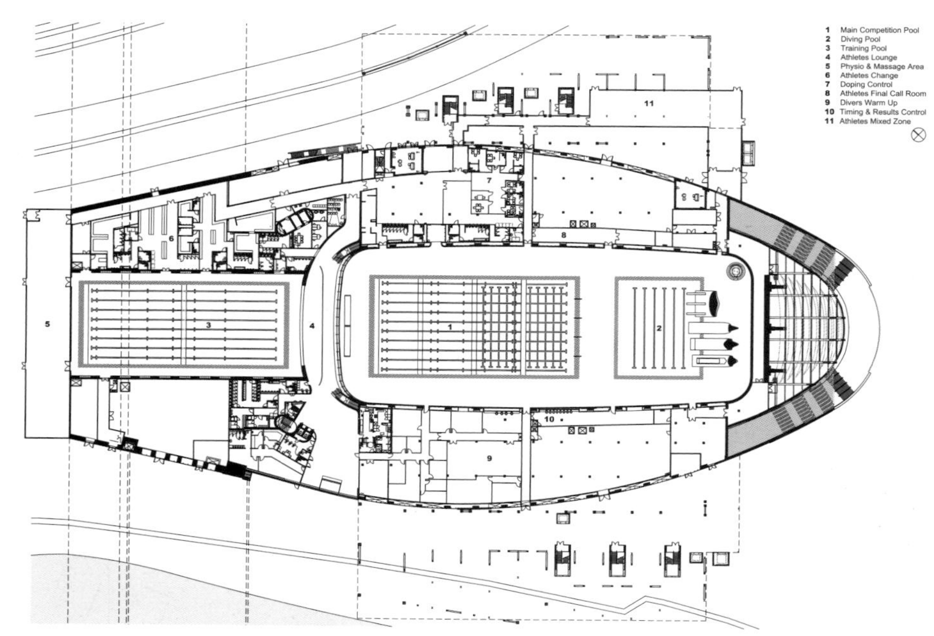

Ground Floor Plan (Olympic Mode)

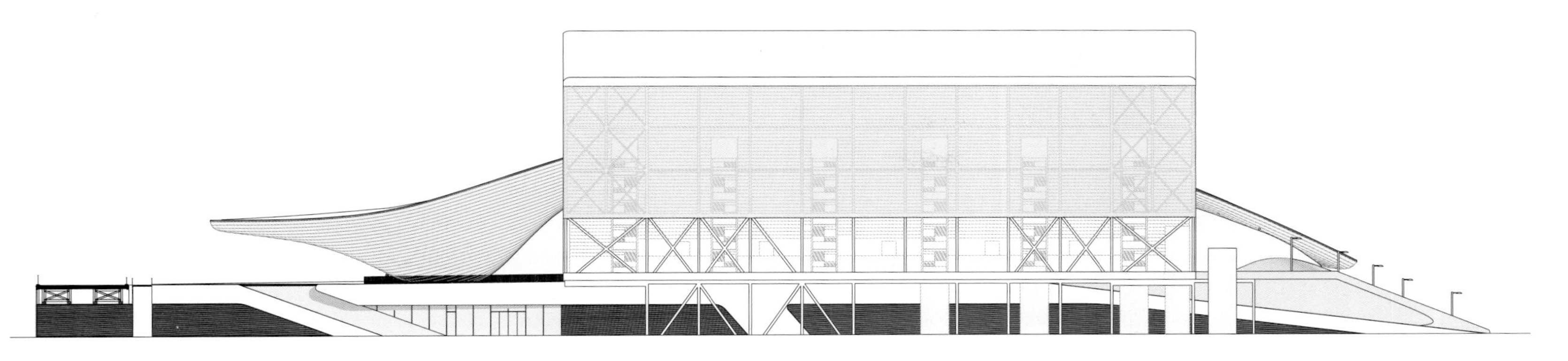

West Elevation (Olympic Mode)

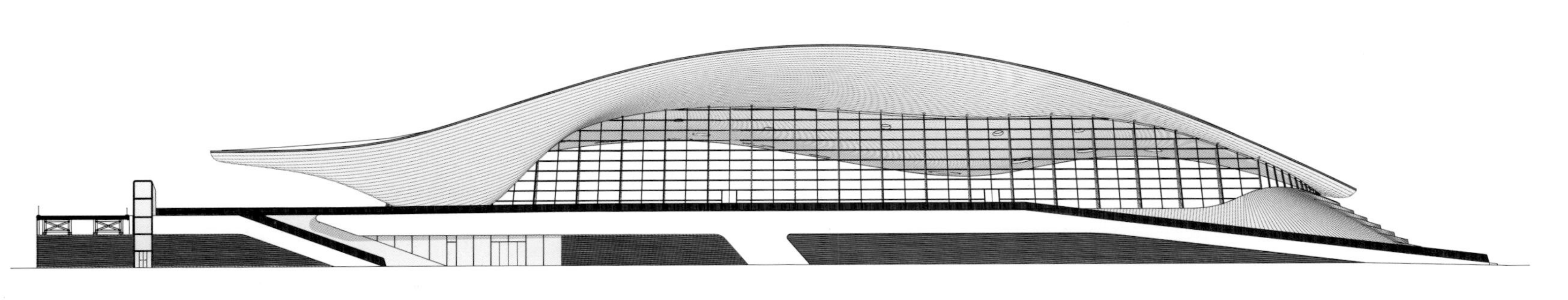

West Elevation (Legacy Mode)

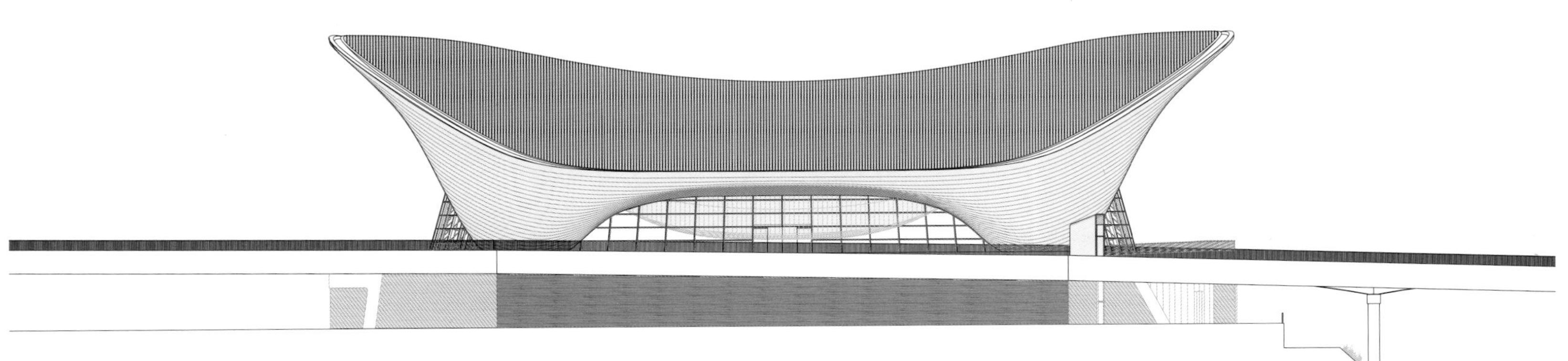

North Elevation (Legacy Mode)

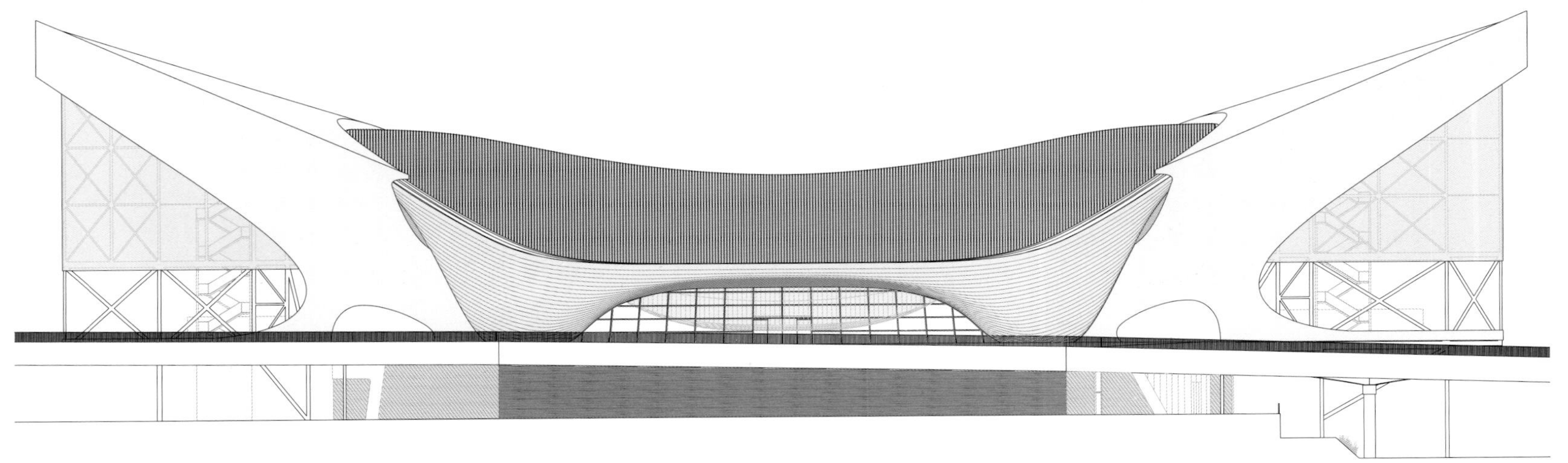

North Elevation (Olympic Mode)

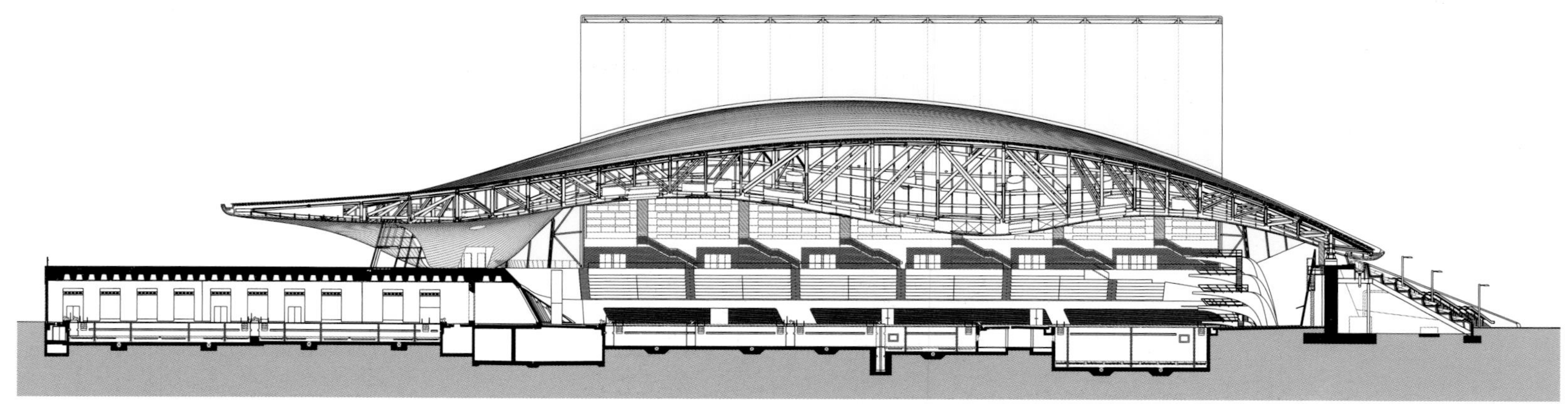

Longitudinal Section (Olympic Mode)

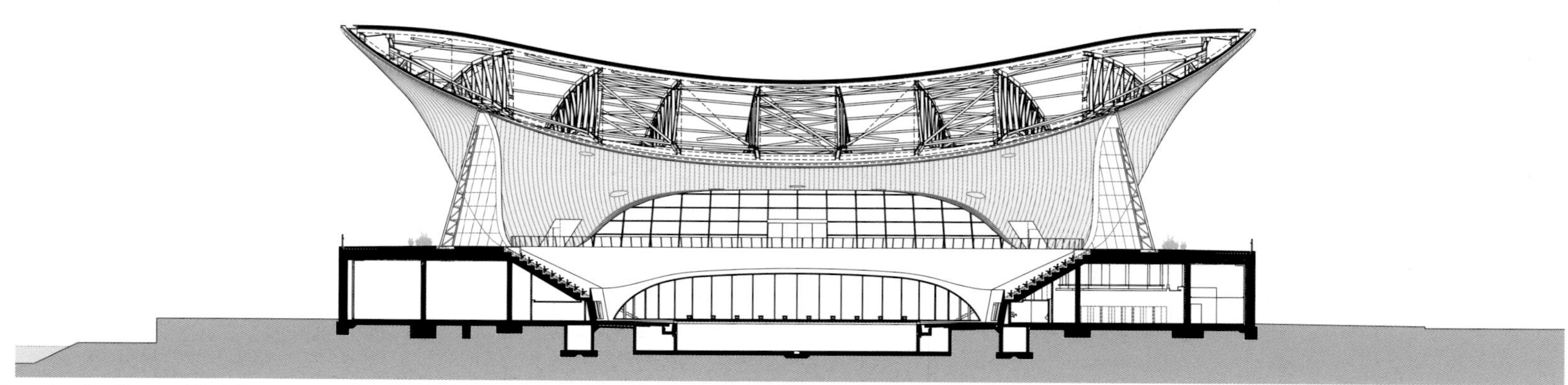

Cross Section (Legacy Mode)

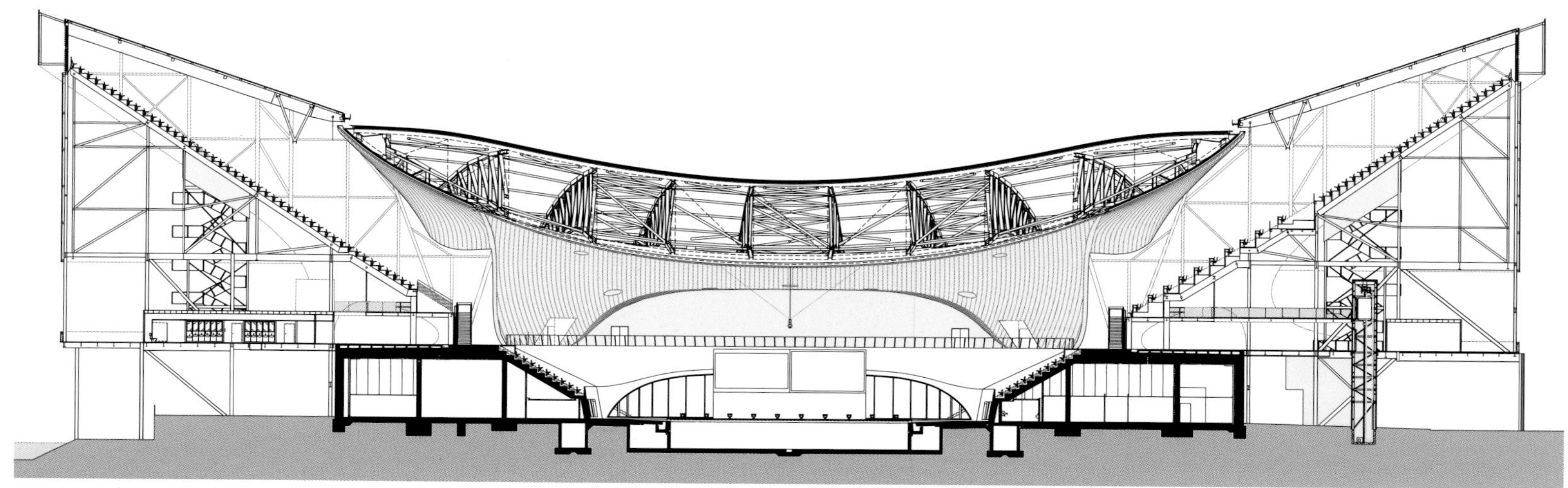

Cross Section (Olympic Mode)

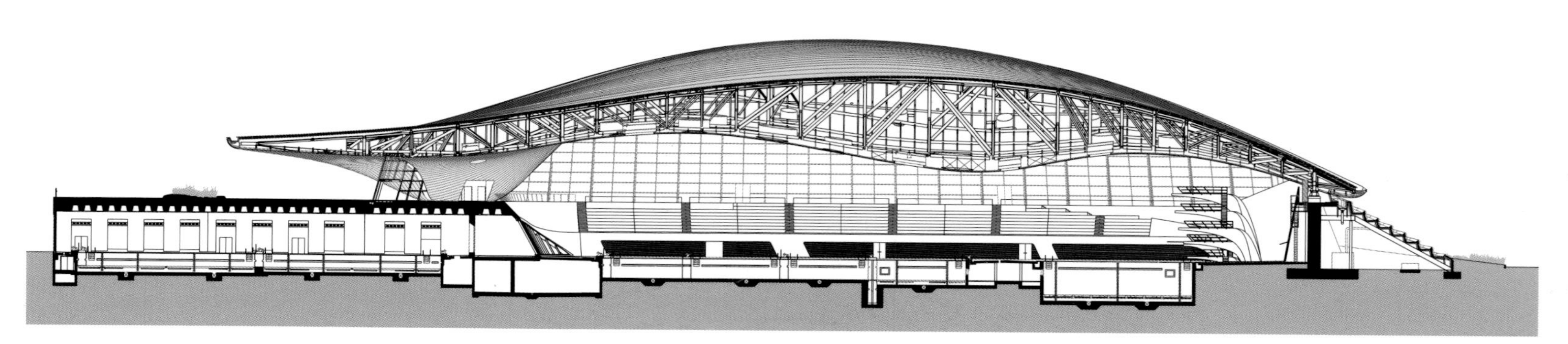

Longitudinal Section (Legacy Mode)

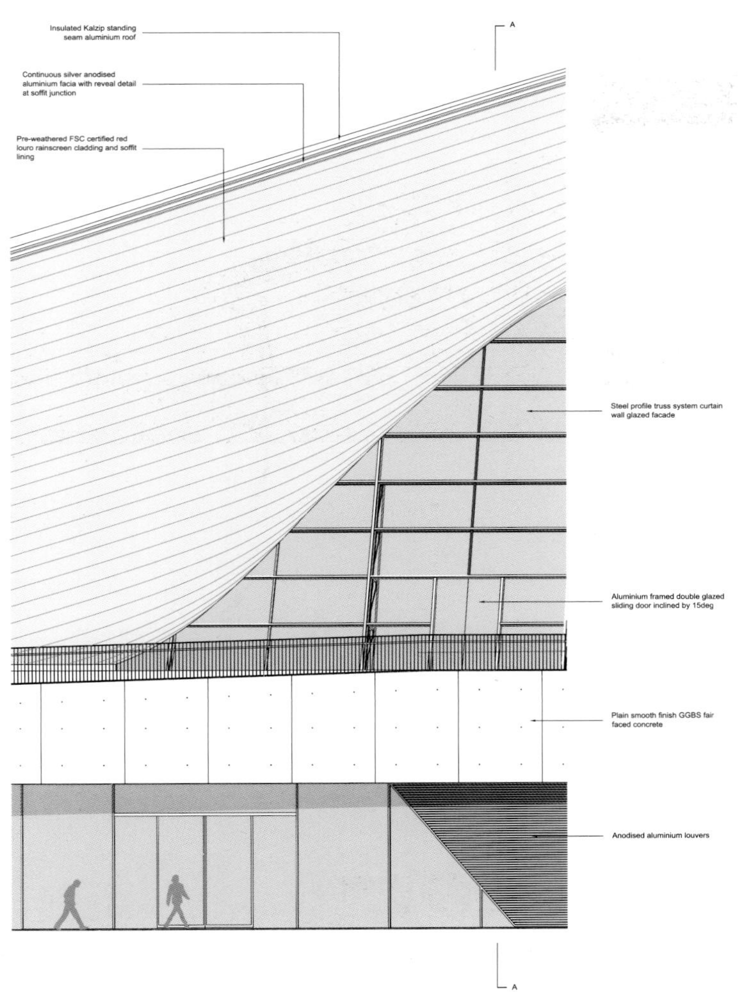

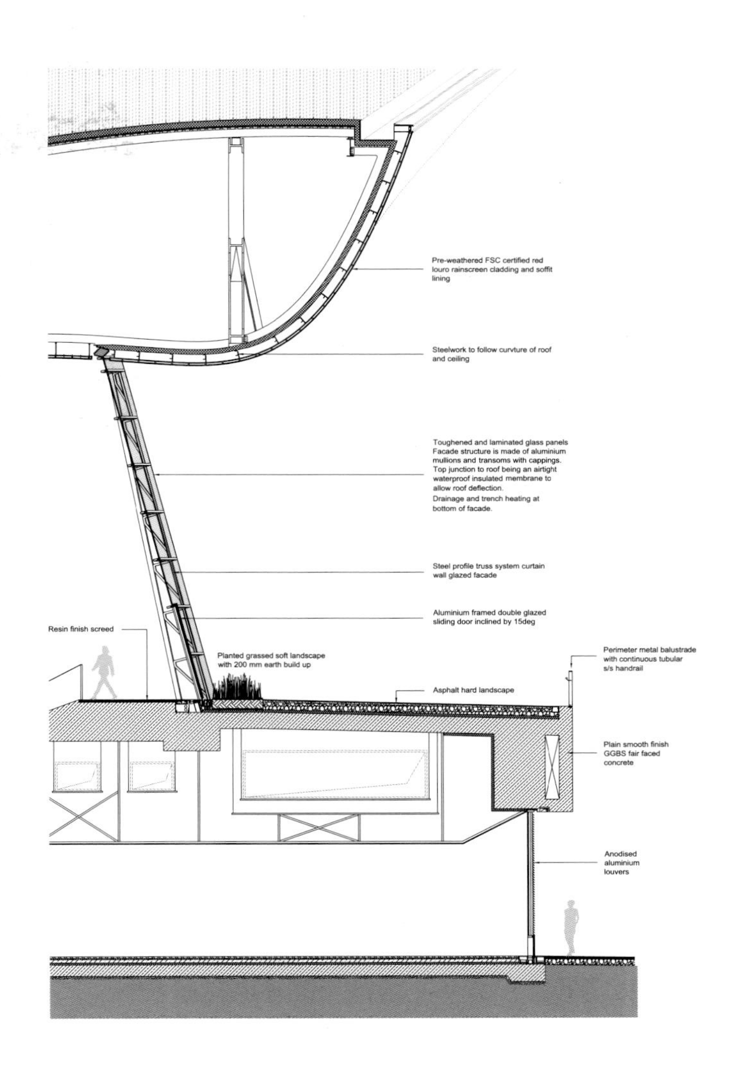

Detail 1 (Legacy Mode, With Text)

Continuous silver anodised aluminium facia with reveal detail at soffit junction

Continuous aluminium gutter incorporating insulated downpipes and overflow wiers

Insulated Kalzip standing seam aluminium roof

Pre-weathered FSC certified red louro rainscreen cladding and soffit lining

Steel profile stick system curtain wall glazed facade

Medium beam projector luminaires fixed to lighting truss and grouped in lighting 'bubble'

Soft landscaping areas with precast concrete surrounds

Aluminium framed double glazed sliding door inclined by 15deg

ENTRANCE LOBBY

PLAZA ENTRANCE

Plaza asphalt hard landscape

Detail 2 (Legacy Mode, With Text)